D1601646

EVOLUTION, TIME AND SPACE: THE EMERGENCE OF THE BIOSPHERE

Proceedings of a symposium on biogeography organized with the British Museum (Natural History), London 6–10 April 1981, on the occasion of the centenary of the museum.

THE SYSTEMATICS ASSOCIATION
SPECIAL VOLUME No. 23

EVOLUTION, TIME AND SPACE: THE EMERGENCE OF THE BIOSPHERE

Edited by

R. W. SIMS

Department of Zoology, British Museum (Natural History), London

J. H. PRICE

Department of Botany, British Museum (Natural History), London

P. E. S. WHALLEY

Department of Entomology, British Museum (Natural History), London

1983

Published for the

SYSTEMATICS ASSOCIATION

by

ACADEMIC PRESS
LONDON NEW YORK
PARIS SAN DIEGO SAN FRANCISCO SÃO PAULO
SYDNEY TOKYO TORONTO

ACADEMIC PRESS INC. (LONDON) LTD
24/28 Oval Road
London NW1

United States Edition published by
ACADEMIC PRESS INC.
111 Fifth Avenue
New York, New York 10003

British Library Cataloguing in Publication Data

Evolution, time and space.—(The Systematics
Association special volume ISSN 0309-2593; v. 23)
I. Biogeography—Congresses
I. Sims, R.W. II. Price, J.H.
III. Whalley, P.E.S. IV. Series
574.9 QH84

ISBN 0–12–6444550–8

LCCCN 82–73806

Printed in Great Britain at the Alden Press, Oxford

Contributors

Adams, C.G., *Department of Palaeontology, British Museum (Natural History), Cromwell Road, London SW7 5BD, UK*

Ball, I.R., *Institute for Taxonomic Zoology, University of Amsterdam, Postbus 20125, 1000 HC Amsterdam, The Netherlands*

Greenwood, P.H., *Department of Zoology, British Museum (Natural History), Cromwell Road, London SW7 5BD, UK*

Grove, A.T., *Geography Department, University of Cambridge, Cambridge CB2 3EN, UK*

Hessler, R.R., *Scripps Institution of Oceanography, La Jolla, California 92093, USA*

Humphries, C.J., *Department of Botany, British Museum (Natural History), Cromwell Road, London SW7 5BD, UK*

Jablonski, D., *Department of Ecology and Evolutionary Biology, University of Arizona, Tucsan, Arizona 85721, USA*

Keast, J.A., *Department of Biology, Queen's University, Kingston, Ontario, Canada K7L 3N6*

Milner, A.R., *Department of Zoology, Birkbeck College, University of London, London WC1 7HX, UK*

Nelson, G., *Department of Ichthyology, American Museum of Natural History, New York, NY 10024, USA*

Owen, H., *Department of Palaeontology, British Museum (Natural History), Cromwell Road, London SW7 5BD, UK*

Patterson, C., *Department of Palaeontology, British Museum (Natural History), Cromwell Road, London SW7 5BD, UK*

Pielou, E.C., *Biology Department, Dalhousie University, Halifax, Nova Scotia, Canada*

Ross, R., *Department of Botany, British Museum (Natural History), Cromwell Road, London SW7 5BD, UK*

Simberloff, D., *Department of Biological Science, Florida State University, Tallahassee, Florida 32306, USA*

Spoel, S., van der., *Institute for Taxonomic Zoology, University of Amsterdam, Postbus 20125, 1000 HC Amsterdam, The Netherlands*

Valentine, J.W., *Department of Geological Sciences, University of California, Santa Barbara, California 93106, USA*

Wilson, G.D.F., *Scripps Institution of Oceanography, La Jolla, California 92093, USA*

Preface

Despite the implications in Patterson's paper to the contrary (ch. 1), we *did* go into the preparation of the programme of this symposium, the fourth major one involving biogeographic themes within about 26 months, with a firm set of intentions. These intentions were framed, but by no means wholly covered, in our introductory publicity literature. We hoped for a measure of controversy in the discussions, especially from some who have elsewhere occasionally given the impression of total preoccupation with the brilliance of particular biogeographic dogma and of consequent inability to perceive even stimulatory virtue in the beliefs of antagonists. Additional dialogue was expected from biogeographic agnostics, who steadfastly eschew any particular faith. Perhaps this aim of the symposium was rarely realized, necks only infrequently being firmly laid on the block. At least central ideas emerged, moulded by the immense scope of the topics, not least the difficulties facing accepted dogma, thus emphasizing the continuing need for more basic research and for the organization of further symposia.

In addition to the original aim of providing a forum for discussion, there was the corollary and secondary intention to stimulate the involvement of recent (biogeographical) hypotheses in studies on groups scarcely penetrated by these new approaches, either in the sense of testing the concepts or of helping elucidate problems in those groups. This intention was possibly largely realized since, of the 27 speakers, 20 were concerned with biogeographical problems encountered in groups distributed in seven phyla of animals and four major categories of plants.

It would be unjust in the extreme to give the impression that the meeting developed as a mutual admiration or commiseration society. But it was the clearest indication yet of the sincere, often only groping, movements in the direction of mutual aid and understanding in seeking productive parallels and contributions from *all* available methods and data, from whichever group of organisms. We have progressed beyond infancy in many aspects simply by the recognition that our views and data are still essentially infantile.

For varying reasons, certain papers presented at the symposium are not published here. The excellent entomological contributions of Coope, Holloway and Turner have been collected together in the *Zool. J. Linn. Soc.* (1982), while that by Mattingly together with the masterly botanical expositions of van den Hoek and Miller appeared in the *Biol. J. Linn. Soc.* (1982).

All the papers published here have been ordered within the volume according to the presence of connecting themes with, or general relationships to, the adjacent papers. The opening and closing groups virtually ordered themselves on background data, conceptual or summarizing grounds. We hope that neither contributors nor readers will find anomalies in the sequencing too intrusive or intolerable.

We initially planned to include a much-needed glossary of terms but the scheme was shelved primarily because the definition of terms commonly employed is now available (Lincoln, R.J., Boxshall, G.A. and Clark, P.F., 1982, *A Dictionary of Ecology, Evolution and Systemics*, Cambridge University Press). As to the terms used in different papers throughout the volume, we have made no alterations that were not dictated by the need for clarification. Words such as *endemism* cf. *endemicity*, used as synonyms by different authors, have been left as originally rendered since they are hardly likely either to confuse or offend. This example is chosen to draw attention not only to the need for care with the usage of words, but also to the problem as to whether or not *endemic* and *indigenous* should be regarded as biogeographically synonymous terms. We proffer no definitive solution, but the latter needs consideration as some editors draw a distinction and might cite an example such as: ". . . the *indigenous* rabbit of North Africa and Iberia, *L. cuniculus*, has by dispersal and introduction become *endemic* in northwestern Europe and Australia"!

Our sponsors for the meeting were many and, although the symposium was jointly organized by The Systematics Association and the British Museum (Natural History), it was a pleasure in the centenary year of the museum to acknowledge the facilities and hospitality provided by the Director, Dr R. H. Hedley, and the aid of various members of staff; without these, realization would have been from difficult to impossible. Assistance and financial aid towards speakers' expenses were provided by a number of organizations, including The Systematics Association, the British Council, British Petroleum Co. Ltd, and E. Leitz (Instruments) Ltd; to all these we are very grateful. Academic Press have been helpful on all aspects of the publication of text and illustrations as a volume in The Systematics Association Special Volume Series.

London, 1982

R. W. Sims
J. H. Price
P. E. S. Whalley

Contents

Papers published by the Linnean Society

Published in the Biological Journal of the Linnean Society

Volume 18 No. 2 (1982)

Distribution groups of benthic marine algae in relation to the temperature regulation of their life histories. By C. VAN DEN HOEK
Bryophyte evolution and geography. By H. A. MILLER

Volume 19 No. 2 (1982)

The palaeogeography of mosquito-borne diseases. By P. F. MATTINGLY

Published in the Zoological Journal of the Linnean Society

Volume 76 No. 4 (1982)

Mimetic butterflies and punctuated equilibria: some old light on a new paradigm. By J. R. G. TURNER

Large-scale changes in the geographical distributions of Coleoptera during the Late Quaternary. By R. COOPE

Mobile organisms in a geologically complex area: Lepidoptera in the Indo-Australian tropics. By J. D. HOLLOWAY

1 | Aims and Methods in Biogeography

C. PATTERSON

Department of Palaeontology, British Museum (Natural History), London SW7 5BD, UK

Abstract: Biogeographic aims are of two principal sorts: to use distribution as a guide to earth history; and to explain distribution through application of theories of earth history and evolution. The first aim, investigative, seeks pattern; the second, explanatory, seeks to explain pattern by process. The methods used to meet the aims of biogeography are reviewed, drawing analogies with systematics, on which all biogeography must depend. If our method is to test hypotheses, the limited aims set by cladistic vicariance theory offer the most promising approach to biogeography.

INTRODUCTION

What should be the aims of a symposium on biogeography in 1981? If *aims* are too abstract or intangible, are there *principles* on which biogeographers might agree, and methods or practices which will serve those principles? I ask these questions, and discuss possible answers to them, in order to bring this meeting into focus, and to place it in the long term and short term of the evolution of biogeography.

In the short term, this is, to my knowledge, the fourth symposium on biogeography in less than two years. Their titles give an impression, more or less precise, of their aims: *Vicariance Biogeography: a Critique* (New York, May 1979; Ferris, 1980; Nelson and Rosen, 1981); *Biogeography of Regional Biotas and Communities* (Vancouver, July 1980);

Systematics Association Special Volume No. 23, "Evolution, Time and Space: The Emergence of the Biosphere", edited by R.W. Sims, J.H. Price and P.E.S. Whalley, 1983, pp. 1–28. Academic Press, London and New York.

Alternative Hypotheses in Biogeography (Seattle, December 1980; *Am. Zool.* **22**, 347–471, 1982); now, *Evolution, Time and Space: the Emergence of the Biosphere.* The handbill advertising this meeting described biogeography as "perhaps the most exciting yet least well-defined, derived discipline of contemporary biology" and mentioned four exciting areas: the new palaeogeography; the dogmas of vicariance and cladism; the theory of faunal and floral regions; and dynamic equilibrium theory. The second area recalls the meeting in New York, the third that in Vancouver, and the second to fourth that in Seattle. As for the new palaeogeography, it is not so new if plate tectonics is implied, but there is an alternative new palaeogeography, discussed at the *Expanding Earth Symposium* in Sydney, February 1981. Is our aim to synthesize all these topics?

AIMS

One unquestioned aim of this meeting was to mark the centenary of the British Museum (Natural History), so I begin by taking the centennial perspective and reviewing the aims of biogeography a hundred years ago. In 1881, there were only two books on biogeography in English: Wallace's *Geographical Distribution of Animals* (1876) and *Island Life* (1880). George (1964:100) wrote "The subject of zoogeography in its modern form may be said to have been 'invented' by Wallace". What were his aims? The first stated aim (1876, *1*:8) was that zoogeography "may reveal to us, in a manner which no other evidence can, which are the oldest and most permanent features of the earth's surface, and which the newest". So Wallace's first biogeographic aim was earth history, "an important adjunct to geology". As a subsidiary aim, he proposed (1876, *1*:9) that the study of zoogeography

> will teach us to estimate the comparative importance of the various groups of animals, and to avoid the common error of cutting the gordian knot of each difficulty by vast hypothetical changes in existing continents and oceans—probably the most permanent features of our globe.

There is evident contradiction between these two aims: the first seeks to use life as an adjunct to geology, a guide to earth history; whereas the second denies that it should do so. [Wallace acknowledged that his book contained inconsistencies (Darwin and Seward, 1903, *2*:13).] In any case, his conclusion (1876, *2*:160) was that the Palaearctic and Oriental regions were the workshop, "the source from which all the other regions were supplied with the higher forms of life". This view, says Wallace (1876, *2*:163):

> is not the result of any preconceived theory, but has grown out of a careful study of the facts accumulated, and has led to a considerable modification of the author's previous views. [On this modification, see Fichman (1977).] It may be described, as an application of the general theory of Evolution, to solve the problem of the distribution of animals.

This theme is carried over into *Island Life*, for "so long as the belief in 'special creations' of each species prevailed, no explanation of the complex facts of distribution *could* be arrived at or even conceived" (Wallace, 1880:8). In this book, Wallace described the aim of biogeography as solving the problems of why organisms live where they do—"they are *there*, and others are *not* there" (1880:6), and solutions are found by developing and applying "a clear and definite theory" (1880:499). That theory is one of causes, which are of four kinds, two biological (dispersal, evolution), and two physical (geographical change, climatic change). By these four variables, all distributions may be explained.

In Wallace's two books one can discern the beginnings of two opposite approaches to biogeography. The first approach, mooted and soon discarded by Wallace, uses the distribution of life as factual evidence bearing on the history of the earth. The second, dominant in Wallace's work, uses theories of the history of the earth and of life to explain distribution. These two approaches are nicely contrasted in the Darwin centenary essays by Thiselton-Dyer (1909, on plants) and Gadow (1909, on animals). Thiselton-Dyer was a forthright advocate of the Darwin–Wallace theory of "the more efficient workshops of the north" ("there seems an obvious parallel with the colonialistic spirit of the times"—Nelson, 1978:299). Given that the continents and oceans are fixed, the southward migration theory best explains observed Recent and inferred fossil distributions. Gadow belonged to a different, continental tradition (a student of Gegenbaur and Haeckel, he came to England, to an appointment at the BM(NH) in 1880). He took the view that the distribution of life denies the fixity of the continents—"tortoises . . . have reached the Galapagos by land, not astride a log" (Gadow, 1913:77)—and produced a series of palaeogeographic maps, elucidating in detail the bearing of life on earth history. His maps (Gadow, 1913), though not the first of their kind (e.g. Arldt, 1907), are amongst the most explicit, for "the bolder the hypothetical outlines are drawn, the better" (Gadow, 1909:332), and he hoped that "critics with more expert knowledge will amend them" (1913:vi).

But this hope was not realized. As is well known, the Darwin–Wallace view—fixed continents and northern workshops—was promoted by Matthew (1915), and held office in biogeography until less than twenty years ago (e.g. Mayr *et al.*, 1952; Darlington, 1957, 1965; Simpson, 1965). Our present disrespect for that view is the result of geophysical, not biological evidence. Mayr (1952:255), citing Rensch, reminded biogeographers that if circular reasoning is to be avoided "zoogeographical conclusions must be based on biological (including paleontological) evidence, and geological conclusions on geological evidence". Which suggests that one of the basic faults of the Darwin–Wallace approach was circular reasoning, evident in the conflict between the two aims expressed by Wallace (1876).

Yet those two aims retain an interesting generality. One may view biogeography either as explanation of distribution by applying theories of the history of life and of earth, or as an investigation into earth history. The latter was the aim of the landbridge and, later, the Wegenerian schools. But geology has since found its own guide to earth history, in magnetic lineations, the results of the DSDP, and other ingredients of tectonic theory. There remains one aim which this conference might consider: to use biogeographic data as a test of the two alternative theories of global tectonics, plate theory and expansion theory. For at the most general level, the two make different predictions about the history of life. Plate tectonics implies that in Phanerozoic time both the continents and oceans have had a complicated history of subdivison and coalescence, but in Mesozoic and Cenozoic time, both continents and oceans have, in general, been successively subdivided. Expansion theory implies a contrasting history of continents and oceans, the continents having been successively subdivided whereas the oceans have coalesced. Is there any sign, in the patterns of distribution of oceanic organisms, of one or other of these two different sequences? Expansion implies increasing regionation through time for continental biotas, and decreasing regionation for those of the oceans; plate tectonics implies increasing regionation on the continents and in the oceans, at least since the early Mesozoic (Valentine, 1969, and Schopf, 1979, report increasing regionation of the shelf benthos). And are there still outstanding anomalies between biogeographic data and tectonic theories, such as those which convinced the minority of biologists who insisted on landbridges, and later believed Wegener? Pacifica (Melville, 1981; Nur and Ben-Avraham, 1981) is still a candidate here. The only other useful aim that I see for the earth history aspect of biogeography is to answer Nelson and Platnick's (1981:540) question: "Might there be a single pattern of relationships (a general cladogram of areas) for all groups of organisms?" I return to this below.

Following the acceptance of plate tectonics, one obvious aim amongst biogeographers has been to reconcile their data with the new palaeogeography. That programme has been criticized by Edmunds (1975), Nelson (1976) and Craw (1978), amongst others. I agree with those critics, for this approach can be no more profitable than was fitting biogeography to the fixed continents axiom of Darwin and Wallace.

Wallace's alternative aim, applying evolution and tectonic theory to explain distribution, has been dominant in biogeography since his time. This aim ("nothing less than the history of the evolution of life in space and time in the widest sense"—Gadow, 1909:330; "to succeed in explaining the causes underlying faunal distribution"—Banarescu, 1975:6; "the general understanding of life in time and space"—Brundin, 1981:152) is, I suppose, acceptable to most biogeographers. I make only one comment on it. It is an aim involving or invoking process. Many contributors to the current debate on the aims of systematics have sought to discriminate between pattern and process. Eldredge (1981) makes the same point with regard to biogeography. He writes "scenarios permeate the literature of biogeography every bit as much as the literature of evolution" (p. 36), and by scenarios, he means "explanation of the pattern in terms of ideas about process . . . *just so* stories . . . usually criticized as untestable inductive narratives" (p. 35).

The notion of "narrative biogeography" was first introduced by Ball (1976), who extended it from evolutionary narratives. To Ball a biogeographic (or evolutionary) narrative "was not aimed at the discovery or confirmation of a law, it did not establish any new empirical fact, and it makes no positive predictions about what may be found in future investigations" (1976:409). Ball recommended that biogeography should progress from a narrative phase into an analytical phase, characterized by restrictive, therefore predictive and testable, hypotheses. Morse and White (1979) also comment on narrative and analytical biogeography, and seek to improve Ball's procedure. Eldredge (1981) took a slightly different view: if the aim of biogeography is to explain distribution by theories of process, "we automatically abdicate any position allowing us to improve our theories of process" (Eldredge, 1981:35). He suggested that patterns of geographic distribution be used to test competing theories of process (e.g. dispersal versus vicariance; plate tectonics versus the expanding earth). Of course, if pattern is to be used to test theories of process, the first question of biogeography must be

> What is the exact pattern and how has this pattern changed through time? Clearly we need a methodology for capturing the pattern . . . a survey of existing methods of capturing biogeographical patterns. Surely improvement in this area ... could truly advance biogeography (Eldredge, 1981:37).

These ideas introduce new ramifications of biogeographic aims. To characterize much biogeography as scenario writing—*just so* stories—and to demand testable hypotheses is a radical alternative to the Wallacean programme of biogeography as applied evolution and tectonic theory. Eldredge's alternative, to use biogeographic pattern to test and, perhaps, improve theories of process, can be seen as an extension of the programme of biogeography as a guide to earth history: distribution patterns might be used to test aspects of evolution theory, as well as tectonic theory. If the data of biogeography are to bear the weight of testing general biological theories, it is doubly necessary, as Eldredge said, to ensure that the pattern is captured by the best method available, and by a method that does not depend on those theories the pattern might be used to test (e.g. speciation theory, or more general aspects of evolutionary theory). We need to survey those methods. Even those who reject a programme of testing theories and accept narrative biogeography as a worthwhile aim, need a method, for the kinds of narratives told will be affected, if not constrained, by the methods on which they depend.

METHODS

It might be thought that capturing biogeographic pattern is, at base, pure observation, untrammelled by theory. That idea is opposed by Darwin's dictum "How odd it is that anyone should not see that all observation must be for or against some view if it is to be of any service!" (Darwin and Seward, 1903, *1*:195). Nelson (this volume, Ch. 16) makes this point neatly with the anecdote of Darwin, Fitzroy and the

Galapagos finches. Ball (1976:408) implies that descriptive or empirical biogeography—"establishing the complexity of the distributional patterns"—is within "the world of primary sense data". I disagree, for even the most basic data of distribution rest on hypothesis and assumption, as is shown, for example, by Nelson's comments on the Galapagos finches, and Nelson and Platnick's (1981:488) discussion of levels of endemism in island birds as dependent on taxonomic convention (whether endemic forms are recognized as species or subspecies). It seems that recognition of basic patterns of distribution starts with a problem (as in all science—Ball, 1976:412).

The problems of distributional data first surfaced in the seventeenth and eighteenth centuries, according to Nelson (1978) and Kinch (1980). Two general problems were posed by the early data: disjunct distributions, and "the delineation and enumeration of the regions of creation" (i.e. of endemism; Kinch, 1980:99). Both are subsumed by the general problem of allopatry ("they are *there*, and others are *not* there" in Wallace's words). These problems were, and still are, problems primarily of systematics. For example, recognition of a disjunct distribution demands both an estimate of relationship between the disjunct populations, and an estimate of the taxonomic level at which they form a unique taxon. Delineation of areas of endemism depends on the same sort of taxonomic judgements, at the species level and for higher taxa (e.g. are there genera or families endemic to particular areas?). Such questions do not concern primary sense data, but invoke hypotheses and assumptions.

1. *Wallace's Method*

"Zoogeography in its modern form may be said to have been 'invented' by Wallace, for he was the first to base the subject on the theory of evolution" (George, 1964:100). Wallace's treatment of systematics (especially Wallace, 1876, *1*:83ff.) was evidently influenced by his commitment to evolution, for in *The Geographical Distribution of Animals* he selected genera as his units, writing that species "are systematically disregarded" because they are too numerous, and "because they represent the most recent modifications of form" (1876, *1*:vii). Wallace sought "to estimate the comparative importance of the various groups of animals" (1876, *1*:9). He settled on mammals, because their dispersal depends on the distribution of land and water rather than on accidents; they are adaptable enough to exist all over the globe; their natural classification was mature enough; and their fossil record was good. From this basis in mammalian systematics at the generic level

> Wallace's method of arriving at the degree of relationship of the faunas of the various regions is eminently statistical. Long lists of genera determine by their numbers the affinity and hence the source of colonisation. This statistical method has found many followers, who, relying more upon quantity than quality, have obscured the problem (Gadow, 1913:11).

On quality versus quantity, see under cladistics below; as for statistics, Wallace's method was statistical at two levels, first in his selection of genera as the "average" unit, and secondly in the use he made of those units (Nelson and Platnick, 1981:399).

Wallace's biogeography was also original in its use of fossils:

> To study fossils as though they were equal in importance to living animals was revolutionary and by no means generally acceptable, for it must be remembered that it was only since the recognition of a process of evolution that fossils had been regarded as of any direct significance to the ancestry of present day animals (George, 1964:132).

Thus Wallace developed a method of biogeography which was original in four main ways: its commitment to evolution; its use of statistical comparisons of genera, treated as units; its reliance on mammals; and its appeal to the fossil record. Wallace's method, and those ingredients of it, were endorsed by Matthew (1915), Darlington (1957) and Simpson (1965). The method was the ruling paradigm in biogeography until the middle 1960s. At that time, plate tectonic theory became generally accepted amongst geologists and biologists. This theory carried the implication that Wallace's method had failed to recognize a major factor in biotic history. Acceptance of drift thus provided a favourable climate for new methods and concepts in biogeography. Coincidentally, they were at hand: Brundin (1966) and Hennig (1966) had been long in preparation, whereas MacArthur and Wilson (1967) seem to owe no debt to tectonic theory. And waiting in the wings was Croizat (especially 1964), who for 15 years had been promoting a method, and mounting a many-pronged attack on Darwin–Wallace biogeography. Croizat's ideas had their roots in a different, continental European tradition of biogeography (Nelson, 1973).

2. *Equilibrium Methods*

MacArthur and Wilson's aim was to reformulate biogeography "in terms of the first principles of population ecology and genetics" (1967:175). As a result, they became "unable to see any real difference between biogeography and ecology" (p. v). Their method differs from that of Wallace in every respect except two—their method is statistical (though of course their mathematical models are worlds apart from Wallace's statistics); and they consider chance dispersal and colonization (with extinction) as the major determinants of distribution patterns. Perhaps the most striking and fundamental difference between the equilibrium approach and the Wallacean tradition is that equilibrium theory seems to demand no necessary commitment to evolution (except in the long-term flux of the taxon cycle; Simberloff, 1974).

As an illustration of this point, consider Harper's (1980) non-evolutionary theory to explain the distribution of the Galapagos finches. This theory predicts that species number per island is controlled by target area of the island, and demands nothing more than a mainland source of species, whose origin is immaterial. Harper does not cite equilibrium theory, but his approach seems to match the basic tenets of that theory. As for "capturing the pattern" of distribution, equilibrium theory has nothing to say, except that the pattern be quantifiable, primarily in numbers of species.

Rosen (1978:160) was of the opinion that equilibrium theory and the methods of historical biogeography stand in the same relation as do population genetics and phylogeny construction—"The relation is one of friendly but independent coexistence. A hope for the future reintegration of these divergent interests may be misguided." That equilibrium theory is concerned only with the short time-scale of the ecologist (Vuilleumier, 1978) is no longer true, for the techniques of mathematical modelling developed by equilibrium theorists are now being applied on the geological time-scale, through the fossil record (Carr and Kitchell, 1981; and many other recent papers in *Paleobiology*). But whether there is any possibility of integration with other methods of historical biogeography, discussed below, is still an open question, and one this conference might address.

3. Phenetic Methods

Vuilleumier (1978) sees contemporary biogeography as split into three Kuhnian paradigms or camps, amongst which exchange is limited. One of the three is equilibrium theory; the other two he calls "the theory of faunal regions and of centers of origin" and "the dogmas of vicariance and of cladism". I will cut the cake in a different way, dividing historical biogeography into phenetic and cladistic, in order to emphasize the pattern-gathering aspect rather than the major explanatory principles invoked. Within both phenetic and cladistic biogeography there are those who emphasize either dispersal or vicariance explanations. In phenetic biogeography, Holloway and Jardine (1968) exemplify dispersal explanation, and Croizat (1964) exemplifies the vicariance viewpoint. If Croizat, Holloway and Jardine seem strange bedfellows, perhaps Darlington, Simpson and Keast will be more congenial one to another. The point I mean to make in listing these names is that all see biogeographic pattern in the same way as Wallace—as statistical (or in Gadow's (1913:11) words "eminently statistical . . . relying more upon quantity then quality"). Holloway and Jardine's dendrograms (1968: Figs 1–3, 9–11); Croizat's generalized tracks ("likened to statistical averages", Nelson, 1973:313; "statistical measures of the overall similarity", Ball, 1976:421); Simpson's coefficient of faunal resemblance (1980:197); Wallace's "long lists of genera" (Gadow, 1913:11); Keast's long lists of families (e.g. 1973: Table 2); all are phenetic estimates of pattern. What such estimates of affinity mean, no one is quite sure. If all the taxa involved are natural (monophyletic in Hennig's sense) they may contribute qualitative information (Nelson and Platnick, 1981). But how much noise is introduced by the quantitative method, no one knows. I take this to be the meaning of Gadow's (1913:11) statement that "this statistical method has found many followers, who, relying more upon quantity than quality, have obscured the problem". The obscuring factor is noise, unquantifiable except perhaps by the specialist in each group, to whom the natural and doubtful groups are often plain. Huxley (1888:116) has a relevant comment here:

> I think there is no greater mistake than to suppose that distribution . . . can be studied to good purpose by those who lack either the opportunity or the inclination to go through what they are pleased to term the drudgery of exhaustive anatomical, embryological, and physiological preparation.

That sounds like a counsel of despair, for Huxley implies that the only competent biogeographers will be those specialists who devote a lifetime of drudgery to one group of plants or animals, and each will be competent to deal only with that group. Indeed, this has been the fashion for much biogeography: "biogeographic analyses of particular groups commonly found relegated to the back pages of systematic revisions" (Platnick and Nelson, 1978:15). But there is a way round this problem. It is for specialists to express the fruit of their drudgery in explicit systematics. "To be applied most efficiently a precise biogeographic method requires a non-arbitrary taxonomy that can be correctly comprehended by the generalist" (Rosen, 1976:433).

4. Cladistic Methods

The cladistic approach to the pattern of biogeography has a much shorter history than the phenetic one. Though foreshadowed here and there, e.g. by the cladograms in Arldt (1907), by Rosa (1931), Kinsey (1930, 1936) and Ross (1956), the method was first fully outlined by Hennig (1966). As with phenetic methods, there are cladist biogeographers who prefer dispersal explanations (Brundin, 1966; Nelson, 1969), or who emphasize vicariance explanations (Nelson, 1975; Rosen, 1976), or who mix the two (Ball, 1976; Brundin, 1981). Vicariance explanations have led Nelson and Platnick (Platnick and Nelson, 1978; Nelson and Platnick, 1980a, 1981) to refine and extend cladistic method far beyond its dispersal applications. This is because vicariance prompts enquiries into methods of combining the cladistic patterns of different taxa which are sympatric through all or part of their range. Though these developments are a consequence of the vicariance approach, they do not depend on it, and are therefore appropriately treated as aspects of pattern-gathering rather than explanation.

The most important points made by Nelson and Platnick are the following:

(1) Biogeographic patterns are necessarily comparative. Like statements of relationship in systematics, they must include a minimum of three geographic components.

(2) The smallest unit of biogeographic pattern is an area occupied by at least one endemic species.

(3) Widespread species by themselves contribute no unambiguous information on biogeographic pattern (in this they resemble widespread

characters—symplesiomorphies—in systematics). But widespread species may contribute to the pattern of an area of endemism within their range.

(4) Species cladograms of different, partially or completely sympatric taxa may be combined into a general cladogram of areas. Widespread species, occurring in more than one of the areas, species cladograms which lack representatives in one or more areas, and species cladograms with more than one species endemic to an area, may all be accommodated and combined in area cladograms.

(5) A species cladogram missing one or more areas, or including widespread species, or with redundant species, or with any two of these, or with all three, will specify some subset of the set of all possible cladograms for the areas concerned. Combining several such species cladograms may allow one to select the one or more members of the total set which are common to all the subsets specified.

(6) There are two alternative methods of dealing with widespread species, missing areas and redundant representation of areas in species cladograms. Nelson and Platnick call these alternatives "assumption 1" and "assumption 2". The two methods may give different results; of the two, assumption 2 may be more realistic.

A common criticism of cladistic analysis is that it relies on the assumption that speciation is dichotomous. And a criticism of cladistic biogeography made by several contributors in Nelson and Rosen (1981) is that it depends on the assumption that speciation is allopatric. If the pattern-gathering aspect of cladistic biogeography did rest on such assumptions, these would be serious criticisms. But it is clear (Platnick, 1980; Nelson and Platnick, 1980b; Patterson, 1980) that cladistic analysis of species demands no prior assumption about speciation or even about evolution. As for area-cladogram analysis, as formulated by Nelson and Platnick (1981) it concerns combination of species cladograms, and introduces no more assumptions than they do.

Thus cladistic biogeography need concern pattern alone. The generality and potential of Nelson and Platnick's methods seem to me to make it certain that they will be adopted. I assume that with use they will be further developed and refined. What is lacking at the moment is the raw material—species cladograms. They are the province of those with "the opportunity or the inclination to go through ... the drudgery of exhaustive anatomical, embryological, and physiological preparation", to quote T.H. Huxley again. The interest of Nelson and Platnick's new method is that it will accommodate biological cladograms that those who have tried their hands at area cladograms (Rosen, 1978; Edmunds, 1981; Patterson, 1981a,b; Humphries, 1981; and this volume) would have rejected as uninformative. Even so, those first attempts at combining species cladograms revealed a congruence that was surprising. Nelson and Platnick's (1981:540) question: "Might there be a single pattern of relationships (a general cladogram of areas) for all groups of organisms?" is not yet answerable. But it could be answered, and answered solely in terms of pattern-gathering, without at first invoking any explanatory principle.

EXPLANATIONS

The original, pre-Darwinian problems of biogeography, identified by Kinch (1980), were disjunct distributions and "delineation and enumeration of the regions of creation". The second is an aspect of pattern-gathering, but the first demanded an explanation. Long before Darwin, two alternative natural explanations applicable to disjunct distributions were proposed (Prichard, 1826). One is that they are due to long-range dispersal, for example by seeds in birds' guts or ocean currents. The other is that they are relicts of species which were formerly widespread: "It is possible that a stock of animals which has been exterminated on a continent, may have been preserved in a contiguous island". The disappearance of the former connection between island and mainland may be explained "by referring to the changes, which we have reason to believe the present surface of the earth to have undergone in the course of ages" (Prichard, 1826, *1*:77).

These two alternative explanations are the essence of the distinction between Vuilleumier's (1978) competing paradigms in historical biogeography, i.e. "faunal regions and centers of origin" versus "the dogmas of vicariance". As an explanatory principle, the first paradigm invokes dispersal from a centre of origin, by means which may involve ocean currents or passage through birds' guts. The vicariance paradigm invokes interruption of formerly widespread species by changes in the surface of the earth. The two ideas are by no means mutually exclusive. No Darwin–Wallace centre-of-origin biogeographer would deny that changes in the earth's surface are an important causal factor in biogeography; no vicariance biogeographer would deny that species originate in some area, or that long-range dispersal occurs. If Vuilleumier is correct, that the two schools represent Kuhnian paradigms (and the warmth generated in *Systematic Zoology* and at the New York symposium suggest that there is something in what he says; cf. Ferris, 1980), then there must be more than this between the two schools.

The difference between them may also be expressed in terms of the second pre-Darwinian problem, enumeration and delineation of the regions of creation. In the Darwin–Wallace tradition, there may be as many centres of creation as there are taxa, and delineation of the individual centre for each taxon in time and space is a worthwhile aim. Gadow, by no means a typical Darwin–Wallace biogeographer, put it like this: "the search for generally applicable regions is a mare's nest" (1913:15).

> If all the various groups of creatures had come into existence at the same epoch and at the same place, then it would be possible . . . to construct a map . . . applicable to the whole animal kingdom. But the premises are wrong. . . . The key to the distribution of any group lies in the geographic configuration of that epoch in which it made its first appearance (1913:13).

The research programme implicit in these ideas is to discover the time and place of appearance of each group. This can be contrasted with Croizat's reiterated slogans "earth and life have evolved together" and "nature forever repeats". In other words, it *may* be possible to construct a map applicable to the whole of life; there *may* be a general cladogram of areas. The research programme implicit here is comparison of the distribution of individual groups, looking for congruence. Centres of origin, though they may well have existed, are regarded as relatively unimportant and perhaps unknowable ("the question . . . becomes unnecessary and, perhaps, irrelevant"—Platnick, 1981:149).

"Nevertheless the simplicity of the view that each species was first produced within a single region captivates the mind" (Darwin, 1859:352). The idea of centres of origin seems an inescapable corollary of evolutionary theory and it is natural to regard research into them as a worthwhile aim. If such centres have been discovered or inferred, that would justify the programme and its continuation. In recent years there has been much discussion, most of it critical, of the methods or criteria proposed to localize centres of origin (Croizat *et al.*, 1974; Ross, 1974; Ball, 1976; Craw, 1978; Platnick, 1981; Slater, 1981). It seems to me that only two candidates have survived this criticism: cladograms and the fossil record, preferably combined (this combination, and the fossil record alone, are discussed in detail in Patterson, 1981a).

As an example of the cladogram method, Ross (1974:214) used caddisflies of the genus *Wormaldia*. He explained the method and its interpretation in detail; it is the same as Brundin's (1981:153) "multiple step morphoclines that can be followed over the map". In this instance, Ross's explanation (Fig. 1) is vicariance rather than dispersal. The single disjunct North American species is explained as an isolated relic of a widespread ancestral species. It is legitimate to ask how this explanation might be tested. Because it is a vicariance explanation and specifies no timing for the sequence of events, testing by dating geological barriers or by finding fossils (Platnick and Nelson, 1978:3; Patterson, 1981a:452, 456) is not effective. A test available only to the specialist is to criticize Ross's cladogram by finding new characters which contradict it, or new caddisflies which extend it. Since it is a vicariance explanation, the most feasible test is to follow up the implication that there was a former,

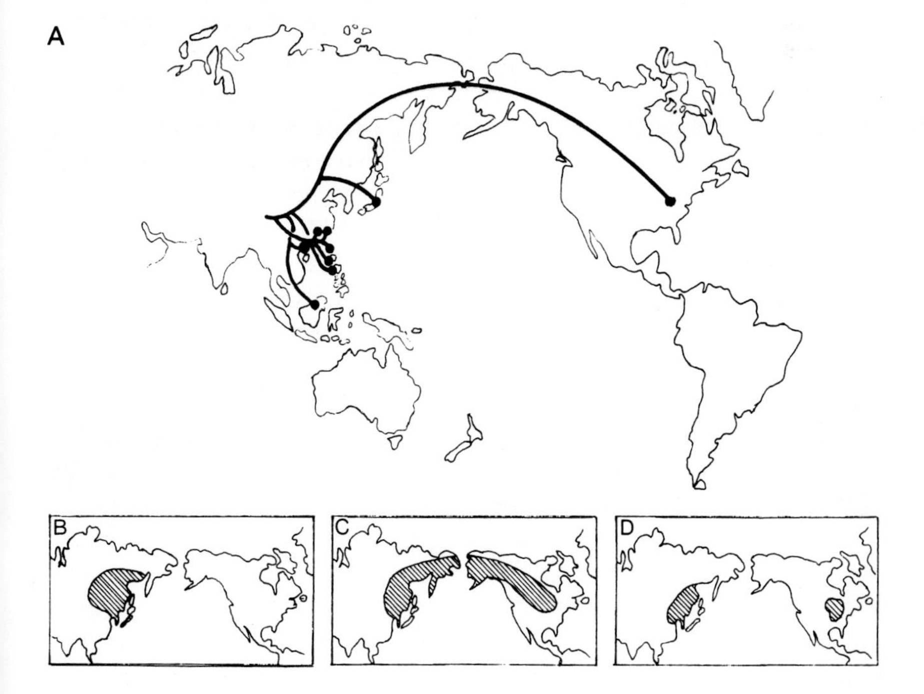

Fig. 1. Ross's (1974) interpretation of the history of the *Wormaldia kisoensis* complex of caddisflies. A, distribution and interrelationships of nine species; B–D, inferred sequence of events responsible for the disjunct Japanese/North American species pair. After Ross (1974: Figs 9.4, 9.5).

widespread east Asian-Nearctic biota. This should have left other traces and the search would be directed towards other groups occupying those areas, to see whether they provided cladograms congruent with Ross's, i.e. towards the question of congruence with a general area cladogram. If such congruence is found, Ross's inference of an Asian centre of origin becomes relatively insignificant. If no congruence is found, a dispersal explanation would be preferable to Ross's vicariance explanation.

As an example of the cladogram combined with fossils, I take Bennett's (1980) analysis of horses. Bennett produced a cladogram, based on skeletal morphology, including all Recent species, nine North American fossil species, and one European and one South American fossil species. By assuming that the fossil record is complete, one may treat all potential ancestors (fossil species without apparent autapomorphies) as actual ancestors, and so produce a phylogram of dispersal (Fig. 2). In this case, multiple dispersals from North America into Eurasia and thence to Africa are resolved. Of this scheme, Bennett (1980:285) wrote

> If various members of *E.* [*Equus*] which are now endemic to Africa migrated through Eurasia to Africa, they now represent relict populations preserved in the once much more widespread savanna environment of Africa.

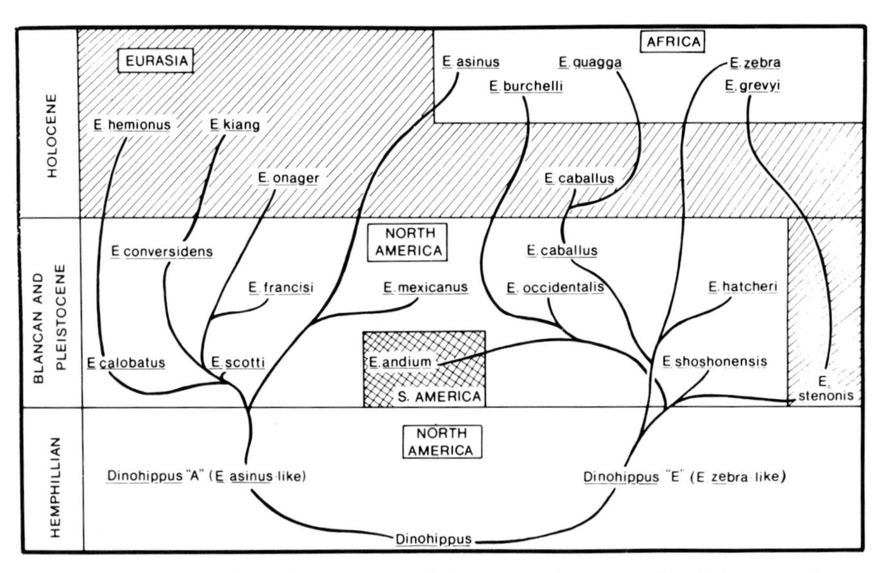

Fig. 2. Zoogeographic phylogram of Recent and certain fossil horses, from Bennett (1980: Fig. 6), by permission.

This statement recalls Prichard's (1826) explanation of island endemics, differential extinction of a formerly widespread species, which in turn implies a vicariance explanation rather than dispersal. Indeed, according to the evolutionary species concept (Wiley, 1978) all five Recent African species in Fig. 2 once also existed in Eurasia and in North America. Why then is dispersal from North American centres postulated, rather than the vicariance explanation of differential extinction in response to change's of the earth's surface? If I understand Bennett correctly, the dispersal interpretation depends on assessments of the fossil record. The North American record is assessed as "excellent and well-documented", whereas that of Eurasia is deficient, for as Bennett says, her analysis predicts that several species of horses are missing from it. But if the fossils are missing, why suppose that what is missing is stratigraphically younger than the American ancestors, rather than of the same age or older, and perhaps more primitive? Bennett's hypothesis does make predictions about the fossil record of North America (complete) and Eurasia (incomplete) that might be tested by further collecting, and by reworking fossils already found (e.g. Davis, 1980). Again, another possible test available to the specialist is to criticize Bennett's cladogram, by finding new characters or new horses.

This has not yet been done, but Eisenmann (1980) has provided an alternative cladogram, including all Recent horses and three Old World fossil species. Her cladogram (Fig. 3) is incongruent with Bennett's in every respect except the sister-group relationship between the African ass (*E. asinus* of Bennett, *E. africanus* of Eisenmann) and the Eurasian

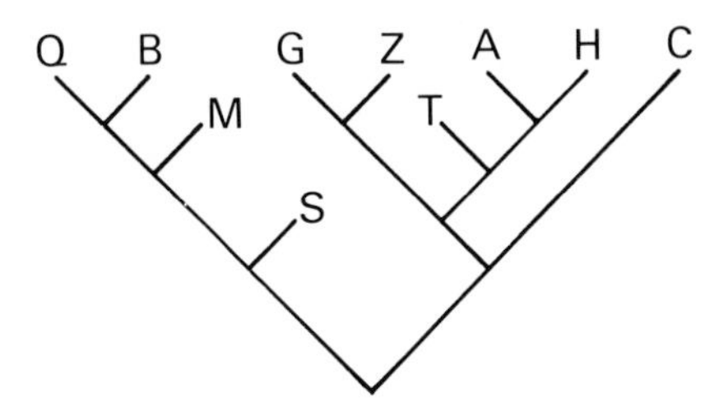

Fig. 3. Cladogram of Recent and certain fossil horses, after Eisenmann (1980: Fig. 67). Nomenclature of species altered to match Fig. 2. A. *Equus asinus*; B, *E. burchelli*; C, *E. caballus*; G. *E. grevyi*; H, *E. hemionus*, *E. kiang* & *E. onager*; M, *E. mauritanicus* (Pleistocene, Algeria); Q, *E. quagga*; S, *E. stenonis* (Pliopleistocene, Europe); T, *E. tabeti* (Pleistocene, Algeria); Z, *E. zebra*.

half-asses (*E. kiang*, *E. onager* and *E. hemionus* of Bennett; *E. kiang* and *E. hemionus* of Eisenmann). Both cladograms cannot be correct. If one assumes that each contains some truth, one can estimate that content only by combining the two (Nelson, 1979, Fig. 4). The result is uninformative, as to both horse interrelationships and biogeography. Horses exist, or have existed, on four continents. Theories of how they got there by dispersal depend on assessments of the fossil record and on the position of fossils in the cladogram (Patterson, 1981a:478). Since Bennett's cladogram is also subject to a vicariance explanation, horses are subject to the vicariance test of congruence with cladograms of other

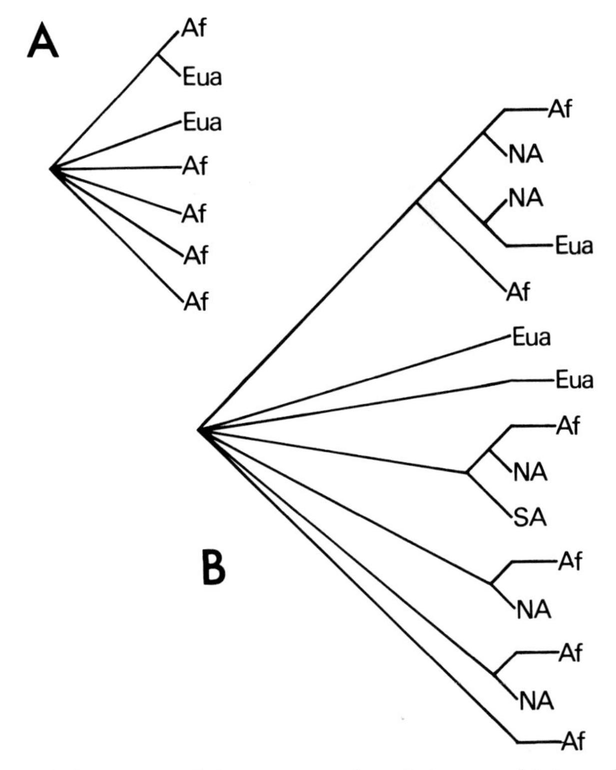

Fig. 4. Area cladograms of horses, produced by combining the species cladograms of Figs 2 and 3. A, Recent species only (the African/Eurasian pairing is A and H of Fig. 3); B, Recent and fossil species, Recent species extended further to the right (North and South American fossils from Fig. 2, African fossils from Fig. 3, the Eurasian fossil *E. stenonis* from Figs 2 and 3). Af, Africa; Eua, Eurasia; NA, SA, North and South America.

groups inhabiting the areas concerned, but this cannot be done until the horse cladogram is resolved in a more informative way.

It seems that the difference between dispersal and vicariance explanations is by no means trenchant. Many apparent dispersals are subsumed under vicariance (Bennett's horses, for example). Random or unique dispersal seems theoretically distinguishable by testing individual taxon cladograms against one another; but mass or concerted biotic dispersal will not be distinguishable from vicariance by this method. Another criterion for dispersal frequently cited (Croizat *et al.*, 1974; Rosen, 1976; Platnick and Nelson, 1978; Brundin, 1981) is sympatry: if two taxa are sympatric, one or both must have dispersed. This criterion depends on the assumption that all speciation is allopatric. Vicariists acknowledge that their assumptions may lead them to underestimate the role of dispersal (Platnick and Nelson, 1978:7). In turn, centre-of-origin biogeographers may overestimate the role of dispersal, assuming it when a vicariance explanation is equally valid (e.g. horses, above; Nelson, 1975). If some speciation is sympatric (as it seems to be in plants, in particular), the criterion of sympatry will also overestimate the role of dispersal.

The difference between dispersal and vicariance explanations may be explored further by analogy between area cladograms and taxon cladograms. Taxon cladograms can be constructed by ordering the characters of organisms into character-state trees, and accepting that cladogram which is conguent with the greatest number of those trees. Having settled on the cladogram which accounts for character distribution parsimoniously (and having checked and criticized the characters as rigorously as possible), one may pass from pattern to process by converting the cladogram into a phyletic tree, concerned with descent and modification rather than the pattern of character distribution. A cladogram represents a set of trees (Harper, 1976; Platnick, 1977), which are of two sorts, those in which one or more terminal taxa in the cladogram are designated ancestors in the tree, and those in which no ancestors are designated ("X" trees). "X" trees stand in one-to-one relationship with cladograms. But "non-X" trees, with specified ancestors, outnumber cladograms (Fig. 5). For example, a three-taxon cladogram is equivalent to at least five "non-X" trees (Harper and Platnick, 1978).

These ideas can be applied to area cladograms in just the same way. An area cladogram is a parsimonious summary of taxon cladograms,

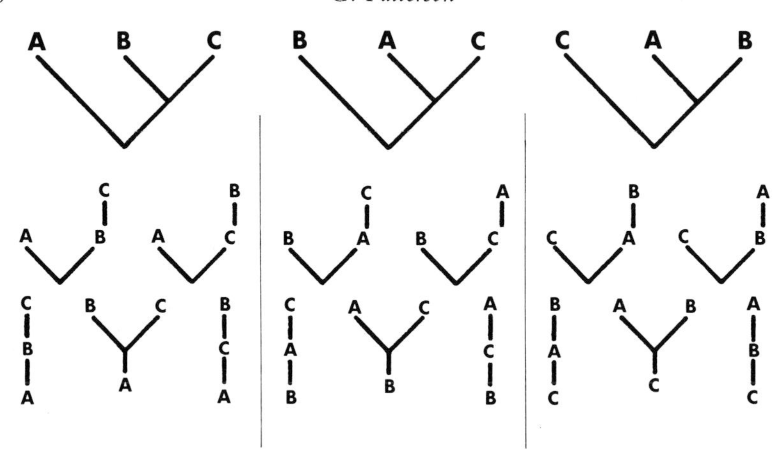

Fig. 5. Cladograms and trees. Top row, the three possible cladograms for three taxa, A, B and C. Beneath each are the five trees compatible with the cladogram. Junctions in trees represent unspecified ancestors ("X"s), and each cladogram is equivalent to an "X tree". A, B and C may also be thought of as areas rather than taxa: junctions in area trees ("X"s) may then represent summation of the descendent areas.

just as a taxon cladogram is a parsimonious summary of character cladograms. An area cladogram may be converted into a tree by invoking process—descent with modification—and the trees will be of the same two sorts as species trees, "X" area trees and "non-X" area trees. "X" area trees, with no designated ancestral areas, stand in one-to-one relationship with area cladograms. The unnamed ancestral areas in "non-X" area trees will be a summation of the descendent areas, and vicariance will be the explanation invoked. "Non-X" area trees, with specified ancestral areas, outnumber area cladograms (Fig. 5). For example, a three-area cladogram is equivalent to at least five "non-X" area trees. In "non-X" area trees, the named ancestral areas may be terminal areas in the cladogram, or other areas, and the explanation invoked will be biotic dispersal.

The problems of designating ancestors in taxon cladograms have been thoroughly explored (Engelmann and Wiley, 1977; Szalay, 1977; Harper and Platnick, 1978; Cracraft and Eldredge, 1979; Patterson, 1981c). They are soluble by making assumptions: that the fossil record is complete, that evolution is irreversible, and that hybridization between species has not occurred. The problems of designating ancestors in area

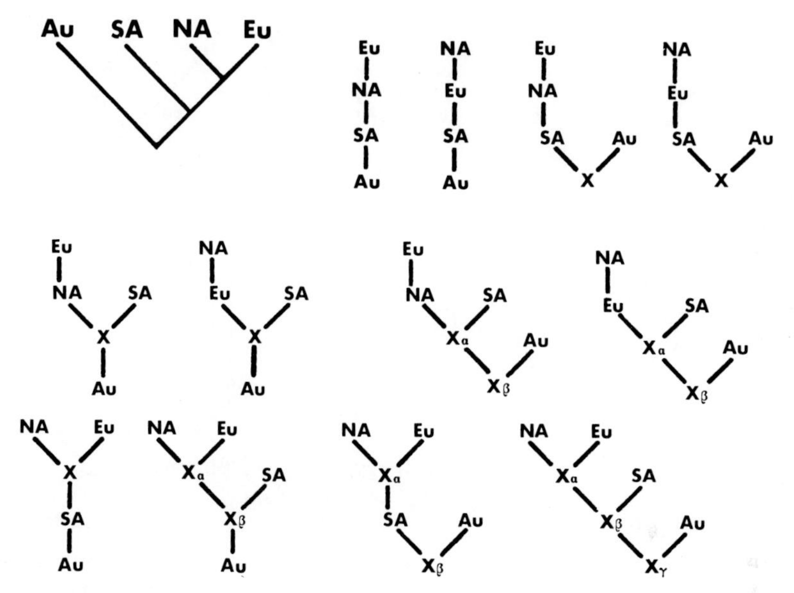

Fig. 6. Area cladograms and trees. Top left, the area cladogram for marsupial mammals (after Patterson, 1981a: Fig. 11.7B), omitting the relationship between North and South America indicated by *Didelphis virginiana*, here assumed to have dispersed into North America recently. Au, Australia and New Guinea; Eu, Europe; NA, SA, North and South America. There are 12 area trees compatible with the cladogram, 11 of which are "non-X area trees", and one is the "X tree" (bottom right).

cladograms have not been explored so thoroughly, though discussions of criteria for recognizing centres of origin concern these problems. I suggest that they are soluble by making equivalent assumptions to those necessary in taxon cladograms: that the geological record is complete, that dispersal is irreversible, and that hybridization between areas has not occurred. Those who postulate centres of origin or biotic dispersal might consider these assumptions, and also the large number of alternative "non-X" area trees compatible with a given area cladogram.

As an example of these numerous alternatives, Marshall's (1980) summary of marsupial historical biogeography reviews and illustrates no less than 12 narratives, three assuming fixed continents and nine based on mobile continents, of which "at least six are still viable" (p. 381). Each of these narratives can be recast as an area tree (Fig. 7). As Fig. 7 shows, all are "non-X" trees, and all introduce named "X"s which are not known to have housed marsupials now or in the past (e.g.

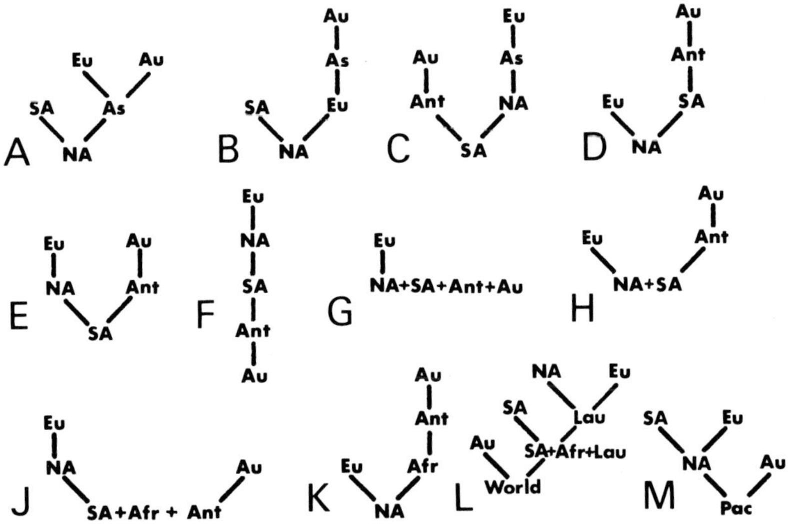

Fig. 7. Theories of marsupial history, adapted from Marshall (1980) as area trees. A–C, "Stabilist Models 1–3" of Marshall (1980: Figs 21.4–6); D–M, "Mobilist Models 1–9" of Marshall (1980: Figs 21.8–16). Afr, Africa; Ant, Antarctica; As, Asia; Au, Australia and New Guinea; Eu, Europe; Lau, Laurasia; NA, North America; Pac, Pacifica; SA, South America.

According to Marshall, A–C are now untenable because marsupials are unknown in Asia, and because of the inferred separation of Australia and Asia in the late Mesozoic and early Tertiary; G is suspect because it implies a widespread origin for marsupials, and continuous land connections between the four ancestral areas; J, K and L are suspect because marsupials are unknown in Africa and because K and L propose that marsupials are older than placentals; and M is supect because of lack of evidence for Pacifica.

Comparison with Fig. 6 shows that only four of the area trees (F, G, J, L) are compatible with the marsupial area cladogram; one of these (L) is equivalent to the 'X tree'. All 12 theories introduce one or more areas in which marsupials are unknown (Antarctica[1], Africa, Asia, Pacifica), so that none of these theories is exactly equivalent to any one of the 12 area trees compatible with the marsupial area cladogram (Fig. 6).

Africa, Antarctica[1], Pacifica). The marsupial area cladogram, as currently resolved (Fig. 6) permits at least eleven "non-X" trees, if the potential ancestral areas are restricted to those in the cladogram. If additional potential ancestors are introduced, such as Africa and Pacifica, there is no limit to the number of "non-X" area trees compatible with the cladogram. From this viewpoint, Marshall's assessment that amongst the alternative stories of marsupial dispersal "at least six are still viable" is less surprising; it might be more interesting to ask how the innumerable other candidates were eliminated.

It is obvious from this discussion that biotic dispersal, as an explanatory principle, is closely analogous to the designation of ancestral species in phylogenies. That practice has been criticized as leading to untestable hypotheses, protected from criticism by a hedge of *ad hoc* assumptions (fossil record complete, evolution irreversible, and so on). In biogeography, equivalent assumptions (geological record complete, dispersal irreversible, and so on) are necessary in designating ancestral areas. A more practical point is the unwillingness of palaeontologists and phyleticists to designate ancestral species with any conviction. Biogeographers seem to have been more ready to designate ancestral areas and centres of origin, perhaps because an area need not be named with the same precision as a species, so that the analogy is more with ancestral groups than ancestral species. Nevertheless, the conviction with which centres of origin are supported seems no greater than for ancestral species.

CONCLUSIONS

I began by asking what aims, principles and methods are appropriate for biogeography in 1981. Biogeography has to begin with systematics, which is the basic tool in capturing pattern. Reporting the 1979 New York meeting, Ferris (1980:68) found the "relative absence of controversy about the cladistic method to be noteworthy". In the hope that past controversies have subsided further during the ensuing two years, I assume that cladistics provides the most efficient "non-arbitrary taxonomy that can be correctly comprehended by the generalist" (Rosen, 1976:433).

Biogeographers, as generalists, scarcely have the opportunity to test or criticize individual taxon cladograms. Nor do they have the opportunity to test assessments of the fossil record. Since the search for fossils still goes on, the only general assessment one can make is that the record is incomplete (Patterson, 1981b:463) and, therefore, that schemes assuming it to be complete are suspect.

If specialists continue to produce taxon cladograms, the most fruitful aim for the biogeographer is surely to compare them one with another, asking to what extent they are geographically congruent. If congruent, they can be combined in area cladograms. Given enough taxon cladograms, biogeographers can answer Nelson and Platnick's (1981:540) question (see above, p. 5). This aim is concerned only with pattern.

Turning to process, or explanation, our aims must depend on answers to questions of pattern. *If* there is only one, or a few, general cladograms of areas, our explanations will be constrained by that simple pattern, and will turn more towards earth history. If there are a myriad area cladograms, or no general pattern, explanations will turn more towards Wallace's aim of applied theories of life. The theories that may be applied will include neo-Darwinism and equilibrium theory. Whatever the answer in terms of pattern, one single pattern or a multitude, the pattern or patterns may be used as a test of theories of earth history (plate tectonics versus earth expansion, for example), and as a test of theories of life (e.g. modes of speciation; dispersal versus vicariance; and more general matters like the randomness predicted by opportunistic neo-Darwinism). But all this is in the future. What we lack still is the raw material, "delineation and enumeration of the regions of creation", and of the relation between them.

ACKNOWLEDGEMENTS

I am most grateful to Drs D.E. Rosen and G. Nelson, American Museum of Natural History, for supplying proof copies of unpublished material, and to Dr Nelson for comments on my manuscript.

REFERENCES

Arldt, T. (1907). "Die Entwicklung der Kontinente und Ihrer Lebewelt". W. Engelmann, Leipzig.

Ball, I.R. (1976). Nature and formulation of biogeographical hypotheses. *Syst. Zool.* **24**, 407–430.

Banarescu, P. (1975). "Principles and Problems of Biogeography". NOLIT, Belgrade. [For National Technical Information Service, Springfield, Virginia.]

Bennett, D.K. (1980). Stripes do not a zebra make, Part I: a cladistic analysis of *Equus*. *Syst. Zool.* **29**, 272–287.

Brundin, L. (1966). Transantarctic relationships and their significance, as evidenced by chironomid midges. *K. svenska VetenskAkad. Handl.* (4) **11**, 1–472.

Brundin, L. (1981). Croizat's Panbiogeography versus phylogenetic biogeography. *In* "Vicariance Biogeography: a Crtitique" (G. Nelson and D.E. Rosen, eds), pp. 94–158. Columbia University Press, New York.
Carr, T.R. and Kitchell, J.A. (1981). Dynamics of taxonomic diversity. *Paleobiology* **6**, 427–443.
Cracraft, J. and Eldredge, N. (eds) (1979). "Phylogenetic Analysis and Paleontology". Columbia University Press, New York.
Craw, R.C. (1978). Two biogeographical frameworks: implications for the biogeography of New Zealand. A review. *Tuatara* **23**, 81–114.
Croizat, L. (1964). "Space, Time, Form: the Biological Synthesis". L. Croizat, Caracas.
Croizat, L., Nelson, G. and Rosen, D.E. (1974). Centers of origin and related concepts. *Syst. Zool.* **23**, 265–287.
Darlington, P.J., Jr (1957). "Zoogeography: the Geographical Distribution of Animals". Wiley, New York.
Darlington, P.J., Jr (1965). "Biogeography of the Southern End of the World". Harvard University Press, Cambridge, Mass.
Darwin, C. (1859). "On the Origin of Species by Means of Natural Selection". John Murray, London.
Darwin, F. and Seward, A.C. (1903). "More Letters of Charles Darwin". 2 Vols. John Murray, London.
Davis, S.J.M. (1980). Late Pleistocene and Holocene equid remains from Israel. *Zool. J. Linn. Soc.* **70**, 289–312.
Edmunds, G.F., Jr (1975). Phylogenetic biogeography of mayflies. *Ann. Mo. bot. Gdn* **62**, 251–263.
Edmunds, G.F., Jr (1981). Discussion of R. Melville's paper. *In* "Vicariance Biogeography: a Critique" (G. Nelson and D.E. Rosen, eds), pp. 287–297. Columbia University Press, New York.
Eisenmann, V. (1980). "Les chevaux (*Equus* sensu lato) fossiles et actuels: crânes et dents dugales supérieures". Centre National de la Recherche Scientifique, Paris.
Eldredge, N. (1981). Discussion of M.D.F. Udvardy's paper. *In* "Vicariance Biogeography: a Critique" (G. Nelson and D.E. Rosen, eds), pp. 34–38. Columbia University Press, New York.
Engelmann, G.F. and Wiley, E.O. (1977). The place of ancestor-descendant relationships in phylogeny reconstruction. *Syst. Zool.* **26**, 1–11.
Ferris, V.R. (1980). A science in search of a paradigm? *Syst. Zool.* **29**, 67–76.
Fichman, M. (1977). Wallace: zoogeography and the problem of land bridges. *J. Hist. Biol.* **10**, 45–63.
Gadow, H. (1909). Geographical distribution of animals. *In* "Darwin and Modern Science" (A.C. Seward, ed.), pp. 319–336. Cambridge University Press, Cambridge.
Gadow, H. (1913). "The Wanderings of Animals". Cambridge University Press, Cambridge.
George, W. (1964). "Biologist Philosopher". Abelard-Schuman, London.

Harper, C.W. (1976). Phylogenetic inference in paleontology. *J. Paleont.* **50**, 180–193.

Harper, C.W. and Platnick, N.I. (1978). Phylogenetic and cladistic hypotheses: a debate. *Syst. Zool.* **27**, 354–362.

Harper, G.H. (1980). Speciation or irruption: the significance of the Darwin finches. *J. biol. Educ.* **14**, 99–106.

Hennig, W. (1966). "Phylogenetic Systematics". University of Illinois Press, Urbana.

Holloway, J.D. and Jardine, N. (1968). Two approaches to zoogeography: a study based on the distributions of butterflies, birds and bats in the Indo-Australian area. *Proc. Linn. Soc. Lond.* **179**, 153–188.

Humphries, C.J. (1981). Biogeographical methods and the southern beeches. *In* "The Evolving Biosphere" (P.L. Forey, ed.), pp. 283–297. British Museum (Natural History) and Cambridge University Press, Cambridge.

Huxley, T.H. (1888). The gentians: notes and queries. *J. Linn. Soc. Bot.* **24**, 101–124.

Keast, A. (1973). Contemporary biotas and the separation sequence of the southern continents. *In* "Implications of Continental Drift to the Earth Sciences" (D.H. Tarling and S.K. Runcorn, eds), vol. 1, pp. 309–343. Academic Press, London and New York.

Kinch, M.P. (1980). Geographical distribution and the origin of life: the development of early nineteenth-century British explanations. *J. Hist. Biol.* **13**, 91–119.

Kinsey, A.C. (1930). The gall wasp genus *Cynips*. A study in the origin of species. *Indiana Univ. Stud.* **84–86**, 1–577.

Kinsey, A.C. (1936). The origin of higher categories in *Cynips*. *Indiana Univ. Publs Sci. Ser.* **4**, 1–334.

MacArthur, R.H. and Wilson, E.O. (1967). "The Theory of Island Biogeography". Princeton University Press, Princeton, N.J.

Marshall, L.G. (1980). Marsupial paleobiogeography. *In* "Aspects of Vertebrate History" (L.L. Jacobs, ed.), pp. 345–386. Museum of Northern Arizona Press, Flagstaff.

Matthew, W.D. (1915). Climate and evolution. *Ann. N.Y. Acad. Sci.* **24**, 171–318.

Mayr, E. (1952). Conclusion. *Bull. Am. Mus. nat. Hist.* **99**, 255–258.

Mayr, E. *et al.* (1952). The problem of land connections across the South Atlantic, with special reference to the Mesozoic. *Bull. Am. Mus. nat. Hist.* **99**, 79–258.

Melville, R. (1981). Vicarious plant distributions and paleogeography of the Pacific region. *In* "Vicariance Biogeography: a Critique" (G. Nelson and D.E. Rosen, eds), pp. 238–302. Columbia University Press, New York.

Morse, J.C. and White, D.F. (1979). A technique for analysis of historical biogeography and other characters in comparative biology. *Syst. Zool.* **28**, 356–365.

Nelson, G. (1969). The problem of historical biogeography. *Syst. Zool.* **18**, 243–246.

Nelson, G. (1973). Comments on Leon Croizat's biogeography. *Syst. Zool.* **22**, 312–320.

Nelson, G. (1975). Historical biogeography: an alternative formalization. *Syst. Zool.* **23**, 555–558.

Nelson, G. (1976). Biogeography, the vicariance paradigm, and continental drift. *Syst. Zool.* **24**, 490–504.

Nelson, G. (1978). From Candolle to Croizat: comments on the history of biogeography. *J. Hist. Biol.* **11**, 269–305.

Nelson, G. (1979). Cladistic analysis and synthesis: principles and definitions, with a historical note on Adanson's *Familles des Plantes* (1763–1764). *Syst. Zool.* **28**, 1–21.

Nelson, G. and Platnick, N.I. (1980a). A vicariance approach to historical biogeography. *Bioscience* **30**, 339–343.

Nelson, G. and Platnick, N.I. (1980b). Multiple branching in cladograms: two interpretations. *Syst. Zool.* **29**, 86–91.

Nelson, G. and Platnick, N.I. (1981). "Systematics and Biogeography: Cladistics and Vicariance". Columbia University Press, New York.

Nelson, G. and Rosen, D.E. (eds) (1981). "Vicariance Biogeography: a Critique". Columbia University Press, New York.

Nur, A. and Ben-Avraham, Z. (1981). Lost Pacifica continent: a mobilistic speculation. *In* "Vicariance Biography: a Critique" (G. Nelson and D.E. Rosen, eds), pp. 341–358. Columbia University Press, New York.

Patterson, C. (1980). Cladistics. *Biologist, Lond.* **27**, 234–240.

Patterson, C. (1981a). Methods of paleobiogeography. *In* "Vicariance Biogeography: a Critique" (G. Nelson and D.E. Rosen, eds), pp. 446–500. Columbia University Press, New York.

Patterson, C. (1981b). The development of the North American fish fauna—a problem of historical biogeography. *In* "The Evolving Biosphere" (P.L. Forey, ed.), pp. 265–281. British Museum (Natural History) and Cam- University Press, Cambridge.

Patterson, C. (1981c). Significance of fossils in determining evolutionary relationships. *A. Rev. Ecol. Syst.* **12**, 195–223.

Platnick, N.I. (1977). Cladograms, phylogenetic trees, and hypothesis testing. *Syst. Zool.* **26**, 438–442.

Platnick, N.I. (1980). Philosophy and the transformation of cladistics. *Syst. Zool.* **28**, 537–546.

Platnick, N.I. (1981). The progression rule or progress beyond rules in biogeography. *In* "Vicariance Biogeography: a Critique" (G. Nelson and D.E. Rosen, eds), pp. 144–150. Columbia University Press, New York.

Platnick, N.I. and Nelson, G. (1978). A method of analysis for historical biogeography. *Syst. Zool.* **27**, 1–16.

Prichard, J.C. (1826). "Researches into the Physical History of Mankind" (2nd edn), 2 vols. J. & A. Arch, London.

Rosa, D. (1931). "L'Ologénèse". Félix Alcan, Paris.

Rosen, D.E. (1976). A vicariance model of Caribbean biogeography. *Syst. Zool.* **24**, 431–464.

Rosen, D.E. (1978). Vicariant patterns and historical explanation in biogeography. *Syst. Zool.* **27**, 159–188.
Ross, H.H. (1956). "Evolution and Classification of the Mountain Caddisflies". University of Illinois Press, Urbana.
Ross, H.H. (1974). "Biological Systematics". Addison-Wesley, Reading, Mass.
Schopf, T.J.M. (1979). The role of biogeographic provinces in regulating marine faunal diversity through geologic time. *In* "Historical Biogeography, Plate Tectonics, and the Changing Environment" (J. Gray and A.J. Boucot, eds), pp. 449–457. Oregon State University Press, Corvallis.
Simpson, G.G. (1965). "The Geography of Evolution". Chilton Books, Philadelphia and New York.
Simpson, G.G. (1980). "Why and How: some Problems and Methods in Historical Biology". Pergamon Press, Oxford.
Simberloff, D. (1974). Equilibrium theory of island biogeography and ecology. *A. Rev. Ecol. Syst.* **5**, 161–182.
Slater, J.A. (1981). Discussion of L. Brundin's paper. *In* "Vicariance Biogeography: a Critique" (G. Nelson and D.E. Rosen, eds), pp. 139–143. Columbia University Press, New York.
Szalay, F.S. (1977). Ancestors, descendants, sister groups and testing of phylogenetic hypotheses. *Syst. Zool.* **26**, 12–18.
Thiselton-Dyer, W. (1909). Geographical distribution of plants. *In* "Darwin and Modern Science" (A.C. Seward, ed.), pp. 298–318. Cambridge University Press, Cambridge.
Valentine, J.W. (1969). Patterns of taxonomic and ecological structure of the shelf benthos during Phanerozoic time. *Palaeontology* **12**, 684–709.
Vuilleumier, F. (1978). Qu'est-ce que la biogéographie? *C.r. somm. Séanc. Soc. Biogéogr.* **54**, 41–66.
Wallace, A.R. (1876). "The Geographical Distribution of Animals", 2 vols. Macmillan, London.
Wallace, A.R. (1880). "Island Life". Macmillan, London.
Wiley, E.O. (1978). The evolutionary species concept reconsidered. *Syst. Zool.* **27**, 17–26.

NOTE

[1] An Eocene polydolopid marsupial has since been reported from Antarctica (Woodburne and Zinsmeister, 1982, *Science* **218**, 284–286).

2 | Spatial and Temporal Change in Biogeography: Gradual or Abrupt?

E.C. PIELOU★

Biology Department, Dalhousie University, Halifax, Nova Scotia

Abstract: Consider a sequence of multivariate observations (e.g. species abundance distributions). The sequence may be spatial or temporal. The pattern of change along the sequence may be shown graphically by an ordered similarity matrix (OSM) whose elements are pairwise similarity (resemblance) coefficients between all possible pairs of observations. The overall appearance of an OSM usually allows one to judge intuitively whether the observations exhibit change along the sequence; and, if they do, whether the change is gradual (continuous) or stepwise (discontinuous). This paper describes a method that allows these judgements to be made objectively rather than subjectively. The "disarray" of an empirical OSM can be measured, and the value compared with the expected disarray in an OSM representing a randomly ordered sequence of observations. One can also compare observed and expected disarrays within parts of a sequence that are thought to represent trend-free stretches between episodes of relatively abrupt change. To illustrate, the method is applied to data on Holocene communities of benthic Foraminifera in the Bay of Fundy. The analysis contributes to an understanding of post-glacial change in a shelf environment affected by rising and falling post-glacial sea levels.

INTRODUCTION

The one theme common to all biogeographic research is change: changing biotas and changing species, changing over space and through

★ Present address: Biology Department, University of Lethbridge, Lethbridge, Alberta, Canada T1K 3M4.

Systematics Association Special Volume No. 23, "Evolution, Time and Space: The Emergence of the Biosphere", edited by R.W. Sims, J.H. Price and P.E.S. Whalley, 1983, pp. 29–56. Academic Press, London and New York.

time. Paleontologists, studying temporal change over thousands or millions of years, are concerned with evolving (i.e. changing) lineages and communities. Neontologists, studying present-day biogeographic patterns, are concerned with spatial variation, some with the morphological or demographic properties of chosen species, others with the changing taxonomic or physiognomic composition of floras and faunas. In all these studies, change, its nature, magnitude, and rate, is the principal theme.

Whenever and wherever change occurs, one is usually less interested in its average rate than in the way the rate itself changes: is change steady and continuous, or does it proceed by fits and starts? Indeed, in the context of temporal change, the problem seems to obsess many students of evolution (see Fig. 1); debate rages between proponents of two contrasting theories, the "gradualistic model" and the "punctuation model" (Gingerich, 1974; Cracraft and Eldredge, 1979; Stanley, 1979).

The problem also arises, if with less sound and fury, in the context of spatial change. As Webb (1954, quoted in Greig-Smith, 1957) wrote: "The fact is that the pattern of variation shown by the distribution of species . . . hovers in a tantalizing manner between the continuous and the discontinuous."

It therefore seems worthwhile to devise a method for discriminating between continuous and discontinuous (or stepwise) change. What is needed is a procedure that can be applied in a wide array of circumstances, to data of very varied kinds. Further, one often wishes to judge

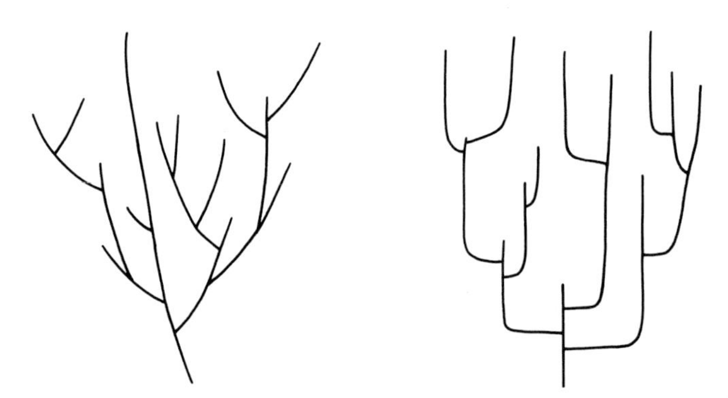

Fig. 1. Two contrasted forms of phylogenetic tree, conforming with the gradualistic model (left) and the punctuational model (right) of evolution. (Adapted from Stanley, 1979).

the degree of correspondence between the modes of change, whether stepwise or continuous, of two different sets of variables, say those specifying the physical environment of a biotic community and those specifying its species composition. The aim of this paper is to describe a way of tackling these problems.

I shall be considering ordered similarity matrices. An ordered similarity matrix (OSM) is a matrix whose (i, j)th element is an index measuring the similarity between the ith and jth entities (e.g. organisms, species, habitats, communities) in an ordered sequence (spatial or temporal) of entities. The whole matrix thus contains a similarity index for every possible pair of entities. For instance, Valentine (1973:353) presents an OSM whose elements are the similarities, as measured by Jaccard's coefficient, between the benthic mollusk faunas of segments of the continental shelf of western North America; the segments are contiguous stretches of the shelf, each one degree of latitude in extent. Other examples appear in Pielou (1979) and later in this paper; they show the similarity in species composition, as measured by the proportional similarity PS (defined below), between foraminiferal assemblages in sequences of sediment samples from cores collected from the sea floor.

Since OSMs are necessarily symmetrical, only their upper right halves will be shown in figures; the cells on the principal diagonal, whose elements are all 100% (the similarity of an entity to itself) are left blank.

OSMs are obviously useful as pictorial representations of change. If the cells of an OSM are shaded, with darker shading denoting closer similarity, the visible pattern is sometimes intuitively interpretable in an obvious way; thus clear examples of stepwise change are readily distinguishable from clear examples of continuous change (see Fig. 2). However, as will be shown below, one can do more with an OSM than merely look at it and judge it intuitively. The patterns (there may be more than one) contained in an OSM can be quantified in a way that permits statistical hypotheses to be tested and different patterns to be compared.

In what follows, the theory underlying the proposed method is discussed first. Then examples are given of its application to the study of foraminiferal assemblages in sediment cores from the Bay of Fundy, and of the light they shed on Holocene changes in the marine environment of the region.

TESTING THE RANDOMNESS OF AN OSM

1. Random OSMs: Two Definitions

An OSM may show an easily interpretable pattern, as do those in Fig. 2, or the pattern may be, in a colloquial sense, random. If a random OSM is to be used as a standard with which nonrandom (or "patterned") OSMs can be compared, it is necessary first to specify precisely what constitutes a random OSM. There are, however, two ways of defining randomness in the context of OSMs. These ways are conceptually entirely different but, as will be shown, the difference is negligible in practice.

According to the first definition, an OSM is random if its elements can be thought of as having been randomly and independently assigned

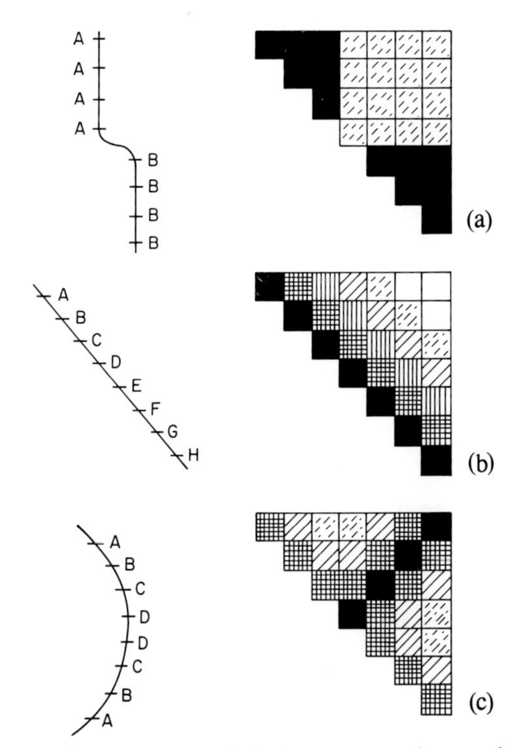

Fig. 2. Three types of sequences and their corresponding ordered similarity matrices. (a) A stepped sequence; (b) a perfectly-graded sequence; (c) a cyclical sequence.

to their positions in the matrix. Such an OSM will be called primitively random. In truth, if one constructs a matrix in this fashion, it may be found to have properties that a real OSM could not have. Consider, for instance, an OSM relating to mollusk communities at a row of sites along a shore. Let s_{ij} denote the similarity (however measured) between the communities at sites i and j. Then if s_{ai} and s_{iz}, say, both have high values, s_{az} must of necessity be fairly high; it cannot be very low. In other words, the elements of a true OSM are not independent of one another as is the case, by definition, in a primitively random OSM. A primitively random OSM could, indeed, more accurately be called a "pseudo-OSM."

According to the second definition of randomness, an OSM is random if its elements are the similarities between entities whose ordering is random. Such an OSM will be called secondarily random. Returning to the example used above, suppose data were available on the mollusk communities at a row of sites. If the data sets were randomly permuted to give a wholly random ordering before an OSM was constructed, the result would be a secondarily random OSM.

It will be demonstrated below that primitively and secondarily random OSMs are indistinguishable as standards of randomness when the pattern of an OSM is measured in the manner now to be described.

2. Measuring the Disarray of an OSM

A sequence of entities may be described as perfectly ordered if its OSM is of the type shown in Fig. 2b. The OSM of an imperfectly ordered sequence exhibits "disarray", and a convenient measure of this disarray may be defined as follows (see Fig. 3).

Consider an OSM of size $n \times n$. It has $n - 1$ diagonals (or super-diagonals) above and parallel with the principal diagonal. Let these diagonals be ranked as shown in Fig. 3. The diagonal r has r elements for $r = 1, 2, \ldots, (n-1)$. Now consider the tth element in diagonal r. Suppose its numerical value is x. Define q_{rt} as the number of elements in diagonals $(r+1), (r+2), \ldots, (n-1)$ with values less than x. Define Q as

$$Q = \sum_{r=1}^{n-2} \sum_{t=1}^{r} q_{rt}$$

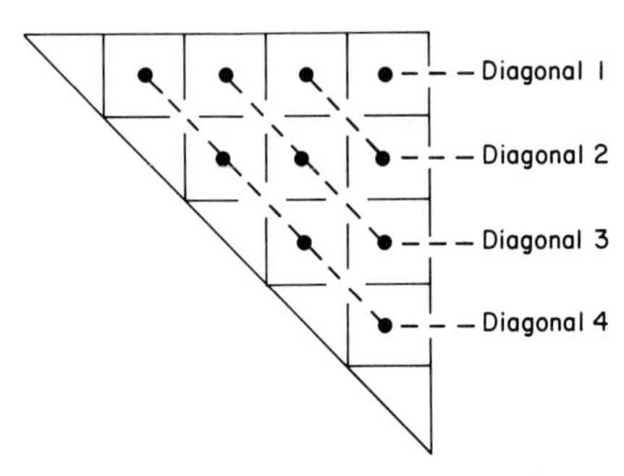

Fig. 3. To illustrate the computation of Q; see text.

Thus Q is the total number of different pairs of elements for which the larger member of the pair is in a lower ranking diagonal than the smaller member of the pair. Tied pairs each contribute one-half to the sum Q.

It should be noticed that it is the ranks of the similarity indices in the OSM, not their absolute magnitudes, that determine the value of Q. If these ranks, from 1 for the smallest to $n(n-1)/2$ for the largest, are substituted for the similarity indices in an OSM, Q can be calculated using the same formula. This makes it easy to construct a primitively random OSM; it is done by randomly permuting the integers 1, 2, . . . , $(n-1)/2$ and placing them in the OSM's cells.

3. The Distribution of Q in Random OSMs

To test whether an observed OSM differs significantly from one that might have been yielded by a randomly ordered sequence of entities, we need to know the distribution of Q in random OSMs.

First, consider primitively random OSMs. As explained above, these are "pseudo-OSMs", but the justification for considering them first is that they are easy to construct, as just described; and further, as will be demonstrated below, they yield a distribution of Q indistinguishable from that yielded by secondarily random OSMs.

Given an OSM of size $n \times n$, the possible values of Q range from a minimum of $Q = 0$ if the OSM has zero disarray (as in Fig. 2b) to a maximum, Q_m, which is easily found to be

$$Q_m = \binom{n}{2} - \sum_{j=1}^{n-1} j\binom{j+1}{2}.$$

This maximum is attained by an OSM in which the pattern is the exact reverse of that having $Q=0$; that is, the similarity indices steadily decrease from the top right corner (diagonal 1) to the longest super-diagonal (diagonal $n-1$).

Table I shows values of Q_m for $n=4, 5, \ldots, 31$. A useful recurrence relation for Q_m is

$$Q_m(n+1) = Q_m(n) + n^2(n-1)/2.$$

In earlier work (Pielou, 1979), the distribution of Q in primitively random OSMs was estimated by computer simulation. Six values of n were tried, namely $n=4$, 6, 8, 11, 16 and 21. For each n, a batch of primitively random OSMs, with ranks in place of similarity indices,

Table I. Values of Q_m for $n=4, 5, \ldots, 31$.

n	Q_m	n	Q_m
4	11	18	10812
5	35	19	13566
6	85	20	16815
7	175	21	20615
8	322	22	25025
9	546	23	30107
10	870	24	35926
11	1320	25	42550
12	1925	26	50050
13	2717	27	58500
14	3731	28	67977
15	5005	29	78561
16	6580	30	90335
17	8500	31	103385

was constructed. The value of Q for each simulated OSM was determined.

The results showed the distribution of Q to be symmetrical. They also permitted estimates to be obtained of the 5th, 10th, 90th and 95th percentiles of the distribution, for each n. Since the distributions are symmetrical, it follows that the median of Q (equivalently, the 50th percentile) is $Q_m/2$. Thus the median for any n can be determined exactly instead of being estimated, as were the other percentiles, from computer simulations.

Now consider secondarily random OSMs. As already explained, empirical OSMs can only be secondarily, not primitively, random. Strictly speaking, therefore, to test the randomness of an empirical OSM, one should ask: What is the distribution of Q in secondarily random OSMs?

There is, however, no unique answer to this question. In calculating the elements of an OSM, one is free to choose any of the multitude of similarity indices that have been defined over the years by a host of research workers; (for a list of indices used in ecology, see Goodall, 1978; for indices used in biostratigraphy, see Simpson, 1960, and Cheetham and Hazel, 1969; for indices used in taxonomy, see Sneath and Sokal, 1973). It cannot be assumed, although it seems intuitively likely, that the distribution of Q in secondarily random OSMs is independent of the functional form of the similarity index used to construct the OSM. In theory one could obtain, by computer simulation, the distribution of Q in secondarily random OSMs for each different similarity index treated separately. But this is not feasible.

It therefore seems best to treat the distribution of Q in primitively random OSMs as, so to speak, the fundamental criterion of randomness. One can check on an *ad hoc* basis whether it adequately represents the distribution of Q that would be strictly appropriate for the similarity index selected for use in any given case. An example of such a check is given below.

First, however, note the analogy between this approach and the one customarily taken by users of standard statistical tests that require the variables concerned to have normal distributions. An investigator either assumes that the variable of interest is normally distributed or, if he is conscientious, tests the data for normality before proceeding with a test. Suppose the result of such a test is affirmative. Then what has been shown is that the variable is "virtually" normal, in other words, indistinguishable from normal by the test being used. But it is probably never true that a variable of practical interest really has a normal distribution, for which the range, in theory, is always from $-\infty$ to $+\infty$.

The definition of proportional similarity (PS) is as follows. Suppose the species abundance distributions of two ecological collections are available and PS is to be used to measure their similarity. Let p_{1j} be the proportion of collection 1 that belongs to species j; let p_{2j} be defined analogously. Then

$$\mathrm{PS} = \sum_{j=1}^{s} \min(p_{1j}, p_{2j})$$

where s is the total number of species in the two collections.

Figures 4 and 5 were constructed in the following manner. The step functions were obtained using real data, on foraminiferal assemblages in sequences of samples from sediment cores. Data from four cores were used, having $n = 6$ and 7 (Fig. 4) and $n = 12$ and 16 (Fig. 5) samples

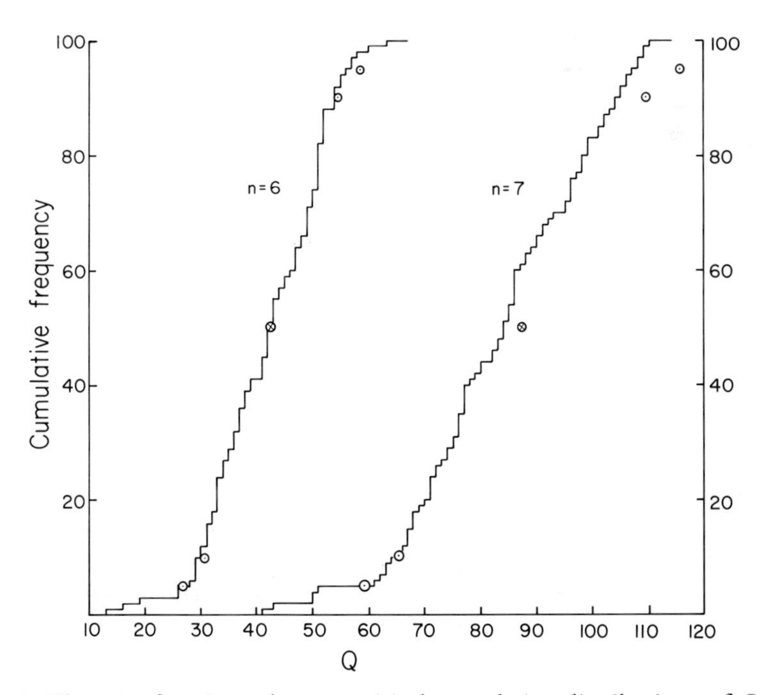

Fig. 4. The step functions show empirical cumulative distributions of Q for $n = 6$ (left) and $n = 7$ (right). They were obtained by randomly permuting real data on foram assemblages in samples from sediment cores. Each set of samples was randomly permuted 100 times. The ⊙ and ⊗ symbols show percentiles of the corresponding "primitive" distributions of Q, inferred by randomly assigning ordinal ranks to the cells of an OSM. (⊙ symbols are the 5th, 10th, 90th and 95th percentiles; ⊗ symbols are medians.)

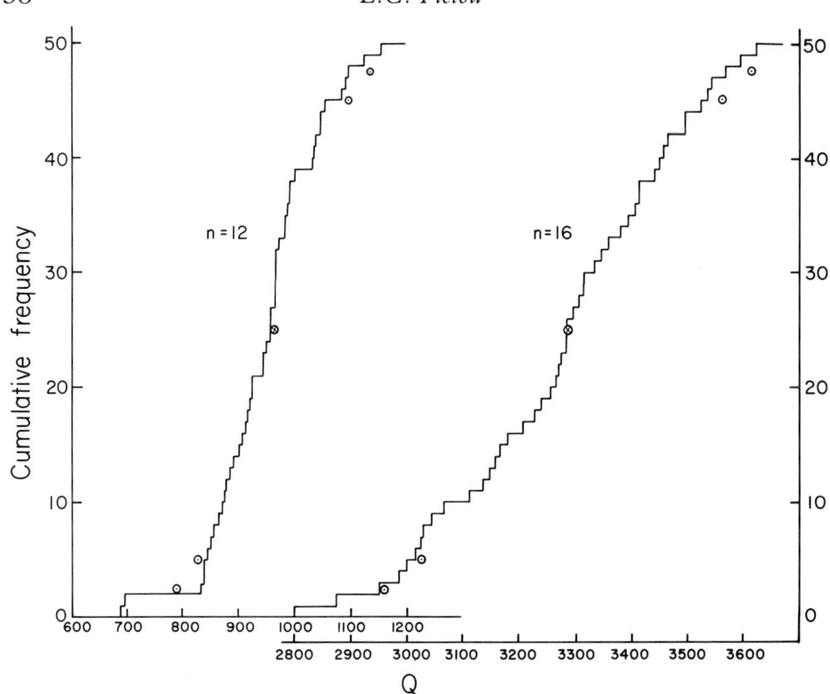

Fig. 5. As for Fig. 4, with $n = 12$ and $n = 16$. The empirical distributions were obtained from 50 random permutations of the sample data. Note the two scales on the abscissa. That on the left is for $n = 12$; that on the right for $n = 16$.

respectively in the sequence. To simulate the distribution of Q in a secondarily random OSM of size $n \times n$, using data from an n-sample core, the n different species abundance lists were randomly permuted a large number of times (100 times for $n = 6$ and 7; 50 times for $n = 12$ and 16) and, at every replication, an OSM was constructed using PS as similarity index. The Q for each OSM was obtained and the observed cumulative distribution of the Qs was plotted. Thus the step functions shown are simulated distributions of Q in secondarily random OSMs of size $n \times n$ for $n = 6$, 7, 12 and 16.

These distributions are now to be compared with the distributions of Q in primitively random OSMs of the same sizes. In Figs 4 and 5 the hollow dots show the 5th, 10th, 50th, 90th and 95th percentiles of Q for the four values of n now being considered. These percentiles (except for the 50th) were estimated by computer simulation as described in an earlier paper (Pielou, 1979). As described above, the 50th percentile is known exactly and is $Q_m/2$.

As Figs 4 and 5 show, there is very close correspondence between the distribution of Q, for given n, in primitively and secondarily random OSMs. This justifies the use of primitively random OSMs as a standard of randomness against which to judge empirical OSMs, at least for OSMs whose elements are PS values. In what follows, all OSMs are constructed from PS values. And the distinction between primitively and secondarily random OSMs is disregarded unless explicitly mentioned.

4. *Testing an Empirical OSM for Randomness*

Testing an empirical OSM for randomness can be done using Fig. 6. Suppose an OSM has been constructed for the data sequence of interest, and that the value of Q for this OSM has been obtained (a FORTRAN program for doing this is given in an Appendix). The desired test statistic is the ratio Q/Q_m which, in words, is the disarray of the observed OSM as a fraction of the maximum possible disarray for an OSM of the same size. The required value of Q_m (for $n = 4, 5, \ldots, 31$) is found in Table I.

Now enter the point $(n, Q/Q_m)$ on the graph in Fig. 6. Accept or reject the null hypothesis of randomness according as the point falls above the curve or below it. The probability of a type I error (mistakenly rejecting a true null hypothesis) is 10% at most (see below). A one-tailed test is used since the only departures from randomness that one can realistically expect are those yielding improbably small values of Q.

Two points to notice are the following:

(1) Although, for clarity, a smooth curve has been drawn in Fig. 6, in reality Q/Q_m is defined only for integer n.

(2) Specifying the significance level of the test is not straightforward. Suppose we knew the distribution of Q in secondarily random OSMs to be truly identical with that in primitively random OSMs; then the distribution would be symmetrical and the 5th percentile would be the critical level for a 5% one-tailed test or a 10% two-tailed test. However, intuition (backed up by the locations of the upper pair of hollow dots in all four distributions shown in Figs 4 and 5) suggests that the distribution of Q in secondarily random OSMs probably departs from

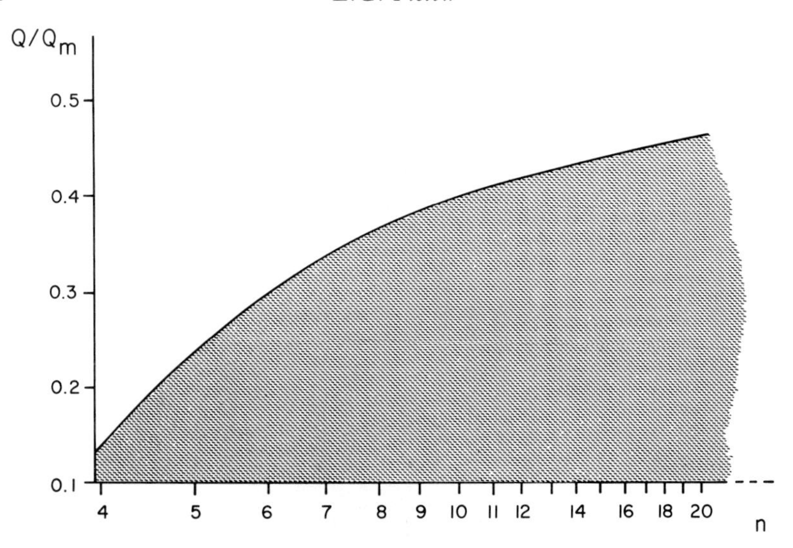

Fig. 6. The curve shows how the 5th percentile of Q/Q_m in primitively random OSMs varies with *n*. Note that the scale of the abscissa is not arithmetic. The chosen scale is such that *n* is proportional to log Q_m; the scale flattens out the curve. The shaded area below the curve is the critical region for a 10% (approximately) one-tailed test.

the "primitive" distribution at the extreme end of its right tail. Thus, while it is obviously true that $\Pr(Q = Q_m) > 0$ for primitively random OSMs, presumably $\Pr(Q = Q_m) = 0$ for empirical OSMs. Conceivably a similarity index could be devised for which $Q = Q_m$ would not be impossible, but none of the similarity indices in common use has this peculiar property. It follows that the area under the right tail is slightly less for the secondarily random distribution than for the "primitive" distribution; hence when, as suggested here, the 5th percentile is used as critical level for a one-tailed test, the probability of error is certainly less than 10%, although somewhat more than 5%.

To illustrate use of the test, it will be applied to data presented by Kellogg *et al.* (1979). They counted benthic Foraminifera in sediment cores from the Ross Sea, Antarctica. In a sequence of 30 samples from sedimentary Unit B in one of their cores (Core EL 32–8), the relative proportions of the six most abundant species★ "apparently show random fluctuations". From this and much other evidence, the authors

★ The species are: *Ehrenbergina glabra*, *Trifarina earlandi*, *Globocassidulina subglobosa*, *Cassidulinoides porrecta*, *Cibicides refulgens* and *Globocassidulina biora*. See Figs 3 and 4 in Kellogg *et al.* (1979).

conclude that the poorly sorted sediments of Unit B probably suffered intense mechanical reworking, under grounded sea ice, during the last one or more Pleistocene glaciations.

Using their data, we now test the hypothesis that the sequence of foraminiferal assemblages is indeed randomly ordered. The OSM of the sequence, of size 30 × 30, was first constructed (Fig. 7 shows the 15 × 15

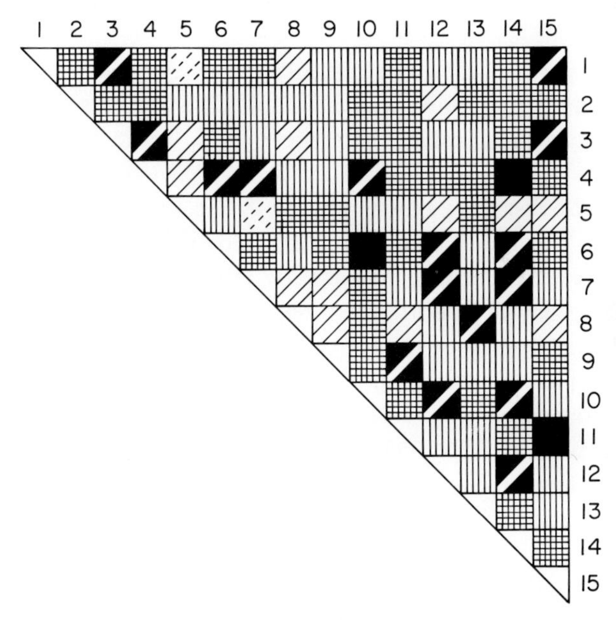

Fig. 7. An OSM based on data from Kellogg *et al.* (1979). The complete OSM for their data is of size 30 × 30. For clarity, a 15 × 15 OSM is shown here, constructed by considering only every second sample in their sequence. The PS values of the different shadings are shown in Fig. 12.

OSM obtained by taking account of the 15 even-numbered samples only). The 30 × 30 OSM was found to have $Q = 46965$. Table I shows that $Q_m = 90335$, when $n = 30$. Thus $Q/Q_m = 0{\cdot}520$ which, being greater than 0·5, obviously accords with the null hypothesis. The authors' intuitive supposition is therefore supported.

INTERPRETING OSMS

1. Q/Q_m as Test Statistic and as a Measure of Disarray

The preceding section showed how Q/Q_m can be used as a test statistic in testing the hypothesis that a sequence of entities is randomly ordered. In the example given, evidence from a variety of different kinds of data (from grain-size analysis and observations on diatoms as well as on Foraminifera) had suggested the hypothesis that the sediments in a sequence of samples from the core in question were randomly ordered. The test, performed by constructing an OSM from the data on benthic Forminifera and determining its Q/Q_m value, led to acceptance of the hypothesis.

Often, however, one uses an OSM to suggest an hypothesis. Obviously one cannot then use the same OSM to test the hypothesis that it has itself suggested. But the Q/Q_m value is still useful in providing a measure of the disarray of an OSM or (inversely) of the complement of disarray which may be called the "gradedness" of the OSM. Thus, like many statistics calculated from raw numerical data, Q/Q_m can serve, according to context, either as a test statistic (when there is a pre-existing hypothesis to test) or as a quantitative measure of an interesting property of numerical observations.

Because it depends partly on n, an observed value of Q/Q_m can only be directly compared with another if both relate to sequences with the same n. However, a convenient dichotomous categorization of all OSMs is possible, using Fig. 6. A sequence can be described as well-graded or ill-graded according as the point $(n, Q/Q_m)$ describing it falls below or above the 5th percentile curve. In other words, the curve in the figure can be used to define the critical region for a test in cases in which a test is appropriate; or as a convenient boundary between two classes of OSM (and hence sequences) in cases in which a test is not appropriate. A sequence that turns out to be ill-graded (i.e., with its point in the acceptance region for a test) can be thought of as being, so far as one can judge, devoid of any trend.

It should be re-emphasized that it is the arrangement, not the magnitudes, of the similarity indices in an OSM that determines Q/Q_m. Thus to say that a sequence is well-graded does not necessarily imply that adjacent entities are very similar, although this may be so; it implies only that pairwise similarities decrease as the distance between pair members increases; in other words, there is a trend. Likewise, in a random sequence, similarities may be all high, or all low, or cover a wide range of values from high to low.

2. Sediment Cores from the Bay of Fundy: the Data

To illustrate these concepts, I now consider some OSMs relating to sequences of samples from sediment cores collected in the Bay of Fundy between Nova Scotia and New Brunswick.

Figure 8 is a map of the Bay of Fundy and shows the sites of the three cores described below. They are cores 127, 136 and 145 from a larger collection of cores collected by Vilks (unpublished). Cores 127 and 145 are each 3·5 m long and yielded $n = 15$ samples, taken at 25 cm intervals; core 136 is 4 m long and yielded $n = 17$ samples.

Each sample gave two data sets. The first data set is a species-abundance distribution for the assemblage of benthic Foraminifera in the sample. The proportional abundances were estimated from counts of

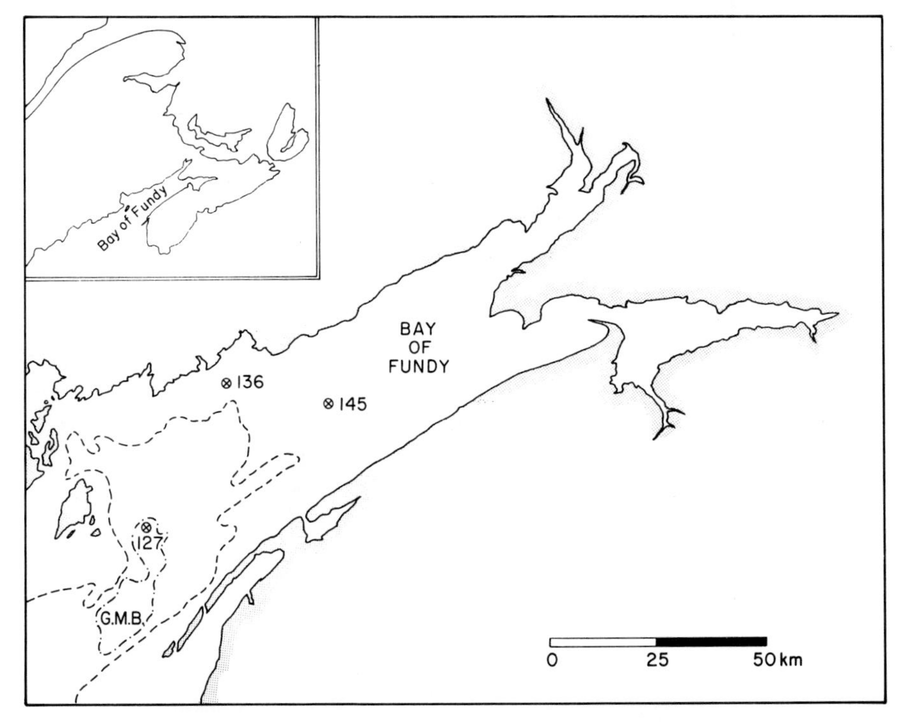

Fig. 8. Map of the Bay of Fundy showing the sites of cores 127, 136 and 145. The bathymetric contours are 50 fathoms (91·4 m), dashed; and 100 fathoms (182·8 m), dash-dot. G.M.B. = Grand Manan Basin.

300 individual foram tests. The total number of tests per sample ranged from 842 to 54600.

The second data set is a grain-size distribution for the sediment particles of each sample. The unit of grain size is $\phi = -\log_2 d$, where d is grain diameter in millimetres. Observed values ranged from $\phi = -3$ representing pebbles of 8 mm diameter, to $\phi = 10$ representing very fine clay particles of less than 0·001 mm diameter. Figure 9 shows the grain-size distributions for the coarsest and finest sediments sampled; these are from the surface sample of core 145 and the deepest sample (at 3·5 m) of core 127, respectively.

Each core thus yielded two OSMs, one relating to the sequence of species-abundance distributions of the Foraminifera, the other to the corresponding sequence of grain-size distributions. The elements used to construct all OSMs are PS values.

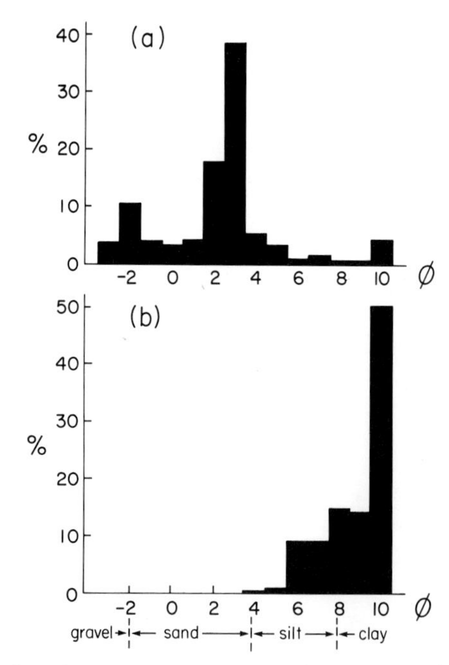

Fig. 9. Grain-size distributions in two strongly contrasted sediment samples, from (a) core 145 at the surface; (b) core 127 at 3·5 m below the surface. Grain size is measured as $\phi = -\log_2 d$ where d is grain diameter in mm. The range of values of ϕ for gravel, sand, silt and clay are, respectively, <-2, -2 to 4, 4 to 8 and >8.

3. Probable Post-glacial History of the Bay of Fundy

The post-glacial history of the Bay of Fundy is complicated, but in broad outline it is believed to be as follows (Grant, 1970; Fader *et al.*, 1977; Rashid and Vilks, 1977; Vilks and Rashid, 1977; Amos, 1978; Amos *et al.*, 1980; Vilks, 1981). At the height of the final glacial advance of the Pleistocene, about 20000 y BP, the bay was ice-filled. Since the melting of the ice there have been two transgressions of the sea separated by a regression (Amos *et al.*, 1980). The first transgression resulted from the eustatic rise in sea level that accompanied melting of the ice; relative to the land, sea level reached a height of 45 m or more above the present mean sea level (MSL) about 14000 y BP. Then, although eustatic rise continued, its effect was offset by crustal rebound and for about 4000 to 6000 y, sea level regressed until it was 30 m below present MSL. Then, 8000 to 10000 y BP, the second transgression began, as a result of crustal downwarping, and sea level has been rising, relative to the land, continuously ever since.

However, conditions changed abruptly about 6300 y BP (Grant, 1970; Amos, 1978). Before that time, while the banks on either side of the mouth of the bay (see Fig. 8) were above or only just below sea level, the flushing rate of the bay was low. It was a sheltered, almost landlocked body of water, cooled and diluted by the cold, fresh water of inflowing rivers. Then, as sea level rose, the water over the shallow banks deepened; by about 6300 y BP the entrance to the bay had attained a cross-sectional area large enough to permit ready influx of the warmer, more saline, ocean waters of the continental shelf and slope to the southwest. At the same time, the tidal amplitude began to increase. As the depth and volume of the bay steadily increased, the tides underwent increasing resonant amplification so that, today, tidal amplitudes range from 3·5 m at the mouth of the bay to over 16 m at its head (Amos *et al.*, 1980). The modern bay is therefore a strongly tidal body of water, kept relatively warm and fully saline by constant influx from the open ocean.

The history of the bay is based on many lines of evidence; these include data on old shoreline positions both above and below present MSL (Grant, 1970; Amos, 1978), and studies of echograms, seismic profiles and samples of surface sediments (Fader *et al.*, 1977), as well as studies of sediment cores.

The most recently collected cores, including the three to be discussed below, are therefore expected to confirm, add to, and possibly modify the details of, existing knowledge about events since the current transgression of the sea began.

4. Analysis of the OSMs of Three Cores

We now consider these cores in turn.

(a) CORE 127. This core was collected in 183 m deep water in Grand Manan Basin, a depression in the sea floor at the entrance to the Bay of Fundy, where fine sediments have accumulated rapidly (Vilks and Rashid, 1977; Rashid and Vilks, 1977). Figure 10 shows the two OSMs derived from it. Inspection of them shows immediately that the grain-size distributions are much more uniform than the foraminiferal assemblages; this follows from the greater magnitudes of the PS values, as shown by darker shading, in the sediment OSM than in the foram OSM.

For the sediment OSM, $(n, Q/Q_m) = (15, 0{\cdot}427)$. If this point is entered on the graph in Fig. 6, it is found to fall just within the shaded region, implying that the sediments probably exhibit a weak trend.

The low PS values between sample 10 and all other samples show that sample 10 is in some way anomalous. Examination of the grain-size distributions shows that in sample 10 the mode is at $\phi = 9$, whereas in all other samples it is at $\phi = 10$. Thus the difference is not great. The sediments are silty clay (see Fig. 9) belonging to the LaHave formation, which consists of material winnowed from glacial till elsewhere in the bay and deposited where current velocities are low (Fader *et al.*, 1977).

Now consider the assemblages of benthic Foraminifera sampled by the core. A total of 39 species was found, disregarding species that formed less than 1% of the total; the number of tests per sample ranged from 4200 to 54 600. The OSM has $(n, Q/Q_m) = (15, 0{\cdot}218)$ which is very strong evidence for nonrandomness (see Fig. 6). The pattern of the OSM resembles that in Fig. 2a rather than that in Fig. 2b, suggesting that the core may consist of two homogeneous segments grading into each other at the level of samples 8 and 9. To test this supposition, data from samples 1 though 8, and from samples 9 through 15, were analysed separately. This is done by considering, individually, the two small OSMs (labelled "upper segment" and "lower segment" in Fig. 10)

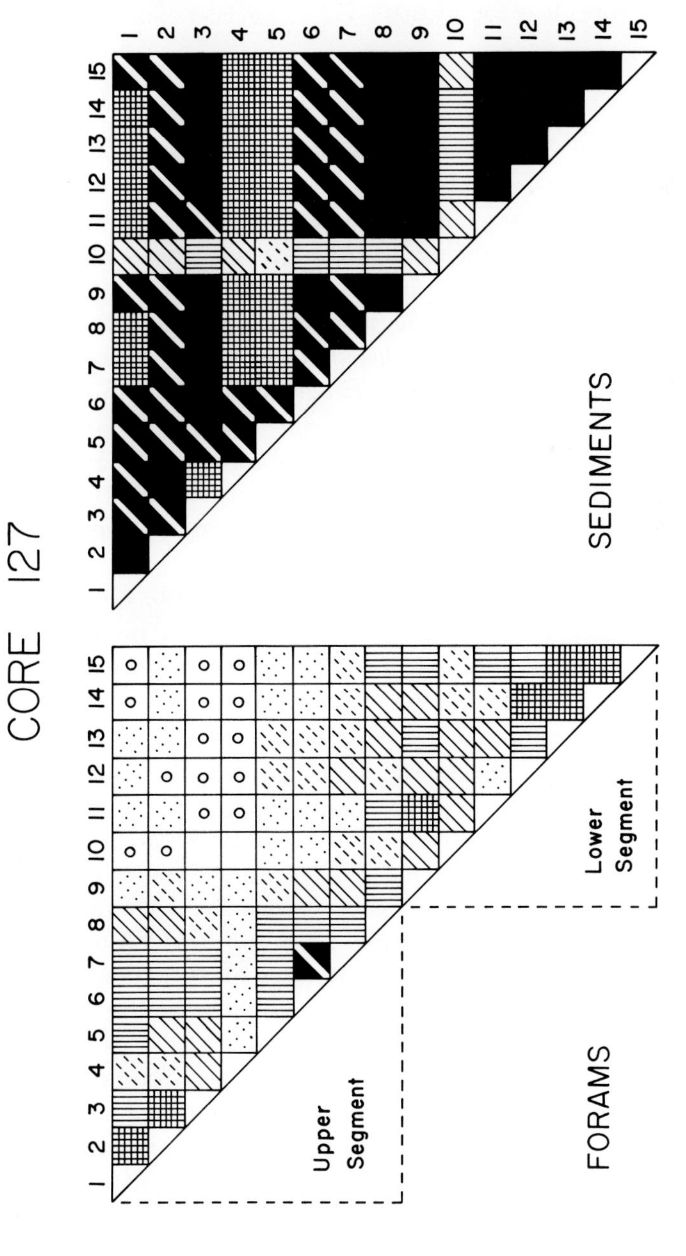

Fig. 10. The OSMs for core 127. The PS values of the different shadings are shown in Fig. 12.

that are imbedded in, and form part of, the single large OSM pertaining to the whole core. It is found that

$$(n,\ Q/Q_m) = \begin{cases} (8,\ 0{\cdot}413) \text{ for the upper segment;} \\ (7,\ 0{\cdot}394) \text{ for the lower segment.} \end{cases}$$

Reference to Fig. 6 shows that each of these segments can be regarded as "ill-graded" or trend-free. The foraminiferal OSM thus provides strong supporting evidence that a fairly abrupt environmental change occurred while the sediments were being deposited. The species-abundance data show that there is indeed a contrast between the foraminiferal assemblages of the upper and lower segments. Lower segment assemblages contain a large proportion of *Elphidium excavatum* f. *clavata* Cushman, whereas this taxon is scarce in upper segment samples. *E. excavatum* f. *clavata* is an ecophenotype of *E. excavatum* and is indicative of cold, somewhat dilute water (Miller *et al.*, unpublished). Presumably therefore, the lower and upper segments consist of sediments laid down, respectively, before and after the sudden change in conditions at 6300 y BP described above.

The upper segment is 2 m thick, so that if it began to form 6300 y BP, the average sedimentation rate was 32 cm/1000 y. This is somewhat slower than the minimum rate of 54 cm/1000 y reported for Grand Manan Basin by Rashid and Vilks (1977). Another possibility is that the base of the upper segment is no older than 4000 y, as hypothesized by Grant (1970); this gives an average sedimentation rate of 50 cm/1000 y.

There are two possible causes for the within-segment randomness inferred from the OSMs. The lack of trend could indicate that environmental conditions, and hence foraminiferal communities, remained constant while the sediments were laid down. Alternatively, it could indicate that the sediments had been reworked so that even if there had been a trend, all evidence of it has been obliterated. It is not possible to decide between these competing explanations from examination of the OSM. However, other evidence (Vilks, pers. comm.) makes it improbable that the sediments have been reworked. The within-segment randomness thus suggests trend-free environmental conditions during the periods in which the sediments were deposited.

(b) CORE 136. This core was collected in water of 79 m depth. Its location is shown in Fig. 8 and its two OSMs in Fig. 11.

The sediment OSM shows that, except for the anomalous surface

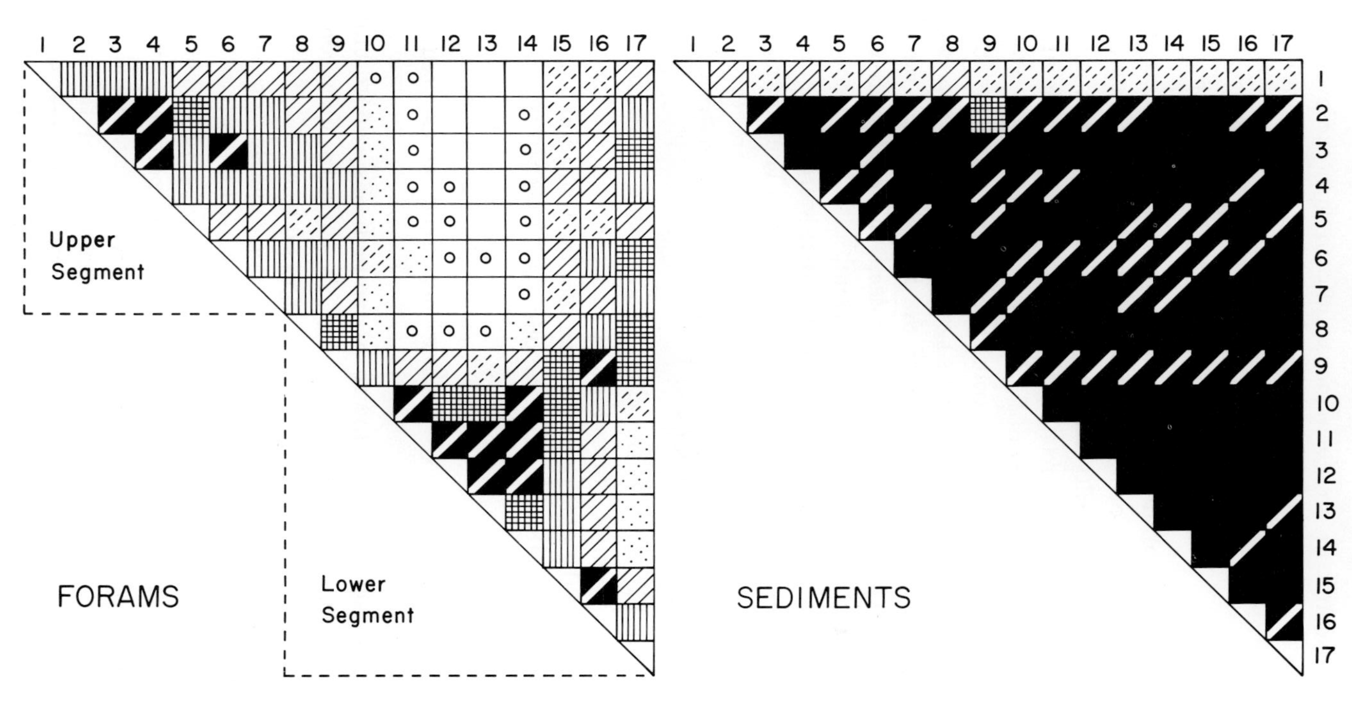

Fig. 11. The OSMs for core 136. The PS values of the different shadings are shown in Fig. 12.

layer, the sediments are even more uniform than those of core 127. Excluding the surface layer from consideration, it is found that for the rest of the core

$$(n, Q/Q_m) = (16, 0{\cdot}491).$$

From the position of this point in Fig. 6, it is reasonable to conclude that the sediments are uniform and devoid of any trend. The raw data show these sediments to be silty clays with grain-size distributions not very different from those in core 127; the sediments contain a slightly larger, but still negligible, fraction of fine sand.

The anomalous surface layer consists of coarser material with a modal grain size of $\phi = 3$ (fine sand). This sandy surface layer is presumably a lag sediment from which some of the fines have been removed by currents. Unlike core 127, which is in a deep, sheltered basin, core 136 is at a site having appreciable bottom current.

Now consider the Foraminifera data. There were 31 benthic species, and the number of tests ranged from 840 to 7300. The non-randomness of the OSM is obvious at a glance. For the whole core, the results are:

$$(n, Q/Q_m) = (17, 0{\cdot}304).$$

As in core 127, there is an evident discontinuity in the foraminiferal assemblages, without corresponding discontinuity in the grain-size distributions of the sediments. In core 136, the transition is at about the level of sample 8 (180 cm). For the upper segment

$$(n, Q/Q_m) = (7, 0{\cdot}377)$$

which gives no reason to assume any trend in the sequence of assemblages.

The lower segment (the lower 10 samples) has a more interesting OSM. It closely resembles that in Fig. 2c and constitutes convincing evidence that the foraminiferal assemblages underwent a cyclic change. Unfortunately, the statistic Q/Q_m is not suitable for diagnosing departures from randomness of this kind. Its usefulness is as a measure of the nonrandomness of an OSM along an axis leading from a completely haphazard pattern at one extreme to perfect grading (as in Fig. 2b) at the other. A cyclical OSM, like those in Fig. 2c and the lower segment of the foram OSM in Fig. 11, does not lie on this axis.

The lack of an appropriate measuring statistic or test statistic for cyclical OSM's is immaterial in the present case, however. The cyclical pattern of the OSM for samples 8 to 17 in core 136 is too conspicuous to be doubted. It implies that, for the first half of the period in which these sediments were deposited, the environment underwent steady, gradual change; the trend then reversed itself so that the starting conditions were restored. The nature of these trends may possibly be inferred from the original data, which show that *Elphidium excavatum* f. *clavata*, the foraminifer believed to indicate cold, dilute waters, was at first abundant (70% in sample 17), then sparse (6% in sample 13), and finally abundant again (53% in sample 8).

It is interesting to notice that the foraminiferal assemblages in the upper segments of cores 127 and 136 differ markedly. In core 136, *E. excavatum* f. *clavata* dominated in all samples of the upper segment, always forming at least 45% of the sample assemblages. By contrast, it will be recalled that this species was scarce in the upper segment (but not in the lower segment) of core 127.

(c) CORE 145. This core comes from water of 71 m depth located near the centre of the bay, where there is a strong bottom current. Its OSMs are shown in Fig. 12.

The most noticeable characteristic of the pair of OSMs for this core is that their patterns resemble each other. In both, the upper two samples differ from each other and from all lower samples more than the lower samples differ among themselves. The uniformity of the lower samples is shown by their Q/Q_m values, which reveal no trend, as well as by their high PS values. Disregarding the two uppermost samples, we have:

for the sediment OSM, $(n, Q/Q_m) = (13, 0{\cdot}535)$;

for the foram OSM, $(n, Q/Q_m) = (13, 0{\cdot}455)$.

Indeed, if the pattern of the sediment OSM had led one to hypothesize that the foraminiferal assemblages in the lower 13 samples were free of any trend, the hypothesis would have been rigorously testable, using the relevant Q/Q_m as test statistic.

The sediment in the lower 13 samples is again a silty clay, with a very slightly larger sand fraction than in the sediments of core 136. The surface sediments in the region of this core belong to the Sambro Sands formation (Fader *et al.*, 1977), which consists of silty and clayey sand with some gravel (Fig. 9a). It is a lag sediment from which fine particles have been removed by strong currents. This material forms the

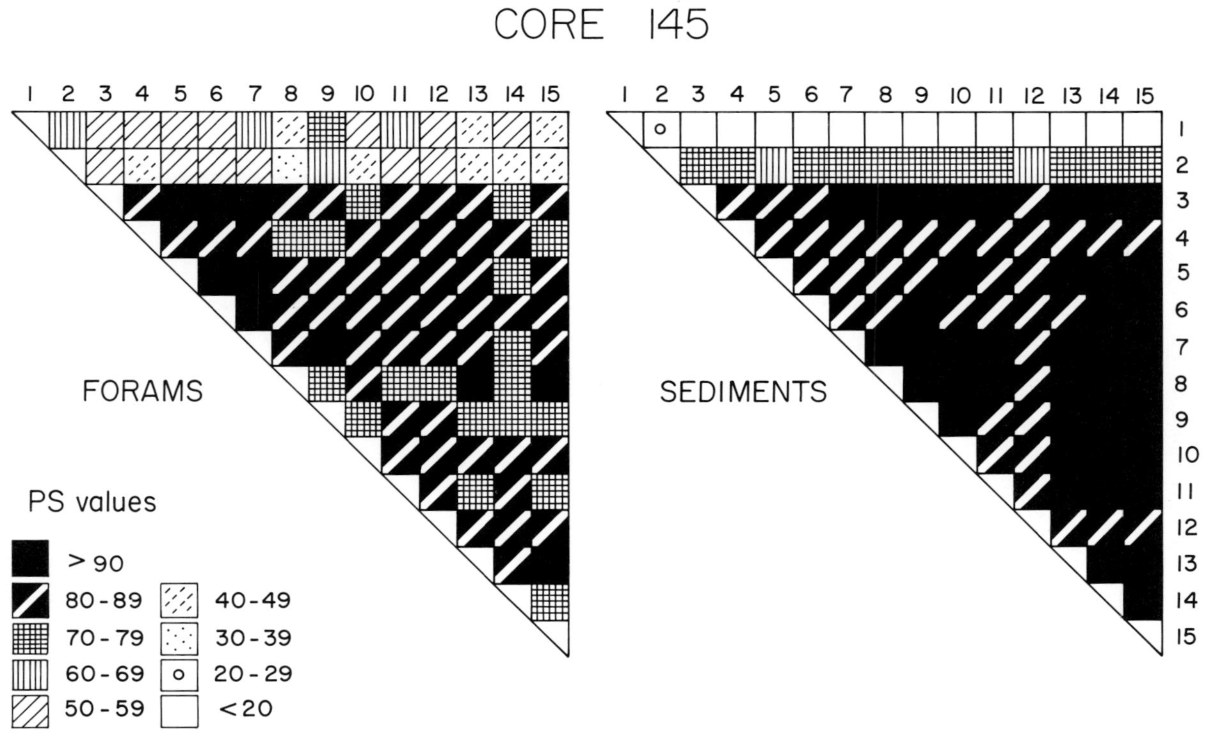

Fig. 12. The OSMs for core 145.

sediment of sample 1, and sample 2 is transitional between this coarse surface material and the clay below.

The foraminiferal samples contained 27 species; the number of tests per sample ranged from 1450 to 6800. The assemblages are dominated at every level by *E. excavatum* f. *clavata*, which suggests that there has been negligible deposition of sediments since tides began to be appreciable 6300 y BP (Amos, 1978). *Cibicides lobatula* is the only other abundant species in the surface sample; it attaches itself to sand grains where tidal currents are strong.

Studies on the OSMs of other cores from the Bay of Fundy are now in progress. The use of the statistic Q/Q_m as a measure of OSM pattern, or as a test criterion, should assist in interpreting the data. Of course, numerical measures of statistical properties rarely (if ever) yield sensational new knowledge all by themselves. At the same time, they can bring objectivity to what would otherwise be wholly intuitive theorizing about the true implications of some very confusing data.

ACKNOWLEDGEMENTS

I thank Gustavs Vilks of the Bedford Institute of Oceanography for the use of his data and for much advice on their interpretation. The work was funded by a grant from the Natural Sciences and Engineering Research Council of Canada.

REFERENCES

Amos, C.L. (1978). The post-glacial evolution of the Minas Basin, N.S. and sedimentological interpretation. *J. sedim. Petrol.* **48**, 965–982.

Amos, C.L., Buckley, D.E., Daborn, G.R., Dalrymple, R.W., McCann, S.B. and Risk, M.J. (1980). Trip 23: Geomorphology and sedimentology of the Bay of Fundy. *Field Trip Guide Book, Geological Association of Canada*, Halifax, 1980.

Cheetham, A.H. and Hazel, J.E. (1969). Binary (presence–absence) similarity coefficients. *J. Paleont.* **43**, 1130–1136.

Cracraft, J. and Eldredge, N. (eds.) (1979). "Phylogenetic Analysis and Paleontology". Columbia University Press, New York.

Fader, G.B., King, L.H. and MacLean, B. (1977). Surficial geology of the eastern Gulf of Maine and Bay of Fundy. *Geological Survey of Canada, Paper* **76–17**.

Gingerich, P.D. (1974). Stratigraphic record of Early Eocene *Hyopsodus* and the geometry of mammalian phylogeny. *Nature, Lond.* **248**; 107–109.

Grant, D.R. (1970). Recent coastal submergence of the Maritime Provinces, Canada. *Can. J. Earth Sci.* **7**, 676–689.

Greig-Smith, P. (1957). "Quantitative Plant Ecology". Butterworth Scientific Publications, London.

Kellogg, T.B., Osterman, L.A. and Stuiver, M. (1979). Late Quaternary sedimentology and benthic foraminiferal paleoecology of the Ross Sea, Antarctica. *J. Foraminiferal Res.* **9**, 322–335.

Pielou, E.C. (1979). Interpretation of paleoecological similarity matrices. *Paleobiology* **5**, 435–443.

Rashid, M.A. and Vilks, G. (1977). Environmental controls of methane production in Holocene basins in eastern Canada. *Organic Geochem.* **1**, 53–59.

Simpson, G.G. (1960). Notes on the measurement of faunal resemblance. *Am. J. Sci.* **358A**, 300–311.

Sneath, P.H.A. and Sokal, R.R. (1973). "Numerical Taxonomy" (2nd edn). W.H. Freeman, San Francisco.

Stanley, S.M. (1979). "Macroevolution, Pattern and Process". W.H. Freeman, San Francisco.

Valentine, J.W. (1973). "Evolutionary Paleoecology of the Marine Biosphere". Prentice-Hall, Englewood Cliffs, N.J.

Vilks, G. (1981). Late glacial-postglacial foraminiferal boundary in sediments of eastern Canada, Denmark and Norway. *Geosci., Canada.* **8**, 48–55.

Vilks, G. and Rashid, M.A. (1977). Methane in the sediments of a subarctic continental shelf. *Geosci., Canada* **4**, 191–197.

APPENDIX

Below is a FORTRAN program for obtaining an OSM with PS values as elements and calculating its disarray, Q. The steps are explained in Comment statements. DIMENSION and FORMAT statements should be modified when JS > 30 or JN > 20. Following the program is a typical output, for a data matrix giving the proportions of JS = 14 species in JN = 13 samples.

```
                    PROGRAM OSM (INPUT = 1000, OUTPUT = 1000)
C  TO CONSTRUCT AN ORDERED SIMILARITY MATRIX AND DETERMINE ITS DISARRAY
C  "Q". "CORE" IS THE DATA MATRIX, GIVING THE PROPORTIONS OF THE JS
C  SPECIES IN EACH OF JN SAMPLES. IT HAS JS ROWS (SPECIES) AND JN
C  COLUMNS (SAMPLES). "SIM" IS THE JN*JN MATRIX OF SIMILARITIES.
C  THESE ARE PS VALUES. "TEMP" IS A JS*2 HOLDING MATRIX FOR THE PAIR
C  OF SAMPLES (COLUMNS) WHOSE PS IS BEING COMPUTED. "T" IS A
C  (JN-1)*JN MATRIX IN WHICH THE ELEMENTS OF "SIM" ARE ARRANGED IN
C  A WAY THAT FACILITATES COMPUTATION OF "Q".
      DIMENSION CORE (30, 30), SIM (30, 30), TEMP (30, 2), T(30, 30)
C  FIRST DATA CARD GIVES VALUES OF JS AND JN. READ IT
      READ 99, JS, JN
   99 FORMAT (3X, 2I4)
C  READ IN THE DATA MATRIX "CORE" ROW WISE.
      DO  1   I = 1, JS
    1 READ 2, (CORE (I, J), J = 1, JN)
    2 FORMAT (20F4.1)
      PRINT 80
      PRINT 81
   81 FORMAT (2X, 19H THE DATA MATRIX IS)
      PRINT 80
      DO   3   I = 1, JS
```

```
  3   PRINT 4, (CORE (I, J), J = 1, JN)
  4   FORMAT (2X, 20F5·1)
      DO  5  I = 1, JN
      DO  5  J = 1, JN
  5   SIM (I, J) = 0·0
      JNN = JN − 1
C   NOW PUT THE 2 COLUMNS OF CORE WHOSE PS IS REQUIRED INTO "TEMP"
      DO  6  I = 1, JNN
      K = I + 1
      DO  7  J = K, JN
      DO  8  L = 1, JS
      TEMP (L, 1) = CORE (L, I)
  8   TEMP (L, 2) = CORE (L, J)
C   NOW FIND PS FOR THIS PAIR OF SAMPLES AND ENTER IT IN SIM (I, J)
      DO  9  L = 1, JS
      IF (TEMP (L, 1) − TEMP (L, 2)) 10, 10, 11
 10   SIM (I, J) = SIM (I, J) + TEMP (L, 1)
      GO TO 9
 11   SIM (I, J) = SIM (I, J) + TEMP (L, 2)
  9   CONTINUE
  7   CONTINUE
  6   CONTINUE
      PRINT 80
      PRINT 82
 82   FORMAT (4X, 25H THE SIMILARITY MATRIX IS)
      DO  12  I = 1, JN
 12   PRINT 13, (SIM (I, J), J = 1, JN)
 13   FORMAT (//4X, 20F5·1)
C   NOW PUT THE ELEMENTS OF "SIM" INTO "T"
      DO  14  I = 1, JNN
      DO  14  J = 1, JN
 14   T (I, J) = 0·0
      DO  15  L = 2, JN
      N = JN − L + 1
      DO  16  K = N, JNN
      KK = K + L − JN
 16   T (K, L) = SIM (KK, L)
 15   CONTINUE
      PRINT 80
 80   FORMAT (///)
C   NOW DETERMINE "Q" FROM "T".
      Q = 0·0
      JM = JN − 2
      DO  18  I = 1, JM
      DO  19  J = 1, JN
      IF (T(I, J) − 0·0) 191, 191, 20
 20   K = I + 1
      DO  21  L = K, JNN
      DO  22  M = 1, JN
      IF ((JN − L) . GE . M) GO TO 221
      IF (T(I, J) − T(L, M)) 221, 30, 23
 23   Q = Q + 1·0
      GO TO 221
 30   Q = Q + 0·5
221   CONTINUE
 22   CONTINUE
 21   CONTINUE
191   CONTINUE
 19   CONTINUE
 18   CONTINUE
      PRINT 24, Q
 24   FORMAT (//11X, 4H Q =, F6·1)
      STOP
      END
```

The following is a representative output, with JS = 14 and JN = 13.

THE DATA MATRIX IS

34·3	0·0	0·0	0·0	7·6	0·0	0·0	0·0	0·0	0·0	0·0	0·0	0·0
15·8	·7	14·0	0·0	0·0	0·0	0·0	3·0	0·0	3·6	1·7	0·0	3·9
2·4	1·4	2·8	0·0	0·0	0·0	·3	1·8	2·0	0·0	1·7	·6	·7
2·0	·5	2·3	·2	0·0	1·6	·5	1·4	·3	·9	·6	·2	1·1
1·0	·5	1·4	·4	·3	1·3	·9	1·1	·7	·6	·6	·6	·9
5·4	1·6	2·8	·4	·3	1·3	2·8	2·6	1·0	1·6	1·3	1·3	1·7
24·6	4·4	3·2	·4	·3	1·3	4·4	4·2	1·3	2·5	2·0	1·8	2·2
4·6	·7	2·8	·2	·3	1·3	3·7	4·0	1·0	2·3	2·0	1·8	2·0
2·5	1·2	3·2	·9	6·3	4·2	3·7	4·9	4·7	4·6	3·5	3·6	2·1
1·5	6·4	8·2	4·4	5·0	5·5	6·3	5·3	4·7	5·8	5·6	6·9	7·0
1·0	12·5	6·6	9·5	10·0	10·6	3·4	10·9	8·6	9·2	11·0	9·4	10·7
·6	12·9	7·8	12·8	10·3	10·2	18·7	9·1	10·6	9·2	10·5	11·1	7·7
1·2	11·6	5·0	11·7	8·3	11·4	7·8	9·3	12·0	9·3	9·5	11·2	10·3
3·3	45·8	40·1	59·1	51·2	51·2	47·7	42·5	53·2	50·6	50·3	51·6	49·8

THE SIMILARITY MATRIX IS

0·0	18·6	38·3	10·1	18·9	16·6	22·7	28·1	16·4	21·6	20·0	16·4	22·2
0·0	0·0	75·7	86·6	81·9	89·0	85·4	87·9	87·7	87·0	90·7	90·2	89·0
0·0	0·0	0·0	66·4	68·9	75·0	76·3	83·9	73·7	80·0	78·2	75·9	81·0
0·0	0·0	0·0	0·0	85·7	89·2	78·6	77·3	91·0	85·2	86·7	90·2	84·2
0·0	0·0	0·0	0·0	0·0	90·1	79·1	81·0	89·0	88·1	88·6	89·0	84·1
0·0	0·0	0·0	0·0	0·0	0·0	83·6	87·4	94·6	93·4	94·7	95·8	91·9
0·0	0·0	0·0	0·0	0·0	0·0	0·0	84·0	82·5	85·1	85·2	85·9	82·6
0·0	0·0	0·0	0·0	0·0	0·0	0·0	0·0	85·0	90·9	90·5	85·5	89·2
0·0	0·0	0·0	0·0	0·0	0·0	0·0	0·0	0·0	91·2	93·0	95·0	88·2
0·0	0·0	0·0	0·0	0·0	0·0	0·0	0·0	0·0	0·0	95·3	93·4	94·8
0·0	0·0	0·0	0·0	0·0	0·0	0·0	0·0	0·0	0·0	0·0	95·1	94·3
0·0	0·0	0·0	0·0	0·0	0·0	0·0	0·0	0·0	0·0	0·0	0·0	92·5
0·0	0·0	0·0	0·0	0·0	0·0	0·0	0·0	0·0	0·0	0·0	0·0	0·0

Q = 1150·0

3 Biogeographic Models, Species' Distributions and Community Organization

D. SIMBERLOFF

Department of Biological Science, Florida State University, Tallahassee, Florida 32306

Abstract: The dynamic equilibrium model of island biogeography has motivated an enormous amount of research, yet evidence bearing on the tenets of the model is equivocal. Correlational studies such as those of species-area relationships are at best consistent with the model, but cannot constitute strong support. Direct assessment of whether an equilibrium species number actually obtains is rarely attempted, and even these attempts are beset by the problem that there is no objective criterion for whether a set of points through time represents an equilibrium. I suggest a few ways of beginning to treat this problem statistically; the major one is based on a runs test. Estimation of turnover rates, which the model envisions as high, are scarce, and in several instances turnover is virtually non-existent. Furthermore, it is unclear to what extent claimed turnover really represents population extinction and not just within-population transient movement. Human activity may also have caused many reported extinctions and immigrations. Much recent research goes beyond analysis of species numbers and turnover rates to focus on island species composition. Neither sizes nor taxonomic affiliations of coexisting species strongly suggest a structuring force, such as competition, that determines which species are found together. This is not to say that species composition is a random matter, only that non-randomness is not manifested by these statistics, and a mechanistic cause remains obscure. Diffuse competition, often said to constitute such a mechanistic cause, is an extremely difficult hypothesis to falsify, except experimentally. One may examine the species compositions of

Systematics Association Special Volume No. 23, "Evolution, Time and Space: The Emergence of the Biosphere", edited by R.W. Sims, J.H. Price and P.E.S. Whalley, 1983, pp. 57–83. Academic Press, London and New York.

islands in an archipelago by clustering either sites or species. The latter mode of analysis is the more common, and has proceeded by seeking extreme dissimilarity between species in which sites they occupy. The existence of species sharing no sites, and therefore forming a "missing combination", has been taken as *prima facie* evidence for competitive exclusion. In fact, such missing combinations would arise even if species were randomly and independently distributed on islands. Simulated archipelago colonization allows statistical statements about when the number of observed missing combinations is extraordinary, and multiway contingency tables with log-linear models are another promising approach to the question of whether competition determines the frequencies of species' co-occurrences.

INTRODUCTION

The attention of ecologically oriented biogeographers has turned increasingly in recent years from how many species coexist at different sites to which species are found at which sites (Simberloff, 1980). It is appropriate to review the dynamic equilibrium model of island biogeography, which motivated and provided a framework for much of the original interest in species richness, and then to examine attempts to move beyond richness to models of species composition in the spirit of the equilibrium model.

THE EQUILIBRIUM MODEL

In its infancy, the equilibrium model (MacArthur and Wilson 1963, 1967) was noted by ecologists and systematists of every stripe (Simberloff, 1974), who sought its confirmation either directly by observing extinction ("turnover") and approximate equilibrium (e.g., Diamond, 1969; Simberloff and Wilson, 1969, 1970), or indirectly by observing phenomena predicted by the model, such as a species–area relationship or depauperization of distant islands (e.g. Vuilleumier, 1970). The model was applied to all manner of insular habitats, not just oceanic islands: caves (Vuilleumier, 1973; Culver *et al.*, 1973), lakes (Barbour and Brown, 1974; Lassen, 1975; Magnuson, 1976), individual plants (Seifert, 1975; Tepedino and Stanton, 1976; Brown and Kodric-Brown, 1977), etc. The sheer number of studies based on the equilibrium model would have sufficed to suggest its veracity to an uncritical observer; in fact the model was widely accepted in spite of rather meagre accomplishments. It has even spawned a subdiscipline, refuge design based on equilibrium theory (Simberloff, 1982), whose precepts have been adopted by the major international umbrella organization for conservation (Anon., 1980).

Evidence on the equilibrium model is equivocal. The myriad studies showing a generally well-defined species–area relationship are at best consistent with the model but cannot be construed as strongly supporting it. The correlative methods by which the relationship is established cannot directly test causality (Simberloff, 1974), and the nearly universally high intercorrelation between two variables that could each potentially affect species number—habitat diversity and area *per se*—makes it especially difficult to interpret multiple regressions. Connor and McCoy (1979) list three non-exclusive forces that could all act to produce a species–area effect similar to that commonly observed. Only one of the forces—decreased population size on small islands—corresponds to the mechanism by which the equilibrium model produces a species–area effect.

Very few data directly assess whether there is an equilibrium species number and, if there is, whether it is maintained by dynamic turnover. In several instances species number is changing in a given direction. Abbott and Grant (1976) examined avifaunas of fourteen islands near Australia and New Zealand and found the number of nonpasserine species to be approximately constant, but the number of passerine species to have been increasing in the past century. For Skokholm Island species number changed erratically and greatly, suggesting that an equilibrium does not obtain even if there is no unidirectional tendency; Abbott and Grant (1976) feel this situation may typify high latitude islands, because of irregularly fluctuating climates. A similar example comes from Haila *et al.* (1979), who find that land bird species richness in the Åland Archipelago (Finland) has increased over the last 50 years.

Other studies (e.g. Diamond, 1972) do not directly observe turnover and changes in species number, but rather infer these from modern distributional data plus more or less firm information on past physical geography. So strong is the desire in some quarters to find nature in accord with the equilibrium model that a *change* in species number can even be viewed as supporting the model. Wilcox (1978) suggested from a species–area regression for lizards of islands off Baja California that species number had fallen over the past 12000 years on all of them, but that this change *confirmed* the equilibrium model. Faeth and Connor (1979) show that Wilcox's data do not support his inference that extinctions have occurred (none were actually observed). Brown (1971) inferred (but did not observe) mammal extinctions on mountaintop "islands" in the Great Basin, and suggested immigration to be non-existent. He construed this as a nonequilibrial fauna.

Several direct observational studies, with censuses at two or more times, report an approximate equilibrium species number. Simberloff and Wilson (1969, 1970) and Simberloff (1976) found for fumigated mangrove islands that original arthropod species number was rather quickly reachieved and subsequently maintained, with some fluctuations about an apparent approximate equilibrium. Diamond for birds of the California Channel Islands (1969) and Karkar (1971) and Terborgh and Faaborg (1973) for birds of Mona Island report similar approximate equilibria, based on fewer censuses. Jones and Diamond (1976) provide more frequent census data confirming the Channel Island result. Diamond and May (1977) claim that 29 annual censuses of birds on the Farne Islands show sufficient constancy to "confirm the basic hypothesis of MacArthur and Wilson, that species number on an island may be in dynamic equilibrium", but the data are equivocal. Mean number of species is 5·86, with a coefficient of variation of 0·20. Species number varied by over 100%, from 4 to 9, and though no study has definitively stated precisely what amount of fluctuation can be accommodated within an "equilibrium" model, these data surely stretch one's notion of constancy.

In addition to constant species number the model envisions frequent turnover, and is thus termed "dynamic". One must always bear in mind the obvious point that estimated extinction rates will be lower the longer the interval between censuses. One can attempt to compensate for this bias statistically (Simberloff, 1969; Diamond and May, 1977) but estimates thus derived cannot be viewed with much confidence, particularly if the only *observed* data points are at the beginning and end, respectively, of a long interval. A second problem with assessing turnover is that the biological significance of even a valid estimate is doubtful without further information on individual movement (Simberloff, 1976). Smith (1975), arguing against the usefulness of the equilibrium model, observed that if one tallied robins arriving at and departing from a tree, one could in principle deduce an extinction rate: every time a bird departed from the tree and no others were left behind, one would record an extinction. But such transient, intrapopulation movement would be unrelated to the population extinction and immigration envisioned by the equilibrium model. The key information, the extent to which population recruitment on an island derives from breeding on the island and not invasion from without, is rarely known (Simberloff, 1976).

Among claimed reports of conformation to the model, turnover estimates vary greatly. The mangrove island arthropod data yield high turnover rates, and even though well over half surely represent transience, a number of undeniable population extinctions occur each year (Simberloff 1976). Diamond's report (1969) of substantial turnover for Channel Island birds was challenged by Lynch and Johnson (1974) primarily on three grounds:

(1) substantial movement between islands, so that defining a population is difficult;

(2) insufficiently frequent, exhaustive, or standardized censuses;

(3) secular environmental change, frequently anthropogenous.

Jones and Diamond (1976) report subsequent censuses that allay one's concern over (2), though (3) and especially (1) are still problematic. Lynch and Johnson (1974) contend on grounds (1) and (2), *contra* Diamond (1971), that turnover on Karkar has not been documented. For Mona Island they concede a large fraction of the turnover reported by Terborgh and Faaborg (1973) but dismiss another fraction as likely to be bound up with human activities. The Farne Islands bird data (Diamond and May 1977) depict substantial yearly turnover, but since the islands are small (80 acres total) bits of land only $1\frac{3}{4}$ mi offshore, one must suspect that these birds are only part of larger, more widespread populations (*cf.* Haila *et at.* 1979).

Abbott and Grant (1976) depict lower, though substantial, turnover for Skokholm Island birds, which they view as not in equilibrium. As this is another small (244 acre), near (2 mi.) island, one again must question the extent to which these birds are parts of larger, more widespread populations. No such reservations must be held concerning the Pacific islands Abbott and Grant studied, and even though only two censuses are reported for each (thus minimizing estimated turnover) an impressive number of extinctions and immigrations are known. But Abbott and Grant attribute at least half of these to man's activities, so that again the assumptions of the equilibrium model cannot be confirmed; they view all these islands as non-equilibrial, for reasons stated above. Slud (1976) for Cocos Island, Abbott (1980) for Christmas Island and Raine Island, and Salomonsen (1976) for the Faroe Islands, by contrast, find constant species number, but virtually no turnover, so that again the model does not apply. Grant and Cowan (1964) report an identical situation for the Tres Mariás Islands.

The equilibrium model, then, has not been generally confirmcd. Indirect evidence, such as species–area relationships, cannot strongly support the model, and infrequent censuses and/or habitat change (usually anthropogenous) weaken inferences that can be drawn from many studies. A key piece of information (where local recruitment derives from) is usually absent (Simberloff, 1976) for examples that appear to conform to the model's tenets, and a similar number of examples do not even conform to the tenets. Gilbert (1980) sees the evidence taken as a whole as sufficiently debilitating to falsify the model as a general depiction of nature. Abbott (1980), examining bird data relevant to the model, is kinder: he views the evidence to date as inconclusive, and observes that northern hemisphere islands at least conform to the model's dynamic component more strongly than Australasian islands do. And Abbott and Grant (1976) suggest that high latitude islands, even if dynamic, are unlikely to have fixed equilibria.

TESTING FOR EQUILIBRIUM

One problem in assessing the validity of the equilibrium model is that there is no generally accepted notion of how constant species number must be in order to be viewed as equilibrial. It must be emphasized that, even in the absence of species interactions, an island will have an equilibrium species number so long as each species α has a fixed probability i_α of having a propagule land on the island in a unit time interval and a fixed probability e_α of being extinguished from non-interactive causes in a unit time interval. Under those circumstances, an equilibrium will obtain (Simberloff, 1969):

$$S = \sum_{\alpha=1}^{P} \frac{i_\alpha}{i_\alpha + e_\alpha}$$

where P is the number of species in the pool. But the probabilistic nature of this formulation insures that even if this model were correct, one would see some fluctuation. And if species interactions were important, there would surely still be chance events, both biotic and abiotic, dictating a certain amount of temporal variation.

Diamond (1969) simply notes that for most Channel Islands, bird species number remained constant within 16% and views this as evidence for equilibrium. Diamond and May (1977), as stated above, construe an over two-fold change in species number of Farne Islands birds as strong evidence for an equilibrium. Other papers in the literature view different percentage changes as either consistent or inconsistent with an equilibrium, and of course there is no objective criterion here (Abbott and Grant, 1976). Diamond and May (1977) also provide the coefficient of variation for the Farne Islands avifauna (0·20), but this is no less arbitrary a criterion for equilibrium than is per cent change in species number. Hence one might ask what coefficient of variation would be taken as *falsifying* the equilibrium.

For an archipelago of islands Abbott and Grant (1976) suggest that if the equilibrium model holds, one would expect equal numbers of islands with increasing and decreasing numbers of species in a specified time interval, and use a sign test to assess deviation from this state, but they do not rationalize the expectation. Could not *all* islands in an archipelago fail to conform to the equilibrium model, with approximately half increasing and half decreasing in species number in a particular interval? S. Hinneri (pers. comm.), in a study of vascular plants of a Finnish archipelago, first constructed a species–area curve, then noted the islands for which observed S was at least 10% greater or 10% less than that predicted from the archipelagic species–area curve. Reasoning that the former should fall disproportionately and the latter rise disproportionately if the number of species on an island were truly a dynamic equilibrium, he observed exactly this state of affairs at a second census. The slope of the species–area curve was virtually unchanged, but with different islands as outliers, both above and below the curve. It is difficult to attach a probability to this event, since even a sequence of bounded random numbers would tend to have high ones followed by large decreases and low ones followed by large increases (see below). But at least the observed trend is consistent with the model.

What is needed is a probabilistic test for a given sequence of species numbers on a single island, and though there is doubtless no single best statistic for this, I now suggest a useful approach. First, it is apparent that two data points are insufficient to establish whether or not an equilibrium obtains, and can at best convey an impression. Diamond (1971) finds 43 lowland bird species on Karkar in 1914, and 49 in 1969. No readily constructed null probability is possible for this change, and whether one views the change as small or large is a matter of taste. With a sequence of N censuses, however, impressions can become stronger even without exact probabilities, and some straightforward statistics are possible. I will use two examples (Fig. 1): Diamond and May's sequence (1977) of 29 annual censuses of birds on the Farne Islands, which they view as clearly equilibrial; and Abbott and Grant's sequence (1976) of 34

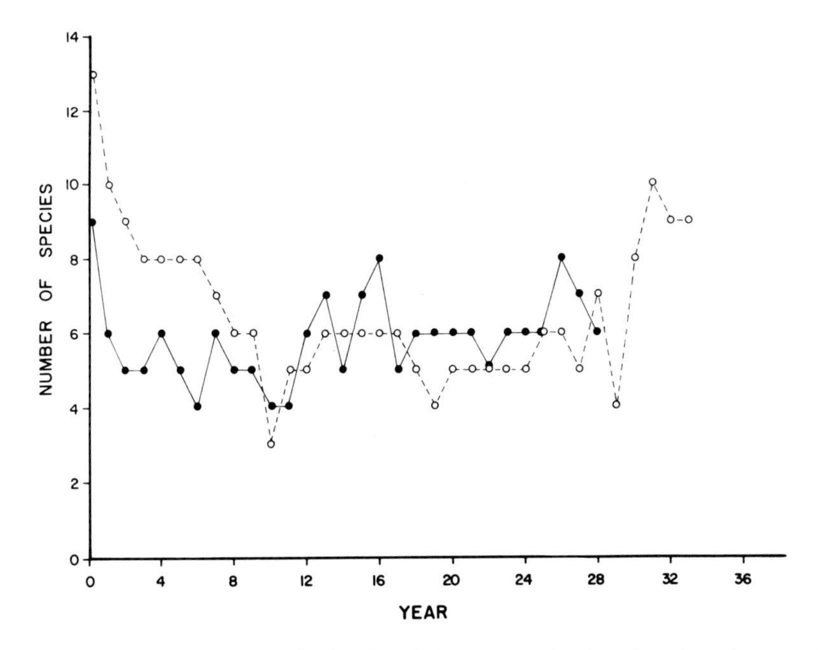

Fig. 1. Colonization curves for birds of the Farne Islands (closed circles, solid line) and passerines of Skokholm Island (open circles, dashed line).

annual censuses of passerines on Skokholm Island, which they view as equally clearly non-equilibrial.

In some sense, since species number must remain within the bounds of O and P, one may view it as always equilibrial. This would render the notion a truism, however, and would obviate all reports of species number equilibria, since an hypothesis of equilibrium could never be falsified. A stronger and more interesting claim would be that species number not only stays within bounds, but tends toward some specific value. As noted above, even in the absence of species interactions such a tendency is possible. One usually views interactions as competitive (see references below) and strengthening a tendency toward an equilibrium by providing a strong impetus for extinction when above-equilibrium numbers obtain. Whatever the possible reasons for species number regulation about a particular equilibrium, we wish to detect such regulation in spite of the fact that we cannot know the exact value of the equilibrium.

First, if species number is regulated about some equilibrium, then one expects that the highest species numbers will tend to be followed by the largest decreases, while the lowest species numbers will tend to be followed by the largest increases. For both of the exemplary sequences (Fig. 2) this is so: for the Farne Islands the correlation coefficient is $-0{\cdot}681$ ($df=26$, $Pr<0{\cdot}01$), while for Skokholm it is $-0{\cdot}535$ (31 df,

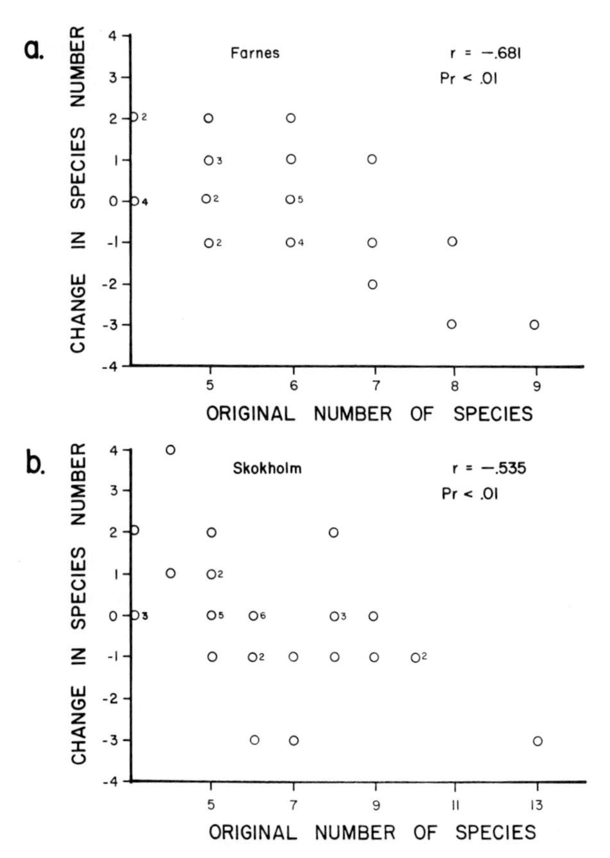

Fig. 2. Changes in species numbers vs. original numbers of species present for Farne Islands birds (a) and Skokholm Island passerines (b).

Pr <0·01). Unfortunately, the probability levels cannot be accepted at face value since, if one randomly drew bounded numbers from a variety of distributions, one would expect a negative correlation between N_i and $(N_{i-1} - N_i)$. This is precisely the difficulty in using population census data to test for density-dependent population regulation (Maelzer 1970, St. Amant 1970), and just as experiment proved critical in assessing that model (Murdoch, 1970; Abbott, 1980), so would experiment likely be most convincing in assessing the equilibrium model (Simberloff, 1978b).

But long strings of censuses can still be useful. If species number is truly regulated about some equilibrium, then the observed number of "runs" (Feller 1968) of increases and decreases ought to exceed that in an unregulated situation. For with regulation, there should be a tendency for an increase to be redressed by a decrease, and vice versa. Observations of "no change" are not tallied, and the sequence of censuses is reduced by eliminating all but one of each string of equal numbers. Of course, a large number of "no changes" would have to be taken as supporting the equilibrium model (at least if species composition changed all the while that species number did not), but no obvious probability attaches to any particular number of observed "no changes". With a "reduced sequence" derived by eliminating "no changes" (Table I), one may ask if the decreases tend to follow the increases and vice versa more frequently than chance would have predicted. For *a* increases and *b* decreases, there are $\binom{a+b}{a}$ distinguishable orderings (Feller, 1968). Any string of successive *a*s (increases) or *b*s (decreases) constitutes a "run", and each ordering has a given number of runs. If we take as a null hypothesis that all orderings are equiprobable, what we require is the tail probability for the observed number of runs or more, since we wish to know if the observed number of runs is extremely high, corresponding to regulatory equilibration. Including the runs on the beginning and end of a sequence of $N - 1$ increases and decreases, formed from a sequence of N censuses, Keeping (1962), following Moore and Wallis (1943), shows that the expected number of runs $E(r) = (2N - 1)/3$, with $\mathrm{var}(r) = (16N - 29)/90$. The tail probability for the observed number of runs *u* or more is easily calculated from Swed and Eisenhart's tables (1943) of the tail probability for an observed number of runs *u* or fewer. Finally, a related group of expectations useful in assessing whether the observed runs are consistent with the null hypothesis that the ordering of increases and decreases is randomly chosen from all orderings consists of the expected numbers of runs of lengths, 1, 2 and greater than 2, respectively. These are (Keeping 1962, following Moore and Wallis 1943) $(5N+1)/12$, $(11N-14)/60$, and $(4N-11)/60$, respectively, and observations may be tested against these expectations by a χ^2 test. For $N > 12$, calculated χ^2 is multiplied by 6/7 and referred to a χ^2 with two degrees of freedom. Levene (1952) addresses the power of different tests based on runs statistics for various alternative hypotheses.

These tests are all applied to both the Farne and Skokholm data (Table I). The key test, whether the number of runs is extremely large, shows that it is not in both examples (# runs = 13 for Farne, Pr = 0·227; # runs = 9 for Skokholm, Pr = 0·704), suggesting that neither avifauna strongly exemplifies the regulatory equilibrium model, at least with a fixed equilibrium. The distributions of run lengths are consistent with this interpretation ($\chi_2{}^2 = 4{\cdot}534$ for Farne, Pr = 0·104; $\chi_2{}^2 = 1{\cdot}868$ for Skokholm, Pr. = 0·393).

SPECIES COMPOSITION

Though an equilibrium number of species would obtain even in the complete absence of species interactions (as I showed in 1969 and reiterated above), the turnover during the equilibration process has been viewed as primarily generated by interspecific competition (Abbott, 1980; cf. MacArthur and Wilson, 1967, p. 121; MacArthur, 1972, pp. 91, 99). Yet the evidence is scant that competitive interactions cause equilibrating extinctions on islands. One-for-one replacement of obvious competitors occurs in none of the examples cited above as supporting the equilibrium model, and the voluminous literature on species introductions on islands also fails to provide evidence for such replacement (Simberloff, 1981). A recent study (Diamond and Veitch, 1981) confirms this trend for New Zealand birds.

An early theoretical attempt to document that competition limits species number on islands focused on species-to-genus ratios; that these were typically lower for islands than mainland was taken as evidence that closely related species are harder pressed to coexist on islands than on mainland (Grant, 1966; MacArthur and Wilson, 1967). However, this difference is probably due to the strong dependence of the ratio on number of species present, and in most cases the species-to-genus ratio on islands is in fact *higher* than that expected for a random draw from the

Table I. Tests of runs for sequences of censuses for birds of the Farne Islands (Diamond and May, 1977) and passerines of Skokholm Island (Abbott and Grant, 1976). Missing censuses on Skokholm (denoted by "—") are ignored. # = number of.

Farne Islands # species			*Sign of change*			
9	6	9				
			−	N = 21		
6	5	6				
			−	a = 9		
5	6	5				
			+	b = 11		
5	6	6				
			−	# runs = 13		
6	6	5				
			−	E(# runs) = 13·667, s.d. = 1·847		
5	8	4				
			+	Pr(# runs > 13) = 1 − Pr(# runs ⩽ 12) = 0·227		
4	7					
		6				
			−		*obs*	*exp*
6	6	5				
			−	# runs length 1	6	8·333
5		4				
			+	# runs length 2	7	3·617
5		6				
			+	# runs length > 2	0	1·217
4	→	7				
			−	6/7 χ^2 = 4·534, Pr = 0·104		

Skokholm Island # species			*Sign of change*			
13	5	13				
			−	N = 19		
10	5	10				
			−	a = 7		
9	5	9				
			−	b = 11		
8	6	8				
			−	# runs = 9		
8	6	7				
			−	E(# runs) = 12·333, s.d. = 1·748		
8	5	6				
			−	Pr(# runs > 9) = 1 − Pr(# runs ⩽ 8) = 0·704		
8	7	3				
			+		*obs*	*exp*
7	4	5				
			+	# runs length 1	4	8·000
6	8	6				
			−	# runs length 2	4	3·250
6	0	5				
			−	# runs length > 2	1	1·083
3	9 →	4				
			+	6/7 χ^2 = 1·868, Pr = 0·393		

4	5	5 9	5
	+		+
6	7	—	6
	+		−
7	8	5	5
	−		+
5	5	6	7
	+		−
7	6	6	4
	−		+
8	5	6	8
	+		+
5	6	6	10
	+		−
6	8	6	9
	−		
6	7	5	
	−		
6	6	4	
		5	
		5	

mainland pool (Simberloff, 1970; Simberloff, 1978a). Another attempt to show the effect of restricted competition on islands was the contention that competition sets a minimum size ratio between related species compatible with coexistence (Hutchinson, 1959), that this minimum ratio is larger on islands than on mainland (Schoener, 1965; Grant, 1968), and that it is larger on small than on large islands (Abbott *et al.*, 1977). But Simberloff and Boecklen (1981) have questioned whether any data indicate the existence of a minimum ratio even on mainland, and Strong *et al.* (1979) have claimed in particular that island ratios tend to exceed mainland ones, and small island ratios tend to exceed large island ones, simply as an artifact of the number of species present. Even a random draw of a proper subset of species from a mainland pool has an expected minimum ratio larger than that in the pool itself. Grant and Abbott (1980) and Hendrickson (1981) dispute Strong *et al.* (1979) on grounds of statistical procedures and inappropriate data, but even using their data and statistical procedures we find little or no evidence that island species are particularly displaced in size (Strong and Simberloff, 1981; Simberloff, 1983).

Perhaps partly in response to the failure of species-to-genus ratios to manifest competitive replacement (Abbott, 1980), MacArthur (1972) and others (e.g. Diamond, 1975) turned to the concept of "diffuse competition" to salvage the notion that competition determines the number of species on an island. Diffuse competition is the phenomenon whereby a group of competitors, not one of which is sufficiently similar to a particular species to exclude it, together reduce resources to an extent that that species cannot exist. MacArthur and Wilson (1967) adduced the banaquit's absence from Cuba as an early example of this phenomenon. Abbott and Grant (1976) pointed out that while an hypothesis of diffuse competition is occasionally plausible, it is usually invoked in an *ad hoc* fashion and is extremely difficult to falsify with normally available data.

The shift in focus to groups of species brought the matter of *composition* of island biotas to the fore, and one now asks not only how many species coexist on an island, but which species these are, and why these. The particular observational data brought to bear on these matters have primarily concerned how many combinations of species are found (and/or how many are not found) on the islands of an archipelago, or how similar island biotas are, and though these matters are amenable to statistical analysis much of the literature on this aspect of diffuse competition is distressingly diffuse, a sad tale of missing or misapplied statistics (Simberloff and Connor, 1979, 1981).

The basic data for an analysis of insular biotic composition consist of an $S \times I$ binary matrix M where S = number of species, I = number of islands or sites, $M(i,j) = 1$ if species i occurs on island j, and $M(i,j) = 0$ if species i is absent from island j. Such matrices can be examined in two ways (Simberloff and Connor, 1979). First, one may classify sites by seeing how similar or different the columns are (analogously to Q-mode analysis in numerical taxonomy), or one may classify species by seeing how similar the rows are (analogously to R-mode analysis). The Q-mode corresponds to assessing site similarity or dissimilarity, the R-mode to evaluating how similar species are in their arrangement over sites.

Q-mode analysis has the longer and older tradition (Simberloff and Connor, 1979, for references) but is beset by the problem that almost all indices used until very recently to measure similarity are *ad hoc* and have unknown probability distributions, so that biological hypotheses cannot be directly tested by them. Nonetheless, causal hypotheses concerning species' biogeographic distributions have occasionally been addressed in this way. Terborgh (1973), for example, calculates the Sørensen similarity index between pairs of West Indian islands to see how similar their avifaunas are, then regresses the value of this index on inter-island distance and concludes that most of the differences between the species compositions of two islands are attributable to area and isolation differences, with interspecific competition responsible for much of the remainder. Yet there is no test for whether two values of the Sørensen index, say 0·37 and 0·49, are significantly different, and the index is sample–size dependent. Power (1975) used a different arbitrary similarity index (Preston's) to examine bird and plant similarity among pairs of Galapagos islands and regressed this index on various physical factors to conclude what forces "explain" what average fraction of the similarity index, or which physical factors are "significant". But since there is no way of telling whether two values of the index are statistically significantly different, one cannot say whether some observed difference in similarity between two pairs of islands is real. And, as in the previous example, the regression that was used to indicate the "important" variables is not an appropriate way to test causal hypotheses, for reasons stated above. A final example is Abbott's study (1977) of plants on small islands of western Australia. For four classes of islands—those with 1–4 species, respectively—he calculated Sørensen's similarity index between all island pairs. He then observed that more pairs of one-species islands have different species than have the same species, and that most pairs of two-species or three-species islands share at least one species, but posed no biological hypotheses to exlain these observations, nor one about the observed distribution of Sørensen similarities.

Recently, a probabilistic similarity index of known distribution suitable for testing an hypothesis about some set of similarities has been proposed (Simberloff, 1978c; Connor and Simberloff, 1978; cf. Grassle and Smith 1976; Smith *et al.*, 1979). The index is a function of number of species shared between sites vs. expected number shared based on some hypothesis about how the species are arranged. For example, Connor and Simberloff (1978) calculate for all pairs of islands in the Galapagos the number of birds and plants shared, and the number that one would have expected if species were randomly and independently arranged among the islands. The similarity index is then the observed number of shared species minus that expected, divided by the standard deviation. The expectation and variance depend on what assumptions one makes about *how* the species are randomly chosen to colonize an island. If all species in the pool are equally likely to colonize any particular island (the "equiprobable" model), the expected number of species shared between an island with m species and another with n is mn/P (Connor and Simberloff, 1978), while the variance is

$$(mn/P) \left[\frac{P^2 - P(m+n) + mn}{P(P-1)}\right]$$

(J. Wright and C. Biehl, pers. comm.). If the species are not equally likely to colonize, the expectation and variance depend on the relative colonization propensities of the different species. Connor and Simberloff (1978) and Simberloff (1978) suggest weighting the species' colonization probabilities by their relative frequencies within the archipelago (the "weighted" model). With this distribution of probabilities, the expected number of species shared and its variance may be calculated by simulation, and for the Galapagos plants and birds the weighted model produces island similarities much closer to observed similarities than does the equiprobable model, for almost all island pairs. J. Wong (pers. comm.) has recently adapted the weighted similarity index to a clustering algorithm, using avifaunas to cluster West Indian islands.

Although the weighted model generally produces similarities far closer to observed ones than does the equiprobable model, still the null hypothesis of random, independent placement of species on islands would be rejected in most instances. However, usually the majority of species shared between islands is accounted for by this model, and one might view it as a baseline, such that any similarity over and above this certainly demands an explanation. Grant and Abbott (1980) object to even this limited interpretation of these results, on the grounds that the entire procedure as a test of an hypothesis of interspecific interaction is "if not circular, . . . at least elliptical", since the distribution of species' frequencies on which their colonization probabilities are based could have been partly determined by interactions. This is certainly true, as was pointed out by Simberloff (1978c), but it seems to me that such an exercise is still useful to indicate what maximum fraction of compositional similarities between islands could plausibly be accounted for by species' acting independently of one another. Of course no such analysis can *prove* what forces do or do not mould insular species compositions. Probably the idea that ought to be tested first, that habitat differences between islands determine how similar their faunas are, requires the most extensive field data. The most parsimonious hypothesis, even if unfalsified, need not be the correct one (Simberloff and Connor, 1981), and probably several lines of evidence (and, one hopes, experiment (Abbott, 1980)) will be brought to bear on these matters. But it is important that testable hypotheses be posed (Popper 1963), if only to move biogeography beyond descriptions or narrative explanations, into the realm of the mature sciences (Ball, 1975).

R-mode analysis, comparing species to see how many sites they share, has proceeded primarily by seeking extreme dissimilarity—that is, pairs or larger groups of species that share *no* sites—and then attributing the existence of such exclusively distributed groups to interspecific competition. Simberloff and Connor (1981) review this literature. For n species there are 2^n possible combinations of species, but one cannot proceed in the most obvious manner, by calculating expected frequencies for all combinations as if the component species colonize independently, then comparing these to the observed frequencies by a χ^2 test. For expectations will be very low, and observed frequencies of most combinations will be zero, even when there are many sites. Askew (1962) was perhaps the first to recognize this problem in his study of the distribution of four gall wasps on oaks.

Pielou and Pielou (1968) proposed an elegant solution—namely, to simulate the colonization of sites by randomly constructing a binary matrix M(i,j), subject only to the constraint that row sums be constant (that is, that species' frequencies be kept as in nature). Species were allowed to colonize independently of one another. From repeated simulations, Pielou and Pielou estimated the expected number of distinct species combinations (and its variance) and compared it to that observed. For most of the bracket fungus populations whose beetles the Pielous were interested in, the observed and expected number of combinations did not differ significantly. But even if they *had* differed, Pielou and Pielou emphasize that species interactions are no more likely an explanation for the discrepancy, *a priori*, than are differences among the sites. The statistics alone cannot demonstrate a biological cause.

Connor and Simberloff (1979) used a similar method to test Diamond's "assembly rules" (1975) for bird communities. These rules, attributed primarily to the workings of interspecific competition, describe patterns of species' distributions in the Bismarck Archipelago. One rule states that some pairs of species occur on no islands; another rule states that some larger groups of species are nowhere found. But since Diamond's data comprise 147 species on only 50 islands, one would have expected some combinations of species not to exist, even if the species had randomly and independently colonized the islands. Interspecific competition would be indicated only if the number of missing combinations exceeds the number that random, independent placement would have produced. To determine these expectations, Connor and Simberloff (1979) simulated the random colonization of the archipelago as the Pielous had, but with column sums (species richnesses) constrained in addition to row sums (species frequencies) to accord with the species–area relationship and other inter-island physical differences that affect richness. For one avifauna, the number of missing combinations was within one standard deviation of expected; for another, observed and expected were almost four standard deviations apart, but the expected number of mutually exclusive pairs (12 448) was 97·6% of the observed number (12 757). For a bat fauna known to have colonized in at least two different periods from two different sources, the observed and expected numbers of missing combinations were not so close. Connor and Simberloff (1979) concluded from this exercise *not* that species colonized any of these archipelagoes randomly and independently, nor that interspecific competition was not crucial in determining which species are found where. Rather, they suggested that the biogeographical data alone do not constitute evidence for the importance of competition.

M'Closkey (1978) suggested that Sonoran desert rodent communities are structured primarily by competition and that the distribution of combinations of species accords with the assembly rules. But no statistical test buttressed these contentions. Grant (1970) studied rodents on islands off the coasts of Canada, Britain, and Denmark, and rejected the hypothesis that the observed distribution of two-genus combinations is rationalized by characteristics of the individual genera because the *Microtus-Clethrionomys* combination is absent at all sites. Instead he concluded that this absence is caused by interspecific competition. However, his statistical approach is suspect (Simberloff and Connor, 1981), since he compares by a χ^2-test the observed frequencies of two-genus combinations to equal expectations for all combinations. Since any combination with a rare species would be expected to be rare (cf. Askew, 1962; Pielou and Pielou, 1968) even if the species had no interactions whatsoever, Grant's high χ^2 values do not impugn an hypothesis of independent colonization.

Abbott (1977) and Abbott *et al.* (1977) take a different approach, using Maxwell-Boltzmann statistics (Feller, 1968) to see if numbers of missing combinations exceed expected for Australian plants and Galapagos finches, respectively. Both studies find for islands with certain numbers of species that too many combinations are missing to sustain the Maxwell-Boltzmann model of balls thrown randomly into boxes, and suggest that a likely cause for the discrepancy would be interspecific competition. But two considerations cast doubt on the assertion (Simberloff and Connor, 1981). First, both papers calculate the Maxwell-Boltzmann probabilities incorrectly. Second, and more important, the Maxwell-Boltzmann model does not constitute an appropriate null hypothesis of species' independent colonization of a set of islands *vs.* an alternative of species' interaction precluding some combinations. For the Maxwell-Boltzmann model views combinations, not species, as the colonizing units, and envisions each island as being assigned to one of these units analogously to balls randomly thrown into boxes, with each box having the same size (or probability of receiving any given ball). But if the species are the colonizing units and act independently of one another, all combinations are not equiprobable: the boxes should be of different sizes. Any combination that includes one or more rare species would *ipso facto* be rare.

Simberloff and Connor (1981) simulated such a model—throwing balls (sites) randomly into different-sized boxes (combinations), with each box's size proportional to the product of the frequencies of its component species—and used it to examine the rodent and plant studies plus that of the Galapagos birds. In all simulations, making box sizes unequal increased the expected number of missing combinations over that of the Maxwell-Boltzmann model. In every instance but one the number of missing species combinations was not significantly different from that expected, if the species had colonized randomly and independently (but had retained their observed frequencies). The single exception was the Galapagos finches, for which islands with three species (and possibly four and five species also, depending on what species were taken to constitute the species pool) had far fewer combinations (only two, for three-species islands) than chance alone would have predicted.

A recent and exciting approach to binary species x site matrices is by Whittam and Siegel-Causey (1981), who use multiway contingency tables and log-linear models to seek evidence for species interactions in Alaskan seabird communities. Four groups of five species each were chosen so that the species in each group all use the same sort of nesting site. The basic log-linear model is:

$$\ln e_{ijklm} = u + u_A + u_B + u_C + u_D + u_E$$

where e_{ijklm} is the expected number of colonies with a given combination of species; i, j, k, l, and m are 0 or 1 depending on whether species A, B, C, D, and E, respectively, are absent or present. The overall mean effect, or average of the logarithms of the expectation of all possible combinations, is u, while the main effect for each species t (u_t, $t = A$ through E) is the average over all combinations including species t, minus u. If the species are mutually independent in their arrangement over sites, the e_{ijklm} can be calculated directly from the total numbers of colonies with and without each species, exactly as one would calculate expectations in a two-way contingency table. If these expectations do not match observations well, additional terms representing species interactions must be added to the model. There are $\binom{5}{2}$ two-way interactions, $\binom{5}{3}$ three-way interactions, etc. The strategy is to achieve an acceptable fit with the fewest interaction terms.

For these four groups, of the total of forty two-way interactions, twenty-four had to be added to the log-linear models to achieve acceptable goodness-of-fit. Of these, twenty-two of the interactions were positive and only two negative, and only these latter two could be consistent with an hypothesis of interspecific competition. Of the forty three-way interactions, only two had to be included in the log-linear models, and both of these constitute "second-order interactions" in that the negative interaction between two of the three species is exacerbated in the presence of the third. Whittam and Siegel-Causey see these as consistent with hypotheses of diffuse competition. Finally, in order to estimate what fraction of these interactions might be statistical artifacts, Whittam and Siegel-Causey constructed a fifth five-species group, no two species of which require the same nest site, and in the log-linear model for this group there are seven significant two-way interactions, of which six are positive. In sum, this study finds very few interactions consistent with a competition hypothesis, and it is as yet unclear whether as many interactions as were seen might be approximately as many as one would expect as statistical artifacts in randomly constructed assemblages.

DISCUSSION

The models discussed in the previous section with which I have been associated, and (I will predict) the multiway contingency model, have been attacked (e.g. Grant and Abbott, 1980) on grounds of lack of realism and probable high type II error rates. That is, the models may not faithfully reflect nature, and it may be difficult to reject the null hypothesis even if one knows it is too highly simplified biologically to be realistic. The latter is certainly a fair charge, and it is precisely the reason why my colleagues and I have been at pains to emphasize that our work does *not* show that geographic or morphological properties of different species are random and independent. Rather, our contention has been that these anlyses show that a matrix of species' locations, or a list of species' sizes, usually cannot by itself reject an hypothesis of species' independence in favour of competition; Pielou and Pielou (1968) propounded a similar message. We have also repeatedly called for an experimental approach to these questions, a plea that Abbott (1980) recently echoed.

But I would be remiss were I not to argue that increasing realism of the models, or at least the source of the evidence on which the purported realism is based, has dangerous pitfalls. This is epitomized by Gilpin and Diamond's study (1976), in which thirteen different versions of the equilibrium model with up to seven parameters were devised to fit, and were tested against, observed distance and area relationships for bird data from 52 islands in the Solomon Islands. Aside from the fact that it is illegitimate to use a single data set both to suggest an hypothesis and to test it (Selvin and Stuart, 1966; Pielou and Pielou, 1968), there is no *a priori* way to determine what the null probability is of finding an acceptable fit given several models and several parameters. That is, even if there were *no* relationship between either distance or area and species number, would it not be possible to concoct a model that came reasonably close to 52 data points, especially given seven parameters? And what null probability attaches to such a model should one be found? Selvin and Stuart (1966) detail the difficulty in assigning significance to such a result. At the least, it would be meaningless unless tested in an entirely different archipelago. I would go further: a model achieved by a survey of this sort is but another "narrative" explanation in Ball's sense. It is at best a plausible explanation, from among many possible, for some observed biogeographic data.

Even if increased "realism" of a model does not stem simply from curve-fitting, but rather from observed biological properties of the component species of a community, I would argue that at some point the increasing realism of the model (added parameters or constraints, etc.) becomes counter-productive. After all, the best-fitting model for a set of points *is* the set of points itself. And the more realistically a model describes a given community, the less likely is it to describe realistically any other community, making a test of the model as an hypothesis increasingly difficult if not impossible. The model then ceases to be a generalized abstraction of elements common to all communities, but rather becomes just an increasingly elaborate narrative description. Once again I am led to conclude with Abbott (1980) that a claim that some biological interaction is biogeographically important ought to rest at the least on several lines of evidence, and at best on controlled experiment.

ACKNOWLEDGEMENTS

Drs O. Järvinen and D. Strong contributed many helpful comments and suggestions.

REFERENCES

Abbott, I. (1977). Species richness, turnover, and equilibrium in insular floras near Perth, Western Australia. *Austr. J. Bot.* **25**, 193–208.

Abbott, I. (1980). Theories dealing with the ecology of landbirds on islands. *Adv. Ecol. Res.* **11**, 329–371.

Abbott, I., Abbott, L.K. and Grant, P.R., (1977). Comparative ecology of Galápagos ground finches (*Geospiza* Gould): Evaluation of the importance of floristic diversity and interspecific competition. *Ecol. Monogr.* **47**, 151–184.

Abbott, I., and Grant, P.R. (1976). Non-equilibrial bird faunas on islands. *Am. Nat.* **110**, 507–528.

Anon. (1980). "World Conservation Strategy". IUCN—UNEP—WWF.

Askew, R.R. (1962). The distribution of galls of *Neuroterus* (Hym: Cynipidae) on oak. *J. Anim. Ecol.* **31**, 439–455.

Ball, I.R. (1975). Nature and formulation of biogeographic hypotheses. *Syst. Zool.* **24**, 407–430.

Barbour, C.D. and Brown, J.H. (1974). Fish species diversity in lakes. *Am. Nat.* **108**, 473–489.

Brown, J.H. (1971). Mammals on mountaintops: Nonequilibrium insular biogeography. *Am. Nat.* **105**, 467–478.

Brown, J.H. and Kodric-Brown, A. (1977). Turnover rates in insular biogeography: Effects of immigration on extinction. *Ecology* **58**, 445–449.

Connor, E.F. and McCoy, E.D. (1979). The statistics and biology of the species–area relationship. *Am. Nat.* **113**, 791–833.

Connor, E.F. and Simberloff, D. (1978). Species number and compositional similarity of the Galápagos flora and avifauna. *Ecol. Monogr.* **48**, 219–248.

Connor, E.F. and Simberloff, D. (1979). The assembly of species communities: Chance or competition? *Ecology* **60**, 1132–1140.

Culver, D.C., Holsinger, J.R. and Baroody, R. (1973). Toward a predictive cave biogeography: The Greenbrier Valley as a case study. *Evolution* **27**, 689–695.

Diamond, J.M. (1969). Avifaunal equilibrium and species turnover rates on the Channel Islands of California. *Proc. natn Acad. Sci. U.S.A.* **64**, 57–63.

Diamond, J.M. (1971). Comparison of faunal equilibrium turnover rates on a tropical island and a temperate island. *Proc. natn Acad. Sci. U.S.A.* **68**, 2742–2745.

Diamond, J.M. (1972). Biogeographic kinetics: Estimation of relaxation time for avifaunas of Southwest Pacific islands. *Proc. natn Acad. Sci. U.S.A.* **69**, 3199–3203.

Diamond, J.M. (1975). Assembly of species communities. *In* "Ecology and Evolution of Communities" (M.L. Cody and J.M. Diamond, eds.), pp. 342–444. Harvard University Press, Cambridge, Mass.

Diamond, J.M. and May, R.M. (1977). Species turnover rates on islands: Dependence on census interval. *Science, N.Y.* **197**, 266–270.

Diamond, J.M. and Veitch, C.R. (1981). Extinctions and introductions in the New Zealand avifauna: Cause and effect? *Science, N.Y.* **211**, 499–501.

Faeth, S.H. and Connor, E.F. (1979). Supersaturated and relaxing island faunas: A critique of the species-age relationship. *J. Biogeogr.* **6**, 311–316.

Feller, W. (1968). "An Introduction to Probability Theory and Its Applications", vol. 1 (3rd edn). Wiley, New York.

Gilbert, F.S. (1980). The equilibrium theory of island biogeography: Fact or fiction? *J. Biogeogr.* **7**, 209–235.

Gilpin, M.E. and Diamond, J.M. (1976). Calculation of immigration and extinction curves from the species–area–distance relation. *Proc. natn Acad. Sci. U.S.A.* **73**, 4130–4134.

Grant, P.R. (1966). Ecological incompatibility of bird species on islands. *Am. Nat.* **100**, 451–462.

Grant, P.R. (1968). Bill size, body size and the ecological adaptations of bird species to competitive situations on islands. *Syst. Zool.* **17**, 319–333.

Grant, P.R. (1970). Colonization of islands by ecologically dissimilar species of mammals. *Can. J. Zool.* **48**, 545–553.

Grant, P.R. and Abbott, I. (1980). Interspecific competition, island biogeography and null hypotheses. *Evolution* **34**, 332–341.

Grant, P.R. and Cowan, I.McT. (1964). A review of the avifauna of the Tres Marias Islands, Nayarit, Mexico. *Condor* **66**, 221–228.

Grassle, J.F. and Smith, W.K. (1976). A similarity measure sensitive to the contribution of rare species and its use in investigation of variation in marine benthic communities. *Oecologia* **25**, 13–22.

Haila, Y., Järvinen, O. and Väisänen, R.A. (1979). Effect of mainland population changes on the terrestrial bird fauna of a northern island. *Ornis scand.* **10**, 48–55.

Hendrickson, J.A., jr. (1981). Community-wide character displacement reexamined. *Evolution* **35**, 794–809.

Hutchinson, G.E. (1959). Homage to Santa Rosalia, or why are there so many kinds of animals? *Am. Nat.* **93**, 145–159.

Jones, H.L., and Diamond, J.M. (1976). Short-time-base studies of turnover in breeding bird populations on the California Channel Islands. *Condor* **78**, 526–549.

Keeping, E.S. (1962). "Introduction to Statistical Inference". Van Nostrand, Princeton.

Lassen, H.H. (1975). The diversity of freshwater snails in view of the equilibrium theory of island biogeography. *Oecologia* **19**, 1–8.

Levene, H. (1952). On the power function of tests of randomness based on runs up and down. *Ann. Math. Stat.* **23**, 34–56.

Lynch, J.F., and Johnson, N.K. (1974). Turnover and equilibria in insular avifaunas, with special reference to the California Channel Islands. *Condor* **76**, 370–384.

MacArthur, R.H. (1972). "Geographical Ecology: Patterns in the Distribution of Species". Harper & Row, New York.

MacArthur, R.H. and Wilson, E.O. (1963). An equilibrium theory of insular zoogeography. *Evolution* **17**, 373–387.

MacArthur, R.H. and Wilson, E.O. (1967). "The Theory of Island Biogeography". Princeton University Press, Princeton, N.J.

Maelzer, D.A. (1970). The regression of log N_{n+1} on log N as a test of density dependence: An exercise with computer-constructed density-independent populations. *Ecology* **51**, 810–822.

Magnuson, J.J. (1976). Managing with exotics—a game of chance. *Trans. Am. Fisheries Soc.* **105**, 1–9.

M'Closkey, R.T. (1978). Niche separation and assembly in four species of Sonoran desert rodents. *Am. Nat.* **112**, 683–694.

Moore, G.H. and Wallis, W.A. (1943). Time series significance tests based on signs of differences. *J. Am. Stat. Assoc.* **38**, 153–164.

Murdoch, W.W. (1970). Population regulation and population inertia. *Ecology* **51**, 497–502.

Pielou, D.P. and Pielou, E.C. (1968). Association among species of infrequent occurrence: The insect and spider fauna of *Polyporus betulinus* (Bulliard) Fries. *J. Theoret. Biol.* **21**, 202–216.

Popper, K.R. (1963). "Conjectures and Refutations: The Growth of Scientific Knowledge". Harper & Row, New York.

Power, D.M. (1975). Similarity among avifaunas of the Galapagos islands. *Ecology* **56**, 616–626.

St Amant, J.L.S. (1970). The detection of regulation in animal populations. *Ecology* **51**, 823–828.

Salomonsen, F. (1976). The main problems concerning avian evolution on islands. *Proc. Internat. Ornithol. Congr.* **16**, 585–602.

Schoener, T.W. (1965). The evolution of bill size differences among sympatric congeneric species of birds. *Evolution* **19**, 189–213.

Seifert, R.P. (1975). Clumps of *Heliconia* inflorescences as ecological islands. *Ecology* **56**, 1416–1422.

Selvin, H.C. and Stuart, A. (1966). Data-dredging procedures in survey analysis. *Am. Statistician* **20**, 20–23.

Simberloff, D. (1969). Experimental zoogeography of islands: A model for insular colonization. *Ecology* **50**, 296–314.

Simberloff, D. (1970). Taxonomic diversity of island biotas. *Evolution* **24**, 23–47.

Simberloff, D.S. (1974). Equilibrium theory of island biogeography and ecology. *Ann. Rev. Ecol. Syst.* **5**, 161–182.

Simberloff, D. (1976). Species turnover and equilibrium island biogeography. *Science, N.Y.* **194**, 572–578.

Simberloff, D.S. (1978a). Use of rarefaction and related methods in ecology. *In* "Biological Data in Water Pollution Assessment: Quantitative and Statistical Analyses". (K.L. Dickson, J. Cairns, Jr., and R.J. Livingston, eds), pp. 150–165. ASTM, Philadelphia.

Simberloff, D.S. (1978b). Colonisation of islands by insects: immigration, extinction, and diversity. *In* "Diversity of Insect Faunas" (L.A. Mound and N. Waloff, eds), pp. 139–153. Blackwell, London.

Simberloff, D. (1978c). Using island biogeographic distributions to determine if colonization is stochastic. *Am. Nat.* **112**, 713–726.

Simberloff, D. (1980). Dynamic equilibrium island biogeography: The second stage. *Proc. Congr. Intern. Ornith. (Berlin)* **17**, 1289–1295.

Simberloff, D. (1981). Community effects of introduced species. *In* "Biotic Crises in Ecological and Evolutionary Time" (M.H. Nitecki, ed.), pp. 53–81. Academic Press, New York.

Simberloff, D. (1982). Island biogeographic theory and the design of wildlife refuges. *Ékologiya*, **4**, 3–13.

Simberloff, D. (1983). Competition among Galapagos and Tres Marias birds: A reconsideration. *In* "Ecological Communities: Conceptual Issues and the Evidence". In press.

Simberloff, D.S. and Boecklen, W. (1981). Santa Rosalia reconsidered: Size ratios and competition. *Evolution* **35**, 1206–1228.

Simberloff, D., and Connor, E.F. (1979). Q-mode and R-mode analyses of biogeographic distributions: Null hypotheses based on random colonization. *In* "Contemporary Quantitative Ecology and Related Econometrics" (G.P. Patil and M.L. Rosenzweig, eds), pp. 123–138. International Co-operative Publishing House, Fairland Md.

Simberloff, D., and Connor, E.F. (1981). Missing species combinations. *Am. Nat.* **118**, 215–239.

Simberloff, D.S., and Wilson, E.O. (1969). Experimental zoogeography of islands: The colonization of empty islands. *Ecology* **50**, 278–296.

Simberloff, D.S., and Wilson, E.O. (1970). Experimental zoogeography of islands: A two-year record of colonization. *Ecology* **51**, 934–937.

Slud, P. (1976). Geographic and climatic relationships of avifaunas with special reference to the comparative distribution in the Neotropics. *Smithson. Contrib. Zool.* **212**, 1–149.

Smith, F.E. (1975). Ecosystems and evolution. *Bull. Ecol. Soc. Am.* **56**, 2–6.

Smith, W., Kravitz, D. and Grassle, J.F. (1979) Confidence intervals for similarity measures using the two-sample jackknife. *In* "Multivariate Methods in Ecological Work" (L. Orloci, R. Rao, and W.M. Stiteler, eds), pp. 253–262. International Co-operative Publishing House, Fairland, Md.

Strong, D.R., Jr., and Simberloff, D. (1981). Straining at gnats and swallowing ratios: Character displacement. *Evolution* **35**, 810–812.

Strong, D.R. Jr., Szyska, L.A. and Simberloff, D.S. (1979). Tests of community-wide character displacement against null hypotheses. *Evolution* **33**, 897–913.

Swed, F.S. and Eisenhart, C. (1943). Tables for testing randomness of grouping in a sequence of alternatives. *Ann. Math. Stat.* **14**, 66–87.

Tepedino, V.J. and Stanton, N.L. (1976). Cushion plants as islands. *Oecologia* **25**, 243–256.

Terborgh, J. (1973). Chance, habitat and dispersal in the distribution of birds in the West Indies. *Evolution* **27**, 338–349.

Terborgh, J. and Faaborg, F. (1973). Turnover and ecological release in the avifauna of Mona Island, Puerto Rico. *Auk* **90**, 759–779.

Vuilleumier, F. (1970). Insular biogeography in continental regions. I. The northern Andes of South America. *Am. Nat.* **104**, 373–388.

Vuilleumier, F. (1973). Insular biogeography in continental regions. II. Cave faunas from Tessin, southern Switzerland. *Syst. Zool.* **22**, 64–76.

Whittam, T.S., and Siegel-Causey, D. (1981). Species interactions and community structure in Alaskan seabird colonies. *Ecology* **62**, 1515–1524.

Wilcox, B.A. (1978). Supersaturated island faunas: A species-age relationship for lizards on post-Pleistocene land-bridge islands. *Science, N.Y.* **199**, 996–998.

4 | Some Principles of Physical Palaeogeography

H.G. OWEN

Department of Palaeontology, British Museum (Natural History), London SW7 5BD, UK

Abstract: Our knowledge of the distribution of seas and land areas at given times in the past is restricted by the lack of adequate syntheses of geological data on a global scale. Certain assumptions are made concerning the evidence for the presence of land and sea-marginal areas in the geological past which may not be correct. It is assumed by some that regions of oceanic crust indicate seaways providing migration routes for marine faunas and barriers to the movement of land animals. Today, areas of oceanic crust, such as Afar, Iceland and the Hawaiian chain are land areas. In making reconstructions of past land-sea distributions, we are faced also with two conflicting hypotheses. On the one hand with the orthodox view of an Earth which has not changed its dimensions during Phanerozoic time in which the ocean-floor spreading data will not coincide with the various reconstructions of the last 200 Ma. On the other, of an expanding Earth in which the various reconstructions accord fully with the ocean-floor spreading patterns during the last 200 Ma. Few major studies have been made to determine the full extent of epicontinental seas at given times in the past. Indeed, most palaeogeographic maps show only the distribution of known marine sediments at the time under consideration. In certain cases, the absence of marine sediments of a given age in a sequence interrupted by a major unconformity, does not indicate the retreat of the sea and the emergence of a land area, but on the contrary, strong current-scour activity within a seaway. This point is illustrated by a brief discussion of the English Channel. If there is difficulty in the determination of the extent of land and sea in the past, the problems of recognizing the presence of upland regions, let alone their altitude and configuration, are prodigious. Nevertheless, a detailed synthesis of

Systematics Association Special Volume No. 23, "Evolution, Time and Space: The Emergence of the Biosphere", edited by R.W. Sims, J.H. Price and P.E.S. Whalley, 1983, pp. 85–114. Academic Press, London and New York.

sediment distribution, fossil content, tectonic activity and oceanic crustal growth can provide a broad picture of the palaeogeography at a given period. A multidiscipline approach is essential in the study of the distribution of faunal provinces and the development of new faunas through time. These pictures, however, have to be viewed with caution.

INTRODUCTION

It is impossible in a relatively short paper to cover all the methods of palaeogeographic analysis now in use and to assess their credibility. The intention of this paper is to review briefly, with examples, those aspects of the methods of determining terrestrial and marine faunal migration routes, and the recognition of faunal provinces, used widely by palaeobiologists at this time. It sounds a note of caution, however, in the manner in which the field evidence is interpreted according to long-held geological principles. During the last 25 years the development of direct sampling and instrument measurement techniques in the seas and oceans have produced results which make necessary reappraisals of certain of our basic assumptions in physical palaeogeographic studies.

Neumayr (1885), not quite a century ago, produced the first palaeogeographic world map using Mercator's projection. This was of the Jurassic and showed the continents in the positions that they are in today, linked by land bridges across the Atlantic and Indian Oceans. This reflected the general view of the day that the Earth's continents and oceanic basins had been formed and fixed in position since the beginning of geological time (e.g. Dana, 1863: 732). Barely twenty years ago, the present paper would have been unacceptable to the scientific establishment of the day because it accepts without question the occurrence of continental "drift", a concept which took 50 years to gain credence. Even now, known inconsistencies between current geological and geophysical hypotheses and the field data will eventually require major revision of thought when they are more widely known. Of particular concern to the palaeobiogeographer is the question of global expansion. At present, many workers would not accept this hypothesis despite the fact that the ocean-floor spreading data support the concept and do not coincide with the geometric requirements of a constant dimensions Earth.

The uniformitarian approach that the present is the key to the past is wrong in major areas of geological thinking, although sedimentary processes have appeared to change but little. The continents today are mostly land and the ocean basins are sea and it is only in the Permo-Triassic that a comparable situation was last seen. The Mesozoic and early Cenozoic witnessed substantial areas of epicontinental seas, the Cretaceous-Tertiary boundary marking the start of a sharp decline in their extent and in their inferred depth. Eustatic changes in sea level due to periods of glaciation and climatic amelioration are documented for the last 2 Ma. The criteria for recognizing such phenomena and the accompanying marine transgressions and regressions in the geological past due to some other geodynamic process, are more equivocal.

I do not intend in this paper to undermine the foundations of palaeogeographic concepts. Rather, to offer alternative interpretations of the field evidence which, if anything, will tend to remove difficult inconsistencies in past faunal migration patterns both marine and terrestrial.

THE GLOBAL VIEW

If 200 million years ago, a time-lapse camera had been placed on the Moon set to take groups of photographs of the Earth at intervals of a million years, the resulting film would show that the Earth's crust is highly mobile. Two hundred million years ago, all the separate continents seen today were assembled together as one supercontinent—Wegener's Pangaea. Our time-lapse film would show Pangaea splitting apart with the development of the Atlantic, Indian Ocean and Arctic Ocean basins. Some workers have come to the conclusion that the film would show that the Earth had expanded in size during this period.

In the continental regions, we would see the development of mountain ranges such as the Alps, the Himalaya and the Verkhoyansk ranges of Asia, in response to the collision and differential motion of one continental fragment against another. We would see also the growth of marginal orogens in the Pacific, such as the Western cordilleran fold belt of North America and the Andes of South America with its continuation into West Antarctica. These marginal fold belts formed in response to the westward displacement of these continents in response to the growing Atlantic and the accompanying overriding of the crust of the eastern Pacific, itself growing by ocean-floor spreading.

The distribution and extent of the relatively shallow seas over the continental crustal regions also showed marked changes during the last

200 Ma. Although these epicontinental seas were very extensive in the past, they have shrunk in area during the Cenozoic to the few seas, such as the North Sea, the Barents Sea, Hudson's Bay and the Arafura Sea in addition to the continental shelves seen today. From the Oligocene, at least in the southern hemisphere in Antarctica, and rather later in the maritime Arctic, we would witness the development of polar ice-caps heralding the various phases of Pleistocene glaciation.

Very few workers would disagree, fundamentally, with this basic picture although the idea of an expanding Earth is an anathema to some. The growth of oceanic crust and, therefore, the break-up and dispersal of the fragments of Pangaea is very well documented over much of the Earth during the last 200 Ma. But, what happened in the Palaeozoic; the preceeding 400 Ma for which no oceanic crust is now preserved to provide a geometric control on reconstructions of continental displacement? Whether we like it or not, the ocean-floor spreading patterns generated during the Mesozoic and Cenozoic do not coincide with those reconstructions which assume an Earth of constant, modern dimensions during this period (e.g. Smith and Briden, 1977; Firstbrook *et al.*, 1980). Those maps which assume an Earth of constant dimensions and in which the ocean-floor spreading evidence appears to fit (e.g. Norton and Sclater, 1979), are the products of computer programmes which do not compensate, or only partially compensate, for the movement of geographic areas relative to the projection pole (Owen, 1983). These maps do not possess cartographic integrity. The ocean-floor spreading patterns and continental geological fit data do support, however, an Earth with a diameter 80% of modern mean value at 180–200 Ma expanding to its present size (Owen, 1976b, 1981).

Reconstructions of Palaeozoic continental displacement which assume an Earth of constant modern dimensions, such as those of Smith *et al.* (1973), Smith and Briden (1977), Ziegler *et al.* (1977) and Scotese *et al.* (1979), have to be treated with caution. Quite apart from the pertinent question of global expansion, the absence of an ocean-floor spreading control and the paucity of palaeomagnetic vector data, render these reconstructions untestable at present. They are carefully considered ideas for which a method of test has still to be found. The few reconstructions published so far of Palaeozoic continental displacement which assume an expanding Earth (e.g. Hilgenberg, 1933, 1966; Creer, 1965; Kremp, 1979) cannot be correct. It is not possible to reduce the diameter of the Earth below 80% of its current mean value without introducing major intra-continental displacements which disrupt the entity of the continents seen today. It is significant that these intra-continental displacements occur along the various Palaeozoic orogenic belts worldwide in the correct chronological sequence, if the diameter of the Earth is increased progressively from 55% of modern value in the Proterozoic to the size and configuration seen in the early Mesozoic (e.g. Fig. 7). But, crustal development in itself does not give a complete picture of the distribution of land and sea during Phanerozoic time.

The broad picture I painted in the first few paragraphs of this section, although correct fundamentally, is already becoming blurred and incomplete in the quest for a meaningful geography of a given time within the last 200 Ma. Further back in time, it is barely possible to see the outlines of the picture. Where were the land areas and where were the seas? More than this, where were the uplands situated and the low-lying marshes, the shallow seas in which life was abundant and the deep unlit oceanic depths? Much information can be obtained from sediment sequences and upon this evidence is based the paleogeographic reconstructions that are published. But what of the oceanic areas which have been deep-drilled at widely scattered sites? We still know very little of the oceanographic conditions in the Mesozoic and Cenozoic over some 70% of the Earth's modern surface. In regions of continental crust, it seems that most palaeogeographic reconstructions are maps of the distribution of marine sediments that have survived erosion and not necessarily maps of the total sea area of the time. Sea-marginal sediments are known in the geological column, but their chance preservation is the exception rather than the rule. By comparison with sedimentation in modern seas in which, of course, we have the benefit of the presence of the sea itself, it is apparent that sediments often considered to be typical of marginal environments may have accumulated well out from the shore in substantial seaways subject to water current activity.

More detailed syntheses of the geological and palaeontological evidence are now being incorporated into regional palaeographic studies. In some of these, the results of sedimentation studies in modern seas are being taken into consideration. But, this work needs to be greatly extended if a meaningful study of marine and terrestrial faunal distributions in the past is to be made world-wide. Throughout much of the geological column, essentially in continental crustal regions, we are able to determine at present only whether the sediments were laid down under marine, brackish or freshwater conditions, and their immediate environment of deposition. More rarely are we able to find sediments that were laid down under terrestrial conditions. We are able to determine marine current directions and sometimes their velocity, but not the depth of the seaway concerned. The distribution of faunas can be important in the determination of sea or land connections and, conversely, the presence of barriers when the sedimentary evidence is equivocal.

IMPLICATIONS OF AN EXPANDING EARTH VS. A CONSTANT DIMENSIONS EARTH FOR PALAEOGEOGRAPHIC RECONSTRUCTIONS

Those geologists who have made any attempt at resolving the inconsistencies of the field data with the so-called "plate-tectonic theory" and its accompanying concept of a constant dimensions Earth, have concluded that the Earth was smaller in the past (e.g. Carey, 1976). Within the group of workers who have advocated expansion, there exist two schools. Those who advocate that all oceanic crustal growth has occurred since the early Jurassic and discount the process of subduction of oceanic crust—the so-called "fast expanders"—are not supported by the ocean-floor spreading data. The other school, to which the writer belongs, advocates a much slower Earth expansion rate strictly in accordance with the ocean-floor spreading and subduction patterns. In the case of the ocean-floor spreading record, the data cannot be made to coincide with the various reconstructions made on an Earth of constant modern dimensions. Spherical triangular voids, or "gores", have to be constructed which widen progressively away from the area of best fit made, usually, at the centre of the reconstruction. In passive-margined oceans, in which a full history of continental break-up and subsequent ocean-floor spreading is preserved, these gores would have to represent continental crust now piled up into a marginal orogen which in fact does not exist, or oceanic crust which has disappeared somehow, because it is not present in the spreading record today. An example of this is the Tethyan Ocean at the time of Pangaea (but not the Tethys Sea) shown in Fig. 4. The geological history of the Himalaya (e.g. Gansser, 1974; Stöklin, 1980; Brunnschweiler, 1974) and even the distribution of dicynodont reptiles (e.g. Cox, 1973) in the Permo-Triassic precludes the wide gap which has to be constructed between eastern Gondwanaland and Asia if the reconstruction of Pangaea is made on an Earth of modern

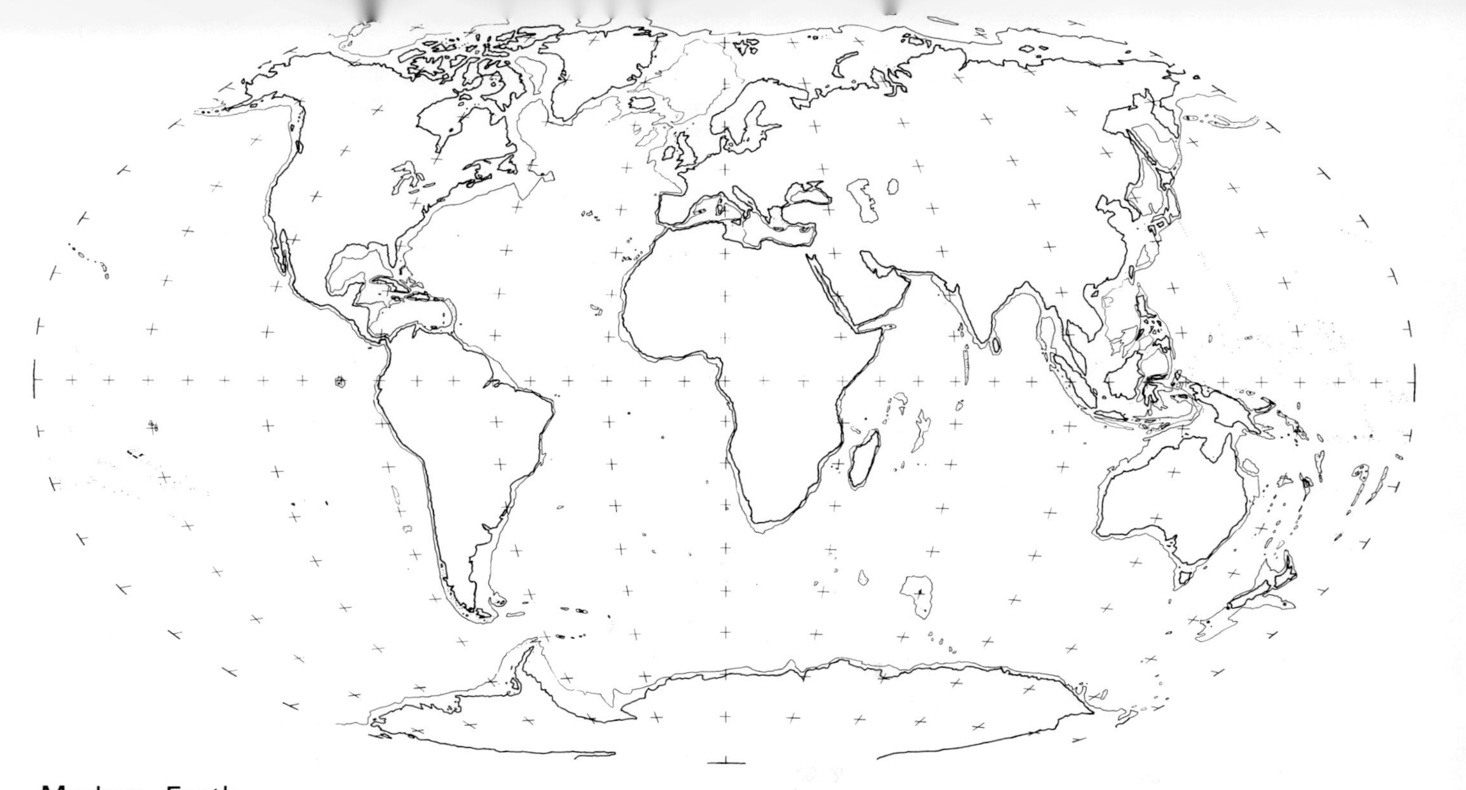

Modern Earth

Fig. 1. Modern Earth outline map showing land margins and the 1000 m isobath. Winkel "Tripel" projection with centre meridian 10°E. longitude. Note that in the series of map Figs 1–4, the areas of anomalous crust in these reconstructions which assume an Earth of constant modern dimensions, decrease to nil at the present day. This is consistent with an expanding Earth.

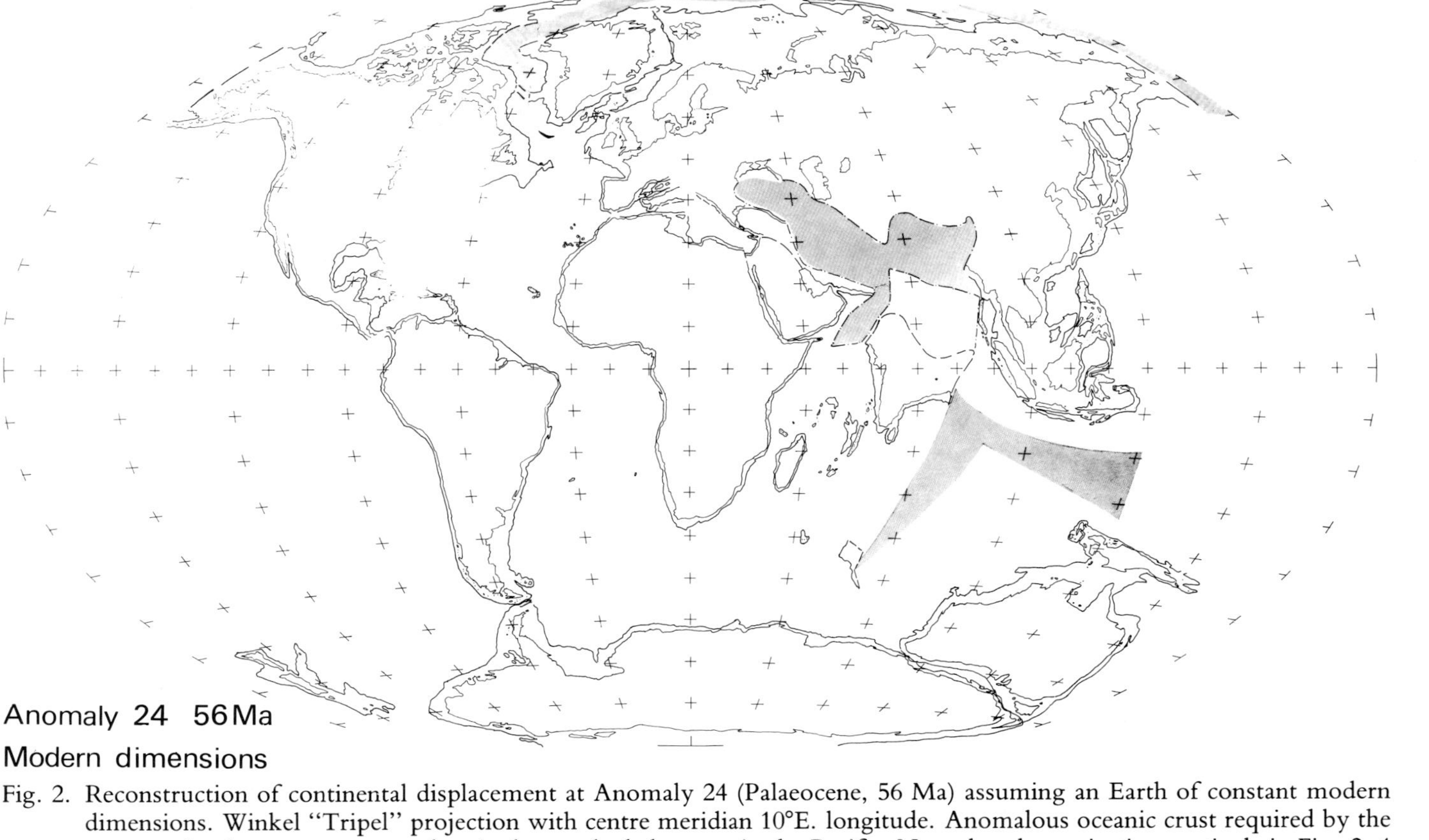

Fig. 2. Reconstruction of continental displacement at Anomaly 24 (Palaeocene, 56 Ma) assuming an Earth of constant modern dimensions. Winkel "Tripel" projection with centre meridian 10°E. longitude. Anomalous oceanic crust required by the construction but not present today, is shown shaded except in the Pacific. Note that the projection graticule in Figs 2–4 coincides with the modern graticule in Fig. 1. The position of continents is the nearest approximation that can be made in modern dimensions Earth reconstructions to the ocean-floor spreading data reflecting the *relative* motions of one continent

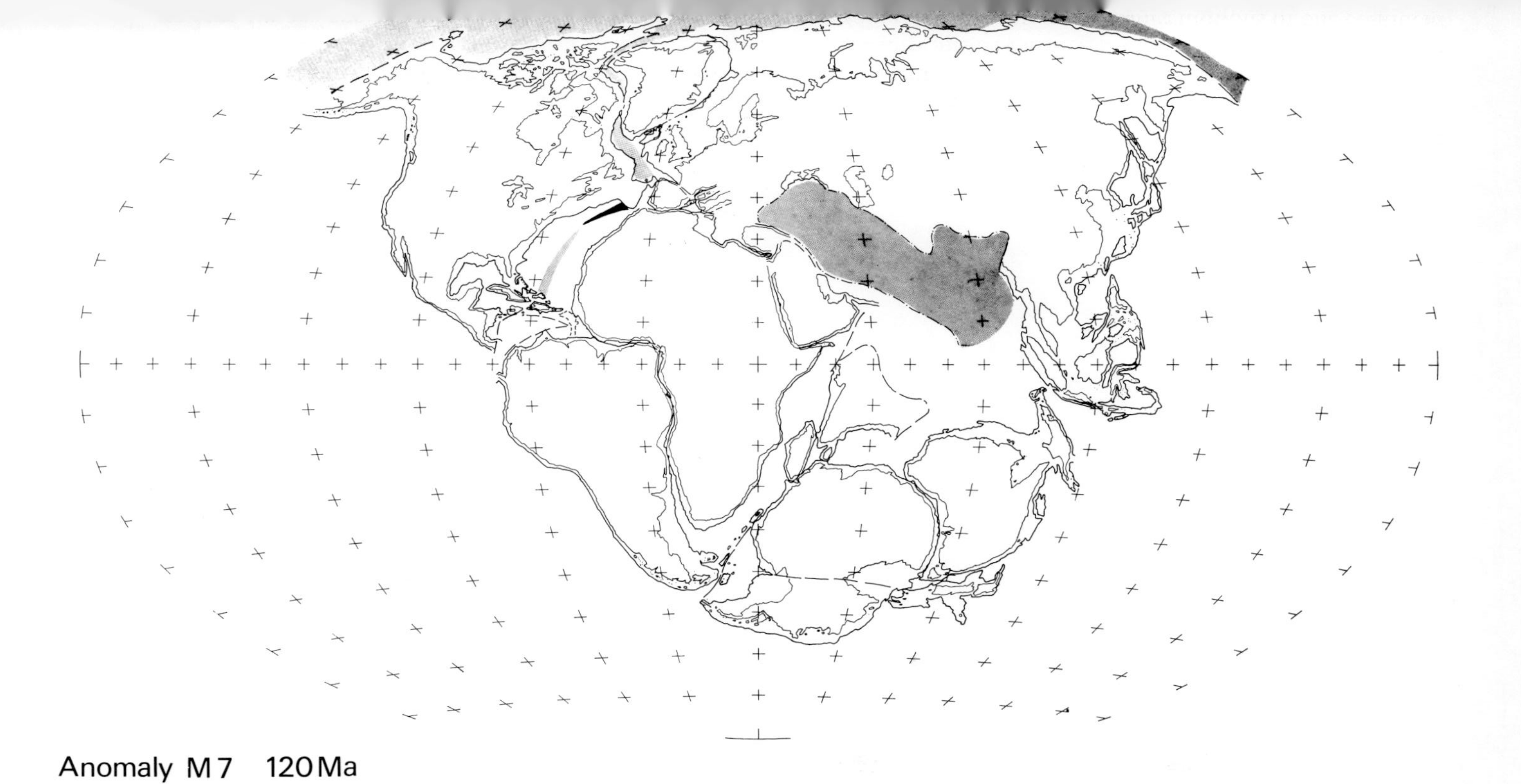

Anomaly M7 120 Ma
Modern dimensions

Fig. 3. Reconstruction of continental displacement at Anomaly M7 (Hauterivian, 120 Ma) assuming an Earth of constant modern dimensions. Winkel "Tripel" projection with centre meridian 10°E. longitude. Anomalous oceanic crust required by the reconstruction, but not present today, is shown shaded except in the Pacific. Area of overlap in the North Atlantic required to provide the nearest fit in the Boreal region is shown black.

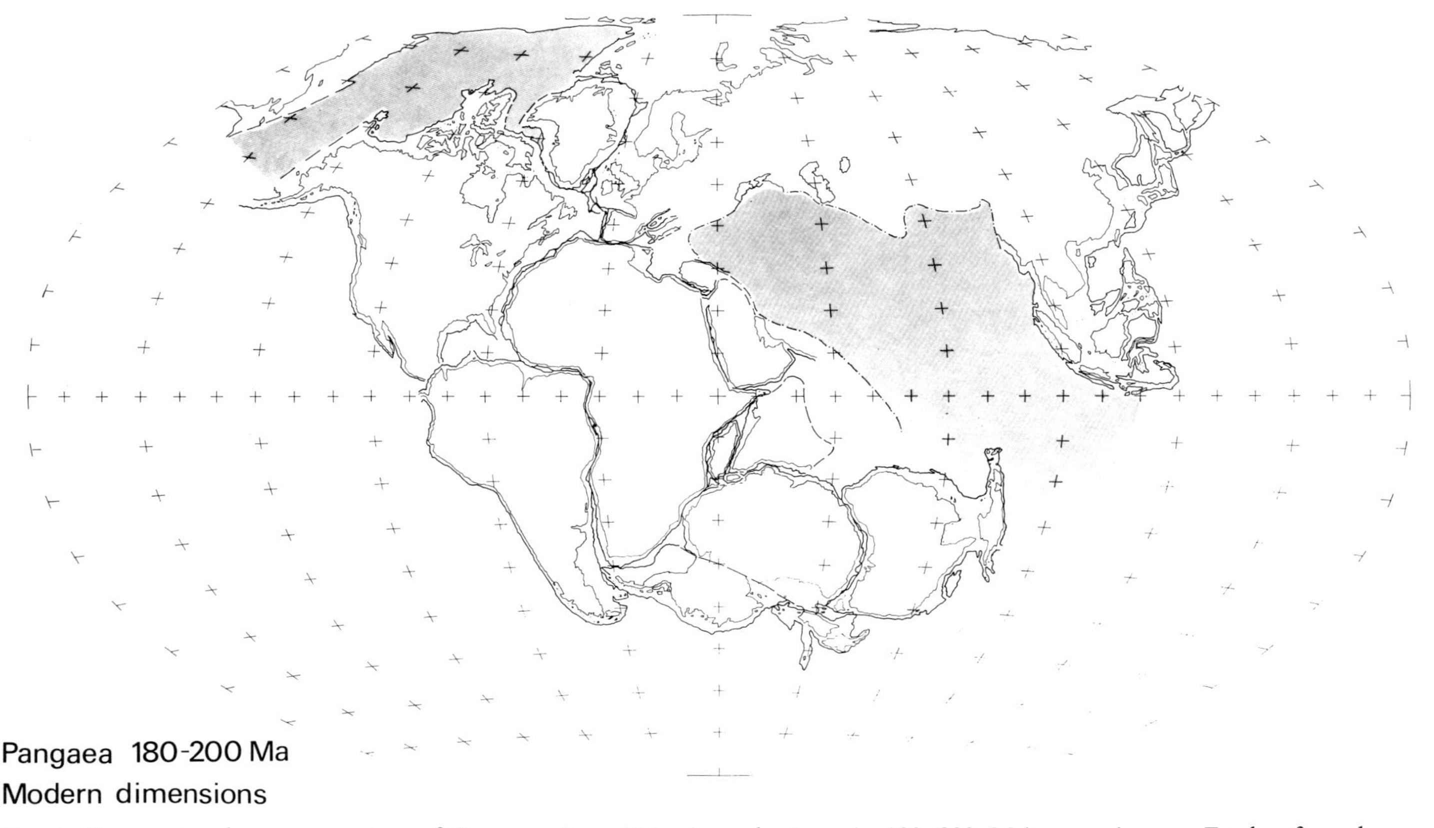

Fig. 4. Conventional reconstruction of Pangaea (Late Triassic–early Jurassic 180–200 Ma) assuming an Earth of modern dimensions. Winkel "Tripel" projection with centre meridian 10°E. longitude. Oceanic crust required to be present by the reconstruction, but of which no evidence exists, is shown shaded except in the Pacific.

dimensions. This problem does not arise in the expanding Earth reconstruction shown in Fig. 7, in which the diameter of the Earth at the time of Pangaea (180–200 Ma) is 80% of modern mean value. The subsequent spreading history agrees with an expanding Earth (Owen, 1976b, 1981, 1983) and coincides with the geological deformation of southern Asia. This rate of expansion, although considerable in terms of the growth of surface area, is much less than is required by those who consider that all expansion and growth of ocean basins has occurred since the early Jurassic (e.g. Shields, 1979). Subduction of oceanic crust by the overriding of ocean-floor by an advancing continental margin (e.g. western North and South America, or by the downthrusting of an older, colder, plate margin within an oceanic crustal region (e.g. at the Marianas–Japan trench system) is required in the expanding model determined by the writer.

Two sets of outline maps, spaced at 60 intervals, are given here for the Mesozoic and Cenozoic. The first series (Figs 1–4) assumes an Earth of constant modern dimensions and the anomalous oceanic crust, anomalous in the sense that it is required by the reconstructions but is absent today, is shown as shaded areas. The second series (Figs 5–7) assumes an expanding Earth and the individual maps coincide with the geological fit and subsequent ocean-floor spreading data.

These maps depict merely, the distribution of continental and oceanic crust at the given times in the past. It is probable that the greater part of the oceanic crustal regions were covered by seas. However, regions such as Iceland and Afar are land areas today and there is evidence from Deep Seas Drilling Project cores of other areas of oceanic crust which were subject to subaerial weathering in the past. It is likely, for example, that the Galapagos Islands were once linked, albeit ephemerally, with the Ecuador coast by means of the now deep-foundered Carnegie Ridge System. Northern North America, Greenland and Europe were linked well into the Cenozoic at the north-west corner of the Barents Shelf and possibly by way of the Greenland–Faroes ridge.

THE WATER PROBLEM

It is the orthodox view among geologists that the volume of water, either in the crustal rocks as ground water, or the seas and oceans and in the atmosphere, is approximately constant and forms a closed system throughout the bulk of geological time (e.g. Donovan and Jones, 1979).

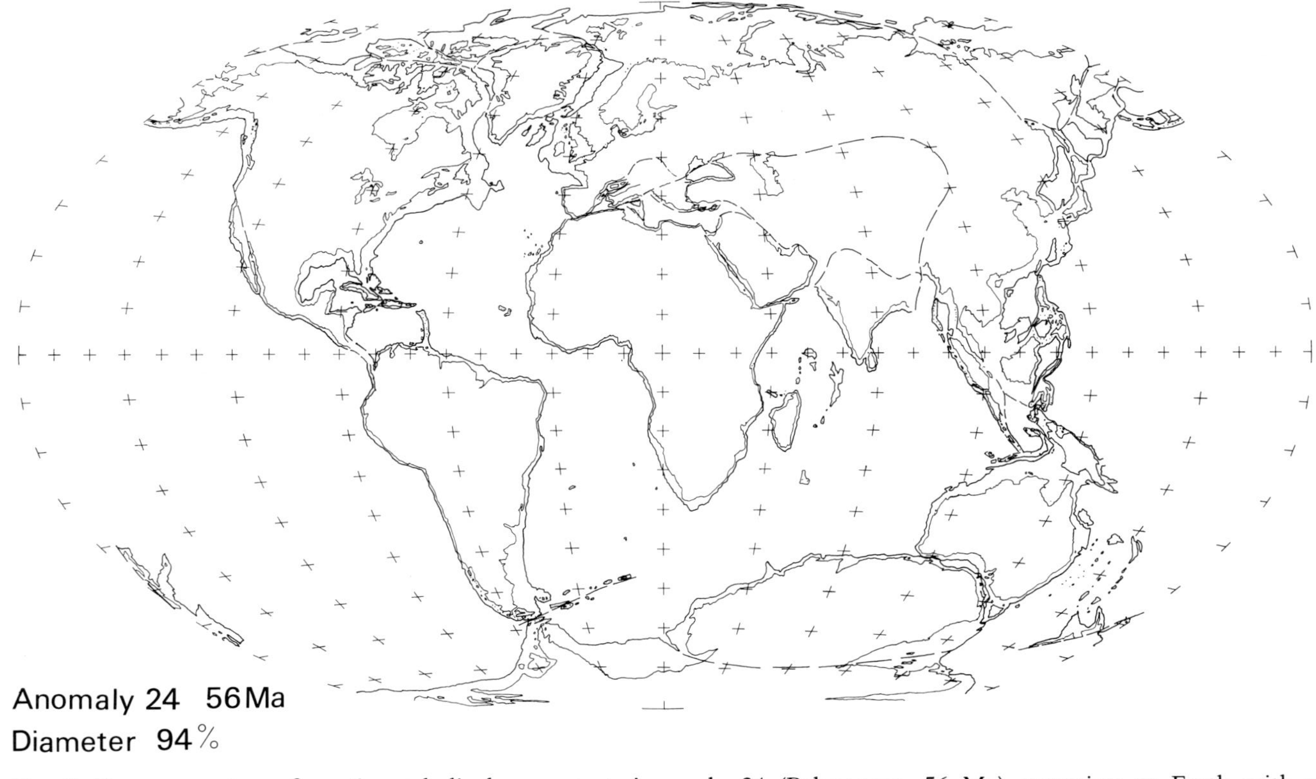

Fig. 5. Reconstruction of continental displacement at Anomaly 24 (Palaeocene, 56 Ma) assuming an Earth with a diameter approximately 94% of modern value. Winkel "Tripel" projection with centre meridian 10°E. longitude. No anomalous areas of oceanic crust are present. Note that the projection graticule in Figs 5–7 has the selected prime meridian (10°E. longitude) and the North geographic pole positions coincident with the Earth's modern graticule in Fig. 1. The length of the meridia increases forward in time to its modern value. The position of each continent accords strictly with the ocean-floor spreading data and, as in Figs 2–4, the motions of the continents

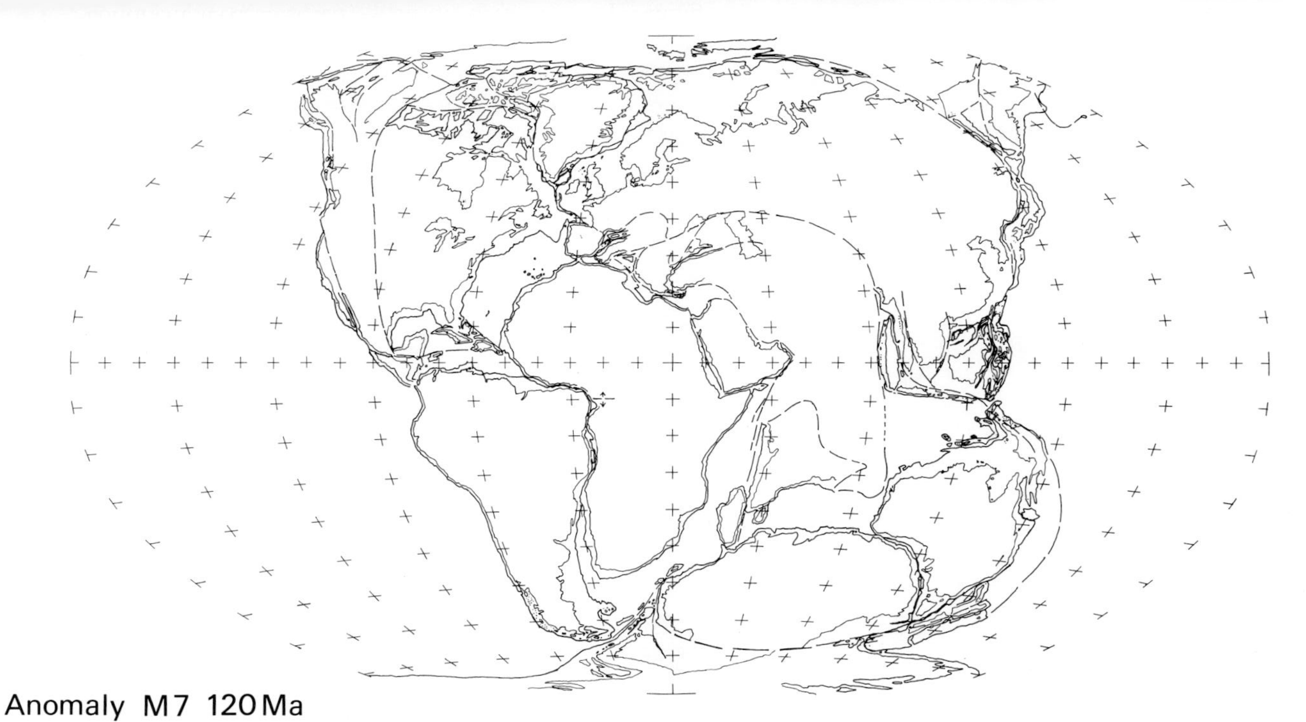

Fig. 6. Reconstruction of continental displacement at Anomaly M7 (Hauterivian, 120 Ma) assuming an Earth with a diameter approximately 87% of modern value. Winkel "Tripel" projection with centre meridian 10°E. longitude. No anomalous areas of oceanic crust are present.

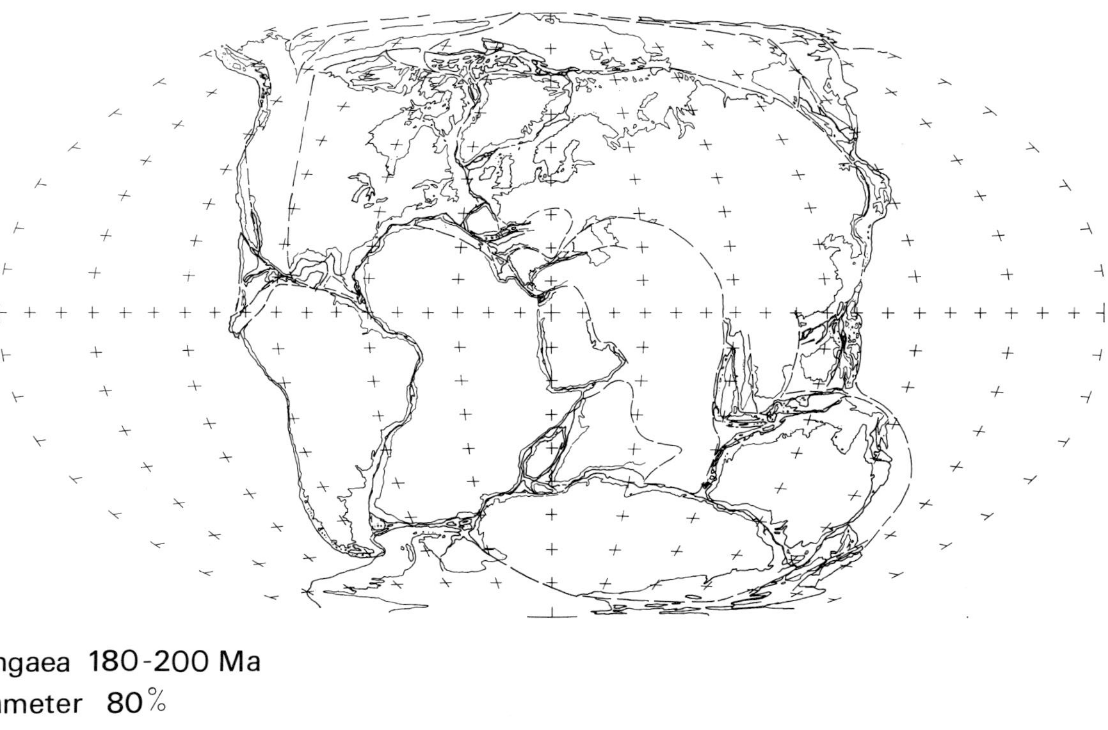

Fig. 7. Reconstruction of Pangaea (Late Triassic–early Jurassic, 180–200 Ma) assuming an Earth with a diameter 80% of modern mean value. Winkel "Tripel" projection with centre meridian 10°E. longitude. The fit of the continents together conforms with the geological data from the continental margins and the subsequent ocean-floor spreading.

This is often used as an argument against global expansion, it being pointed out correctly, that on a smaller diameter Earth such as is shown in Fig. 7, all but the highest mountain peaks would be submerged. Fig. 7 applies as much to the Permo-Triassic in which very extensive areas of continental crust were land, as well as to the lower Jurassic in which extensive epicontinental seas were present. To those who believe in outgassing, the volume of water seen today represents no problem so far as global expansion is concerned.

Is the Earth's water system closed and, as a whole, of great geological antiquity? There is no doubt that water-laid sediments are well represented in the early Archaean. Variations in sea level attributed to eustasy have been advocated throughout the Phanerozoic. Some of these episodes in the Palaeozoic and in the last 2 Ma have been attributed correctly to ice-cap formation and decay (e.g. Fairbridge, 1961). Sea-level variations in periods of climatic optima such as the Mesozoic have been attributed to ocean-basin formation (e.g. Hallam, 1963; Cooper, 1977; Reyment and Mörner, 1977), but Donovan and Jones (1979) have suggested that these are too slow to account for the short duration episodes of regression and transgression, although they might affect long-term cycles. Unfortunately, the spread of DSDP borings is too thin at present to provide the detailed information necessary to test these ideas in respect of the Mesozoic and much of the Cenozoic.

One thing is clear, however, as Egyed (1956) pointed out, epicontinental seas have shrunk in area and decreased in depth, particularly since the Mesozoic despite the assertions of Cogley (1981) to the contrary. If both Modern polar ice caps were melted, the increase in the extent of epicontinental seas would barely reach that of the Miocene. Today, epicontinental seas are confined to the continental shelves and to regions such as Hudson's Bay, the Barents and North Seas, parts of the South China Sea and the Arafura Sea. Albeit that they are thinly spread through the World's oceans, the DSDP/IPOD borehole sequences do not indicate a progressive deepening of the ocean-basins from the Jurassic onward, or a rapidly subsiding oceanic region since the Cretaceous. If such subsidencies had occurred, this would explain the draining of the epicontinental seas. The assumption that surface, ground and atmospheric water has been more or less constant during Phanerozoic time (reiterated in Donovan and Jones, 1979) is not proven. It is apparent that significant quantities of water vapour are dissociated in the upper atmosphere and the Hydrogen ions so released are lost to space (e.g. Walker, 1977). If this occurred in the geological past, the Earth's water system cannot be closed and would require replenishment by outgassing, since free water first appeared on the Earth's surface shortly after the crust was formed.

There is still a tendency to equate continental crustal regions with land areas when faunal distributions are discussed. The present is not the key to the past and the extensive continental land areas of today were last seen as long ago as the Permian. There are problems in the intervening period in determining the true extent of epicontinental seas and in the interpretation of the evidence of regression and transgression. These have now to be considered.

EPICONTINENTAL SEAS: RECOGNITION OF MARINE TRANSGRESSIONS AND REGRESSIONS, FORMER LAND AREAS AND MARINE CONNECTIONS

It is a relatively simple matter to define what is meant by "marine transgression" and "marine regression". Expressed in the simplest terms, if the tide comes in on the flood and does not go out again and this continues as a progressive inundation of a land area, we are witnessing a marine transgression. If the limits of high tide recede progressively off the land area, we recognize regression. The above is obvious, but the processes which cause these phenomena and their geological expression may prove difficult to interpret. For example, are we observing the effect of a global change in sea level due to the advance or contraction of polar ice which we can determine, or in response to the deepening of the ocean basins by some geodynamic process which at present we are unable to determine? On the other hand, are we observing the effect of relatively superficial epeirogenic crustal warping, or folding on the distribution of epicontinental seas over a limited or extensive region of the Earth's crust at any one time? The process of initial rift-faulting and doming along the line of future spreading axes within the supercontinent Pangaea, can be seen in the stratigraphic records of the surrounding regions. The subsequent subsidence of these regions as the source of heat moves away during the process of ocean-floor spreading, can also be seen in the stratigraphic record. The effect of these epeirogenic movements on the regression and transgression of epicontinental seas is seldom considered by those who advocate eustasy to be the essential cause of this phenomenon.

There are problems, however, in the recognition of transgressions and regressions in the geological record (Fig. 8). In Fig. 8A the conventional portrait of a transgressive sequence is shown. Downward movement of the crust relative to sea level produces a progressive inundation of the land area and the migration of the various marine facies landward through time. In Fig. 8B the same crustal movements occur, but the downward movement of the sea-bed relative to sea level normally associated with transgression, occurs at a rate which is slow enough, or sufficiently intermittent, to permit the deposition of terrigenous sediments in fringes extending out from the former coastline. The sedimentary profile in this case can give the appearance of a marine regression. An example of this is provided by the Early Cretaceous sequence in southern England. At the end of the Jurassic, major block and rift-faulting produced a marked regression of the former extensive upper Jurassic seas over Europe and elsewhere. By the early Cretaceous, freshwater deltaic and lacustrine sediments were being built out into a slowly subsiding Wealden trough south of the upfaulted former Jurassic sea-bed area in the region of the London Platform, which was undergoing extensive subaerial denudation (e.g. Allen, 1976; Owen, 1971). At short-lived intervals, the Wealden trough subsided more rapidly allowing marine inundations which are detectable only by micro-faunal and micro-floral incursions. In northern Germany, the Wealden facies of the subsiding basins was short-lived and was followed by a permanent marine invasion.

A third possibility is of considerable importance in the study of faunal migration routes and dispersal (Fig. 8C). In this case, slightly faster deepening of the sea associated with strong current scour and, possibly, a paucity of terrigenous sediment, can produce the classic appearance of emergence above wave-level in the geological record, especially if this is associated with folding. The sedimentary sequence shows an hiatus which may be accentuated into a major non-sequence by folding and strong current and cobble scour. The marine transgression in reality would be interpreted in this instance in the geological column as a major regressive and erosional phase. The marine transgression in orthodox theory would be interpreted as occurring at the base of an immediately overlying sedimentary sequence. In fact, the area was below surface wave-level at the time of erosion and the sedimentary sequence formed only when the velocity of the bottom currents slackened sufficiently to allow sedimentation to occur and to survive.

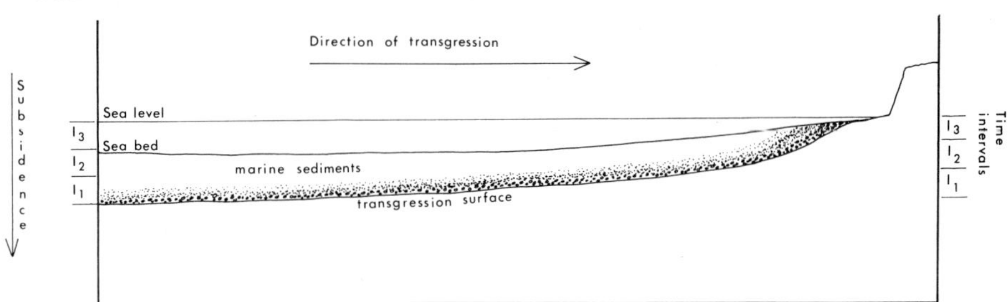

A. Conventional transgressive sequence with progressive onlap

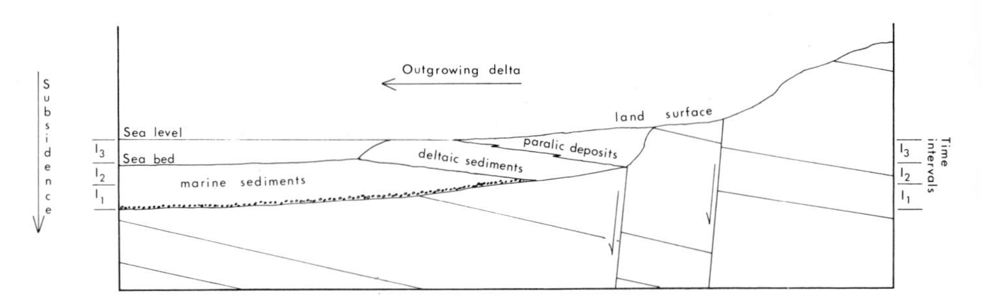

B. Regressive sequence into a subsiding basin

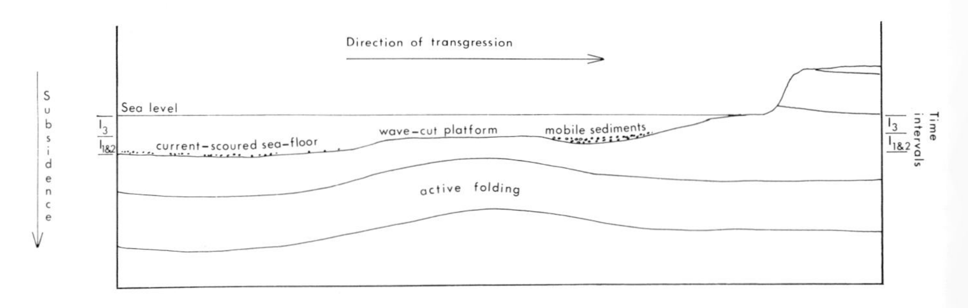

C. Transgression with strong current scour precluding sedimentation

Fig. 8. Diagram of three examples of relative deepening of the sea. A. is the conventional transgression model. B. shows the effect of fast terrigenous sedimentation in a relatively shallow sea. C. is a common example of the situation in many modern shelf-sea areas. I_{1-3} represent arbitrary time intervals of landward erosion; I_3 being the present.

This third possibility can be illustrated by the modern Channel, separating England and France, and the southern North Sea (Fig. 9). The geological history, sedimentation and oceanography of this region of epicontinental sea have been examined in some detail (e.g. Dunham, Smith *et al.*, 1975; Dunham, Gray *et al.*, 1972; CNRS, 1972, and the bibliographies of Smith *et al.*, 1972; Hamilton *et al.*, 1975). In summary, these seas of the last 8000–10 000 years are the product of the Flandrian (or Holocene) marine transgression. This present transgression, which is the latest in a series of essentially climatically-controlled eustatic transgressions and regressions which occurred in the Pleistocene, is due to the combined effect of the major post-Weichselian eustatic rise in sea level and the relatively minor, but geologically fast, crustal warping detected in this region. Synclinal folding, reflecting subsidence of fault-bounded blocks in the structural basement, occurs today in the region of the lower Thames and the Strait of Dover (e.g. Akeroyd, 1972; d'Olier, 1972; Shephard-Thorn *et al.*, 1972:109). Despite the eustatically controlled glacial regressions and interglacial transgressions within the Pleistocene, which are superimposed upon the pattern of regional crustal warping and isostatic readjustments, the sum effect during the Quaternary and Holocene is of a marine transgression, albeit episodic.

Throughout much of the western Channel and in the Strait of Dover, bottom current scour of the bed-rock occurs in response to tide and wave activity (Fig. 9). The comparatively slow bottom currents of between 2 and 4 knots are sufficient to prevent sedimentation except in the slack water of submarine valleys, such as the Hurd Deep in the central part of the western Channel (Hamilton and Smith, 1972), the Fosse Dangeard in the Strait of Dover (Destombes *et al.*, 1975), or where influenced by certain bottom features (e.g. Stride *et al.*, 1972; Lapierre, 1975). In shallow marginal areas undergoing tidal scour, transport of sediment derived from cliffs and rivers is obvious enough, but even at depths of 50–100 m and locally more, strong, storm-wave induced current activity, is sufficient to transport or rework the surface detritus. This surface re-working is just as apparent in areas of the central and eastern Channel in which thick Holocene sediments have accumulated (Stride *et al.*, 1972). On the other hand, in marginal areas of slack water that are now the Walland and Romney marshes, Wissant, the Goodwin Sands and Holland, deposition has built out sediments to form regions of new land (albeit aided by Man), or barely submerged banks, during the Holocene.

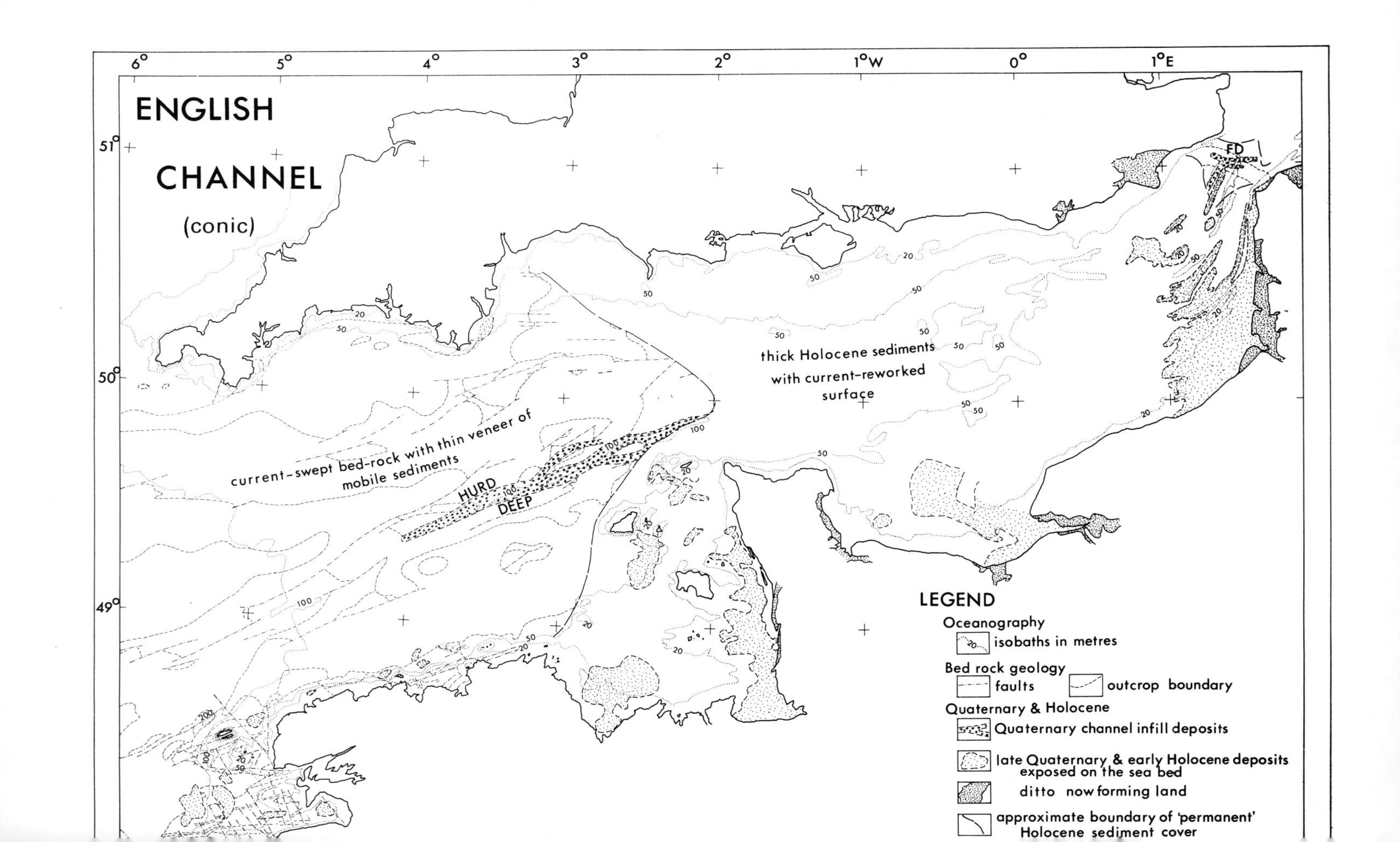
ENGLISH
CHANNEL
(conic)
6°
5°
4°
3°
2°
1°W
0°
1°E
51°
50°
49°
FD
thick Holocene sediments
with current-reworked
surface
current-swept bed-rock with thin veneer of
mobile sediments
HURD
DEEP
LEGEND
Oceanography
isobaths in metres
Bed rock geology
faults
outcrop boundary
Quaternary & Holocene
Quaternary channel infill deposits
late Quaternary & early Holocene deposits
exposed on the sea bed
ditto now forming land
approximate boundary of 'permanent'
Holocene sediment cover

Let us now imagine at some future date, the whole of this region becomes further submerged to a depth where current activity declines at the sea-bottom to permit the overall accumulation of a sedimentary sequence. Initially, there would occur a complex of patchy lenses of conglomerates and current-bedded sands, followed above by more regular bedded and progressively finer grained sediments. These facies would migrate outward from the present region of the central and eastern Channel as the area of the sea increased. Ultimately one would record a sequence similar to that in Fig. 10. The surface of the sediments would be subject to periods of current scour as activity increased for a variety of different reasons, such as, the filling of the basin with sediment up to a point where water body movement would again preclude deposition, or changes in the configuration of the sea occurred sufficient to alter the pattern of water circulation. Today, if one proceeds westward from the Channel into the Western Approaches, Quaternary and Holocene sediments again thicken and, as in the eastern Channel, rest with marked unconformity on an eroded surface of folded and faulted Neogene and older strata. Indeed, in the western Channel, Neogene sediments are more widespread than those of the Quaternary and Holocene (Fig. 9).

These actual erosion and sedimentary features and postulated future sedimentation, if seen in the sense of the geological record without the benefit of the presence of the sea, would be interpreted (Fig. 10) as a post-Neogene regression throughout the Quaternary and Holocene, followed by a post-Holocene marine transgression. Moreover, the outgrowing sedimentary sequences of Walland and Romney Marshes, the French and Belgian coast from Wissant eastward, together with Holland, if preserved, in conjunction with the restricted area and nature of Holocene marine sedimentation would confirm regression. In fact, we are looking at a transgressive situation. It is apparent from Quaternary and Holocene examples, that the early stages of renewed transgression are associated with increased water current activity, which can scour the surface of previously deposited soft sediment reworking

◀ Fig. 9. Sketch map of the English Channel showing areas of exposed bedrock, or bedrock covered by a thin veneer of mobile sediments, and the regions covered by thick later Holocene sediments. Sources: Carte géologique de la Manche (1974), Curry *et al.* (1972), Hails (1975), Lapierre (1975), Larsonneur (1972), Lefort (1975), Shephard-Thorn *et al.* (1972). FD = Fosse Dangeard.

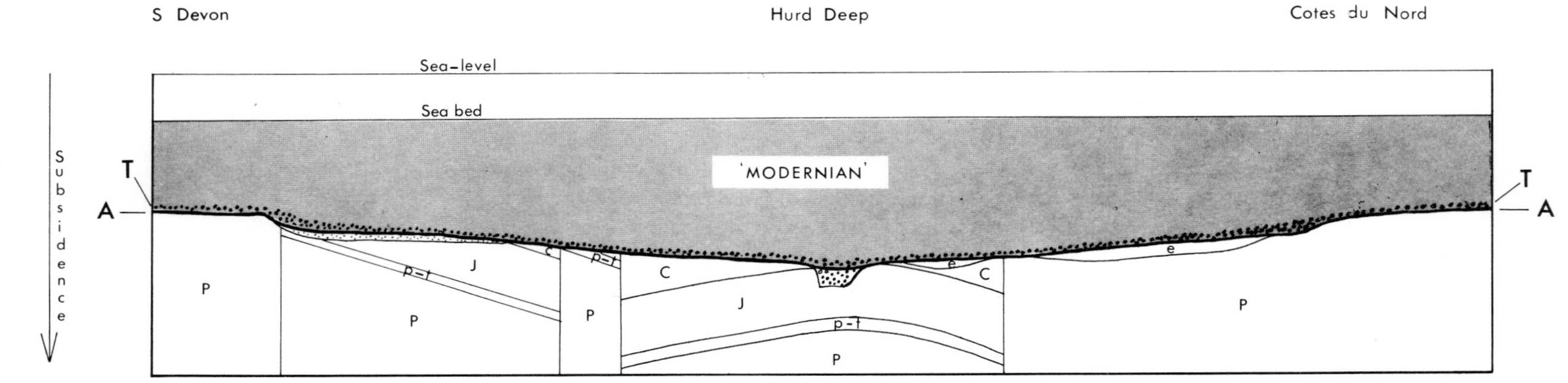

Fig. 10. Cross-section of the western English Channel interpreted as a sub-unconformity surface after subsidence and renewed "Modernian" sedimentation at some future date. A–A, sub-unconformity surface which orthodox interpretation would make a land surface. T–T, base of the "Modernian" sequence which would be interpreted as the initial marine transgression. P, Pre-Cambrian and Palaeozoic; p–t, Permo-Triassic; J, Jurassic; C, upper Cretaceous; e, Eocene. Hurd Deep infilling is of Quaternary age.

it, scour again large areas of bed-rock and prevent the deposition of fine-grained sediment.

The Channel is far from unique among modern epicontinental seas in showing these erosive and sedimentation features (e.g. Friedman and Sanders, 1978). In the sedimentary remains of the far more extensive epicontinental seas of the Mesozoic, for instance, examples comparable to modern shelf seas can be demonstrated, but their description is not possible within the confines of this brief review. In these examples, the postulated "Modernian" of Fig. 10 is represented by actual sediment, with depositional structures preserved, from which the history of formation and the marine environment can be determined by reference to data obtained from modern seas.

TECTONIC DISTURBANCES USUALLY INTERPRETED AS REGRESSION IN THE GEOLOGICAL RECORD

In the foregoing section of this paper, a brief examination has been made of the relatively passive type of transgression where the sea inundates a region in response to eustatic and, or, epeirogenic causes. If on the other hand, the marine transgression is associated with faulting, or if faulting affects a previously stable shelf sea area, marked disturbance of sediments by seismic shock waves in the underlying crust and erosion of them by the water body movements associated with the tsunami-wave occur (e.g. Lisitzin, 1974:198). This activity is sufficient in comparatively shallow epicontinental seas and continental shelves, to scour away the upper surface of previously deposited soft unconsolidated or partly consolidated sediment, or to mobilize it as a whole producing turbidity currents and sediment flows down slopes. This situation is well documented in modern shelf-seas and there is much evidence of it in the past, quite apart from sequences which have the characteristics of turbidity-flow formation.

One such example in the past is to be seen at the top of the middle Albian (Cretaceous) sediments in north-west Europe, where there is a marked erosion surface upon which rests the current-swirled phosphatic debris derived from former clays of middle Albian age (Owen, 1971, 1973, 1976a, 1979). The phosphatic nodules and pebble fauna occur in pockets embedded in clay of early upper Albian age. This erosional episode is used by authors such as Hancock (1975) and Cooper (1977) as part of their evidence to support a eustatically-controlled global marine regression at the end of the middle Albian. However, a different interpretation is suggested for north-west Europe by the contemporaneous tectonic history which affected this region in the early upper Albian. The field evidence indicates marked, but relatively short-lived, fault movements along older Jurassic-Cretaceous boundary fault lines. These movements were associated with strong current scour sufficient to remove over the southern area of the London Platform, for example, a thickness of about 10 m of Lower Gault clay and mud. Rather than indicating a marine regression due to a eustatic change in sea-level, these fault movements heralded increased marine transgression associated with a general subsidence of the depositional basins into which were accumulated substantial quantities of fine-grained sediment during the upper Albian.

A similar "regression" is postulated at the end of the Albian followed by a Cenomanian transgression (see also Matsumoto, 1977:80) by the workers who have attempted to correlate world-wide *sedimentation* events (because that is what they are) in the Cretaceous as evidence of eustatic changes in sea-level. The phase of transgression within the lower Cenomanian is well documented, but the uppermost Albian "regression" is not. If there had been a late Albian marine regression, how is it that sediments of the highest ammonite zone (that of *Stoliczkaia dispar*) are so widespread, particularly in areas considered to be the marginal regions of earlier upper Albian depositional basins? A detailed study of late Albian sedimentation in northern Europe indicates that the shallowing of the depositional region evident in the sediment sequence is due to basin inversion (or isostatic rebound) in response to the earlier accumulation of relatively thick sediment sequences.

There are other examples in the Mesozoic in which the field evidence does not agree with the concept of regular phases of eustatic change in sea level. Faulting and associated erosion of the sea-bed by strong current scour can give the impression of a regression hiatus if the vertical sequence in a limited area is considered on its own. Basin inversion and accompanying shallowing of the sea also can give the impression of regression. If the youngest sediments are geographically more widespread over the depositional area, however, one is still looking at a transgressive situation. This is true, even if wave turbulence and current action causes bottom sediment re-working and precludes continuous sedimentation in the now shallow sea over the more central region of the former subsiding basin.

THE GEOGRAPHY OF THE LAND REGIONS

In the above sections, in which the marine environment has been considered and relatively detailed examples been given, the land areas have received only passing mention. In Fig. 11, an extensive area of land is postulated, because there is no evidence that it was covered by the sea during the upper Albian. But, the true land–sea margins, the nature of the terrain, the pattern of rivers, the physiography of its surface, the

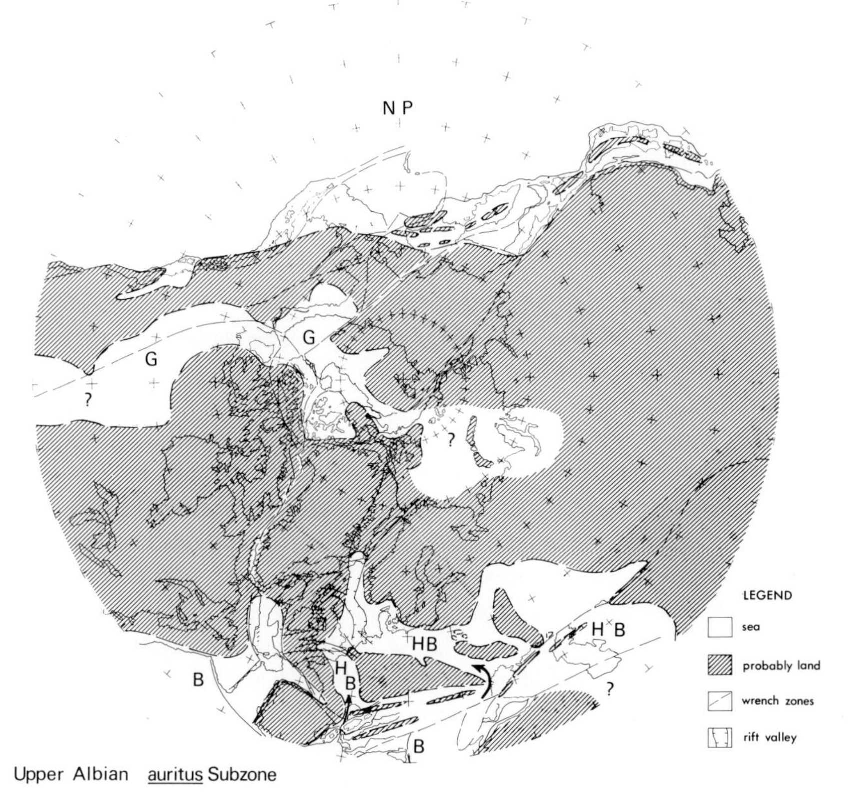

Fig. 11. Palaeogeographic map of the Boreal region in the upper Albian (*c.* 102 Ma) assuming an Earth with a diameter approximately 89% of modern mean value. Ammonite faunal provinces are indicated: NP, North Pacific; G, Arctic or Gastroplitid; H, European or Hoplitinid; B, Brancoceratid or Tethyan-Gondwana. HB denotes the general mingling of the Hoplitinid and Brancoceratid faunal provinces in the European epicontinental seas during the upper Albian.

extent of floral cover and such like are unknown because the evidence for these features has not survived. The major problem with all palaeogeographic analysis is the dearth, or total absence, of a sedimentary record except in limited and relatively local cases of the regions which do not show evidence of marine inundation at a given point in time. Major and minor continental basins filled with intermontane and desert sediment or extensive lake deposits, together with much rarer glacial deposits are known throughout the geological column. But, even in these instances, the picture of the terrestrial environment that they give at any one place at a specific point in time, is severely limited.

CONCLUSIONS

The purpose of this paper has been to sound a note of warning in the way in which some faunal distributions are interpreted against a background of geological orthodoxy. A "Pacifica" has to be invented by some to transport certain faunas and floras between Gondwanaland and the east Pacific margin, because of the huge area problems in reconstructions of the past which assume an Earth of constant modern dimensions. For similar reasons, it is found necessary to route certain land faunas from South America through Antarctica to reach Australia by the early Tertiary, because of the presence of the last vestiges of a wide Tethyan Ocean gap between Australia and South East Asia. These and other problems of faunal distribution, are eased by the expanding Earth model supported as it is by the ocean-floor spreading patterns.

Time is a factor which, for some biogeographers apparently, is difficult to envisage in their cursory examination of the geological record. A few centuries of land contact between two "permanent" terrestrial regions would barely show in the sedimentary sequence of a shallow sea subject to mild epeirogenic warping and not at all in the case of a young, but emergent, spreading ridge. Such a short length of time would be more than adequate for a terrestrial fauna to migrate or disperse from one region to another across ephemeral land bridges. An example of one such ephemeral bridge is provided by Palk Strait between India and Sri Lanka within the Pleistocene stadials. Because a Palaeozoic fauna is not present in contemporaneous sediments each side of what is now a major geological structural boundary, it does not necessarily mean that they were separated by a wide ocean. Today, continental active orogenic belts, whether marginal or not, show major lateral wrench motions and this phenomenon is evident in past orogens such as the Caledonides and the Variscides. The subsequent off-setting of one portion of a formerly continuous sedimentary basin relative to the other portion of it by such wrench motions, has also to be considered.

In the interpretation of the sedimentary record, the detailed work carried out during the last twenty years in the various continental shelf seas of today, makes necessary a re-appraisal of what in fact are sea-marginal sediments as opposed to sediments accumulated at the margins of deposition, which can be two very different things. Moreover, it is now apparent that re-working of sediments can occur at depths well below those thought possible at one time. These distinctions are of fundamental importance in reconstructions of palaeogeography in which we do not have the benefit of the presence of the sea concerned, but merely a sedimentary record with which to interpret its extent and oceanographic features. The possibility of water current activity preventing any sedimentation in sea areas comparable to modern examples, or of producing sand banks and ripple-marked surfaces at moderate depths instead of near-shore, as is normally assumed, must now be taken into account in palaeogeographic environmental reconstructions. Eustatic changes in sea-level might have occurred in the Mesozoic, but their detection is difficult because we can only examine in detail a very small proportion of the Earth's crust, both continental and particularly oceanic, present at this time. Synchronous sedimentary events apparent in the stratigraphic sequence of different areas, have to be examined in the light of the evidence of continental splitting and displacement mechanisms, which have become apparent only in the last decade, as well as eustasy.

The advance in knowledge of geodynamic processes and physical oceanography during the last two decades makes necessary a reappraisal of many long-held concepts used in palaeogeographic analysis. Rather than introducing doubt into the validity of such studies, in the writer's opinion, the new information helps to eliminate previous inconsistencies both in the interpretation of sedimentary sequences and in the distribution of faunas.

ACKNOWLEDGEMENTS

I should like to thank Professor A.J. Smith (Bedford College, London) and Dr E.R. Shephard-Thorn (Institute of Geological Sciences) for useful information about the English Channel and for constructive suggestions.

REFERENCES

Akeroyd, A.V. (1972). Archaeological and historical evidence for subsidence in southern Britain. *Phil. Trans. R. Soc. Lond.* **A272**, 151–169.

Allen, P. (1976). Wealden of the Weald: a new model. *Proc. Geol. Ass.* **86**, 389–437.

Brunnschweiler, B.O. (1974). Indoburman Ranges. *In* "Mesozoic-Cenozoic orogenic belts" (A.M. Spencer, ed.). pp. 279–299. Geological Society, London.

Carey, S. Warren (1976). "The Expanding Earth" (Developments in Geotectonics vol. 10.) Elsevier, Amsterdam.

CNRS (1972?). "Atlas de la Manche, thème 2: connaissance et exploitation des matières minérales et fossiles". Éditions CNEXO, Paris.

Cogley, J.G. (1981). Late Phanerozoic extent of dry land. *Nature, Lond.* **291**, 56–58.

Cooper, M.R. (1977). Eustacy during the Cretaceous: its implications and importance. *Palaeogeogr. Palaeoclimatol. Palaeoecol.* **22**, 1–60.

Cox, C.B. (1973). Triassic Tetrapods. *In* "Atlas of Palaeobiogeography" (A. Hallam, ed.), pp. 213–223. Elsevier, Amsterdam.

Creer, K.M. (1965). An expanding Earth? *Nature, Lond.* **205**, 539–544.

Curry, D., Hamilton, D. and Smith, A.J. (1970). Geological and shallow subsurface geophysical investigations in the Western Approaches to the English Channel. *Repts Inst. geol. Sci.* **70/3**, i–iv, 1–12.

Dana, J.D. (1863). "Manual of Geology". Bliss, Philadelphia.

Destombes, J-P., Shephard-Thorn, E.R. and Redding, J.H. (1975). A buried valley system in the Strait of Dover. *Phil. Trans. R. Soc. Lond.* **A279**, 243–256.

Donovan, D.T. and Jones, E.J.W. (1979). Causes of world-wide changes in sea-level. *J. geol. Soc.* **136**, 187–192.

Dunham, K.C. and Gray, D.A. (eds) (1972). A discussion on problems associated with the subsidence of southeastern England: a symposium. *Phil. Trans. R. Soc. Lond.* **A272**, 79–274.

Dunham, K.C. and Smith, A.J. (eds) (1975). A discussion on the geology of the English Channel: a symposium. *Phil. Trans. R. Soc. Lond.* **A279**, 1–295.

Egyed, L. (1956). Determination of changes in the dimensions of the Earth from palaeogeographical data. *Nature, Lond.* **178**, 534.

Fairbridge, R.W. (1961). Eustatic changes in sea-level. *In* "Physics and Chemistry of the Earth", vol. 4 (L.H. Ahrens, F. Press, K. Rankama, and S.K. Runcorn, eds), pp. 99–185. Pergamon Press, London and Oxford.

Firstbrook, P.L., Funnel, B.M., Hurley, A.M. and Smith, A.G. (1980). "Paleoceanic reconstructions 160–0 Ma." Scripps Institution of Oceanography, University of California.

Friedman, G.M. and Sanders, J.E. (1978). "Principles of Sedimentology". John Wiley & Sons, New York.

Gansser, A. (1974). Himalaya. *In* "Mesozoic-Cenozoic orogenic belts" (A.M. Spencer, ed.), pp. 267–278. Geological Society, London.

Hails, J.R. (1975). Offshore morphology and sediment distribution, Start Bay, Devon. *Phil. Trans. R. Soc. Lond.* **A279**, 221–228.

Hallam, A. (1963). Major epeirogenic and eustatic changes since the Cretaceous and their possible relationship to crustal structure. *Am. J. Sci.* **261**, 397–423.

Hamilton, D., Hommeril, P., Larsonneur, C. and Smith, A.J. (1975). Geological bibliography of the English Channel (Part 2). *Phil. Trans. R. Soc. Lond.* **A279**, 289–295.

Hamilton, D. and Smith, A.J. (1972). The origin and sedimentary history of the Hurd Deep, English Channel, with additional notes on other deeps in the western English Channel. *Mém. Bur. Rech. Géol. miner.* **79**, 59–78.

Hancock, J.M. (1975). The sequence of facies in the upper Cretaceous of northern Europe compared with that of the Western Interior. *Spec. Pap. Geol. Ass. Canada* **13**, 83–118.

Hilgenberg, O.C. (1933). "Vom Wachsenden Erdball". Hilgenberg, Berlin.

Hilgenberg, O.C. (1966). Die paläogeographie der expandierenden Erde vom Karbon bis zum Tertiär nach paläomagnetischen Messungen. *Geol. Rdsch.* **55**, 878–924.

Kremp, G.O.W. (1979). The Earth expansion theory and the climatic history of the lower Permian. *In* "Fourth International Gondwana Symposium 1977, Calcutta, India" (B. Laskar, and C.S. Raja Rao, eds), pp. 3–20. Hindustan Publishing, Delhi.

Lapierre, F. (1975). Contribution à l'étude géologique et sédimentologique de la Manche orientale. *Phil. Trans. R. Soc. Lond.* **A279**, 177–187.

Larsonneur, C. (1972). Le Modèle sédimentaire de la Baie de Seine a la Manche centrale dans son cadre géographique et historique. *Mém. Bur. Rech. géol. miner.* **79**, 241–255.

Lefort, J.-P. (1975). Étude géologique de socle ante-mésozoique au nord du Massif Armoricain: limites et structures de la Domnonée. *Phil. Trans. R. Soc. Lond.* **A279**, 123–135.

Lisitzin, E. (1974). "Sea-level changes" [Elsevier Oceanography Series vol. 8.] Elsevier, Amsterdam.

Matsumoto, T. (1977). On the so-called Cretaceous transgressions. *Spec. Pap. palaeont. Soc. Japan* **21**, 75–84.

Neumayr, M. (1885). Die Geographische Verbreitung der Juraformation. *Denkschr. Akad. Wiss. Wien.* **50**, 57–144.

Norton, I.O. and Sclater, J.G. (1979). A model for the evolution of the Indian ocean and the breakup of Gondwanaland. *J. geophys. Res.* **84**, 6803–6830.

Olier, B. d' (1972). Subsidence and sea-level rise in the Thames Estuary. *Phil. Trans. R. Soc. Lond.* **A272**, 121–130.

Owen, H.G. (1971). The stratigraphy of the Gault in the Thames Estuary and its bearing on the Mesozoic tectonic history of the area. *Proc. geol. Ass.* **82**, 187–207.

Owen, H.G. (1973). Ammonite faunal provinces in the middle and upper Albian and their palaeogeographical significance. *In* "The Boreal Lower

Cretaceous" (R. Casey, and P.F. Rawson, eds). *Geol. Jl, Spec. Issue* **5**, 145–154.

Owen, H.G. (1976a). The stratigraphy of the Gault and Upper Greensand of the Weald. *Proc. geol. Ass.* **86**, 475–498.

Owen, H.G. (1976b). Continental displacement and expansion of the Earth during the Mesozoic and Cenozoic. *Phil. Trans. R. Soc. Lond.* **A281**, 223–291.

Owen, H.G. (1979). Ammonite zonal stratigraphy in the Albian of north Germany and its setting in the hoplitinid faunal province. *In* "Aspekte der Kreide Europas" [I.U.G.S. Series A, No. 6.] (J. Wiedmann. ed.), pp. 563–588. Schweizerbart'sche, Stuttgart.

Owen, H.G. (1981). Constant dimensions or an expanding Earth? *In* "The Evolving Earth" (L.R.M. Cocks, ed.), pp. 179–192. British Museum (Natural History), Cambridge University Press, Cambridge.

Owen, H.G. (1983). "Atlas of continental displacement 200 Ma to Present". Cambridge University Press, Cambridge (in press).

Reyment, R.A. and Mörner, N.A. (1977). Cretaceous transgressions and regressions exemplified by the South Atlantic. *Spec. Pap. palaeont. Soc. Japan* **21**, 247–261.

Scotese, C., Bambach, R.K., Barton, C., Van der Voo, R. and Ziegler, A. (1979). Paleozoic base maps. *J. Geol.* **87**, 217–277.

Shephard-Thorn, E.R., Lake, R.D. and Atitullah, E.A. (1972). Basement control of structures in the Mesozoic rocks in the Strait of Dover region, and its reflexion in certain features of the present land and submarine topography. *Phil. Trans. R. Soc. Lond.* **A272**, 99–113.

Shields, O. 1979. Evidence for initial opening of the Pacific Ocean in the Jurassic. *Palaeogeogr. Palaeoclimatol. Palaeoecol.* **26**, 181–220.

Smith, A.G. and Briden, J.C. (1977). "Mesozoic and Cenozoic Paleocontinental Maps". Cambridge University Press, Cambridge.

Smith, A.G., Briden, J.C. and Drewry, G.E. (1973). Phanerozoic world maps. *In* "Organisms and Continents through time" (N.F. Hughes, ed.), pp. 1–42. *Palaeontology Spec. Pap.* **12**.

Smith, A.J., Hamilton, D., Williams, D.N. and Hommeril, P. (1972). Bibliographie géologique de la Manche. *Mém. Bur. Rech. Géol. miner.* **79**, 233–240.

Stöklin, J. (1980). Geology of Nepal and its regional frame. *J. geol. Soc.* **137**, 1–34.

Stride, A.H., Belderson, R.H. and Kenyon, N.H. (1972). Longitudinal furrows and depositional sand bodies of the English Channel. *Mém. Bur. Rech. géol. miner.* **79**, 233–240.

Walker, J.G. (1977). "Evolution of the Atmosphere". Macmillan, New York and London.

Ziegler, A.M., Scotese, C.R., McKerrow, W.S., Johnson, M.E. and Bambach, R.K. (1977). Paleozoic biogeography of continents bordering the Iapetus (pre-Caledonian) and Rheic (pre-Hercynian) oceans. *In* "Paleontology and plate tectonics with special reference to the history of the Atlantic Ocean" [Milwaukee Public Museum Spec. Publ. No. 2.] (R.M. West, ed.), pp. 1–22. Milwaukee Public Museum, Milwaukee, Wis.

5 Evolution of the Physical Geography of the East African Rift Valley Region

A.T. GROVE

Geography Department, University of Cambridge, Cambridge CB2 3EN, UK

Abstract: The evolution of the physical geography of the region is considered as having taken place in four periods of diminishing magnitude approaching the present. In the Mesozoic and Tertiary eras, the continental outline and rift system were roughed out by coastal movements. During the Late Pliocene and Pleistocene, a number of volcanic, tectonic and climatic events have been dated. Sequences of climatic changes in the Late Quaternary are traced in the four drainage basins, Zambezi, Zaire, Nile and Awash; the possible range of temperature and rainfall fluctuations is discussed. Finally, mention is made of the results of man's activities in the region over the last few centuries.

INTRODUCTION

The ancient history of East Africa is mainly recorded in the lake basins and at their margins, especially in the northern part of the rift system where volcanic rocks are widespread and where faulting and erosion in Quaternary times have revealed fossil and archeological remains varying widely in age. The results of recent researches in the north have recently been presented in two volumes, *The Geological Background to Fossil Man* (Bishop, 1978) and *The Sahara and the Nile* (Williams and Faure, 1980).

The rift valleys lie along two intersecting lines running from the Red Sea to the Kalahari and from the Sudd to the Zambezi delta (Fig. 1).

Systematics Association Special Volume No. 23, "Evolution, Time and Space: The Emergence of the Biosphere", edited by R.W. Sims, J.H. Price and P.E.S. Whalley, 1983, pp. 115–155. Academic Press, London and New York.

From the Red Sea coast the Afar depression narrows to a trough that runs SSW across Ethiopia to Chew Bahir (Lake Stefanie). Lake Turkana (Rudolf) is offset to the west and the rift valley extends south from Turkana as the Gregory Rift across the dome of the Kenya Highlands to

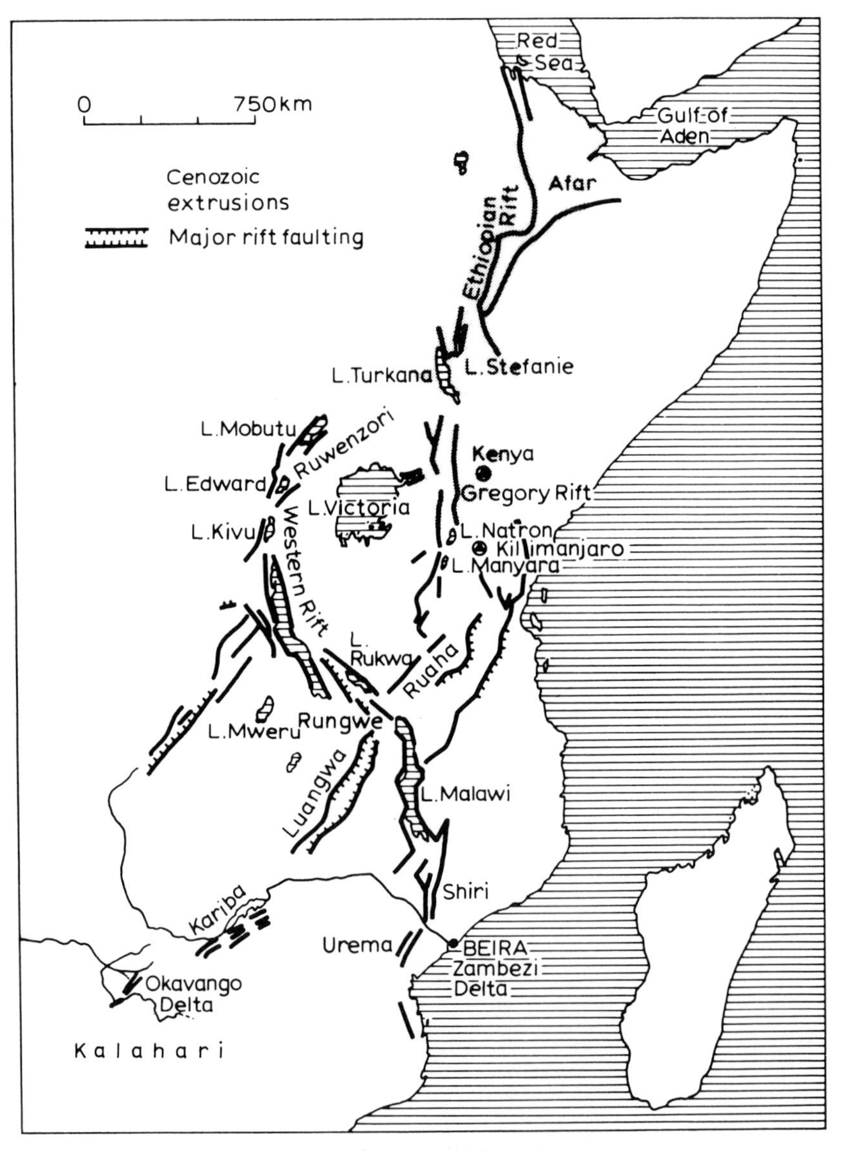

Fig. 1. The East African Rift Valley Region.

Lake Natron and then splays out to accommodate Lakes Eyasi and Manyara. Fault scarps and tilted blocks extend south across Tanzania and follow the Ruaha valley south-west to the Rungwe Mountains, an isolated volcanic mass at the north end of Lake Malawi. The Luangwa valley prolongs the system further to the south-west across Zambia, and the line of the rift can then be traced uncertainly via the Kariba section of the Zambezi valley to the linear NE–SW lower extremity of the Okavango delta in Botswana. Another set of rift valleys is followed by the eastern frontier of Zaire from Lake Mobutu (Albert) via lakes Edward and Kivu to Lake Tanganyika. The Rukwa basin, offset to the east of the Tanganyika trough, extends south-east to the foot of the Rungwe mountains and then deepens to run south via Lake Malawi and the Shire valley to the coast.

For the most part East Africa consists of upland plains at elevations between 1000 and 2500 m. The coast approaches high ground in the vicinity of Tanga where the Usumbara Mountains are only about 100 km inland (Kent, 1971). To the north and south of Tanga, lowlands broaden to extend from Somalia across Turkana to eastern Equatoria in the Sudan Republic and from coastal Mozambique up the lower Zambezi, the lowlands being backed by escarpments rising to the high plains of the interior. The rift valleys are generally 50–100 km wide with lakes occupying hollows on the floors of the troughs, and escarpments rising as much as 1500 m to the uplifted shoulders of the rifts. Associated with the rifts, but sometimes 100–200 km away from them, are several great volcanoes of which the highest are Kilimanjaro, 5895 m, and Kenya 5200 m. Both these mountains and Ruwenzori, 5110 m, support glaciers. The Ruwenzori range on the border of Uganda and Zaire is remarkable in that it consists of a block of crystalline rocks, uplifted some 5 km vertically along the line of the Western Rift. Centrally placed between the Gregory Rift and the Western Rift is Lake Victoria at a height of 1134 m, resting in a gentle depression and extending over an area of 70 000 km^2.

Environmental conditions are largely controlled by the climate which is related in detail to the topography (Griffiths, 1972). Over most of the region the mean annual rainfall exceeds 750 mm but a tongue of greater aridity extends south-west from Somalia across northern and eastern Kenya into the heart of Tanzania. Rainfall totals generally increase with altitude to reach a maximum at about 2500 m and then decline, so that the wettest areas are the crests of the highlands and the middle slopes of the high mountains facing the rain-bearing winds. These winds are generally from the south-east in the eastern part of the region, and from the south-west in the west. The rift floors and rain shadow areas are dry, evaporation values are high and the lakes depend on streams rising at the higher levels. The equatorial part of the rift experiences two rainy seasons which are unreliable and unpredictable. Towards the tropics the seasonal rainfall peaks draw close together giving single rainfall peaks; in July/August in southern Ethiopia, in January/February in Botswana.

Savanna woodland dominates the scene on the upland plains. The vegetation is most varied and complex on the great mountains and the highlands at the rift margins. Semi-arid grasslands and woodland on the plains are overlooked by montane forest at 1500 to 3000 m, then by moorland and by Alpine zones approaching the highest summits. Almost everywhere the plant cover has been greatly modified by cultivation, grazing and burning.

The balance between precipitation and evaporation losses in the rift basins is quite delicate. The lakes in the north-eastern rift valleys are small and saline as compared with those in the west and south. Confined to closed basins they vary in level markedly from season to season and over periods of years, according to the rainfall (Fig. 2). The lakes in the west and south are drained to the sea by three of the world's great rivers, but their outflow is not reliable. Lake Malawi normally overflows via the Shire river to the Zambezi, but ceased to do so between 1915 and 1935 when rainfall was below normal and sediment accumulated at the

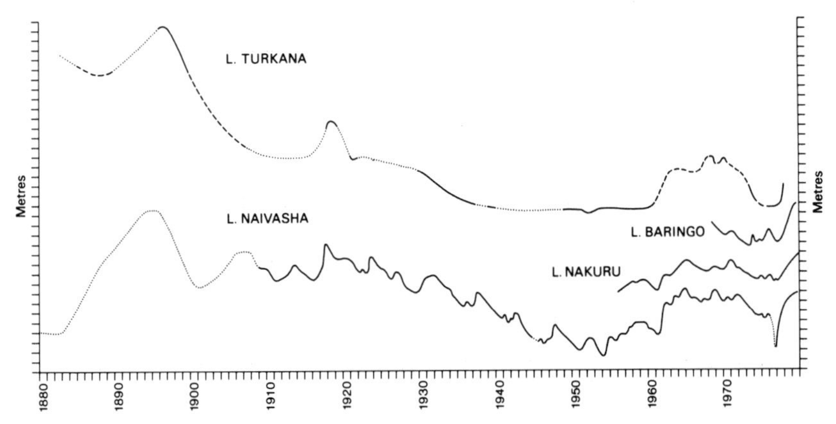

Fig. 2. Fluctuations of lake levels; broken lines inferred; from Butzer (1971) and Hydrology Department, Nairobi, Kenya.

outfall (Fig. 3). Lake Tanganyika's supplies to the Zaire are similarly interrupted from time to time. If it were not for the flow of the Ruzizi river from Lake Kivu, Tanganyika would be a closed basin, and the Ruzizi contributes to the Zaire only because volcanic lavas dammed back a headwater of the Nile to form Lake Kivu at some stage probably late in the Pleistocene.

The presence of lavas and other volcanic rocks in the vicinity of the rift valleys allows tectonic events and former land surfaces to be dated by isotopic methods. The troughs of the rift system have trapped volcanic and lacustrine sediments, preserving them and their fossil remains. Glacier tongues on the highest mountains and vegetation zones arranged altitudinally have moved up and down as temperatures have oscillated in the course of the Late Quaternary, leaving a record in moraines and pollen sequences. Strandlines above the present lakes present a record of the changing balance of precipitation and evaporation; the lake sediments preserve diatom and molluscan remains and pollen. Taken together these allow the outlines of the physical evolution of East Africa to be traced through time thereby providing a back-

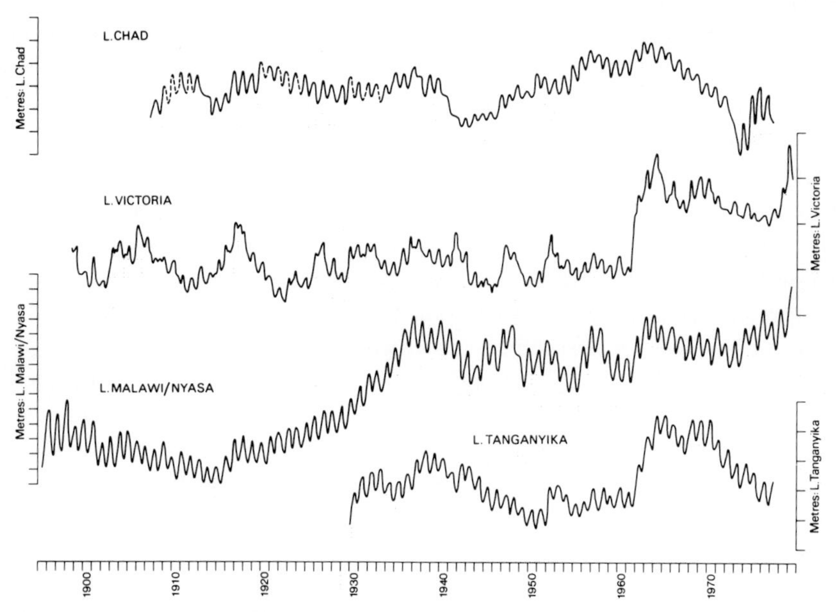

Fig. 3. Fluctuations of lake levels in certain of the Great Lakes. Note that all of them were high in the 1960s.

ground against which past and present biological distributions may be viewed.

Here the sequence of events is discussed as having taken place in four time periods of diminishing magnitude as we approach the present. First the continental outline and the rift system were roughed out by crustal movements operating over more than 100 million years. Then, during the Late Pliocene and most of the Pleistocene, over a period lasting about 3 million years, a succession of volcanic, tectonic and climatic events occurred which is being disentangled by studies of sedimentary basins in the northern and eastern parts of the rift valley system. Thirdly in the last 30000 years of the Late Quaternary, which comes within the range of 14C dating, a sequence of climatic changes can be traced in some detail. Finally, attention is given to the current episode, with man's activities dominating the changing scene, and climatic fluctuations continuing to be significant for man and his environment.

THE ROUGHING OUT OF THE RELIEF

Former land surfaces lie preserved as unconformities at the base of sedimentary formations. In some cases the sedimentary rocks have been stripped away to exhume, for example, in South Africa, landscapes glaciated in Permo-Carboniferous times or, in East Africa, plains surmounted by inselbergs. Some of the old surfaces surviving as upland plains may never have been deeply overlain by sediments but have been protected by resistant superficial layers of lateritic or calcrete material (Pallister, 1956). Dixey (1956) regarded the upland plains of Africa as having originated as subaerial peneplains, each the outcome of long-continued erosion and related to a particular marine base-level. The erosion surfaces, it seemed to him, offered opportunities to fill in the gaps in the geological record of the continental interior where uplift and erosion have long been dominant and fossils useful for dating purposes are lacking. The age of an erosion surface lies between that of the rocks into which it is cut and the time when dateable rocks overlying it were laid down. Usually the bracketing dates are far apart and the ages of the erosion surfaces are consequently very imprecise.

Dixey attributed great importance in the blocking out of the relief of south-east Africa to large-scale faulting early in Mesozoic times, faults with throws of several kilometres letting down Karoo rocks into troughs running parallel in some areas, to trend lines of pre-Cambrian mobile belts and following the boundaries of old cratons, notably the Tanzanian shield. Elsewhere the Karoo rocks were laid down in pre-existing depressions. These rocks were removed from upstanding areas by Jurassic planation and were left preserved within the basins and downfaulted troughs which have since been the scene of repeated cycles of sedimentation, erosion and faulting. Dixey saw the Zambezi as continuing to flow east across the rising rim of the continent, a northward extension of the Lebombo monocline, thereby providing a large-scale example of antecedent drainage. Excavation of Karoo rocks from the Malawi and Rukwa troughs was followed by the accumulation in them of rocks containing dinosaur remains which were of Cretaceous age. These beds were used as time markers for the surfaces on which they rest, as were the Miocene mammalian beds of Rusinga Island in the Winam (Kavirondo) Gulf of Lake Victoria. But such fossiliferous beds are relatively rare in the African interior and in the absence of other diagnostic features the dating and correlation of erosion surfaces was commonly on a basis of their height and disposition; there was consequently uncertainty and disagreement between different scientists working in the same area as to the age of the surfaces.

King's studies of the African erosion surfaces (L.C. King, 1962, 1978) are probably the best known. Like du Toit he was convinced that Africa had once formed a part of Gondwanaland, the super-continent that had broken up and drifted apart in the manner described by Wegener. He considered that the erosion surfaces provided a means of identifying the stages involved in the process and proposed a mechanism for their formation involving the extension of pediplains inland as escarpments were formed and successively retreated from the coast following intermittent uplifts of the continental margin. He claimed to recognize a staircase of erosion surfaces stepping down to the coast of southern and eastern Africa, with interior structural basins, notably the Kalahari, bordered by wider surfaces at lesser vertical intervals. These surfaces were seen as falling into five groups. The oldest consisted of the remnants of surfaces that had originated before the break-up of Gondwanaland. The next group, belonging to the post-Gondwana cycle, was regarded as having been initiated in early Cretaceous times. The most extensively developed surfaces, which evolved in the early Cenozoic and were called "African", were said to provide the key to the recognition of the other surfaces. They were inscribed by Miocene and late Cenozoic cycles and finally dissected by valleys formed as a result of uplift and drainage rejuvenation in Plio-Pleistocene times.

Du Toit and King have been justified in their belief in continental drift at a time when the interest in the subject of other earth scientists had waned, even though the process of ocean-floor spreading that has been elaborated and the associated movements of crustal plates carrying the continents differs considerably from the scheme as they had envisaged it. For over a decade it has now been generally accepted that the continents were all united at the end of the Palaeozoic into one landmass surrounded by a world ocean, and a remarkably well-authenticated and detailed history of the movements of the continents since that time is emerging as a result of oceanographic, geophysical and paleomagnetic studies. The development of techniques for coring the sea-floor and the employment of methods of isotopic analysis have been mainly responsible for the new knowledge that has been acquired.

The opening of the Atlantic began in the north with North America breaking away from north-west Africa. By the early Mesozoic a Tethys seaway linked the embryo Atlantic to the world ocean and eastern Gondwanaland, including Madagascar and India, began to break away from western Gondwanaland. About this time, in the Jurassic period, the gymnosperms evolved and expanded their range so as to dominate the plant world in the Cretaceous. The flowering plants evolved, diversified and in their turn became dominant. The modern freshwater fish groups began to appear and the dinosaurs, though suffering successive waves of partial extinction in the course of the Mesozoic, also continued to diversify. Throughout this time it seems, elements of Gondwanaland's flora and fauna were able to reach Eurasia and North America. But by the Oligocene, about 40 million years ago, Africa, still joined to Arabia, had become an island, and it remained an island for 25 million years (Fig. 4).

In the course of the middle Cenozoic, the African island was moving slowly northwards from a latitudinal position about 15° south of the present day. Climates were probably less strongly differentiated than is now the case; equatorial forests and grasslands may not have differed very greatly from those of the present day, and faunal assemblages evolved that were adapted to them. By 30 million years ago, updoming, faulting and the eruption of basalts had already begun in the area which was to become the junction of the Red Sea, the Gulf of Aden, and Afar (Baker *et al.*, 1972). The Winam or Kisumu rift came into existence and,

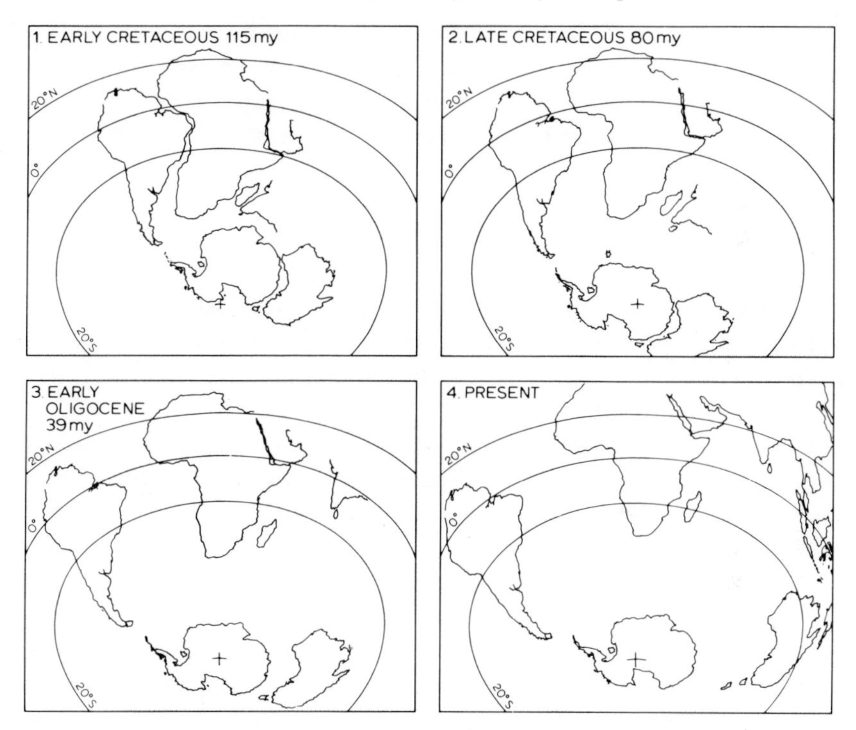

Fig. 4. The break-up of Gondwanaland, from Norton and Sclater (1979).

from 23 million years ago, was filled by lavas and tuffs from the Kisingini and Tinderet volcanoes on the south side of the gulf. These volcanic eruptives, and others thrown out by volcanoes along the Kenya–Uganda border, Elgon, Moroto, Napak and Kadam, and also in southern Turkana, have been dated by K/Ar techniques and give minimum ages for the erosion surfaces and intervening scarps lying beneath them (Bishop and Trendall, 1967). Furthermore, the volcanic materials overlie sediments containing vertebrate remains, and being rich in carbonates are excellent preservatives of bony material, as Bishop (1971) pointed out. The earliest mammalian and hominoid assemblages known from Africa were buried 22 to 24 million years ago by the first eruptions of Mount Elgon; others were preserved beneath and between later tuffs and lava flows and together they provide a remarkably complete record of life in the area in the Miocene.

As Arabia began to separate from Africa, great volumes of basaltic lavas were extruded over the Ethiopian region and by 20 million years ago, when the Indian ocean still spread across northern Kenya, faulting had initiated the Turkana depression and basalts were extruded over its western parts. About 17 million years ago Arabia and the north-east corner of Africa came into contact with this Iranian and Turkish portion of South-West Eurasia. The African fauna, including elephants and hominoids, were able to disperse north and east, while creatures such as the ancestors of the zebra, hyaena and hippopotamus, and probably also the cyprinid fishes, moved into Africa from Asia. Whereas, according to the van Couverings (1976), the East African mammalian assemblages had persisted with scarcely any evolutionary change for 7 million years at least, within the next two or three million years they may have changed rather rapidly.

The Miocene, between 17 and 13 million years ago, saw changes of global importance. There are indications of strong polar cooling. India was encroaching northwards on Eurasia, elevating the Himalayan ranges and thereby modifying climates throughout the northern tropics. About 14 to 12 million years ago the Iberian portals of the Mediterranean closed and, as evaporation exceeded the supply of water by precipitation plus the inflow of rivers from the surrounding land areas, the Sea shrank, eventually drying up almost completely about 6½ million years ago (Hsu *et al.*, 1977). An enormous desert basin, 3000 m deep, persisted to the north of Africa until the beginning of the Pliocene when a seaway to the Atlantic was re-established and a new fauna colonized the reconstituted Mediterranean.

In East Africa, the site of the Gregory Rift, about 15 million years ago, seems to have been a north-south trending arch of crystalline rocks which formed the continental watershed, with rivers draining east to the Indian Ocean and west towards the Atlantic across the line of the present Western Rift. The crest of the arch was an area of moderate relief, surmounted by the remnants of more than one erosion surface and with the crest itself occupied by a linear depression (B. King, 1978; Fig. 5). In the late Miocene, about 15 million years ago, great volumes of fluid phonolite erupted from a small number of centres, giving low-angled volcanoes about 70–80 km in diameter. The phonolite filled the crestal rift depression and about 12 million years ago overflowered eastwards across Laikipia, westwards as far as the Winam rift and northwards over the Kitale Plateau (Shackleton, 1978). Faulting began to block out the Elgeyo escarpment and the Tugen Hills and was repeated from time to time successively further east as phonolites continued to be erupted

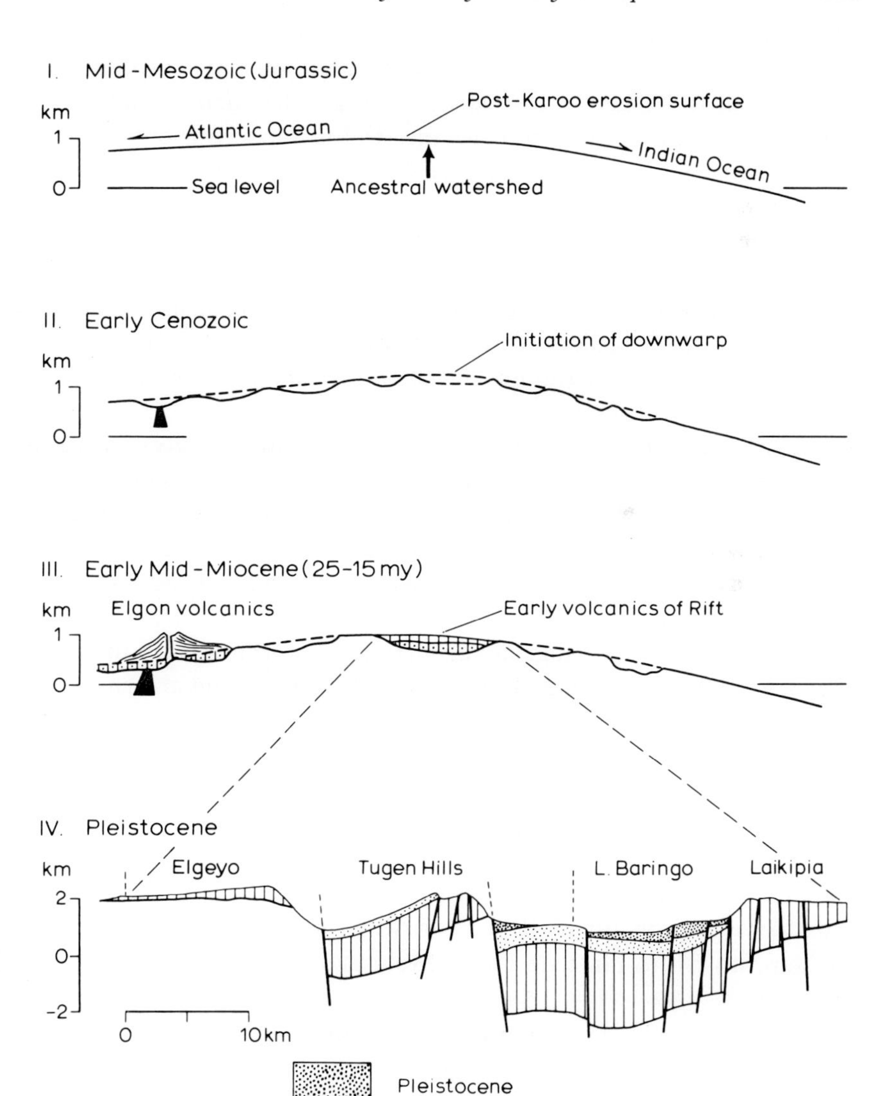

Fig. 5. The evolution of the Gregory Rift, after King (1978).

intermittently. Sediments accumulated in the intervening periods and those with an age believed to be between 12 million and 9 million years have proved to be very informative. The faunal remains point to the existence at this time of mammals ancestral to modern groups in East Africa and resembling rather closely faunas known to have existed in southern Eurasia about the same time. Pickford (1978a,b) found indications that during this part of the Miocene the landscape of north-west Kenya was comparable to that of the present day, though the relief may have been more subdued, with habitats ranging from lacustrine, and riverine with gallery forest, to open woodland and possibly grassland. The climate, though never arid, seems to have approached semi-aridity from time to time.

The structural history of the Lake Mobutu rift was very different from that of the Baringo area. Volcanic rocks are much less important and tectonic activity has been more restricted to repeated movements along the same boundary faults, resulting in the formation of a deep trough, its floor tilted down to the south-south-west. Clays, silts and fine sands of the Kaiso formation, derived from remodelling of the Early Cenozoic or African group of surfaces have accumulated to depths of at least 1200 m and possibly to more than 2500 m and are much less well-exposed by faulting and erosion than in the case of the Baringo basin.

In Ethiopia, flood basalts of Oligocene and early Miocene age are widely concealed by more silica-rich volcanics and ignimbrites which have given K/Ar dates of 9·5 to 2 million years. In addition great shield volcanoes of trachyte, the Arussi Mountains, Kakka, Badda, Chilalo and Enkwolo, probably mark an abortive phase of crustal extension east of the proto-rift in Pliocene times (Mohr and Potter, 1976).

PLIOCENE AND PLEISTOCENE TIMES

Although it would seem inevitable that the Rift System is related in some way to the global system of mid-ocean-ridges there is little or no evidence of spreading taking place within the rift valleys. In the north-east of the region, where the abundance of volcanic material has allowed faulting and eruptions to be dated, it is apparent that the modern troughs as topographical features are the outcome of faulting in Pleistocene times. Between 2 and 1·5 million years ago a surface in Ethiopia and Kenya that had become deeply weathered was strongly rifted; tectonic activity has continued from time to time since then with the formation of step faults and backward tilting blocks, as if movement was continuing at intervals on major dislocations at depth (King, 1978).

Large calderas have formed along the rift floor in the course of the Pleistocene and have divided it into separate basins. The Gademotta ridge overlooking Lake Ziway in southern Ethiopia is the western rim of a very large caldera dated 1·05 million years by Laury and Albritton (1975). The Alutu Caldera lies between Ziway and Langano, the Corbetti caldera between Shala and Awasa. Lake Shala itself occupies a caldera that may be of Late Pleistocene age. In Kenya, a volcanic complex rising from the rift floor south of Lake Turkana and blocking the Suguta valley rests on lacustrine sediments standing some 150 m above the present lake. The summit of the barrier has collapsed to form a caldera over 3 km in diameter (Williams, 1978). Further south, the Nakuru-Naivasha basin is defined by Menegai caldera to the north and by Longonot to the south (Scott, 1980). These calderas and others such as Rungwe in southern Tanzania have emitted lava flows and ignimbrites; the ash and pumice thrown out by them have settled in or been washed into the rift troughs, modifying drainage patterns and affecting soil conditions over extensive areas.

East of the Gregory Rift, volcanic activity in Pliocene and Pleistocene times has included the building of Mount Kenya and Kilimanjaro. Mount Kenya has long been extinct (Baker, 1967) but lavas have erupted from the Kibo vent of Kilimanjaro from time to time in the Quaternary. The growth of Kilimanjaro and mudflows from Meru, a cone to the west of it, have interrupted southward drainage forming the Amboseli basin (Downie and Wilkinson, 1972).

The Pliocene and early Pleistocene records are best known for those areas where volcanic ash and lake sediments have accumulated and have lately been uplifted and dissected to reveal the sedimentary sequences and the remains of creatures that lived in the lakes and their vicinity. The impetus for much of the research has been provided by interest in the hominid remains but much information has also been obtained about contemporary faunas and their environments.

The succession of events in Olduvai Gorge has been worked out by Hay (1971). The oldest sediments dating from 2 to 1·75 million years ago, were laid down in a perennial lake about 25 km wide. Sediments which accumulated about 1 million years ago include an aeolian tuff which separates a very diverse older fauna that includes extinct elements such as deinotheres and calichotheres from a later fauna that lacks some of the archaic elements. This change, associated with the appearance of Acheulian implements, may be attributable to a change to drier steppe conditions, though the region seems to have been semi-arid throughout the period.

The Turkana basin has been the scene of much scientific activity since international expeditions in the late 1960s began to investigate deltaic sediments laid down by the Omo during the Pliocene and Pleistocene. Investigations have extended to the east side of the lake and also to the south-west in the vicinity of Lothagam Hill (Patterson *et al.*, 1970). K/Ar dating of tuff layers and paleomagnetic reversal sequences have allowed marker horizons to be identified and traced over extensive areas. Each tuff layer defines an ancient landscape. Fifteen are distinguishable in the area east of Lake Turkana, 120 in the Omo sediments (Fitch and Vondra, 1976). Difficulties have arisen in relating the two chronologies but in time there is little doubt that this region will provide the chronological base for anthropoid studies at least in East Africa. Already it has been possible to study time slices of the paleoenvironment with the associated macro and microfossil remains; about 50000 identifiable fossil mammals have so far been unearthed. The lake seems to have reached its greatest extent about 3·9 million years ago. Fossils, mineralogy and chemistry of the sediments show, according to Cerling *et al.* (1977) that it was constantly fresh during the Pliocene and early Pleistocene and had its own endemic molluscs. From pollen studies it could seem that the vegetation of the lower Omo 2 million years ago was not very different from that of the present day, with shrubby and treed grassland on the plains, forest near the mountains. Changes in the pollen and faunal composition point to the possibility of a shift towards drier conditions about 1·8 million years ago (Bonnefille, 1976a). The last major faulting in the basin seems to have taken place in early to mid-Pleistocene times with renewed rifting along the Suguta, South Turkana, Chew Bahir axis. In the Late Pleistocene and Holocene, when the Kibish formation accumulated, the lake has been intermittently saline.

The Afar depression was subject to violent earth movements throughout much of the Quaternary. Lake Abhé, lying in an early Quaternary trough fault, is the terminal lake of the Awash river which rises west of Addis Ababa and has at times received an overflow from the Ziway-Shala basin. Below the rift scarp the Awash flows north below the western escarpment and then north-east, cutting down through massive fans of bouldery gravel and into the fluvial and lacustrine Plio/Pleistocene Hadar formation. A basalt flow which has given an age of 3 million years is succeeded by sediments which may have accumulated under conditions like those around present-day Lake Chad. The fauna consisted of creatures typical of savanna or light forest; more than 30 hominid-bearing beds have been found. Other lakes formed in central Afar at intervals during the Plio-Pleistocene; their diatoms indicate that before the main faulting gave the present horsts and graben, conditions were cooler than they are now (Gasse *et al.*, 1980). One lake near the escarpment, which had been fairly deep according to its diatom flora, dried up probably about 2 million years ago. At Melka Kontouré, above the scarp crest at 2000 m and 150 km downstream of the sources of the Awash, montane forest species such as *Olea*, *Juniperus* and *Hagenia* were present round 1·75 to 1·6 million years ago. But conditions became less humid; the montane forest trees totally disappeared, hippos became rarer and remains of antelope more common (Bonnefille, 1976b).

The Olorgesailie basin, 55 km south-west of Nairobi, though small in comparison with the Turkana and Afar depressions, has received much attention in recent years (Shackleton, 1978, Isaac, 1977, Owen and Renaut, 1981). Covering about 100 km^2 it was probably one of a series of lakes occupying the rift floor between Longonot and Lake Magadi in the Middle Pleistocene. The sediments, 60 m in thickness and brought in by rivers possibly originating at times in the Naivasha basin, include much volcanic debris probably thrown out by Longonot. They preserve a rich and abundant archeological record of early man and the creatures living alongside him. Mollusca and pollen are rare but diatoms give some indication of variations in lake level. At times when the lake in the basin was deep the climate must have been markedly wetter than at present, but the sediments were disturbed by faulting, with dislocations of as much as 15 m in the course of deposition, and so the sequence cannot be interpreted solely in climatic terms. Dating is uncertain and the sedimentary record may span only a few tens of thousands of years. Eventually southward tilting and through-drainage resulted in dissection and exposure of sedimentary sections and former living floors.

By degrees fragments of information are accumulating about the early and middle Pleistocene, "the Muddle in the Middle" as Isaac (1975) has called it. The Pleistocene climatic record has been very complicated;

ocean cores and loess studies point to many oscillations of temperature and therefore, in all likelihood, many shifts in precipitation/evaporation ratios in East Africa. Probably the best potential sources of information are the sediments on the floors of the rift valley lakes, especially Malawi and Tanganyika which, judging from their fauna, have been in existence for a very long time. Already a deep core obtained from Lake Abiyata in the course of geothermal prospecting is being examined for its diatom and pollen content.

THE LATE QUATERNARY

A climatic succession correlating African pluvials and Alpine glacial advances put forward in 1947 in connection with the first Pan-African Congress on Prehistory, was undermined by Cooke (1958) and Flint (1959), who showed that the evidence for the early pluvials was quite inadequate. Bishop and Posnansky's (1960) exposition of the origins of Lake Victoria disposed of the basis for the first or Kageran pluvial. The sediments supposed to indicate the second Kamasian pluvial were shown by Bishop (1971) to include beds varying in age from the Pliocene to the Holocene. Tricart (1956) controverted the glacial/pluvial hypothesis in West Africa and, in 1963, radio-carbon dates obtained by Faure *et al.* (1963) from lacustrine sediments in the Chad basin pointed to the last lacustrine period on the south side of the Sahara as having been post-glacial. A reassessment of the Late Quaternary in East Africa rapidly took place. Celia Washbourn (1967) put the Gamblian pluvial in a new time context, showing that Lakes Nakuru and Elmenteita had been united to form a single lake *c.* 9500 years ago, standing at 180 m and overflowing into the Menengai caldera.

Three kinds of investigation were involved. One group, mainly consisting of pollen analysts, studied the content of organic cores obtained from the floors of montane lakes and the swampy fills of rivers reversed by faulting in western Uganda. A second group constituted mainly of geomorphologists was concerned with the levels of moraines on the glaciated mountains, and raised strandlines around lake basins. The third group, mainly consisting of pupils of Professor Livingstone (1978, 1980) took deep cores from the lakes and analysed their pollen, micro-flora and sediment chemistry.

The high mountains of East Africa, Elgon, Kenya, Kilimanjaro and Ruwenzori and also the Balé Mountains of southern Ethiopia, supported ice fields each covering several hundreds of square kilometres at certain stages in the Pleistocene. Nilsson's (1940) early studies of the glacial features are remarkable for their scope and detail at a time when air photographs were unobtainable, maps sketchy and transport difficult. His work provided a useful starting point though many of his conclusions have been shown to be untenable, for he had no good means of dating the features he was studying. In recent years it has been confirmed that there were several glaciations of Kilimanjaro (Downie, 1964), Ruwenzori (Osmaston, 1965) and Mt. Kenya (Mahaney, 1978).

During the last glaciation, which was in general somewhat less extensive than two or three predecessors, the lowering of the snowline as calculated from the positions of moraines indicates an average lowering of temperature of about 5°C, supposing precipitation to have been the same as it is now (Table I). Whether this degree of cooling would have extended to lower altitudes is uncertain. It might be argued that, given drier conditions, lapse rates would have been steeper than today and temperatures near sea-level not very different from those of the present. Certainly the CLIMAP figures for ocean surface temperatures around Africa at 18000 BP (Gardner and Hays, 1976) indicate a cooling of nearer 2° or 3°C than 5°C. On the other hand if, as seems to be the case, precipitation was less than now at 18000 BP, temperatures on the glaciated mountains must have been more than 5°C lower to give the required accumulation of snow. Livingstone (1980), adopting Harvey's (1976) figure for a reduction in precipitation of 29 per cent, calculated a mean temperature lowering of about 7·5°C at high altitudes. Lancaster (1980) made a study of the current variation in temperature with altitude in Malawi which did not indicate very strong departures from the mean from place to place or season to season at the present day,

Table I. Fernline lowering in the Pleistocene on East African mountains (from Livingstone, 1980).

Glaciation	*Ruwenzori* (m)	*Kilimanjaro* (m)	*Mt. Kenya* (m)
Lake Mahoma (the last)	640	884	762
Rwimi basin	731	914	
Katabarua	640		1067

and this would suggest that the temperature lowering towards the end of the late Pleistocene was 7·5°C at low altitudes as well as at the snowline.

The timing of the last glacial advance on the East African mountains is uncertain. Reliance has to be placed on radio-carbon dates of organic material resting in hollows known to have been occupied by the ice and which give minimum dates for its retreat. A small cirque on Mount Badda in Ethiopia was abandoned before 11 500 ± 200 BP (Hamilton, 1977); ice had retreated from the caldera of Mount Elgon before 11 000 BP (Hamilton and Perrott, 1978); Lake Mahoma on Ruwenzori was ice-free shortly before 14790 ± 290 BP (Livingstone, 1962); Coetzee's (1964) date of 15862 ± 185 BP near Sacred Lake on Mount Kenya points to deglaciation having occurred before that time. The possibility should, perhaps not be dismissed that the last glacial maximum occurred in the East African mountains before the onset of aridity, possibly over 20000 years ago when conditions may have been both cooler and more humid than at present and thus more favourable to glaciation than at the time of the glacial maximum in middle latitudes.

The altitudinal zonation of the vegetation on the mountains of East Africa and Ethiopia is complicated and varies from one mass to another (Flenley 1979). Broadly one can distinguish montane rain forest at 1500–2500 m, a bamboo zone at 2500–3000 m, *Hagenia-Hypericum* forest at 3000–3500 m, an *Erica* belt above the treeline and an alpine belt above about 4000 m. Forests are confined to the wetter sides of the mountains but the exact nature of the climatic and edaphic controls of the vegetation patterns is still not clear. Zinderen Bakker (1962) was inclined to interpret pollen diagrams from cores taken at about 3000 m, showing an increase in arboreal pollen at the end of the Pleistocene, in terms of rising temperatures and the consequent replacement of the *Erica* belt by forest. However, Livingstone (1967) saw the greater proportion of grass pollen before 12500 BP as indicating greater aridity in the late Pleistocene; the different kinds of grasses cannot be deduced from the pollen and he considered the non-arboreal pollen was probably derived from dry tropical grassland, not from temperate high-level grassland. Hamilton's (1972) studies of modern pollen rain and the contrasts in the mobility of pollen from different species has allowed reinterpretation of some of the older pollen diagrams and a re-appraisal of the Late Quaternary climatic evidence from tropical Africa. It is now clear from the evidence both from the mountains and from the lake basins that East

Africa and Ethiopia were on the whole cooler and drier for some thousands of years before 12000 BP than they are now.

The different kinds of evidence are complementary. Glacial moraines are good evidence of past cooler conditions but are difficult to date, and calculation of former temperatures from them requires information about former precipitation. High lake strandlines provide good evidence of past wetter conditions (Fig. 6), but calculation of former mean rainfall figures demands values of past temperatures to be fed into the hydrographic expressions. As evidence accumulates, the opportunities grow to improve solutions to the simultaneous equations.

The biogeographic consequences of climatic changes in the Late Quaternary have probably been greatest for freshwater ecosystems, and lake basins have yielded most information about the timing and scale of climatic sequences. It is therefore convenient to consider the evidence for Late Quaternary climatic change in each of the main drainage basins in turn.

1. Zambezi Basin

The upper Zambezi and even the Luangwa may conceivably have carried sediments into the Kalahari basin in Cenozoic times. Following uplift in the late Cenozoic the lower Zambezi cut down strongly, the Victoria Falls came into existence and exploiting vertical planes in Batoka lavas of Karoo age had retreated, according to archeological evidence, 8 km since middle Pleistocene times, exposing a sequence of duricrusts, terrace gravels and redeposited Kalahari-type sands (Bond, 1967).

South-west of the Victoria Falls, in an area that is very gently sloping vertical movements along east north-east to west-south-west lines have modified the lower courses of the Okavango and Cuando (Wright, 1978). At present most of the Cuando's discharge finds its way to the Zambezi via the Chobe swamps which are also fed by the Makwegana spillway from the Okavango (Fig. 7). Most of the Okavango's annual discharge of 11 km^3 is lost by evaporation from its own "delta" but a small flow of water usually escapes to Lake Ngami and via the Botletle river to terminal lake Dow in the Makgadikgadi depression (Grove, 1969). The swamp region is known to be very active seismically and gravity surveys have shown that rift structures extend from the middle Zambezi to the lower end of the "delta" (Reeves, 1972).

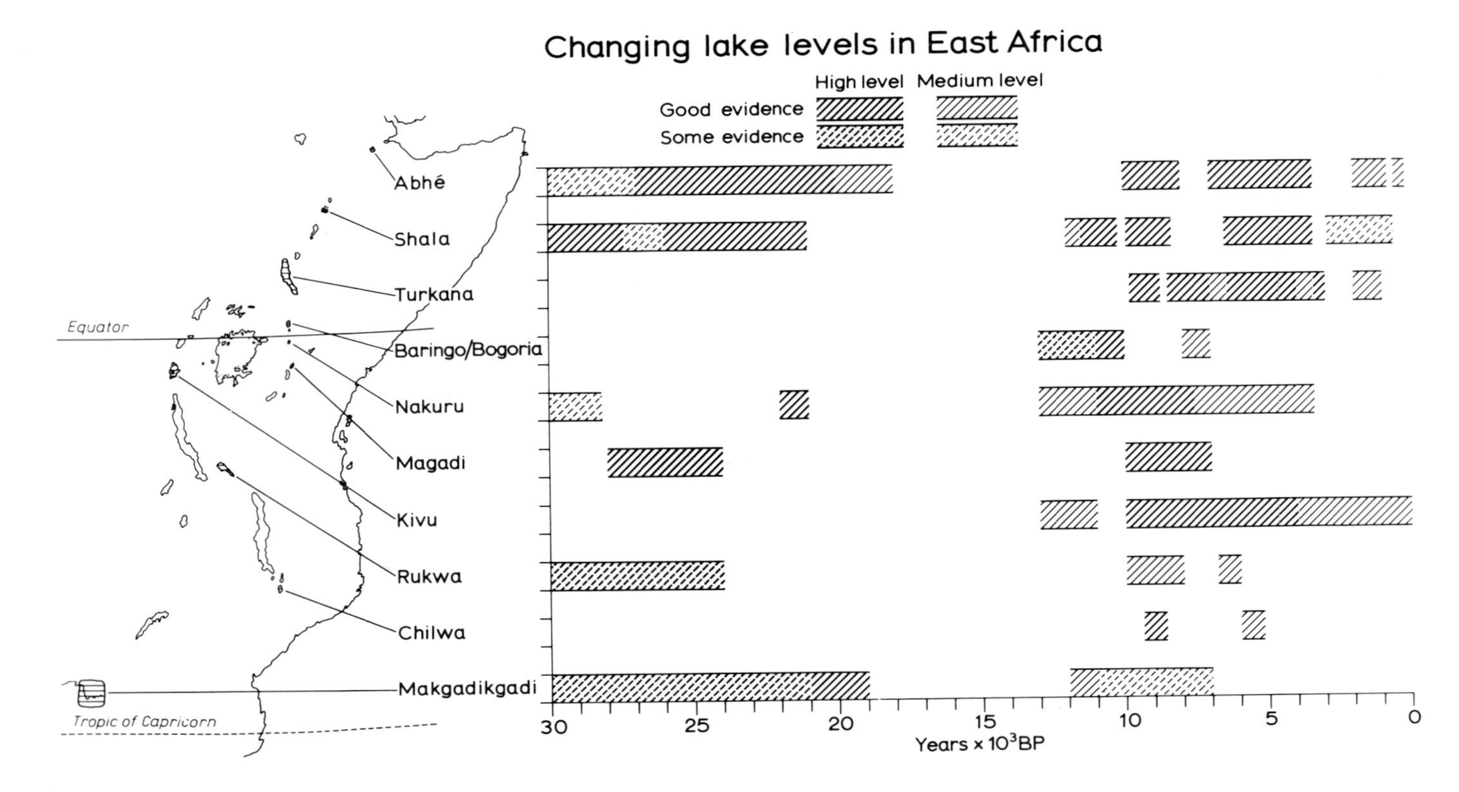

Fig. 6. Changing levels in certain East African lakes during the Late Quaternary.

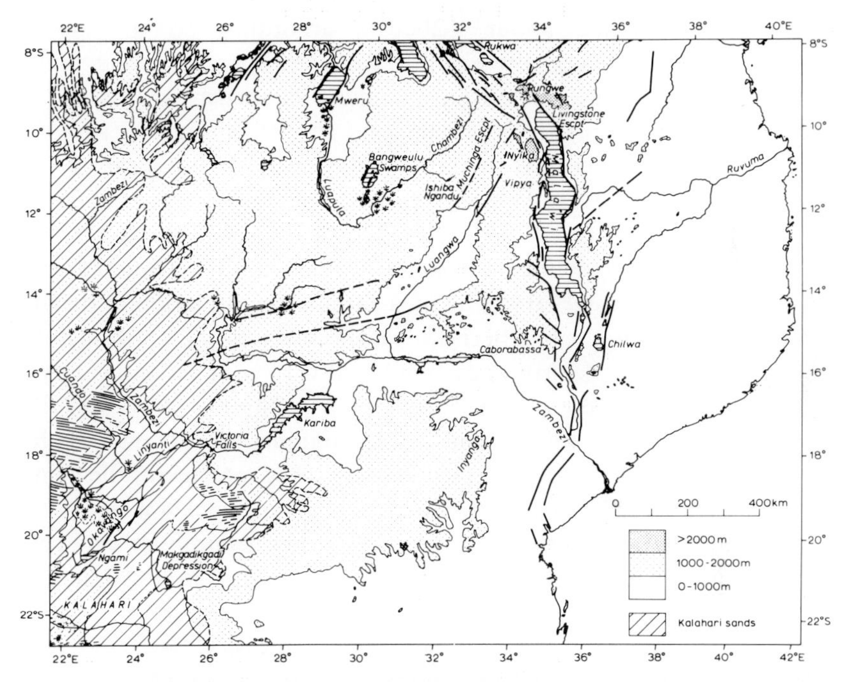

Fig. 7. The Zambezi basin. Unevenly spaced lines running approximately East–West in the vicinity of the Okavango delta and Makgadikgadi depression represent ancient linear dunes now fixed by vegetation.

A photogeological map of Botswana produced by the Institute of Geological Sciences (1980) showed an extensive area of lake shorelines and alluvium in the Makgadikgadi depression surrounded by linear dunes fixed by vegetation, running east-west in south-west Zimbabwe and diverging west of the Makgadikgadi to run west-north-west and also curving round to the south-west. According to radio-carbon dating of lacustrine carbonates and mollusca, there have been at least two periods of high lake levels in the depression. An earlier lake, probably existing between about 30000 and 19000 years ago, occupied some 60000 km^2, an area almost as large as Lake Victoria, and could conceivably have overflowed by the Savuti channel to the Chobe and Zambezi (Grey and Cooke 1977). Tentative estimates of the rainfall required to support such a lake, supposing temperatures were no lower than now, are three times that of the present (Ebert and Hitchcock, 1978). A lake with an area of about 35000 km^2 (Grove, 1969) may have come into existence about 12000 BP according to a date of 11920 ± 1630 BP obtained by Heine (1978). This smaller lake would scarcely have been maintained by an inflow equivalent to that of the modern Zambezi. Whether or not faulting allowed an inflow from that source, numerous dry river channels, apparently of no great age are unconnected with the Zambezi and climatic change seems to provide the best overall explanation.

The fixed dunes are evidence of conditions having been much drier at times in the Late Quaternary. The mean annual precipitation of the area under fixed dunes in the northern Kalahari is in excess of 400 mm and in some places reaches 700 mm; nearly all of it is treed grassland. Only in the extreme south-west of the Kalahari where the rainfall is less than 150 mm is sand driven by the wind and even under these conditions it is only the dune crests that are affected. This is some measure of the degree of former aridity. It would seem that in their original state the fixed dunes resembled the active linear dunes of the Namib desert; bare, sharp-crested ridges of sand, a hundred meters high instead of the 25 m high linear mounds to which they are now reduced. Some barchanoid dunes on the floor of the Makgadikgadi depression, and the inner members of lunate pairs of dunes on the leeward sides of pans post-date the Holocene lake in the Makgadikgadi (Lancaster, 1979); massive linear dunes running across the valleys of the Okavango and Zambezi formed in the late Pleistocene, probably as the 60000 km^2 lake dried up.

The evidence for important changes in climate in the northern Kalahari in late Quaternary times is clearly very strong and one might expect the consequences of such changes to be reflected in the geomorphology and vegetational history of the adjacent regions. Bond (1963) described sections in dambo sediments in Zimbabwe which he interpreted in terms of climatic oscillations involving variations in rainfall from about 50–150% of modern means. However, Tomlinson (1973, 1974), was unable to distinguish any very marked changes in the vegetation over the last 11000 years from the pollen content of dambo sediments collected from the Inyanga highlands of eastern Zimbabwe. Meadows (1980) examined pollen samples from cores extracted from dambos on the Nyika Plateau of Malawi. Small clumps of forest occupy valley heads bordering the dambo floors and White (1978) suggested they are the remnants of more extensive forest destroyed by burning. Shroder (1976) found charcoal in sediments exposed by land-slipping in one of the valley heads and dating to 9300 BP, but Meadows (1980) was unable to find evidence of forest having been much more extensive than it is now, either in the last 5000 years of peat accumulation, or in peaty clays 11 000 to 9300 years old.

2. Zaire Basin

Two pollen studies in northern Zambia have failed to show any important vegetation changes over the last 20000 years (Lawton, 1963; Livingstone, 1971). At Ishiba Ngandu where gentle crustal warping dammed back the Mansha stream, 200 km west of the Nyika plateau, Livingstone found the only significant change to be a decline in evergreen forest pollen from low to even lower levels. The pollen spectra at all levels, as Meadows found in Nyika, and Tomlinson in Zimbabwe, are dominated by grasses.

The lack of change in the pollen assemblages does not necessarily signify that climate has remained unchanged. Savanna species of trees do not seem to be abundant producers of pollen, many of them being insect pollinated, and so the pollen may not represent the vegetation very satisfactorily. However the absence of Chenopodiaceae and Amaranthaceae at Ishiba Ngandu probably does indicate that northern Zambia has not been extremely arid within the last 20000 years; the absence of montane and lowland forest pollen does not necessarily mean that the rainfall has been no higher than at present. It has be be recognized that forest trees are at present mainly confined to sites such as valley heads where seepage water is available to their roots throughout the year; a more extensive forest cover might well require a year-round supply of water. The limiting factor may be not mean annual rainfall but seasonality of the rainfall. If a markedly dry season has persisted as a feature of the climate even at times when lakes have benefited from higher annual totals only limited numbers of trees, as individuals or species, may have been able to grow away from existing forest sites, even in the wetter parts of the Holocene.

The southern ends of Lake Tanganyika and Lake Rukwa both lie within 400 km of the three sites last mentioned where pollen studies have been made in Zambia and Malawi. Both lakes lie in ancient basins; gravity surveys indicate that sediments in the Rukwa basin are 6 km thick. (Sowerbutts, pers. comm. 1981). Lake Tanganyika has a very large number of endemics and the mollusca with their resemblances to marine species have long attracted attention; they probably acquired their characteristic shell forms as defences against specialized molluscan predators amongst the fishes. The floor of the lake, which goes well below sea-level, was probably faulted down through 1200 m in Plio-Pleistocene times. For much of the time since then it has probably been a closed basin, its level much lower than now though not necessarily saline. At present the discharge of the Ruzizi from Lake Kivu not only causes Lake Tanganyika to spill via the Lukuga into the Lualaba, it also dominates the great lake's water chemistry.

The history of Lake Kivu's Late Quaternary oscillations in level are much better known than those of Lake Tanganyika, largely because of a study made by Hecky and Degens (1973). Echo-sounding of the sediments has shown scores of reflecting layers. Pebbles and shell fragments dated at 13700 BP are taken to be a former beach deposit at a depth of 310 m; assuming a temperature at the time 3°C lower than now and reconstructing the paleo-hydrological conditions the authors suggest rainfall was about half its present value. By 12000 BP the lake was probably 86 m below its modern level. A shift from aragonite to monohydrocalcite in sediments 10000 years old was probably the result of freshening of the water as the lake overflowed, and the continuance of high calcium contents in the sediments after 9300 BP points to high levels persisting for several thousands of years. At the same time diatom assemblages point to a general warming in the climate, *Stephanodiscus astraea* declining as *Nitzschia fonticola* and then *N. spiculum* increased. About 5000 years ago carbonate deposition suddenly ceased, probably as a result of the lake becoming strongly stratified as dense spring water accumulated in the lowermost layers, and then mixing was re-established about 2000 years ago.

Lake Rukwa's history is much less well known, in spite of thick Quaternary beds being exposed in deep sections especially in the south of the basin. It has long been suspected that Rukwa overflowed to Tanganyika by a spillway near Karema (Stockley, 1938). Lacustrine flats at about 150 m above the present lake and at present undated probably mark this stage. It may correspond with dates of 32830 ± 930 BP (SMU–296) and a minimum of 24600 BP (A–946) obtained by Clark *et al.* (1970) from lake marls in the Songwe valley north-west of Mbeya (Fig. 8). Shell beds further downstream near Galula towards the top of terraces overlooking the river and 60 m above the present lake have given dates of 9740 ± 130 and 8060 ± 120 BP and a bed at 50 m a date of 6340 ± 50 BP. (Fig. 9) Shells from this last bed include several species which are either

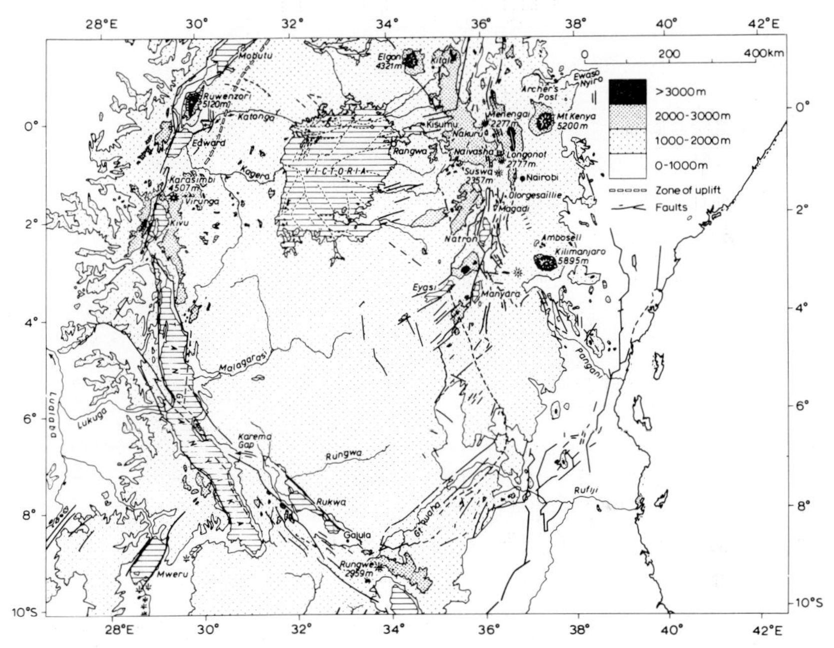

Fig. 8. The Zaire headwaters and the East African Plateau.

living in Lake Tanganyika at the present day or did so in the past. Further investigations of the sections in the lacustrine beds are likely to be productive, especially because dateable ash layers from Rungwe and Ngozi craters are widely distributed over the southern part of the basin.

3. Nile Basin

Drainage from the western shoulders of the Gregory Rift was towards the Atlantic in the Late Cenozoic, with the Kagera, Kafu and Katonga rivers continuing to flow into the Albertine rift well into Quaternary times (Fig. 8). Eventually, at some stage which has yet to be ascertained, uplift of the eastern shoulder of the Albertine rift exceeded the rate of downcutting by the three rivers and they were reversed. Lake Victoria was formed by the water ponded back and, as uplift of the rift shoulder continued, Victoria's western shore was displaced eastwards in the latter part of the Pleistocene generating a series of tilted strandlines (Bishop and Posnansky, 1960). Three sets of shorelines, at 18–20, 12–14 and 3-4 m, are horizontal, formed when tilting had ceased and an overflow to

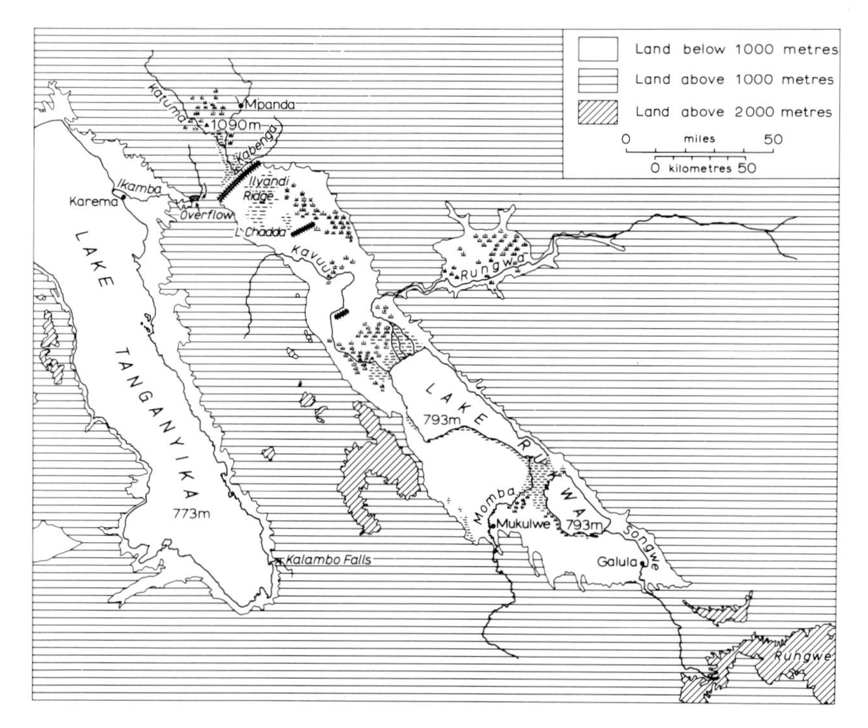

Fig. 9. The Rukwa trough. A rise in the level of Lake Rukwa since 1960 has resulted in the two lakes shown here combining to inundate the Momba delta. Late Quaternary shorelines are marked by barrier beaches in the north-western part of the Rukwa trough and by terraces containing sub-fossil mollusca overlooking the lower Rungwa and Songwe rivers. The Ilyandi ridge, and terraces at about 150 m above the lake south of Mukulwe, may be associated with a former overflow to Lake Tanganyika near Karema.

the north by way of a tributary of the Kafu was cutting down intermittently. The lowest shoreline has been dated to 3720 BP. At this stage Lake Nabugabo was formed as a lagoon on the west side, and since that time five species of *Haplochromis* endemic to Nabugabo, it is claimed, have diversified from the parent stock (Greenwood, 1965).

Livingstone (1975) found evidence that the level of Victoria was at least 75 m lower than now prior to 12500 BP. At that time the lake would have been no more than 26 m deep, probably resembling Lake Kyoga at the present day, but being a closed basin. Several cores from Pilkington Bay, just to the south of the exit of the White Nile from Lake Victoria, were examined by Kendall (1969). The longest core, was dated in 28 places giving an unusually well controlled record. Below 15·4 m, sediments in the core dating from more than 12000 BP are rich in bases indicating the lake contained a high concentration of dissolved salts, the basin was closed and that it had been for several centuries. It was open for the next 2000 years, ceased to discharge for a brief period about 10000 BP and has remained a source of Nile discharge since then. Analysis of the pollen content from Victoria cores shows a change from savanna to forest vegetation of the evergreen type corresponding with the initial rise of the lake and then, about 6000 BP, semi-deciduous species becoming more prevalent. In the past few thousands of years, forest trees have decreased in abundance, probably as a result of increased burning and cultivation, possibly combined with reduced precipitation.

Leaving Lake Victoria the White Nile enters Lake Kyoga, which is clearly the flooded headwaters of the Kafu, and flows up one of the Kafu's former tributaries on its way to the Karuma and Kabelega (Murchison) Falls and Lake Mobutu. Cores from Lake Mobutu examined by Harvey (1976) showed that it was open after 12500 BP and thus, presumably, transmitting water received from Lake Edward and Lake Victoria northwards to the White Nile and the Sudd region. Reduced alkalinity and the presence of a diatom flora between 18000 BP and 14000 BP indicated that the lake was also open at this time when most other inter-tropical African lakes seem to have been at low levels (Street and Grove, 1976; Fig. 6). It seems to have been closed from 25000 to 18000 BP and open for 2000 years before that.

Lake Manyara, which lies in a closed basin in northern Tanzania, has produced a record which is also rather anomalous. Examined by Holdship (1976), cores from the lake show alternations of zeolites and diatomites, the zeolites indicating drier conditions the diatomites wetter ones. Here it would seem that a Late Pleistocene wet period began about 22000 BP, later than elsewhere, and seems to have lasted until about 16000 years ago. Conceivably lakes within a few degrees of the equator have experienced a different climatic sequence from those in somewhat higher latitudes, or perhaps the records provided by lake cores bring out different fluctuations from those indicated by lake strandlines.

At the present day, run-off from much of south-west Ethiopia and north-west Kenya reaches Lake Turkana by the Omo and the Turkwell and Kerio rivers (Fig. 9). Earlier in the Holocene, Lake Turkana received a greater inflow and its catchment area was still more extensive. Chew Bahir, at 500 m, is normally dry at the present day but vertebrae of large Nile Perch are common along raised strandlines and in shingle spits stretching across the lake floor. Occasionally it is flooded by water carried down by the Saggan river which rises near the shores of Lake Abaya and at times it also receives an overflow from that lake. Stromatolites on spits and headlands, and algal limestone coatings on rock faces are found at heights up to 20 m above the floor of Chew Bahir and a 14C date from shells in this material has given its age as 5600 ± 100 BP (Grove *et al.*, 1975). It is likely that a lake at this level overflowed to Lake Turkana by a channel clearly visible on satellite imagery, thus extending its catchment to the source of the Billate river.

Lake Turkana's level of about 375 m has fluctuated through a range of 20 m in the course of this century (Fig. 2). To the north-west of the lake a spillway can be readily traced from near Sanderson's Gulf westwards into Equatoria Province of the Sudan Republic and then northwards to the Pibor tributary of the White Nile (Fig. 10). A rise of about 80 m would allow Lake Turkana to contribute to the Nile's discharge and judging by the incision of the channel north-west of the Lotagipi

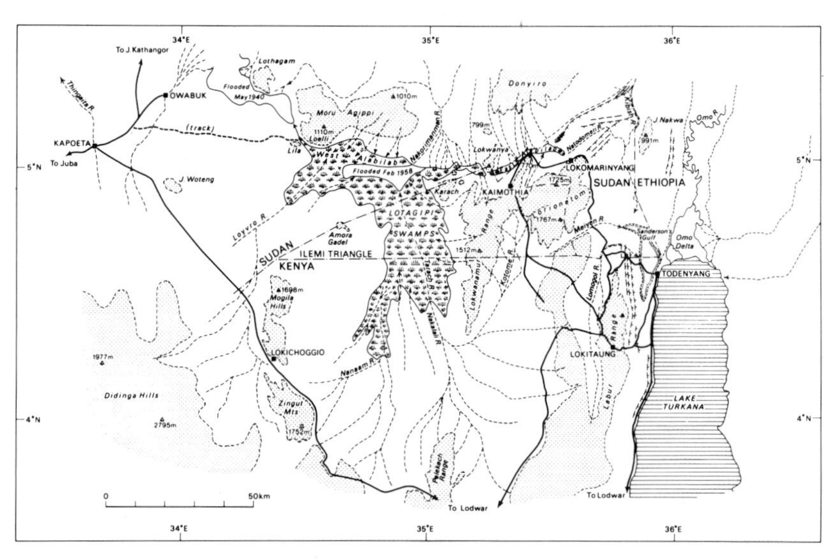

Fig. 10. The overflow from Sanderson's Gulf in the Turkana basin to the Pibor tributary of the White Nile was by way of Eastern and Western Alabilab and Jebel Kathangor.

Swamps, it may have done so over a lengthy period in the past. Butzer's (1971 and 1980) surveys of alluvial features in the vicinity of the lower Omo and Kibish show it did so during the period from 10 000 to 7000 BP, again from 6500 to 4000 BP and, briefly, about 3250 BP. Conditions seem to have hovered about the modern level between 7000 and 6500 BP and during the last 2500 years. Neither Butzer nor other investigators have identified a high stand of Turkana during the last 30 000 years of the Pleistocene.

To the south of Turkana, the Teleki volcanic barrier blocks the entry of the Suguta river to the lake. A terminal lake, Logipi, expanded far up the Suguta valley in the early Holocene and overflowed to reach Turkana by the Kerio valley (Truckle, 1976). Baringo and Bogoria were united into a single large Late Pleistocene lake and were again at high levels about 10 500 BP; there is no evidence of an overflow to the Suguta, though water may have found its way northwards underground, and may continue to do so today. Further south still, Washbourn-Kamau (1971) has calculated the various combinations of precipitation, run-off and evaporation losses, that could have been associated with the diminished area of Lake Nakuru in the Late Pleistocene and with its increased area in the Early Holocene when it stood at 180 m above the modern lake and overflowed into the Menengai caldera. The results (Table II) can be compared with those of Richardson and Richardson (1972) from Lake Naivasha. They have interpreted cores as showing high lake levels beginning about 12 500 BP and lasting until 5650 BP with the lake possibly overflowing at times to the south through Hell's Gate Canyon. The lake dried up completely for over a thousand years but formed again and has persisted through the last three millennia. At Naivasha's highest level the precipitation is calculated to have been 65% above the modern mean. There are variations in the radio carbon dates for high levels in these lakes in the eastern rift but the pattern is reasonably consistent in showing high stands in the early and middle Holocene, with an intervening drier phase and then, after a second dry phase, a few millennia less arid than the present (Fig. 6).

4. Awash Basin and Afar

Lake level fluctuations and associated environmental changes have been studied more carefully in Ethiopia than anywhere else in the Rift Valley (Fig. 11). Cores from Abhé were analysed by Gasse (1977) in association

Table II. Comparison of paleoprecipitation estimates from lakes in Eastern Africa (from Street 1979)

	Source	*Assumed temperatures*	*% Present precipitation*
Terminal Pleistocene			
Ziway-Shala	Street (1979)	3°C lower	91
		6°C lower	68
Nakuru-Elmenteita	Washbourn (1967)	3°C lower	80
Kivu	Hecky and Degens, (1973)	3°C lower	54
Victoria	Harvey (1976)	5°C lower	71
Early Holocene			
Ziway-Shala	Street (1979)	Present	147
		2°C lower	128
Nakuru-Elmenteita	Butzer *et al.* (1972)	2–3°C higher	165
Naivasha	Butzer *et al.* (1972)	Present	125
Manyara	Holdship (1976)	Present	133

with strandlines around that lake and others in Afar including Afrera. She discovered evidence for three Late Pleistocene wetter periods. The second of these came to an end about 30000 BP when the level of the Abhé lake fell and its water became more saline though it never dried up completely and by 27000 BP was high again. It reached a maximum between 25000 and 23000 BP, fell rapidly from 20000 BP onwards and by 17000 BP was dry. A lake in the Ziway-Shala basin also reached high levels between 26500 and 21000 BP, overflowing to the Awash and ultimately into Lake Abhé. The diatom flora indicates that conditions were probably somewhat cooler as well as wetter than at present.

After an arid period lasting until 12000 BP and possibly until 10000 BP, lake levels in the Ziway-Shala basin rose to the overflow level 112 m above Shala in the early Holocene and Abhé rose 160 m above its present level. At this time the area of lakes in the Awash basin exceeded 10000 km^2. The high levels were maintained for over a thousand years until 8400 BP and then there was a regression for another one or two millennia before a second maximum immediately preceding 6500 BP in the case of Abhé, and immediately following that date in the case of

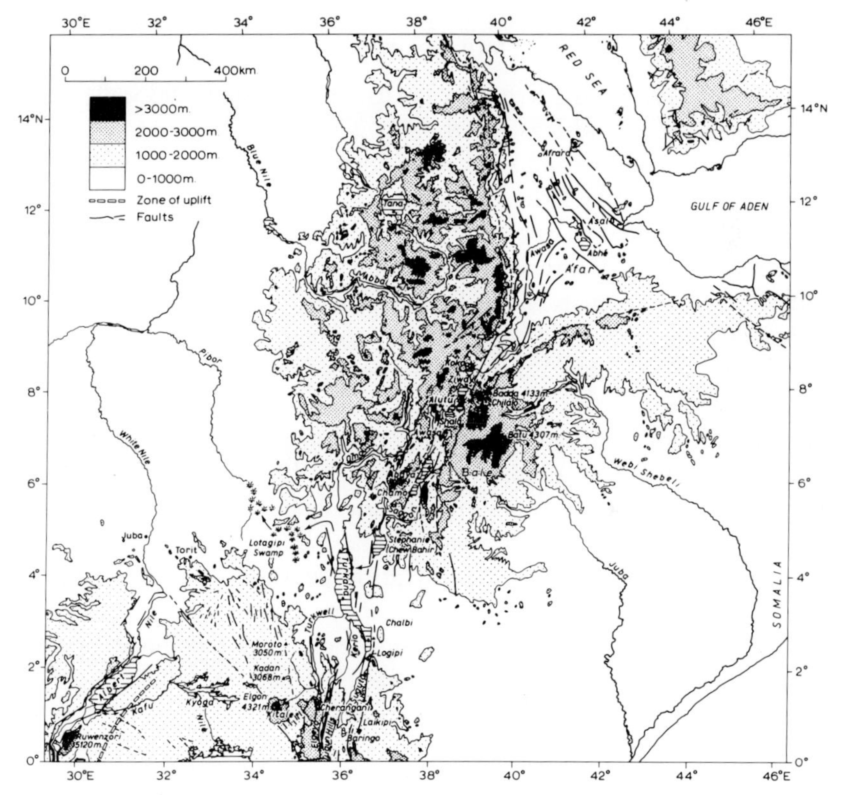

Fig. 11. Ethiopia; Afar, the Awash basin and the headwaters of the Nile.

Ziway-Shala. The late Holocene was a dry period with only short humid pulsations.

Street (1979) made careful reconstructions of the paleohydrology of the Ziway-Shala basin for different stages of the lakes. For the terminal Pleistocene minimum lake level, assuming a temperature 3°C lower than at present gave a value of 91% of the modern precipitation or for a temperature 6°C lower, precipitation of 68% of the present mean. For the early Holocene high stand, assuming modern temperatures gave a precipitation of 147% of the current mean and similarly for the Late Holocene (about 2000 BP) a value of 122%. These calculations do not take into account the possible effects of cloud cover and wind strength changing but they are conservative, in the sense that they do not take into account the losses of water from the basin to the Awash by overflows which may at times have been quite large.

MODERN TIMES

Hominoids have been present in the rift region since Miocene times, walking on their hind legs for the last 3 million years. At what stage they began to use fire and thereby modify the plant cover and run-off characteristics of very extensive areas is unknown, but probably as far back as the Middle Pleistocene. By the end of the Pleistocene men were becoming efficient hunters and fishermen but they were still not very numerous. A great increase in their effectiveness came at the end of the Holocene wet period some 5000 years ago when the megalithic monuments of western Europe were being erected; the floodwaters of the Nile were controlled for irrigation in Egypt, and pastoralists entered the area around Turkana, bringing pottery with them. They were followed about two thousand years ago by iron-working people needing charcoal, and cultivators clearing fresh areas of land annually.

The pace of change accelerated with the coming of the Europeans. Portuguese calling at East African ports en route to and from India introduced South American crops new to Africa, notably maize and cassava which greatly extended the variety and reliability of staple foodstuffs and must have allowed a great increase of populations when they were adopted in the Rift Valley region. The nineteenth century saw the introduction of firearms by Arab ivory traders based at Bagamoyo and by European hunters especially on the fringes of the Kalahari and in northern Kenya. Towards the end of the century the rinderpest epidemic decimated game animals as well as cattle, and human populations were reduced by smallpox, measles and other infectious diseases.

The recent history of land use and migration has been influenced by continued fluctuations in the climate. In the mid-eighteenth century, though Nile floods deriving from the Blue Nile were high, low-level stages in Egypt were below normal, implying that dry conditions in the Sahel at the time may have been shared by parts of East Africa (Nicholson, 1976:130). After low levels of Lake Tanganyika, and below normal low stages of the Nile in the early nineteenth century, the last three decades, until 1895, were on the whole wet; the levels of the Great Lakes were high and Lake Tanganyika commenced to overflow via the Lukuga in the 1870s. From 1895 rainfall totals were reduced for a quarter of a century with particularly severe droughts about 1898 in southern Ethiopia. During the first 60 years of this century Lake

Victoria never reached the levels it had attained towards the end of the nineteenth century. Lake Malawi ceased to overflow from 1915 to 1935. The discharge of the Nile as measured at Aswan in the first half of the century was on average less than 80% of its discharge in the last three decades of the nineteenth century. Greater aridity in the first half of the twentieth century was experienced by many tropical areas as Kraus (1955) has shown. Then, quite suddenly, after unusually heavy rains at the end of 1961, all the great lakes in East Africa rose by a metre or two and continued to rise with above average rains until 1964. All the lakes reached their highest levels this century and then declined somewhat, though generally remaining higher than they had been in the first half of the century. They fell in the early 1970s but recovered later and in 1979 Lake Malawi was higher than had ever been recorded this century (Fig. 3).

The higher lake levels since the middle of the century have not been an unmitigated blessing. In the Rukwa basin, for example, the greatly increased size of the Lake has involved the submergence of grassland and a great reduction in the antelope population (Gunn, 1973). The main feeding grounds of flamingoes have shifted in the wet periods from Nakuru to Bogoria and Baringo (Vareschi, 1978). Droughts have continued, those of the early 1970s in Tanzania and Ethiopia, and of 1980 in Turkana were particularly severe. They have accentuated the processes which are involved in the desertification syndrome.

The degradation of the environment in the dry parts of the world which has been the subject of so much attention in recent years is the outcome of increased human populations and consumption of materials of all kinds. In Kenya the population is increasing at 4% annually and in the Rift Region as a whole one might guess that there are now about 5 times as many people as there were at the beginning of the century and that domestic animals have increased in a similar proportion. Pereira (1962) brought together the hydrological effects of the changes in land use in some East African catchment areas. Some of the results of erosion for Tanzania have been assembled by Rapp *et al.* (1972). Three symposia held in Gaberone (1971, 1976, 1979) have considered the problems of managing the Kalahari and Okavango environments.

In addition to inadvertent influences on the environment there have been deliberate interventions by man. Kariba and Caborabassa dams, both on the Zambezi, are amongst the most important instances: the Owen Falls dam on the White Nile and the Shire barrage below Lake

Malawi are less conspicuous but still important influences on the hydrology of the rivers and lakes they control. Many reservoirs formed by smaller dams are filling with sediment resulting from accelerated erosion in deforested headwaters.

The high mountain environments are particularly precious and susceptible to change. Ever since the highest peaks were first climbed less than a century ago their glaciers have been in retreat and several have been reduced to fragments (Messerli, 1980). For a short time about 1961 there were reports of advances (Temple, 1968), and between 1963 and 1974 the upper reaches of the Lewis Glacier on Mount Kenya showed signs of thickening (Young, 1980), but retreat has been dominant, probably as a result of global warming in the first half of the century. Much more serious has been the destruction of forest below about 3000 m for timber and for agriculture. Patterns of population and associated distributions of plants and wildlife are taking place as people concentrate in urban areas and as villagization transforms the rural landscape of entire tribal areas and countries, Tanzania in the 1970s, Kikuyuland in the 1950s, within a decade. Game parks have been established, but reserves are far from being sacrosanct and the disruption and population movements accompanying civil wars have affected great tracts of country since independence came within the last quarter of a century.

BIOGEOGRAPHICAL IMPLICATIONS

Broad similarities between life forms in Africa and some of the other continents, especially South America, are no doubt associated with their having been joined together until towards the end of Mesozoic times. The isolation of the continent from the rest of the world's land areas for 25 million years, when the mammals were rapidly evolving in the middle of the Cenozoic, allowed the continent to acquire its own fauna by the Miocene, at a time when it was less desertic than now and rain forests occupied vast low-level plains. The creatures involved included our hominoid ancestors.

Contact with Eurasia in the middle of the Miocene coincided with violent tectonics, increased relief and more diversified climates and vegetation. Evidence of the faunal changes associated with immigrants from Europe and Asia was preserved in volcanic and lacustrine sediments within the rift system and in some cases has been revealed by renewed faulting in the Plio/Pleistocene and Pleistocene. The rift valleys with their semi-arid floors and forested shoulders provided routeways for the migrations of plants and animals southwards from temperate latitudes and northwards from southern Africa.

There is little indication that the Pliocene was a time of extraordinary aridity in the Rift region. Signs of climatic fluctuations appear in the Plio-Pleistocene but become more apparent towards the end of the Quaternary period. There is some evidence that lakes were more extensive in the period 30000 to 20000 BP than at any stage since. Late Pleistocene aridity must have extinguished the freshwater fauna of tropical latitudes and reduced the area under rain forest quite drastically. When conditions wetter than those of the present were re-established at the very end of the Pleistocene rain forest may have been able to recolonize certain areas with two rainy seasons such as southern Uganda from forest refuges. In other areas, the continued long dry season may have prevented expansion of forest into areas receiving high mean annual rainfall totals. Lakes expanded in the Early and Middle Holocene integrating drainage systems, and causing certain areas to drain to the sea instead of to closed basins. This explains many features of freshwater fish and mollusc distributions.

The return of drier conditions about 5000 years ago was accompanied by the extension of pastoralism and agriculture. Climates ameliorated for a time about 3000 years ago and have continued to fluctuate since then, but the main changes in the landscape over the last century or two have been caused by growing human populations, deterioration of the plant cover and increased erosion.

ACKNOWLEDGEMENTS

I wish to thank R.B. Owen, R.W. Renaut and J.W. Pallister for their useful comments, and A. Shelley for drawing the maps and diagrams. Some of the data on which this survey is based have been obtained in the course of research financed by the Natural Environment Research Council, the Royal Society and Royal Geographical Society.

REFERENCES

Baker, B.H. (1967). "Geology of the Mount Kenya Area". [Report no. 79, Geological Survey Kenya.] Ministry of Natural Resources, Nairobi.

Baker, B.H., Mohr, P.A. and Williams, L.A.J. (1972). Geology of the Eastern Rift System of Africa. *Geol. Soc. Am. spec. Paper* **136**, 1–67.

Bishop, W.W. (1971). Late Cenozoic History of East Africa in Relation to Hominoid Evolution. *In* "Late Cenozoic Glacial Ages" (K.K. Turekian, ed.), pp. 493–528. Yale University Press, New Haven and London.

Bishop, W.W. (ed.) (1978). "The Geological Background to Fossil Man". Scottish Academic Press, Edinburgh.

Bishop W.W. and Posnansky, M. (1960). Pleistocene environments and early man in Uganda. *Uganda. J.* **24**, 44–61.

Bishop, W.W. and Trendall, A.F. (1967). Erosion surfaces, tectonics and volcanic activity in Uganda. *Q. Jl. geol. Soc. Lond.* **122**, 385–420.

Bond, G. (1963). Pleistocene environments in southern Africa. *In* "African Ecology and Human Evolution" (F.C. Howell and F. Bourlière, eds), pp. 308–334. Viking Fund Publications in Anthropology, Aldine Press, Chicago.

Bond, G. (1967). River valley morphology, stratigraphy and paleoclimatology in southern Africa. *In* "Background to Evolution in Africa" (W.W. Bishop and J.D. Clark, eds), pp. 303–312. University of Chicago Press, Chicago.

Bonnefille, R. (1976a). Palynological evidence for an important change in the vegetation of the Omo basin between 2·5 and 2 million years ago. *In* "Earliest man and environments in the Lake Rudolf basin" (Y. Coppens, F.C. Howell, G.H. Isaac and R.E.F. Leakey, eds, pp. 421–31. University of Chicago Press, Chicago.

Bonnefille, R. (1976b). Vegetations et climats des temps olduwayens et acheuléens à Melka—Kunturé (Ethiopia). *In* "L'Ethiopie avant l'histoire", vol. 1 (J. Chavaillon, ed.), pp. 55–71).

Botswana Society (1971). Proceedings of the Conference on sustained production from semi-arid areas with particular reference to Botswana. *Botswana Notes and Records, Special Edition* No. 1. [Mimeographed.]

Botswana Society (1976). "Proceedings of the Symposium on the Okavango Delta and its future utilisation". National Museum, Gaborone.

Botswana Society, (1978). "Symposium on Drought in Botswana" (M.T. Hinchey, ed.). Clark University Press, Worcester, U.S.A.

Butzer, K.W. (1971). "Recent history of an Ethiopian delta". [Department of Geography Research Paper No. 136.] Chicago University.

Butzer, K.A. (1980). The Holocene Lake Plain of North Rudolf, East Africa. *Physical Geogr.* **1**, 42–58.

Cerling, T.E., Hay, R.L. and O'Neil, J.R.O. (1977). Isotopic evidence for dramatic climatic changes in East Africa during the Pleistocene. *Nature, Lond.* **267**, 137–138.

Clarke, J.D., Haynes, C.V., Mowby, J.E. and Gautier, A. (1970). Interim report on palaeoanthropological investigations in the lake Malawi Rift. *Quaternaria* **13**, 305–354.

Coetzee, J.A. (1964). Evidence for a considerable depression of the vegetation belts during the upper Pleistocene on the East Africa mountains. *Nature, Lond.* **204**, 564–566.

Cooke, H.B.S. (1958). Observations relating to Quaternary environment in east and southern Africa. *Bull. geol. Soc. S. Africa* **20** (Annex), 1–74.

Dixey, F.H. (1956). The East African Rift System. *Bull. Colonial Geol. Mining Research, Suppl.* **1**, 1–71.

Downie, C. (1964). Glaciations of Mount Kilimanjaro, northeast Tanganyika. *Bull. geol. Soc. Am.* **75**, 1–16.

Downie, C. and Wilkinson, P. (1972). "The Geology of Kilimanjaro". University Department of Geology, Sheffield.

Dunne, T. (1979). Sediment yield and land use in tropical catchments. *J. Hydrology* **42**, 281–300.

Ebert, J.I. and Hitchcock, R.K. (1978). Ancient lake Makgadikgadi, Botswana; mapping, measurement and paleoclimatic significance. *Palaeoecology* **10**, 47–56.

Faure, H., Manguin, E. and Nydal, R. (1963). Formations lacustres du Quaternaire supérieur du Niger oriental: diatomites et âges absolus. *Bull. Bur. Rech. géol. min. Paris* **3**, 41–63.

Fitch, F.J. and Vondra, C.F. (1976). Tectonic and stratigraphic framework. *In* "Earliest Man and environments in the Lake Rudolf basin" (Y. Coppens, E.C. Howell, G. Ll. Isaacs and R.E.F. Leakey, eds), pp. 8–12. University of Chicago Press, Chicago.

Flenley, J.R. (1979). "The equatorial rain forest: a geological history". Butterworths, London.

Flint, R.F. (1959). On the basis of Pleistocene correlation in East Africa. *Geol. Mag.* **96**, 265–84.

Gardner, J.W. and Hays, J.D. (1976). Responses of sea surface temperature and circulation to global climatic change during the past 200,000 years in the eastern equatorial Atlantic. *Mem. geol. Soc. Am.* **145**, 221–246.

Gasse, F. (1977). Evolution of Lake Abhé (Ethiopia and TFAI) from 70,000 B.P. *Nature, Lond.* **264**, 42–45.

Gasse, F., Rognon, P. and Street, F.A. (1980). Quaternary history of the Afar and Ethiopian R.H. lakes. *In* "The Sahara and the Nile." (M.A.J. Williams and H. Faure, eds), pp. 361–400. Balkema, Rotterdam.

Greenwood, P.H. (1965). The cichlid fish of Lake Nabugabo, Uganda. *Bull. Br. Mus. nat. Hist., Zool.* **12**, 315–57.

Grey, D.R.C. and Cooke, H.J. (1977). Some problems in the Quaternary evolution of the landforms of northern Botswana. *Catena* **4**, 123–133.

Griffiths, J.F. (ed.) (1972). "Climates of Africa". Elsevier, Amsterdam.

Grove, A.T. (1969). Landforms and climatic change in the Kalahari and Ngamiland. *Geogrl. J.* **135**, 191–212.

Grove A.T., Street, F.A. and Goudie, A.S. (1975). Former lake levels and climatic change in the Rift Valley of Southern Ethiopia. *Geogrl. J.* **141**, 177–202.

Gunn, D.L. (1973). Consequences of cycles in East African climate. *Nature Lond.* **242**, 457.

Hamilton, A.C. (1972). The interpretation of pollen diagrams from highland Uganda. *Palaeoecology of Africa* **7**, 45–149.

Hamilton, A.C. (1977). An upper Pleistocene pollen diagram from highland Ethiopia. *Abstracts INQUA Congress, Birmingham* **10**, 193.

Hamilton, A.C. and Perrott, A. (1978). Aspects of the glaciation of Mt. Elgon, East Africa. *Palaeoecology of Africa* **11**, 153–161.

Harvey, T.J. (1976). "The paleolimnology of Lake Mobutu Sese Seko Uganda-Zaire: the last 28000 years". Doctoral dissertation, Duke University.

Hay, R.L. (1971). "Geology of the Olduvai Gorge". University of California Press.

Hecky, R.E. and Degens, E.T. (1973). "Late Pleistocene—Holocene chemical stratigraphy and paleoliminology of the rift valley lakes of central Africa". Woods Hole Oceanographic Institution.

Heine, K. (1978). Radiocarbon chronology of late Quaternary lakes in the Kalahari southern Africa. *Catena* **5**, 145–149.

Holdship, S.A. (1976). "The paleoliminology of Lake Manyara, Tanzania: a diatom analysis of a 56 metre sediment core". Doctoral dissertation, Duke University.

Hsu, K.J., Montadert, L., Bernoulli, D., Cita, M.B., Erickson, A., Garrison, R.E., Kidd, R.B., Melières, F., Muller, C. and Wright, R. (1977). History of the Mediterranean salinity crisis. *Nature, Lond.* **254**, 207–209.

Isaac, G. Ll. (1975). Sorting out the Muddle in the Middle. *In* "After the Australopithecines" (K.W. Butzer and G.L. Isaac, eds), pp. 875–87. Mouton, The Hague.

Isaac, G. Ll. (1977). "Olorgesailie; archeological studies of a middle Pleistocene Lake Basin in Kenya". Chicago University Press, Chicago.

Kendall, R.L. (1969). An ecological history of the Lake Victoria basin. *Ecol. Monogr.* **39**, 121–76.

Kent, P.E. (1971). "The geology and geophysics of coastal Tanzania". [*Geophysical Paper No. 6. Institute of Geological Sciences.*] HMSO, London.

King, B.C. (1978). Structural and volcanic evolution of the Gregory Rift Valley. *In* "Geological Background to Fossil Man" (W.W. Bishop, ed.), pp. 29–54. Scottish Academic Press, Edinburgh.

King L.C. (1962). "The Morphology of the Earth". Oliver and Boyd, London.

King, L.C. (1978). The geomorphology of central and southern Africa. *In* "Biogeography and ecology of southern Africa" (M.J.A. Werger, ed.), pp. 3–17. Junk, The Hague.

Kraus, E.B. (1955). Secular changes in tropical rainfall regions. *Q. J. R. meteorol. Soc.* **81**, 198–210.

Lancaster, I.N. (1979). Quaternary environments in the arid zone of southern Africa. [*Environmental Studies, Occasional Paper* no. 22.] University of Witwatersrand, Johannesburg.

Lancaster, I.N. (1980). Relationships between altitude and temperature in Malawi. *South African Geogrl. Jl.* **62**, 90–97.

Laury, R.L. and Albritton, C.C. Jnr. (1975). Geology of Middle Stone Age archeological sites in the Main Ethiopian Rift Valley. *Bull. geol. Soc. Am.* **86**, 999–1011.

Lawton, R.M. (1963). Palaeoecological studies and ecological studies in the Northern Province of Northern Rhodesia, *Kirkia* **3**, 46–77.

Lind, E.M. and Morrison, M.E.S. (1974). "East African Vegetation". Longman, London.

Livingstone, D.A. (1962). Age of deglaciation in the Ruwenzori range, Uganda. *Nature, Lond.* **194**, 859–60.

Livingstone, D.A. (1967). Postglacial vegetation of the Ruwenzori Mountains in Equatorial Africa. *Ecol. Monogr.* **37**, 25–52.

Livingstone, D.A. (1971). A 22000 year pollen record from the plateau of Zambia. *Limnol. Oceanogr.* **16**, 349–356.

Livingstone, D.A. (1975). Late Quaternary climatic change in Africa. *A. Rev. Ecol. Syst.* **6**, 249–280.

Livingstone, D.A. (1980). Environmental changes in the Nile headwaters. *In* "The Sahara and the Nile" (M.A.J. Williams and H. Faure, eds), pp. 339–360. Balkema, Rotterdam.

Mahaney, W.C. (1978). Quaternary stratigraphy of Mount Kenya: a reconnaissance. *Palaeoecology of Africa* **11**, 163–170.

Meadows, M.E. (1982). "The past and present day environments of the Nyika Plateau, Malawi". Ph.D. Dissertation, Cambridge University.

Messerli, B. (1980). Mountain glaciers in the Mediterranean area and in Africa. *World Glacier Inventory IAHS Publ.* No. **126**, 197–211.

Mohr. P.A. and Potter, E.C. (1976). The Sagatu ridge dike swarm, Ethiopian rift margin. *J. Volcanol. geotherm. Res.* **1**, 55–71.

Nicholson, S.E. (1976). "A climatic chronology for Africa". D.Phil. dissertation, University of Wisconsin.

Nilsson, E. (1940). Ancient changes of climate in British East Africa and Abyssinia. Stockholm. *Geogr. Ann.* 22:1–79.

Norton, J.O. and Sclater, J.G. (1979). A model for the evolution of the Indian Ocean and the break-up of Gondwanaland. *J. Geophys. Res.* **84**, 6803–6830.

Osmaston, M.A. (1965). "The past and present climate and vegetation of Ruwenzori and its neighbourhood". Ph.D. thesis, Oxford University.

Owen, R.B. and Renaut, R.W. (1981). Paleoenvironments and Sedimentology of the Middle Pleistocene Olorgesailie Formation, southern Kenya Rift Valley. *Palaeoecology of Africa* **13**, 147–174.

Pallister, J.W. (1956). Slope development in Buganda. *Georg. Jl.* **122**, 80–87.

Patterson, B. Behrensmeyer, A.K. and Sill, W.D. (1970). Geology and fauna of a new Pliocene locality in north western Kenya. *Nature, Lond.* **226**, 918–21.

Pereira, H.C. (ed. 1962). Hydrological effects of changes in land use in some East African catchment areas. *E. Afr: Agr. For. J. Special Issue* **27**. Nairobi.

Pickford, M.H.L. (1978a). Geology, palaeoenvironments and vertebrate faunas of the mid-Miocene Ngorora Formation, Kenya. *In* "Geological Background to Fossil Man" (W.W. Bishop, ed.), pp. 237–262. Scottish Academic Press, Edinburgh.

Pickford, M.H.L. (1978b). Stratigraphy and mammalian palaeontology of the late-Miocene Lukeino Formation, Kenya. *In* "Geological Background to

Fossil Man" (W.W. Bishop, ed.), pp. 263–278. Scottish Academic Press, Edinburgh.

Rapp, A., Berry, L. and Temple, P.H. (1972). Soil erosion and sedimentation in Tanzania. *Geograf. Annlr* **54A**, 105–379.

Reeves, C.V. (1972). Rifting in the Kalahari? *Nature, Lond.* **237**, 95–96.

Richardson, J.L. and Richardson, A.E. (1972). History of an African Rift lake and its climatic implications. *Ecol. Monogr.* **42**, 499–534.

Scott, S.C. (1980). Geology of Longonot volcano, central Kenya. *Phil. Trans. R. Soc.* **296**, 437–465.

Shackleton, R.M. (1978). Geological map of the Olorgesailie Arca, Kenya. *In* "Geological Background to Fossil Man" (W.W. Bishop, ed.), pp. 171–172. Scottish Academic Press, Edinburgh.

Shroder, J.F. (1976). Mass movements on the Nyika Plateau, Malawi. *Z. Geomorph.*, N.F. **20**, 56–77.

Stockley, G.M. (1958). "The geology of parts of the Tabora, Kigoma and Ufipa Districts, N.W. Lake Rukwa". Geological Survey, Tanganyika.

Street, F.A. (1979). "Late Quaternary lakes in the Ziway-Shala Basin, Southern Ethiopia". Ph.D. Dissertation, University of Cambridge.

Street, A.F. and Grove, A.T. (1976). Environmental and climatic implications of Late Quaternary lake level fluctuations in Africa. *Nature, Lond.* **261**, 285–327.

Temple, P.H. (1968). Further observations on the glaciers of the Ruwenzori. *Geograf. Annlr.* **50A**, 136–150.

Tomlinson, R.W. (1973). The Inyanga Area. An essay in regional biogeography. *Occ. Pap. Univ. Rhodesia.* **1**, 1–67.

Tomlinson, R.W. (1974). Preliminary biogeographical studies on the Inyanga Mountains, Rhodesia. *African Geogrl. Jl.* **56**, 15–26.

Tricart, J. (1956). Tentative de corrélation des périodes pluviales africaines et des périodes glaciares. *Comptes rend. Summ. Soc. geol. Fr.* **9–10**, 164–167.

Truckle, P.H. (1976). Geology and late Cainozoic lake sediments of the Suguta Trough, Kenya. *Nature, Lond.* **263**, 380–383.

Van Couvering, J. and van Couvering J. (1976). Early Miocene mammal faunas from East Africa: aspects of geology, faunistics and palaeoecology. *In* "Human Origins" (G. Isaac and E, McCown, eds), pp. 155–207. W.A. Benjamin, Menlo Park, Calif.

Vareschi, E. (1978). The ecology of Lake Nakuru (Kenya). The abundance and feeding of the lesser flamingo. *Oecologia* **32**, 11–35.

Washbourn, C.K. (1967). Lake levels and Quaternary climates in the eastern Rift Valley of Kenya. *Nature Lond.* **216**, 672–3.

Washbourn-Kamau, C.K. (1971). Late Quaternary lakes in the Nakuru—Elmenteita Basin, Kenya. *Geogr. J.* **137**, 522–35.

White, F. (1978). The Afro-montane Region. *In* "Biogeography and Ecology of Southern Africa" (M.J.A. Werger, ed.), pp. 463–513. Junk, The Hague.

Williams, L.A.J. (1978). Character of Quaternary volcanism in the Gregory Rift Valley. *In* "Geological Background to Fossil Man" (W.W. Bishop, ed.), pp. 55–70. Scottish Academic Press, Edinburgh.

Williams, M.A.J. and Faure, H. (eds) (1980). "The Sahara and the Nile". Balkema, Rotterdam.

Wright, E.P. (1978). Geological studies in the northern Kalahari. *Geogrl. J.* **144**, 235–249.

Young, J.A.T. (1980). The glaciers of East Africa. *World Glacier Inventory. AIHS Publ.* no. **26**, 213–217.

Zinderen Bakker, E.M. van (1962). A late-glacial and post-glacial climatic correlation between East Africa and Europe. *Nature, Lond.* **194**, 202–3.

6 Endemism and Cosmopolitanism in the Diatom Flora of the East African Great Lakes

R. ROSS★

Department of Botany, British Museum (Natural History), London SW7 5BD, UK

Abstract: The known diatom flora of the East African Great Lakes consists of about 380 species. 100 of these are not known to occur outside Africa and 36 have been recorded from the Asiatic or American tropics or from both; the remainder, almost 250 species (64%), extend into the temperate regions, the majority, and probably almost all, being cosmopolitan. None of the currently recognized genera are confined to Africa. The proportion of the species present that are confined to Africa is much higher in the Surirellaceae than in any other group, except for a few genera represented in the East African Great Lakes by very few species. The distribution patterns show that many species of diatom have very great ecological tolerance, that they are capable of long-distance dispersal, and that speciation in general is slow. It is suggested that these characteristics are a consequence of their relatively low level of complexity. The reasons for the much greater rate of speciation in the Surirellaceae and, to a lesser extent in *Nitzschia*, are obscure; adaptive radiation does not seem to be the explanation.

INTRODUCTION

The biogeographical patterns exhibited by the diatoms are very different from those of many of the groups considered in this symposium, and

★ *Present address*: Garden House, Evesbatch, Bishop's Frome, Worcester WR6 5BE, UK.

Systematics Association Special Volume No. 23, "Evolution, Time and Space: The Emergence of the Biosphere", edited by R.W. Sims, J.H. Price and P.E.S. Whalley, 1983, pp. 157–177. Academic Press, London and New York.

those of the East African Great Lakes present a marked contrast to the fishes of that region dealt with in another paper. In this contribution, the data that are available about the distribution of the species and genera constituting the diatom flora of those lakes are presented and analysed, and the implications considered. As a basis, I have listed all the species of diatom known to occur in Lakes Malawi (Nyassa), Tanganyika, Kivu, Edward, Mobuto Sese Seko (Albert), Victoria and Turkana (Rudolf), using the published records (Dickie, 1879; Müller, 1904a,b, 1905, 1910; West, 1907, 1909; Ostenfeld, 1908, 1909; Schröder, 1911; Virieux, 1913; Woloszynska, 1914; Hustedt, 1921, 1949; Erlandsson, 1928; Bachmann, 1933, 1939; Rich, 1933; Zanon, 1938; Cholnoky, 1954; Van Meel, 1954; Kufferath, 1956, and those in Schmidt, 1874–1959, and Huber-Pestalozzi, 1942), with my own studies of material from Lake Victoria and, to a lesser extent, Lake Tanganyika. I have classified each species into one or other of three biogeographical groups: "cosmopolitan", "pan-tropical" and African. They have also been sub-divided ecologically into planktonic, attached, and benthic forms, and the Surirellaceae have been segregated from the last group because their distribution patterns are very different.

LIMITATIONS OF THE DATA

1. *Taxonomy*

In compiling the list, it has been necessary to make a number of decisions about taxonomy on a basis that one would not normally consider adequate. Thus, I have considered the diatom described by Müller (1904, b) as *Melosira distans* var. *africana* O. Müll. as a species separate from *M. distans* (Ehrenb.) Kütz., and hence one confined to Africa, on the basis of his description and figure without seeing a specimen. Similarly, I have recorded *Cymbella grossestriata* O. Müll. as a species confined to Africa because Frenguelli's (1938) description and figure of *C. grossestriata* var. *recta* Freng. from Argentina indicate, in my opinion, that it is not conspecific with the African diatom.

Furthermore, the general state of diatom taxonomy is such that failure to recognize the extent of variability within species may have led to some inflation of the list by the inclusion of a species under two or more synonyms, or to the listing of species as from Africa only under a name that is a synonym of a species with wider distribution. Had it not been for very recent studies of one section of the larger genus *Nitzschia* (Lange-Bertalot, 1976; Lange-Bertalot and Simonsen, 1978) 76 species instead of 59 would have been recorded as occurring in the East African Great Lakes, with 28 instead of 23 confined to Africa. *Nitzschia*, however, is taxonomically a particularly difficult genus, and it is most unlikely that any other genera are in need of such drastic revision.

It is also possible that some diatoms from the East African Great Lakes have been misidentified as species occurring elsewhere, and hence that a few that I have classed as cosmopolitan or pan-tropical are really confined to Africa. Even if this is so, however, the number involved is undoubtedly very small.

In the material from Lake Victoria that I have examined there are some 30 species that I have not yet identified. In the case of a very few of these I am certain that they are undescribed, and I have included them in the list. Some of the remainder may be extremes of variation of species recorded from the lake, or be representatives of species described from eslewhere, whilst others may well be undescribed and confined to Africa, but the proportions in these various categories are very uncertain.

2. Geographical Distributions

The diatom floras of the world are far from adequately known. Europe, North America, Japan, South Africa and Argentina are the only areas the freshwater diatom flora of which can be regarded as reasonably well covered. It is not unusual for a species of diatom to occur in a limited number of localities widely distributed over the world, and hence remarkable extensions of known ranges are recorded from time to time. For this reason it has seemed justifiable to class as "cosmopolitan" not only the species known to have a world-wide distribution but also the small number found in the East African Great Lakes and also in a few temperate localities outside Africa. Amongst these latter, there may be a few species that do not in fact extend to America or Australasia, or may even be confined to Africa and Europe, but the failure to distinguish these few from those that are truly cosmopolitan is not of any great significance for the analysis undertaken here. It is certain that no particular genus or higher taxon is predominant among them.

For similar reasons I have grouped together as "pan-tropical" 19 species known only from Africa and tropical Asia, seven known only from Africa and tropical America, and ten known from the tropics of all three continents. The diatom floras of tropical Asia, and even more tropical America, are so poorly known that one can have little confidence that species known from the tropics of two continents do not occur in the third. Some, too, of the species currently known only from Africa will no doubt be found to have ranges extending outside that continent.

As far as the East African Great Lakes themselves are concerned, our knowledge of their diatom flora is manifestly incomplete. The most extensive published lists are those for Lake Malawi (Müller, 1904a,b, 1905, 1910), Lake Kivu (Erlandsson, 1928; Zanon, 1938; Hustedt, 1949) and Lake Edward (Hustedt, 1949). There are only very short lists for Lake Mobuto Sese Seko (West, 1909; Bachmann, 1933) and Lake Turkana (Rich, 1933; Bachmann, 1939), whilst the published lists for Lake Tanganyika (van Meel, 1954, which includes all previous records; Kufferath, 1956) and Lake Victoria (West, 1907; Ostenfeld, 1908, 1909; Virieux, 1913; Woloszynska, 1914; Hustedt, 1921; Bachmann, 1933; Rich, 1933) are intermediate in length. In spite of the considerable number of papers listing diatoms from Lake Victoria, of the 188 species that I have identified in my own material, 125 (66%) had not been recorded previously from the lake. That many more species will be found in the East African Great Lakes is certain, but to what extent, if any, the distribution patterns of these will be found to alter the picture given by those already known cannot be predicted.

3. *Ecological Categories*

The main difficulty that arises in placing species of diatom into one of the three broad ecological groups adopted in this discussion is in deciding for some species whether they should be classed as benthic or planktonic. Both benthic and attached species are often found in plankton samples, having been raised into the water column by turbulence. A number of these can not only survive suspended but also multiply considerably, to such an extent that they can form a prominent part of the content of the sample. It is for the most part not difficult, in fresh waters, to decide whether the primary habitat of such a species is or is not the plankton, and this is made easier by the fact that, in almost every genus, the species all belong to the same habitat group. The principal exception to this is *Nitzschia*, many species of which are apparently truly planktonic whilst others are definitely benthic. This is the only genus in which, in the present study, the species have been divided between two ecological classes, but the information available about a number of them is not sufficient for the allocation to be made with any real certainty.

Although the large genus *Melosira* contains benthic as well as planktonic species, all those found in the East African Great Lakes belong to the group that has been segregated as a separate genus under the name *Aulacoseira* by Simonsen (1979), and all members of this taxon are planktonic. The correct application of the generic name *Aulacoseira* is currently in doubt, hence the wording of the above sentence.

Many of the species of *Surirella* found in the East African Great Lakes have been classed as planktonic (Huber-Pestalozzi, 1942), but it would seem that all species of this genus are primarily benthic forms, although many of them can maintain themselves and multiply when suspended by turbulence.

The only other likely error in the ecological classification is that a few species of *Cymbella* should probably be classed as benthic rather than attached, but this possibility only arises for about four of the 21 species found in the East African Great Lakes.

DISTRIBUTION PATTERNS

The distribution patterns of the species found in the East African Great Lakes are summarized in Table I. From this it will be seen that, of the 382 so far recorded from the lakes, almost two-thirds have been classed as cosmopolitan and that only about one-quarter are confined to Africa. The proportion of species not found outside that continent is, however, higher amongst the planktonic species (39%) than amongst the attached (19%) or the benthic other than the Surirellaceae (17%). The Surirellaceae, on the other hand, have a very different pattern of distribution, with two-thirds of the species confined to Africa. The number of species classed as "pan-tropical" is small in all ecological groups; this group is more strongly represented in the large genus *Navicula* than elsewhere, but even in that genus it only reaches 16% against an overall proportion of 9%.

The species confined to Africa are listed in Table II, which gives details of the Great Lakes in which they have been recorded, and also of the other parts of Africa where they are known to occur. Many of them

Table I. Distribution patterns of diatoms occurring in the East African Great Lakes: numbers of species in biogeographical groups.

	Cosmopolitan	*Pan-tropical*	*Africa*	*Total*
Plankton				
Actinocyclus	1			1
Asterionella	2			2
Cyclotella	7			7
Melosira	3	2	5	10
Meridion	1			1
Nitzschia	11	2	12	25
Odontella	1			1
Rhizosolenia	1		2	3
Stephanodiscus	3		1	4
Thalassiosira			2	2
Total	30(54%)	4(7%)	22(39%)	56
Attached				
Achnanthes	9	2	1	12
Berkeleya	1			1
Cocconeis	3			3
Cymbella	19	2	2	23
Cymbellonitzschia			1	1
Diatoma	3			3
Epithemia	4	1		5
Eunotia	8	1		9
Fragilaria	11		3	14
Gomphocymbella			2	2
Gomphoneis	1			1
Gomphonema	10	1	5	16
Gomphonitzschia			2	2
Hannaea	1			1
Mastogloia	2			2
Rhoicosphenia	1			1
Rhopalodia	3		4	7
Synedra	8	1	2	11
Tabellaria	2			2
Total	86(74%)	8(7%)	22(19%)	116
Benthos				
Amphora	5	1		6
Anomoeoneis	1			1
Brachysira	2			2
Caloneis	6	1		7
Capartogramma		1	3	4
Denticula	1			1
Diploneis	2	1		3

Table I (*cont.*)

	Cosmopolitan	*Pan-tropical*	*Africa*	*Total*
Frustulia	2		1	3
Gyrosigma	4			4
Hantzschia	1	1		2
Navicula	48	11	9	68
Neidium	4		1	5
Nitzschia	22	1	11	34
Pinnularia	18	2	2	22
Stauroneis	3	1	1	5
Total	119(71%)	20(12%)	28(17%)	167
Surirellaceae				
Campylodiscus			1	1
Cymatopleura	2		2	4
Stenopterobia	1			1
Surirella	8	4	25	37
Total	11(26%)	4(9%)	28(65%)	43
Overall Total	246(64%)	36(10%)	100(26%)	382

have also been found in smaller lakes, rivers and streams in the general area of the Great Lakes from Malawi and Zambia to Ethiopia, but this information has not been separately recorded. The information about distribution in Africa has been drawn from a variety of sources, the most important being Forti (1914), Zanon (1941), Cholnoky (1953, 1954, 1957a,b, 1958a,b,c, 1966, 1968), Woodhead and Tweed (1956, 1960), Compère (1967, 1975, 1980) and Maillard (1977). Forty-one of the 95 species confined to Africa occur outside the region of the Great Lakes, the proportion of the planktonic species amongst these being high (59%) and of the Surirellaceae low (28%).

A third of the species confined to Africa belong to genera represented in the Great Lakes primarily by cosmopolitan species. It is the remainder that call for comment.

1. *Plankton*

Melosira is represented in the East African Great Lakes by ten species, all belonging to the group segregated by Simonsen (1979) as a separate genus, under the name *Aulacoseira* (see above, p. 161). This taxon consists entirely of freshwater planktonic species. Five of these ten species are confined to Africa, with three of them extending outside the area of the Great Lakes.

Table II. Distribution in Africa of species occurring in the East African Great Lakes and confined to that continent.

	Lake Malawi	*Lake Tanganyika*	*Lake Kivu*	*Lake Edward*	*Lake Mobuto Sese Seko*	*Lake Victoria*	*Lake Turkana*	*Elsewhere in Africa*
Plankton								
Melosira								
argus O. Müll.	+	−	+	−	−	−	−	11
ikapoensis O. Müll.	+	−	−	−	−	+	−	1,5,9
nyassensis O. Müll.	+	−	−	−	+	+	−	1,5,8
pyxis O. Müll.	−	+	−	−	−	−	−	
sp. (*distans* var. *africana* O. Müll.)	−	−	−	+	−	+	−	
Nitzschia								
aequalis Hust.	−	+	−	+	−	+	−	5
amphioxoides Hust.	−	−	−	+	−	+	−	1,2,11
asterionelloides O. Müll.	+	+	−	−	−	−	+	
consummata Hust.	−	−	−	+	−	−	−	5
nyassensis O. Müll.	+	+	−	−	−	+	−	14
pelagica O. Müll.	+	−	−	−	−	−	−	
spiculoides Hust.	−	−	−	+	−	+	−	1,2
spiculum Hust.	−	−	+	+	−	+	−	1,5
stricta Hust.	−	−	−	+	−	−	−	1,5
subregula Hust.	−	+	−	−	−	−	−	
vanoyei Choln.	−	−	+	−	−	−	−	2
sp. (*fonticola* sensu Hustedt 1949)	−	−	+	+	−	+	−	
Rhizosolenia								
curviseta Hust.	−	−	−	−	−	+	−	
victoriae B. Schröder	−	−	−	−	−	+	−	
Stephanodiscus								
damasii Hust.	−	−	+	+	−	−	−	
Thalassiosira								
faurei (Gasse) Hasle	−	−	+	−	−	−	+	6
rudolfi (Bachm.) Hasle	−	−	−	+	−	−	+	2,5
Attached								
Achnanthes								
congolensis Hust.	−	−	−	+	−	+	−	
Cymbella								
grossestriata O. Müll.	+	+	+	+	−	−	−	
vanoyei Choln.	−	−	+	−	−	−	−	

Table II (*cont.*)

	Lake Malawi	*Lake Tanganyika*	*Lake Kivu*	*Lake Edward*	*Lake Mobuto Sese Seko*	*Lake Victoria*	*Lake Turkana*	*Elsewhere in Africa*
Cymbellonitzschia								
minima Hust.	−	+	−	−	−	−	−	
Fragilaria								
aethiopica G.S.West	−	+	−	−	−	−	−	
africana Hust.	−	+	−	+	−	+	−	1,2
longissima Hust.	−	−	−	−	−	+	−	
Gomphocymbella								
beccarii (Grun.) Forti	+	+	+	+	−	+	+	4,5,8,11,14
gracilis Hust.	−	+	−	−	−	+	−	
Gomphonema								
aequatoriale Hust.	−	−	−	+	−	+	−	1,5,11
africanum G.S. West	−	+	−	+	+	−	−	11
apicetrapezon Kuff.	−	+	−	−	−	−	−	
brachyneura O. Müll.	+	−	+	−	−	+	−	1
navicella O. Müll.	+	−	+	−	−	−	−	
Gomphonitzschia								
ungeri Grun.	−	−	+	+	−	+	−	4,11,14
sp.	−	−	−	−	−	+	−	
Rhopalodia								
gracilis O. Müll.	+	+	+	+	−	+	+	
hirudiniformis O. Müll.	+	+	+	+	−	+	+	
rhopala (Ehrenb.) Hust.	+	−	+	+	−	+	+	3,4,5,6,8,10, 13,15,16,
stuhlmannii O. Müll.	+	−	−	−	−	+	−	
Synedra								
cunningtonii G.S. West	−	−	−	−	+	+	−	5
nyansae G.S. West	−	−	−	−	−	+	−	
Benthic								
Capartogramma								
amphoroides R. Ross	−	+	−	−	−	−	−	
karstenii (Zanon) R. Ross	+	+	−	−	−	+	−	2
rhombicum R. Ross	−	+	−	−	−	−	−	
Frustulia								
sp. (*Vanheurckia africana* G.S. West)	−	−	−	−	+	−	−	
Navicula								
barbarica Hust.	−	−	−	+	−	+	−	

Table II (*cont.*)

	Lake Malawi	*Lake Tanganyika*	*Lake Kivu*	*Lake Edward*	*Lake Mobuto Sese Seko*	*Lake Victoria*	*Lake Turkana*	*Elsewhere in Africa*
damasii Hust.	–	–	–	+	–	+	–	1,2,5
faceta Hust.	–	–	–	–	–	+	–	1
molestiformis Hust.	–	–	–	+	–	–	–	1,2
nyassensis O. Müll.	+	+	+	+	–	+	–	1,2,5,8,9,10
pseudokrasskei Choln.	–	–	+	–	–	–	–	
vanmeelii Kuff.	–	+	–	–	–	–	–	
zanonii Hust.	–	+	+	+	–	+	–	1,2,5,6,9,11
sp.	–	–	–	–	–	+	–	
Neidium								
tanganyikae (G.S. West) F.W. Mills	–	+	–	–	–	–	–	
Nitzschia								
adapta Hust.	–	–	–	+	–	+	–	1,2,6
epiphytica O. Müll.	+	+	–	–	–	+	–	1,5
epiphyticoides Hust.	–	–	+	+	–	+	–	1,2,5,12
erlandssonii Choln.	–	–	+	–	–	–	–	
intermissa Hust.	–	–	–	+	–	+	–	2
kapimbiense Kuff.	–	+	–	–	–	–	–	
lacustris Hust.	–	+	–	–	–	+	–	5
obsoleta Hust.	–	–	–	+	–	+	–	
profunda Kuff.	–	+	–	–	–	–	–	
pseudosubrostrata Kuff.	–	+	–	–	–	–	–	
sp. (*epiphytica* sensu Hustedt 1949)	–	–	–	+	–	+	–	
Pinnularia								
scaettae Zanon	–	–	+	–	–	–	–	
tropica Hust.	–	–	–	–	–	+	–	1,2,5,11,15
Stauroneis								
subobtusa Hust.	–	–	+	–	–	–	–	
Surirellaceae								
Campylodiscus								
tanganicae Hust.	–	+	–	–	–	–	–	
Cymatopleura								
calcarata Hust.	–	+	–	–	–	–	–	
nyansae G.S. West	–	–	–	–	–	+	–	5

Table II (*cont.*)

	Lake Malawi	Lake Tanganyika	Lake Kivu	Lake Edward	Lake Mobuto Sese Seko	Lake Victoria	Lake Turkana	Elsewhere in Africa
Surirella								
aculeata Hust.	–	+	–	–	–	–	–	
acuminata Hust.	–	+	–	–	–	–	–	
brevicostata O. Müll.	+	+	–	–	–	+	–	
debesii Hust.	–	+	–	–	–	–	–	2,5,14
effusa Hust.	–	+	–	–	–	–	–	
engleri O. Müll.	+	+	+	+	+	+	–	7,11,12
fasciculata O. Müll	–	–	–	+	–	–	–	
fuellebornii O. Müll.	–	+	+	+	–	+	–	2,5,10
gradifera Hust.	–	+	–	–	–	–	–	
heidenii Hust.	–	+	–	–	–	–	–	
lancettula Hust.	–	+	–	–	–	–	–	
latecostata Hust.	–	+	–	–	–	–	–	
margaritacea O. Müll.	+	–	–	–	–	+	–	
margaritifera Hust.	–	+	–	–	–	–	–	
nyassae O. Müll.	+	–	–	–	–	+	–	3
obtusiuscula G.S. West	–	+	–	–	–	–	–	5
plana G.S. West	–	+	–	–	–	+	–	
reicheltii Hust.	–	+	–	–	–	–	–	5
sparsipunctata Hust.	–	+	–	–	–	–	–	
spiraloides Hust.	–	+	–	–	–	–	–	
striolata Hust.	–	+	–	–	–	–	–	
subcontorta Hust.	–	+	–	–	–	–	–	
subrobusta Hust.	–	+	–	–	–	–	–	
turbo O. Müll.	+	–	–	–	–	+	–	
vasta Hust.	–	+	–	–	–	–	–	7

Key to right-hand column:

1: South Africa
2: Namibia
3: Angola
4: Zimbabwe
5: Lake Chad
6: Niger Republic (Aïr Mountains)
7: Nigeria
8: Dahomey [= Benin]
9: Ghana
10: Upper Volta
11: Sierra Leone
12: Mali
13: Egypt
14: Libya
15: Algeria
16: Madagascar

Nitzschia is the genus with the greatest number (25) of planktonic species in the East African Great Lakes; of these, 12 are confined to Africa, but only four of them have not been found outside the region of the Great Lakes. The genus is particularly well represented in the plankton of Lake Edward (15 species, seven of them confined to Africa) and of Lake Victoria (13 species, six of them confined to Africa). Cholnoky (1968) states that *N. vanoyei* occurs in a number of the East African Great Lakes, but I have found no published record of it from anywhere other than Lake Kivu. Lange-Bertalot (1976) has pointed out that the species from Lakes Kivu and Edward identified by Hustedt (1949) as *N. fonticola* Grun. is not so.

Rhizosolenia is a large genus which is primarily marine but has a very small number of freshwater species. Lake Victoria is the only one of the East African Great Lakes in which it occurs and there it is represented by three species, one cosmopolitan and two known only from that lake.

Thalassiosira is another large genus whose species are mostly marine. Two species occur in the East African Great Lakes, both of them confined to Africa. Hasle (1978) recorded *T. faurei* from Lakes Kivu and Turkana and from smaller lakes in Ethiopia and Tanzania, and *T. rudolfi* from Lake Edward and from two smaller lakes in Tanzania, in one of which *T. faurei* also occurred. *T. rudolfi* was originally described from Lake Turkana but, according to Hasle, who was unable to locate type material, the description and figure match the species she found in material from Lake Edward and not that which she found in the material examined from Lake Turkana. Whilst I have followed Hasle, there is obviously the possibility that the original *T. rudolfi* is the species that she called *T. faurei*, and that it is only this species that occurs in Lake Turkana. There must also be some doubt about the identity of the diatoms on which the records from Namibia and Lake Chad are based. Indeed Compère (1980), when recording *T. faurei* from the Aïr mountains, remarks that his record from Lake Chad (Compère, 1975) of *T. rudolfi* (as *Coscinodiscus rudolfi* Bachm.) may have been based on specimens of *T. faurei*. Furthermore, the two species are so similar that Hasle (1978) suggests that they may not be distinct.

Kufferath (1956) described a centric diatom from Lake Victoria under the name *Coscinodiscus limifixus*. His description and figure make it plain that it is not a *Coscinodiscus* and it is impossible to be certain that it is not one of the species of *Stephanodiscus* or *Thalassiosira* already known from the the East African Great Lakes. It has been omitted from the list.

2. Attached Diatoms

Cymbellonitzschia is a genus of only three species, one known only from Lake Tanganyika and the other two, which may not be distinct from one another, from the coastal regions of western Europe. This genus is asymmetric about the apical axis, the valve having one margin convex, the other straight. In *C. minima*, the African species, the raphe lies along the convex margin, but in the European species it lies along the straight margin. This indicates that *C. minima* has been derived from a symmetric form quite independently of the European species and is not at all closely related to them. Whether it should be placed in *Hantzschia* or *Nitzschia* or treated as generically separate from these is doubtful, but if not included in one of those genera it constitutes a monotypic genus confined to Africa.

Gomphocymbella is a very similar case. There are two species, neither of which occurs outside Africa, in the East African Great lakes. Although clearly distinct, they are closely related to one another, and there is only minor asymmetry about the transapical axis to separate them from *Cymbella*. They clearly do not belong to the same genus as the other species that have been included in *Gomphocymbella*, most of which are closer to *Gomphonema* than to *Cymbella*.

Gomphonitzschia is also represented in the East African Great Lakes by two species, one of them undescribed and neither recorded from outside Africa. Other species of the genus occur in India and in North America. *Gomphonitzschia* differs from *Nitzschia*, not only in being asymmetric about the transapical axis but also in other characters, and it is a truly distinct genus.

Rhopalodia has four species, in the East African Great Lakes, that are confined to Africa and three that are cosmopolitan; one of these latter is more frequent in brackish than in fresh waters. This is a comparatively small genus with a number of its species in brackish and marine habitats. Two of the four species confined to Africa differ from all the others in the genus by being asymmetrical about the transapical axis, the character that separates *Gomphocymbella*, or, at least its African species, one of which is the type of the genus, from *Cymbella*. *Rhopalodia* presents an interesting contrast with the closely related *Epithemia*, a genus confined to fresh waters and having, like *Rhopalodia*, a small number of cosmopolitan species that are very common throughout much of their range. Four of these and one pan-tropical species are present in the East African Great Lakes, but there are no species of the genus confined to Africa. *R. rhopala* and *R. hirudiniformis*, the two species of *Rhopalodia* asymmetric about the transapical axis, form colonies on rocks in the spray zone around the shores of the lakes there, the individual cells borne on the tips of branched gelatinous stipes. *Epithemia turgida* (Ehrenb.) Kütz., present but rare in all the East African Great Lakes, forms similar colonies in comparable habitats throughout much of its range but apparently does not occupy this ecological niche in East Africa.

3. Benthos

Capartogramma is a genus of only four species, all of which occur in Lake Tanganyika, two of them not having been found elsewhere. *C. karstenii* is confined to Africa and has not been found in the living state outside the Great Lakes, although it was present in Nigeria during the Pleistocene. The fourth species, *C. crucicula* (Grun. ex Cleve) R. Ross is more widespread in Africa and occurs in America both in the tropics and in eastern North America. This is an undoubtedly distinct genus and the only one for which there are any definite indications of an African origin.

Nitzschia is represented in the benthos of the East African Great Lakes by more species than in the plankton, but a smaller proportion of its benthic species than of its planktonic ones is confined to Africa. Nevertheless, the proportion of its benthic species not found outside that continent is about double that for the benthic species as a whole.

4. Surirellaceae

This is the most advanced family of diatoms. It consists of two large genera, *Campylodiscus*, which is primarily marine, and *Surirella*, which has many marine and many freshwater species, and four small genera, two marine and two, *Cymatopleura* and *Stenopterobia*, freshwater. Two-thirds of the species of this family in the East African Great Lakes are confined to Africa, a much larger proportion than of any other group. The great majority of the species confined to Africa, 23 out of 28, occur in Lake Tanganyika; 19 of them have not been found in any other of the Great Lakes, although four of these have been recorded from other areas of Africa. The remainder are confined to Lake Tanganyika and are not known from smaller bodies of water in East Africa. This is in considerable contrast to the distribution patterns of the 64 other species not known outside Africa, only 12 of which are known from no other of the Great Lakes than Tanganyika.

Campylodiscus is represented in the East African Great Lakes by one species confined to Lake Tanganyika and *Stenopterobia* by a single cosmopolitan species known only from Lake Victoria, among the Great Lakes. The two cosmopolitan species of *Cymatopleura* occur in the East African Great Lakes, one of them, *C. librile* (Ehrenb.) Pant., being very widespread there. There are in addition one species confined to Lake Tanganyika and one occurring in Lake Victoria and recorded also from Lake Chad. It is *Surirella* that provides the bulk of the species of this family present in the East African Great Lakes, including the great majority of those confined to Africa.

Summarizing the distribution patterns of the diatoms occurring in the East African Great Lakes, one can say:

(a) two-thirds of the species have a distribution extending into the temperate regions, almost all of these being definitely cosmopolitan, as are all the genera represented there except for *Capartogramma* (which however extends to America) and the doubtfully distinct groups represented by *Cymbellonitzschia minima* and by *Gomphocymbella beccarii* and *G. gracilis*:

(b) only one quarter of the species are confined to Africa, with about 60% of these not known outside the region of the Great Lakes;

(c) the proportion of species confined to Africa is higher amongst the planktonic ones (39%) than amongst the attached (19%) or the benthic (17%), the Surirellaceae apart;

(d) two-thirds of the Surirellaceae are confined to Africa, and a high proportion of these are endemic to Lake Tanganyika, from which few other endemic species are known.

DISCUSSION AND CONCLUSIONS

When one comes to consider what has led to this state of affairs, so different from that in the fishes of the East African Great Lakes, with no species and virtually no genera occurring outside the African continent, there are three main points that can be made. The first of these is that the diatoms have a much greater ecological tolerance than the fishes. This is emphasized by the fact that at least 150 of the 246 East African species classed as cosmopolitan have also been recorded from Greenland. Whilst there may be amongst these a few cases where the same name has been applied to two different diatoms, in the great majority there is no morphological difference between the diatoms in the two areas.

The second point is that the diatoms have a much greater ability to disperse. The cosmopolitan distribution of so many species cannot be explained by their having had a continuous distribution on Pan-Gaea before the break-up of the continents. There are no fossil records of freshwater diatoms earlier than the Miocene, except that specimens of *Eunotia*, a genus and a member of a family otherwise known only from fresh waters, occur rarely in the very late Eocene marine deposit from Oamaru, New Zealand. This deposit was laid down in shallow water and the specimens of *Eunotia* in it may have been derived from the adjacent land. Marine diatoms are known from the Cretaceous, and their fossil record indicates that the Pennales, to which the great bulk of freshwater species belong, did not arise until the Paleocene. Many species of diatoms can resist desiccation and they are light enough to be carried in the air for great distances; fossil specimens of diatoms from Pleistocene deposits exposed to the south of the Sahara come down in rain on London from time to time. Also, whilst the experimental investigations by Atkinson (1980) give little support to the idea that they might be transported in the gut of migrating birds, the possibility of their being carried in dried mud on their legs cannot be ruled out.

Thirdly, with a few exceptions dealt with below, there has been little speciation in the East African Great Lakes. Whilst the diversity of the diatoms suggests that there must have been periods of rapid diversification in their history, their rate of speciation seems in general to have been low. Although we know less about Miocene and Pliocene freshwater diatoms than we do about modern ones, what is known indicates that many present-day species are indistinguishable from Miocene ones.

It seems reasonable to suppose that these three characteristics of the diatoms, ecological tolerance, great dispersability and a slow rate of speciation, are all a consequence of their being unicellular organisms with a much lower level of morphological, physiological and behavioural complexity than advanced multicellular groups. The connexion between their small size and their dispersability is obvious. It is much more difficult to see a reason why ecological tolerance should be a consequence of a comparatively low level of complexity, but the very wide distribution of many unicellular organisms, not only diatoms, suggests that it is. If this is the case, then it is reasonable to suppose that selection pressures, the extent to which changes in the environment can be accommodated only by changes in the genome, are greater in more complex organisms and hence that simplicity of organization results in greater stability of the genome.

Another factor is that there are presumably in the same area more different ecological niches for organisms with a more complex behaviour than there are for simpler ones. If your life style is just to sit or crawl about taking in nutrients from the ambient medium and absorbing sunlight as a source of energy, you cannot specialize as you can if your diet can be algal crusts or planktonic Crustacea or mud-inhabiting worms.

Whilst these may be the reasons for the absence or small amount of speciation in most genera, the problem remains of why there has been so much more speciation in the Surirellaceae and, to a lesser degree, in some other genera, in particular in *Nitzschia*, especially the planktonic species. It is difficult to believe that 26 species of *Surirella* living on the bottom mud of Lake Tanganyika (there are three cosmopolitan and two pan-tropical species in the lake), many of them together on the same patch, occupy different ecological niches, and the same is true of the 15 species of *Nitzschia*, seven of them confined to Africa, that occur in the plankton of Lake Edward. If adaptive radiation must be regarded as an unlikely explanation, there is no other that is obvious. The Nitzschiaceae and the Surirellaceae are regarded as the most advanced families, but they have fossil histories extending back to the upper Eocene and the upper Oligocene respectively. They were thus long past their period of initial differentiation when the East African Great Lakes came into existence.

There are examples in other groups of organisms of similar and more spectacular bursts of speciation. The fish genus *Haplochromis* in Lake Victoria is one. The flowering plant genus *Erica* in the Cape region of South Africa is another. That genus, which seems to have had a Mediterranean origin and to have spread through Africa from north to south, indulged in such a burst of speciation when it reached the tip of the continent that it is now represented there by some 600 species, very diverse and including a few groups so distinct from the rest that they are regarded as constituting separate genera. Whilst adaptive radiation may have played some part in these two cases, it is difficult to believe that a sufficient number of different ecological niches is present for that alone to have been the determining factor.

The only other explanation for rapid speciation that has, to my knowledge, been advanced is the suggestion by Gillett (1962) that it might be caused by pest pressure. There is, however, no evidence that I am aware of to suggest that this applies more strongly to *Nitzschia* and *Surirella* than it does to, for instance, the Naviculaceae, which show very low rates of speciation in Africa.

From this analysis of the distribution patterns of the diatoms in the East African Great Lakes, two points of general interest arise. One is that, whilst vicariance is the probable explanation of discontinuous distributions in many groups of more advanced organisms, in the diatoms at least, and probably in many other groups of small forms, long-distance dispersal occurs. The other is that rates of speciation within a major group can vary very considerably, proceeding slowly in many lineages but very rapidly, resulting in bursts of speciation, in others. The reasons for these bursts of speciation are not obvious; adaptive radiation would appear not to be an adequate explanation. If the great bulk of the species arising in such bursts are of short duration and without descendants, they are presumably of little significance in the general progress of evolution. On the other hand, if many of them survive for long periods, changing and diverging more slowly, then the resultant large group, with both a common ancestry and characters in common unique to the group, would probably be regarded as a distinct sub-family or even family. Speculations about how groups of this level have arisen have little in the way of facts as a basis, and the possibility that they have arisen in this way could well be examined.

REFERENCES

Atkinson, K.M. (1980). Experiments in dispersal of phytoplankton by ducks. *Br. phycol. J.* **15**, 49–58.

Bachmann, H. (1933). Phytoplankton von Victoria Nyanza, Albert Nyanza und Kiogasees. *Ber. schweiz. bot. Ges.* **32**, 705–717.

Bachmann, H. (1939). Mission scientifique de l'Omo. Beiträge zur Kenntnis des Phytoplanktons ostafrikanischer Seen. *Revue Hydrol.* **8**, 119–140.

Cholnoky, B.J. (1953). Studien zur Ökologie der Diatomeen eines eutrophen subtropischen Gewässers. *Ber. dt. bot. Ges.* **66**, 347–356.

Cholnoky, B.J. (1954). Neue und seltene Diatomeen aus Afrika. *Öst. bot. Z.* **101**, 407–427.

Cholnoky, B.J. (1957a). Beiträge zur kenntnis der südafrikanischen Diatomeenflora. *Port. Acta biol.* **6**, 53–93.

Cholnoky, B.J. (1957b). Neue und seltene Diatomeen aus Afrika. III. Diatomeen aus dem Tugela-Flusssystem, hauptsächlich aus dem Drakensberg in Natal. *Öst. bot. Z.* **104**, 25–99.

Cholnoky, B.J. (1958a). Hydrobiologische Untersuchungen in Transvaal II. Selbstreinigung im Juksei-Crocodile Flusssystem. *Hydrobiologia* **11**, 205–266.

Cholnoky, B.J. (1958b). Beiträge zur Kenntnis der südafrikanischen Diatomeenflora. II. Einige Gewässer in Waterberg-Gebiet, Transvaal. *Port. Acta biol.* **6**, 99–160.

Cholnoky, B.J. (1958c). Einige Diatomeen-Assoziationen aus Südwest-Afrika. *Senckenberg. biol.* **39**, 315–326.

Cholnoky, B.J. (1966). Die Diatomeen im Unterlaufe des Okawango-Flusses. *Beih. Nova Hedwigia* **21**, 1–102.

Cholnoky, B.J. (1968). "Die Ökologie der Diatomeen in Binnengewässern". J. Cramer, Lehre:

Compère, P. (1967). Algues du Sahara et de la région du lac Tchad. *Bull. Jard. bot. natn. Belg.* **37**, 109–288.

Compère, P. (1975). Algues de la région du lac Tchad. IV. Diatomophycées. *Cahiers O.R.S.T.O.M., ser. Hydrobiol.* **9**, 167–192.

Compère, P. (1980). Algues de l'Aïr (Niger). *Bull. Jard. bot. natn. Belg.* **50**, 269–329.

Dickie, G. (1879). Algae from Lake Nyassa, E. Africa. *J. Linn. Soc., Bot.* **17**, 281–283.

Erlandsson, S. (1928). Diatomeen aus Afrika. *Svensk Bot. Tidskr.* **22**, 449–461.

Forti, A. (1914). Terza contribuzione alla flora algologica della Libia. *Atti R. Ist. Veneto Sci.* **73**, 1441–1551.

Frenguelli, J. (1938). Diatomeas del Querandinense Estuarino del Rio Matanza en Buenos Aires. *Revta Mus. La Plata*, n.s., *Paleont.* **1**, 291–314.

Gillett, J.B. (1962). Pest pressure, an underestimated factor in evolution. *In* "Taxonomy and Geography" (D. Nichols, ed.), pp. 37–46. Systematics Association, London.

Hasle, G.R. (1978). Some freshwater and brackish water species of the diatom genus *Thalassiosira* Cleve. *Phycologia* **17**, 263–292.

Huber-Pestalozzi, G. (1942). Das Phytoplankton des Süsswassers. Systematik und Biologie. Diatomeen. *Binnengewässer* **16** (2), 367–549.

Hustedt, F. (1921). Bacillariales. *In* B. Schröder, Zellpflanzen Ostafrikas gesammelt auf der Akademischen Studienfahrt 1910. *Hedwigia* **63**, 117–173.

Hustedt, F. (1949). Süsswasser-Diatomeen. *Explor. Parc natn. Albert Miss. H. Damas* **8**, 1–119.

Kufferath, H. (1956). Organismes trouvés dans les carottes de sondage et les vases prélevées au fond du Lac Tanganyika. *Résult. scient. Explor. Hydrobiol. Lac Tanganyika* **4** (3), 1–74.

Lange-Bertalot, H. (1976). Eine Revision zur Taxonomie der Nitzschiae lanceolatae Grunow. Die "klassichen" bis 1930 beschriebenen Süsswasserarten Europas. *Nova Hedwigia* **28**, 253–307.

Lange-Bertalot, H. and Simonsen, R. (1978). A taxonomic revision of Nitzschiae lanceolatae Grunow. 2. European and extra-European fresh water and brackish taxa. *Bacillaria* **1**, 11–111.

Maillard, R. (1977). Diatomées d'eau douce du Mali, Afrique. *Bull. Mus. natn. Hist. nat. Paris*, sér. 3, **433**, 1–45.

Müller, O. (1904a). Berichte über die botanischen Ergibnisse der Nyassa-See- und Kinga-Gebirgs-Expedition der Hermann- und Elise- geb. Heckmann-Wentzel-Stiftung. VII. Bacillariaceen aus dem Nyassalande und einige benachtbarten Gebieten. Erste Folge: Surirelloideae—Surirelleae. *Bot. Jb.* **34**, 9–38.

Müller, O. (1904b). Berichte über die botanischen Ergibnisse der Nyassa-See- und Kinga-Gebirgs Expedition der Hermann- und Elise- geb. Heckmann-Wentzel-Stiftung. VII. Bacillariaceen aus dem Nyassalande und einige benachtbarten Gebieten. Zweite Folge: Discoideae—Coscinodisceae. Discoideae—Eupodisceae. *Bot. Jb.* **34**, 256–301.

Müller, O. (1905). Berichte über die botanischen Ergebnisse der Nyassa-See- und Kinga-Gebirgs Expedition der Hermann- und Elise- geb. Heckmann-Wentzel-Stiftung. VII. Dritte Folge: Naviculoideae—Naviculeae—Gomphoneminae—Gomphocymbellinae—Cymbellinae. Nitzschioideae—Nitzschieae. *Bot. Jb.* **36**, 136–205.

Müller, O. (1910). Berichte über die botanischen Ergebnisse der Nyassa-See- und Kinga-Gebirgs-Expedition der Hermann- und Elise- geb. Heckmann-Wentzel-Stiftung. VIII. Vierte Folge (Schluss): Naviculoideae—Naviculeae—Naviculinae. Fragilarioideae—Fragilarieae—Fragilarinae. Fragilarioideae—Fragilarieae—Eunotinae. *Bot. Jb.* **45**, 69–122.

Ostenfeld, C.H. (1908). Phytoplankton aus dem Victoria Nyanza. Sammelausbeute von A. Bogert, 1904–5. VIII. Abhandlung. *Bot. Jb.* **41**, 330–350.

Ostenfeld, C.H. (1909). Notes on the phytoplankton of Victoria Nyanza, East Africa. *Bull. Mus. comp. Zool. Harv.* **52**, 171–181.

Rich, F. (1933). Scientific results of the Cambridge expedition to the East African Lakes. 7. The Algae. *J. Linn. Soc., Zool.* **38**, 249–275.

Schmidt, A. (1874–1959). "Atlas der Diatomaceenkunde". Aschersleben: Schlegel, Siever; Leipzig: Reisland; Berlin: Akademie-Verlag.

Schröder, B. (1911). *Rhizosolenia victoriae* n.sp. *Ber. dt. bot. Ges.* **29**, 739–743.

Simonsen, R. (1979). The diatom system: ideas on phylogeny. *Bacillaria* **2**, 9–71.

Van Meel, L. (1954). Le phytoplankton. État actuel de nos connaissances sur les grands lacs est-africains et leur phytoplankton. *Résult. scient. Explor. hydrobiol. Lac Tanganyika* **4** (1), 1–681.

Virieux, J. (1913). Plancton du lac Victoria Nyanza. *In* "Voyage de Ch. Alluaud et R. Jeannel en Afrique orientale (1911–1912). Résultats scientifiques". 23 pp.

West, G.S. (1907). Report on the freshwater algae, including phytoplankton, of the Third Tanganyika Expedition conducted by Dr. W.A. Cunnington, 1904–1905. *J. Linn. Soc., Bot.* **38**, 81–197.

West, G.S. (1909). Phytoplankton from the Albert Nyanza. *J. Bot., Lond.* **47**, 244–246.

Woloszynska, J. (1914). Studien über das Phytoplankton des Victoria-Sees. *Hedwigia* **55**, 184–223.

Woodhead, N. and Tweed, R.D. (1956). A check-list of tropical West African Algae. *Hydrobiologia* **11**, 299–395.

Woodhead, N. and Tweed, R.D. (1960). Freshwater Algae of Sierra Leone. 3. The algae of Rokupr and Great Scarcies River. *Revue algol.*, n.s. **5**, 116–150.

Zanon, V. (1938). Diatomee della regione del Kivu (Congo Belga). *Commentat. pontif. Acad. Scient.* **2**, 535–668.

Zanon, V. (1941). Diatomee dell'Africa Occidentale Francese. *Commentat. pontif. Acad. Scient.* **5**, 1–60.

7 The Zoogeography of African Freshwater Fishes: Bioaccountancy or Biogeography?

P.H. GREENWOOD

Department of Zoology, British Museum (Natural History), London SW7 5BD, UK

Abstract: The major lakes of east and central Africa are characterized by their highly endemic faunas of cichlid fishes. The explosive speciation undergone by these animals, and the often extreme and apparently convergent morphological differentiation undergone by certain taxa further compound the problem of their biogeographical history (or histories).

Apparently the cichlid fishes (and to a certain extent the non-cichlid fishes) of each lake can be taken as an example of multiple and temporally truncated vicariant speciation events, albeit vicariance on a geographically circumscribed scale. But, question marks hang over the more distant histories and interrelationships of these fishes, the value of the fossil record, the information which may be derived from the suspected geomorphological history of the lake basins and associated river systems, and over what information can be gained from the fishes of those rivers. Many answers have been proposed to explain the histories of the different lake faunas and their interrelationships. These will be reviewed in the light of new information, particularly that stemming from a fresh look at the phylogenetic relationships of the cichlid fishes involved. Some thought will also be given to the way in which different philosophical approaches have influenced both the questions asked by various investigators, and the way in which answers have been formulated. In a wider context, there is the question of how this particular (and in some respects unusual) problem in biogeography can be related to, or assist in the solution of other biogeographical problems.

Systematics Association Special Volume No. 23, "Evolution, Time and Space: The Emergence of the Biosphere", edited by R.W. Sims, J.H. Price and P.E.S. Whalley, 1983, pp. 179–199. Academic Press, London and New York.

> In Africa there is a comparatively greater variety of distinct Freshwater types [than in India], imparting to the study of its fauna an unflagging pleasure. . . . Albert Günther on the distribution of freshwater fishes (1880)

A hundred and three years later, Günther's remarks still hold true but the problems associated with that study are now far more complex. Whereas the known African freshwater fishes in Günther's time numbered some 255 species in 15 families, the modern student is faced with about 2500 species and 30 families. On current taxonomic criteria these primary freshwater fishes* show a high level of endemicity, with only about 16 of the 270 genera, and probably no more than three or four species, occurring outside Africa.

Perhaps this very complexity, both in taxonomic and ecological terms, and this high degree of endemicity have exerted an almost inhibitory effect on detailed zoogeographical studies of the fauna. Whatever the cause, there have been few major zoogeographical studies on the African freshwater ichthyofauna during the last century, and our basic understanding of African freshwater fish biogeography is little different from that outlined by Günther (1880).

There have, of course, been several intracontinental studies at what may be termed the bioaccountancy level (e.g. Boulenger, 1905, Poll, 1957, 1973; Matthes, 1964; Roberts, 1975), and some progress has been made towards unravelling certain intercontinental relationships, but in general there has been little improvement in matters of detail. As yet there has not been enough time to detect a change in conceptual thinking on the subject; the influence of vicariance biogeography is apparent only in a few of the more recently published papers (Nelson, 1969; Patterson, 1981).

The opportunities for deriving unflagging pleasure from the freshwater fishes of Africa are now, as they were in Günther's time, unlimited, and the unresolved problems awesome.

The African freshwater ichthyofauna, as Günther (1880) recognized, shows intercontinental affinities with that of Asia and that of South America, particularly the former, while the apparent relationships of some taxa also bring Australia and Europe into the picture.

* Primary freshwater fishes: this term is used to include Patterson's (1975) two categories, archaeolimnic and telolimnic, respectively groups originating in fresh water and always so confined, and groups confined to fresh waters at present but less closely restricted in the past. It does not include peripheral species (e.g. anguillids, salmonids and gasterosteids) which move freely between fresh and salt water.

The following may be cited as typical examples of the observations on which intercontinental affinities are based; these affinities were used by several authors to propose schemes of dispersal and evolution (Regan, 1922; Darlington, 1948, 1957; Menon, 1951, 1964; Gosline, 1975; Patterson, 1975; Novacek and Marshall, 1976; Briggs, 1979; Fink and Fink, 1981), but still have troublesome elements.

(1) The ostariophysan order Characiformes occurs today in Africa and America (tropical South America to Texas), but only one of its families (the Characidae) is common to both continents. During the Eocene, however, characid fishes occurred in Britain and France (Patterson, 1975). Teeth from deposits in the latter countries have morphological features which strongly suggest their derivation from fishes belonging to an African subgroup of the family, the alestines.

(2) Some genera in the ostariophysan family Cyprinidae (carps and their allies), namely members of the bariliine, barbine, garrine and labeine subdivisions of that family, occur in Africa and Asia (with one genus, *Barbus*, also distributed in Europe; see Howes (1980) for details). The Cyprinidae, it should be noted, do not occur in South America, but are widespread in North America.

(3) Three families (Bagridae, Schilbeidae and Clariidae) of the ostariophysan order Siluriformes (catfishes) occur only in Africa and Asia; one clariid genus, *Clarias*, occurs in both regions but there are no other genera shared intercontinentally. Neither Africa nor Asia shares any primary freshwater siluriform family with South or North America.

(4) Three percomorph families (Channidae, Anabantidae and Mastacembelidae) are found only in Africa and Asia. At least nominally, no anabantid and no channid genera are shared by the two continents, but one of the two recognized mastacembelid genera occurs in both regions and the resemblance between the Asian and Africa channid genera is so close that their distinction is not accepted by all authorities.

(5) The percomorph family Cichlidae is widely distributed in South and Central America, and in Africa; it also occurs in peninsular India, Sri Lanka and Madagascar. No genera are common to any of the regions except India and Sri Lanka. Could dispersal through the sea play an important role in cichlid biogeography?

(6) A fifth percomorph family, the Synbranchidae (swamp-eels), has one genus (*Ophisternon*) with species in Africa, South America, Mexico, Cuba, Indo-Malaysia and Australia, and a second genus (*Monopterus*)

with species in Africa, south-east Asia (including India and Malaysia), China and Japan, but not in the neo-tropical region. Could marine dispersal be involved here?

(7) The Osteoglossomorpha (bony-tongues), an archaic and primitive group of bony fishes (Greenwood, 1973a) has a distribution that could be described as Pangeaean since it involves Africa, North and South America, Asia and Australia.

One family (Osteoglossidae, suborder Osteglossoidei) occurs in Africa, South America and Australia, being represented in each area by different genera; fossil osteoglossids are found in Asia, Australia and North America. A related family (Pantodontidae) is endemic to Africa, as was the entirely extinct family Singididae (Greenwood and Patterson, 1967). The Notopteridae (suborder Notopteroidei) occurs only in Africa and South East Asia. Its sister family, the Mormyridae, is endemic to Africa but the sister group of both families is represented by an extant endemic North American family (Hiodontidae) and an extinct family (Lycopteridae) known only from the Jurassic of China (Greenwood, 1970).

(8) In contrast to the seven families common to Africa and Asia alone, there are but two (Characidae [see above] and Lepidosirenidae [lungfishes]) restricted only to Africa and South America. Again, no genera are shared by the two continents. A second lungfish family (Ceratodontidae) has a single living representative in Australia, but fossil ceratodontids have a much wider distribution.

A discussion of these distribution patterns (and others like them) is perhaps most fruitful if it centres around the question of *exactly* what information can be gained from them on the biogeographical history of the African primary freshwater ichthyofauna. That question is equally relevant to dispersalist and to vicariance biogeographers. Its answer is particularly critical in relation to hypotheses about areas of origin, as well as hypotheses of dispersal. The latter can relate either (i), to the definition of primitive cosmopolitan distributions preceding the vicariance events which led to current distribution patterns; or, (ii), alternatively to the formulation of serial acts of dispersal and speciation thought to explain the same patterns.

What, then, do the examples cited above contribute to our historical interpretation of the observed distributions and especially to our interpretation of the possible processes underlying them? For most examples the answer, I believe, is "disappointingly little".

The reason for that answer is simple to give but difficult to remedy. It lies principally in the lack of adequate hypotheses for the phylogenetic relationships of the fishes involved. Such schemes of relationship are needed at various categorical levels, not just at the highest ones, and must be so argued that any particular taxon's primitive or derived status, relative to those of its sister taxa, is readily ascertainable.

Without sound phylogenies even the most carefully contrived scenario-style biogeographical schemes lack credibility, no matter how well substantiated by ecological and physiological data (or surmises) they may appear to be. They lack credibility since all are based primarily on assumed phylogenetic relationships (often a reflection only of overall similarity or dissimilarity) and on assumed degrees of primitiveness or specialization amongst the taxa concerned. Furthermore, these schemes often are supported by untestable *ad hoc* hypotheses invoking various ecological parameters to account for routes of dispersal or the absence of taxa from certain regions.

The Ostariophysi (examples 1–3 above) are a case in point (see Gosline's, 1975, account of the group). Long an important element in establishing relationships between the freshwater ichthyofaunas of Africa and South America on the one hand, and of Asia and Africa on the other, the broad outlines of their historical biogeography have been reconsidered in four recent papers (Gosline, 1975; Patterson, 1975; Novacek and Marshall, 1976; Briggs, 1979). Unfortunately none of these authors could draw on a detailed phylogenetic analysis of the group, and their conclusions are at variance for this and other reasons.

Fink and Fink (1981) have now laid the basis for that much needed analysis and have thereby provided a sounder foundation for future biogeography studies, an aspect of the Ostariophysi reviewed but briefly in their paper.

Even with the foundation laid by the Finks, we will, at first, only be able to approach the wider aspects of the problem, and much work remains to be done before we can start to investigate the finer points.

Questions at that level are many and complex, as for example those involving the relationships of the three endemic African catfish families (Amphiliidae, Mochokidae and Malapteruridae); do they lie with South American, with Asian, with Eurasian or with other African families; what, indeed, are the interrelationships of all the catfish families?

Then there are problems involving the precise relationships between the characin fishes of South America and Africa, and in particular whether the one shared family (the Characidae, example 8 above) is really a holophyletic unit (a current assumption recently queried by Vari, 1979). The presence during the Eocene of alestine characids in Europe—the only record of the Characiformes occurring outside Africa and America south of Texas—is another problem. This, however, is perhaps most parsimoniously explained as a dispersal event from Africa during the Palaeocene (see maps in Adams, 1981). Why the characids spread no further into Europe is an open question and one probably related to the reason why they did not move into (or become established in) Asia or America north of Texas.

There are questions too concerning the relationships of Asian and African cyprinoid ostariophysans. The information we do have is certainly insufficient to warrant statements about a ". . . largely one-way migration of south-east Asian fishes into Africa" (Gosline, 1975). That there has been some dispersal between the two continents cannot be gainsaid, and it seems probable that the extant freshwater representatives of the primitive sister group within the Ostariophysi (the Anotophysi of Rosen and Greenwood, 1970) are restricted to Africa. A primary problem here is to determine at what level or levels of phyletic differentiation the dispersal or dispersals took place, a problem which involves all ostariophysan groups and virtually all continents, not just a possibly recent exchange between Africa and Asia of cyprinoid and siluriform taxa. A start on the problem of Asian–African cyprinid relationships has recently been made by Howes (1980) but, as he admits, there is still a long way to go.

We turn now to other groups shared by Asia and Africa, in particular to the shared percomorph groups (examples 4–6 above). Again, one finds that there is little or no information on the inter- and intrarelationships of the taxa involved. Consequently few hard biogeographical conclusions can be drawn for these groups either, especially with regard to their probable areas of origin.

Within the families Channidae, Anabantidae and Mastacembelidae (example 4 above), African and Asian species and genera seem, from the evidence available, to be closely related. Thus, these species might be considered as the end-products of vicariance events following a relatively recent (i.e. early Caenozoic) freshwater dispersal from one continent to the other.

There is little critical evidence for evaluating the wider interrelationships of the three families. It has been suggested that the

Channidae and Anabantidae are related, but the evidence is inconclusive (see Nelson, 1976, for a review of this question; also Liem, 1963). The sister family of the Mastacembelidae, the Chaudhuridae, is confined to Burma.

A few members of the family Cichlidae (example 5 above) are known to tolerate and even breed in sea- and brackish-water, and the nearest living relatives of the family are entirely marine in habitat (Liem and Greenwood, 1981). Thus there are grounds for thinking that some phase or phases of marine dispersal may have been involved in the biogeographical history of an otherwise predominantly freshwater (and speciose) group with the rather improbable hallmarks of a Gondwandian distribution (see below). Nothing is known about intrafamilial relationships amongst cichlids, but the question is under review (Stiassny, 1982; Greenwood, 1983 a and b, and work in progress).

Like the cichlids, a few members of the Synbranchidae (example 6 above) are tolerant of marine and brackish water conditions but, also like the cichlids, most of the species are confined to truly freshwater habitats. A partially resolved phylogeny of the Synbranchidae is available (Rosen and Greenwood, 1976) but it is not sufficiently refined to be of particular value in detailed zoogeographical research. The most primitive synbranchid species occurs in Asia (Malaysia). One African taxon has some of its most closely related species in Asia and others in South America, whilst the second African taxon has its sister species (all more derived) widely distributed in Asia.

There is no fossil record for the Synbranchidae, and that for the Cichlidae extends, with certainty, no further back than the Oligocene. Taking into account the known fossil fish record for the Mesozoic, it seems improbable that either group, or the immediate common ancestors of either group, were present in Gondwana times. The questions of synbranchid and cichlid dispersal routes, and the ecological nature of those routes must, therefore, remain open.

Finally we must consider the two "archaic" groups (example 7 and 8 above) the Osteoglossomorpha and the Dipnoi (lungfishes), representatives of which were known to exist in Gondwana times.

A fairly complete phylogeny is available for the Osteoglossomorpha (Greenwood, 1973a; Patterson, 1981) and, unusually for freshwater fishes, there is a critical fossil record as well (see Patterson, 1981). In a series of papers, Nelson (1969) and Patterson (1975, 1981) have reconstructed a broad and reasonably complete biogeographical history for the group. But, even here, doubts remain and these preclude the writing of a fully detailed story. There are, for example, uncertainties about the history of the Afro-Asian family Notopteridae, the immediate relationships of the endemic African family Mormyridae, and uncertainties about the relationships of some recently described fossils from China (see Patterson, 1981).

Despite these difficulties, however, we have for the Osteoglossomorpha probably the most complete biogeographical history of any freshwater fish group in Africa. In brief, the biogeographical patterns of osteoglossomorph history were established, intracontinentally, in Pangea (up to 145 my BP), and were further developed by the break-up of Gondwanaland, with possibly some intercontinental exchange between Africa and Asia during the Caenozoic.

The Dipnoi present a different and less well-documented history. The four species of African lungfishes and the single species from South America are more closely related to one another than any is to the Australian species. This relationship is reflected in the current classification of lungfishes. The Afro-American species are placed in one family (Lepidosirenidae) and the Australian species in another (Ceratodontidae), the latter a family to which the majority of fossil taxa can be referred, and one with a long history.

The present intercontinental distribution of the lepidosirenids is probably to be ascribed to the dispersal of their common ancestor during a late Mesozoic Afro-South American interconnection (Patterson, 1975). The earlier history of the group, and thus of the relationship between Afro-America and Australian lungfishes, is more complex. Patterson (1975) has argued that the extant lungfishes should not be considered a primary freshwater group (that is one originating in, and thereafter strictly confined to, fresh waters). The nearest relatives of the living Australian lungfish, all now extinct, are marine species and thus it is most parsimonious to conclude that the common ancestor of the Australian and the Afro-American lungfishes (i.e., the common ancestor of the Ceratodontidae and the Lepidosirenidae) was also marine. If that is so (and the evidence is persuasive, see Patterson, 1975) then it follows that, in general terms, the Dipnoi contribute little towards elucidating the history of primary freshwater fish biotas.

The discussion and evidence reviewed so far can be summarized quite simply: the primary freshwater fish fauna of Africa is related in part to that of Asia and in part to that of South America. This conclusion, as we know, was reached by Günther in 1880 on virtually the same evidence. Certainly for some of the groups involved the evidence has been strengthened, but in general we have learned little more and the picture we have is still a coarse-grained one.

That less progress has been made in refining that picture is, I think, mainly due to the fact that we are without a detailed scheme of phylogenetic relationships for the majority of taxa involved. Denied that information we lack the data on which to build sound biogeographical hypotheses, in particular those relating to probable areas of origin and hence to the directions of dispersal routes taken by the evolving lineages. Patterson's (1981) account of the Osteoglossomorpha discussed above, and Wiley's (1976) study of the garpikes (Lepisosteidae), demonstrate the value of this approach to biogeography.

Lack of phylogenetic data also prevents us from evaluating the relationships of endemic African taxa, taxa generally omitted from estimates of biota relationships just because they are endemic (a status usually conferred on them because of their relative morphological isolation). Here, as in taxonomy, merely acknowledging an apparent morphological gap by some nomenclatural means can be tantamount to donning blinkers. Unravelling the phylogenetic relationships of Africa's endemic taxa would add substantially to the sum of evidence available for biogeographical research.

I would, in this context, query the total validity and value of such sweeping remarks as "African rivers and swamps harbor an extraordinary assortment of archaic and phyletically isolated fish groups, most of them endemic and several bizarrely modified" (Roberts, 1975). True, several are bizarrely modified (but then so are several Asian and South American species) and some are archaic (as are some South American and Asian taxa), but is Roberts justified in writing, in such general terms, of "phyletic isolation" when the data for that conclusion simply are not available?

Ironically, it is for many of these archaic groups that the best phylogenetic data are available and have enabled us to show that, with one exception (the Polypteridae), they are not phyletically isolated (see p. 185–186 above).

I foresee little progress in the broader study of African freshwater fish biogeography until much more research has been devoted to the higher level taxonomy of those fishes. In that endeavour I would favour the use of a cladistic methodology, the value of which is apparent in the recently improved understanding of osteoglossomorph biogeography (see above) and in the biogeography of other fish groups.

Obviously an increased knowledge of phylogenesis alone will not provide ready answers to all biogeographic problems. The absence or presence of particular taxa in particular regions will still, on occasion, be inexplicable. But, need we provide untestable *ad hoc* hypotheses masquerading as explanatory scenarios to account for these enigmas? It is an empirical observation that no cyprinid fishes occur in South America today. Do we gain anything by postulating some unsuccessful past competition with the characins to "explain" that absence? By doing so, I believe, we run the risk of inhibiting more critical re-analyses of the data available, or even of accepting the fact that certain data are not yet available.

Similar comments and criticisms can be levelled at the state of intracontinental biogeographical studies on African freshwater fishes, the most recent of which (Roberts, 1975) reviews its predecessors in the field.

These several studies, beginning with Boulenger's paper of 1905, have been concerned mainly with delimiting faunal provinces. Roberts (1975) recognized ten such regions. Provincial boundaries have been modified somewhat over the years; new provinces were defined as more taxa were described, the taxonomic (but not necessarily the phylogenetic) status of others was reassessed, and as more information was obtained on the distribution of various species. Overall, however, the changes have not been profound.

Faunal provinces are, in effect, areas of endemicity, usually at the generic level but reinforced by the presence of endemic species from more widely distributed genera. The recognition of faunal provinces can be considered a stage in the production of faunal inventories, their use a means of recognizing areas of cladogenic activity. As such their importance must be acknowledged, but they tell us little about the interrelationships of biotas. For that, as for intercontinental studies, we require detailed phylogenetic studies; so far few of these have been forthcoming. I cannot, therefore, agree with Roberts' (1975:281) sentiment that recognition of faunal provinces *per se* "... provide[s] much more insight into zoogeographical relationships". Rather, as noted before, their value lies in pinpointing probable areas of past cladogenic activity.

In this respect, it is instructive to superimpose the outlines of Roberts' ten faunal provinces (Fig. 1) on a map showing presumed hydrographical features at the Mio-Pliocene transition (Fig. 2).

The coincidence of many faunal provinces with the major drainage basins at that time is most noteworthy. However, some restraint should be exercised before drawing conclusions from this congruity of areas. In preparing the Mio-Pliocene map, some account was taken of present-day fish distributions (see Howell and Bourlière, 1963:651), as well as of information derived from purely geomorphological and palaeogeographical studies. There may, therefore, be an element of circularity underlying the coincidences involved.

Fig. 1. The ichthyofaunal provinces of Africa; compounded from figures in Roberts (1975). The faunal boundaries for the Upper and Lower Guinea provinces are taken from the text of that paper and not from the figures.

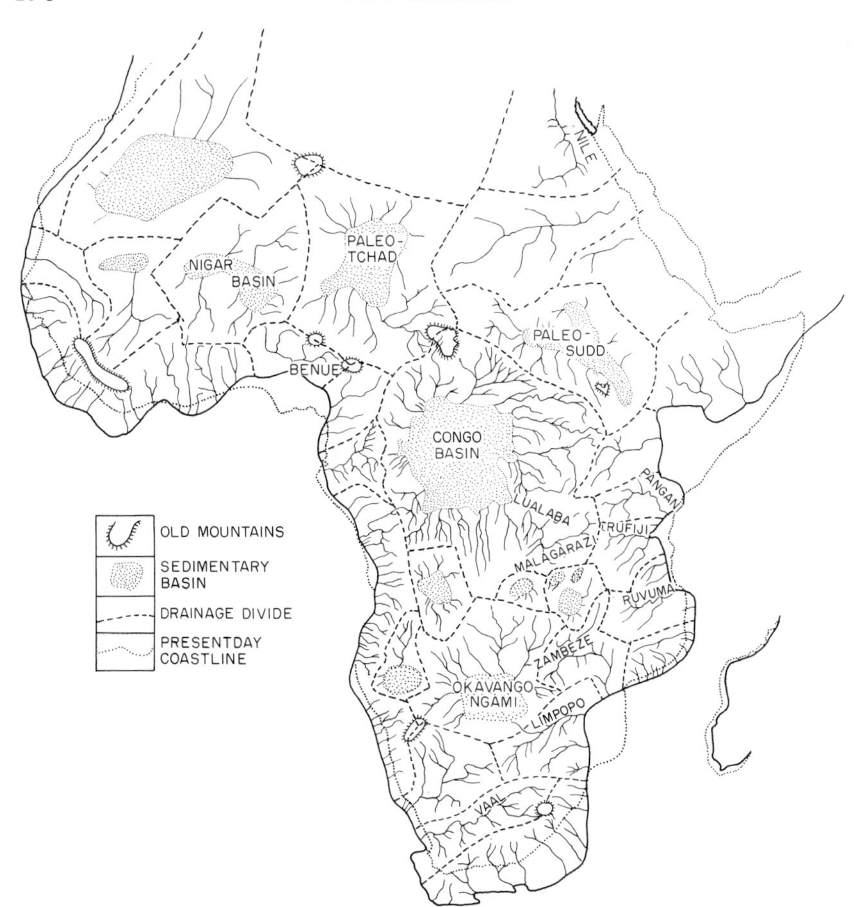

Fig. 2. Hydrogeographical map of Africa at the Miocene-Pliocene transition. Modified from Roberts (1975, after the original in Howell and Bourlière, 1963).

To some extent, as remarked previously, it is possible to interrelate provincial biotas on the basis of shared taxa. This is, however, only meaningful when the shared taxa are of limited general distribution. A widely distributed taxon, like the plesiomorph taxon in systematics, is only of value when determining relationships at the higher levels of universality. The question of shared species is also confounded by the problem of geographically circumscribed differential extinction (see below, p. 191, 194).

Using taxon distributions, Poll (1973) has suggested a relationship between the Nilo-Sudanian and Zairean provinces (nomenclature after Roberts, 1975), and Roberts (1975) has indicated "an early faunistic relationship" between the upper Guinean, lower Guinean and Zairean provinces.

A relationship between the European fish fauna and that of the Maghreb province (Roberts, 1975) is clearly indicated, but it is difficult to establish other related biotas on the present evidence, although Roberts' suggestion of affinities between the Zambesi and East Coast provinces seems reasonable.

The history of the Maghreb province (Tunisia, Algeria and Morocco) is an interesting one, showing as it does the extent to which faunal affinities can change in relatively recent geological times. Today, the Maghreb fish fauna has a noticeable European component. For example, the genera *Phoxinus* (or *Phoxinellus*), *Cobitis*, *Aphanius*, *Salmo* and *Gasterosteus* are present (the two latter being peripheral and not primary freshwater elements) but do not occur elsewhere in Africa, whilst at the species level some of the Maghreb *Barbus* are probably conspecific with European taxa. The remaining elements of its generally depauperate ichthyofauna occur in Africa, in both Africa and the Middle East, or are apparently related to African species (see Roberts, 1975, for details).

A critical phylogenetic study of the *Barbus* species and those of related genera occurring in the Maghreb should throw more light on the overall relationship of the fauna, but for once, even without that information its mixed affinities cannot be denied.

The Miocene freshwater ichthyofauna of the Maghreb (Greenwood, 1973b) stands in marked contrast to the present-day one. Amongst the fishes known from fossils, there are at least two exclusively African primary freshwater taxa (*Polypterus* and *Clarotes*) and a species of the common African genus *Clarias*, none of which is present in the Maghreb today; also present is the genus *Lates* which, although it is a peripheral element, is very widely distributed in Africa but absent from the present-day Maghreb ichthyofauna.

None of the Maghreb fossils is identifiable with any of the present-day European elements in that region. In its totality, this Miocene ichthyofauna is, at the generic level, comparable with a segment of its contemporary Nilotic fauna, and indeed with a segment from the Nilo-Sudanian fauna of today.

Although a detailed phylogenetic study was by no means essential to assay the change in the Maghreb fauna, nor to establish contemporary faunal affinities, such as study would help to clarify two other intriguing problems in African zoogeography and hydrology. These are the postulated but apparently incongruous former drainage of the upper Lualaba river (now part of the Zaire river system) into the Nile, and the relationships of several taxonomically isolated elements in the Cape of Good Hope ichthyofaunal province (see Roberts, 1975:297 and 307 [on Lualaba and Cape], Banister and Bailey, 1979 [on Lualaba]).

According to Poll (1957, 1963), a number of species occurring in the upper Lualaba river are disjunctly distributed Nilo-Sudanic elements and are therefore indicative of a former direct connection between the upper Lualaba and the Nile. Banister and Bailey (1979) have investigated this claim in some detail and find little to support it.

From Poll's original list of eight presumed Nilotic species, Banister and Bailey were able to show that two had been included in error, that for a third species there are no grounds for assuming it has Nilotic affinities, that the record of a fourth species was based on a misidentification, and that two others occur elsewhere in the Zaire system. Thus, of the eight species, only two, *Polypterus senegalus* and *Polypterus bichir* remain as undisputed Nilo-Sudanian elements.

In challenging an earlier suggestion (Greenwood, 1976) that the Lualaba "Nilotics" were relics of a formerly more widely distributed fauna, Poll (1976) expressed doubts that the species could have reached the upper Lualaba river in former times without leaving traces of their passage in other Zairean rivers. We now know (see above) that two species do occur elsewhere in the Zaire system, so that Poll's doubts apply only to the *Polypterus* species *P. senegalus* and *P. bichir*. Traces of their passage might be found in the form of those *Polypterus* species today recognized as Zairean endemics. A phylogenetic study of extant polypterids would test that hypothesis and also test the conspecificity of the Nile and Lualaba populations.

The Cape of Good Hope ichthyofaunal province (totalling about 50 species, mostly Cyprinidae) is characterized by the presence of a distinctive group of *Barbus* species (the redfins), by the high level of endemicity amongst its other cyprinid fishes, by an endemic anabantid genus (*Sandelia*), and by the poor representation of species from the neighbouring Zambesi province. A phylogenetic analysis of the endemic cyprinids (especially the redfins and their relative, the near-extinct, orophilic *Oreodaimon quathlambae*; see Greenwood and Jubb,

1967; Skelton, 1976) and the endemic *Sandelia* could well indicate the relationships, and possibly the origin, of this distinctive assemblage.

The Great Lakes of Africa (and some of the minor ones too) have long fascinated ichthyologists and evolutionary biologists. The centres of this attraction are the adaptively multiradiate and speciose flocks of cichlid fishes, each flock endemic to a particular lake. Fishes from families other than the Cichlidae also show a fairly high level of endemicity, but few have produced the adaptations exhibited by the cichlids (Greenwood, 1981).

Using non-endemic species (especially non-cichlid taxa) as indicators, individual lakes have been associated with different ichthyofaunal provinces (see Roberts, 1975). But here again caution must be urged until the phyletic relationships of different elements in the lake faunas are better understood (or known at all), and until we have a better appreciation of the "indicator" value assignable to those species whose distributions extend beyond the lakes. Some of the latter species could be members of what are, effectively, primitively widespread faunas.

For example, Roberts (1975) puts Lake Victoria in the East Coast province, but places Lakes Edward and George, together with Lakes Albert and Turkana, in the Nilo-Sudanian province. At the same time, he notes that Lake Victoria also has some Nilo-Sudanian affinities, and indirectly through that link, affinities with Lakes Edward and George as well. This rather mixed bag of relationships needs further sorting.

Recent phylogenetic studies on the haplochromine cichlid fishes of all five lakes (Greenwood, 1979, 1980) suggest that the majority of species in Lakes Victoria, Edward and George share a recent common ancestry not in turn shared with the haplochromines of Albert and Turkana. The latter apparently are derivable from a recent common ancestor shared with the haplochromines of the Nile and with certain species in the Zaire river. One species from Lake George is, however, a member of this lineage (Greenwood, 1979), and thus raises questions about possible past or even recent contacts between that lake and the Nilo-Zairean systems. Many non-endemic species belonging to families other than the Cichlidae, and common to all or most of these five lakes, have a wide distribution in the Nilo-Sudanian region; some even extend into the Zaire province as well. These species should probably be treated as suspect "primitive cosmopolitans".

Lake Victoria, and to a lesser extent Lakes Edward and George (but not Lakes Turkana and Albert) certainly do share some species exclusively with the East Coast province (Whitehead, 1962). The presence of these species may truly reflect the affinities of the three lakes with the East Coast province.

Contrary to Roberts (1975), I would, therefore, mainly on the evidence of the cichlid fishes, include Lakes Edward and George in the same faunal region as Lake Victoria. But, I would agree with Roberts in considering that region to be part of, or at least to have strong affinities with, the East Coast province. In reaching that conclusion I have placed particular emphasis on the shared East Coast provincial species, and have ignored the supposed evidence provided by those taxa with a much wider distribution. The latter I consider to be remnants of an earlier preprovincial fauna of tropical Africa.

Present-day Lakes Edward and George lack several so called Nilotic (i.e. Nilo-Sudanian) elements in their fish faunas, elements which, in contrast, are present in Lake Victoria (Greenwood, 1959). The good Caenozoic fossil record for Lake Edward shows that these absentee "Nilotics" (with other Nilo-Sudanian species absent from present-day Lake Victoria) were once present in Lake Edward. Some, indeed, were present until a mere 8000 years ago (Greenwood, 1959).

The rapidity, in geological terms, with which the Edward fish fauna lost what are often considered marker species of faunal affinity, should serve to emphasize the frailty of the data on which relationships can be erected.

Studies concerned with the inter- and intrarelationships of the cichlid fishes in Lakes Tanganyika and Malawi are still in their infancy. It is therefore impossible to say what contribution these fishes can make towards identifying the faunal affinities of the two lakes.

Preliminary research on the cichlids of Lake Malawi (Greenwood, in progress) do not bear out the oft repeated suggestion (see Roberts, 1975) that two Zambesian species, *Astatotilapia calliptera* (formerly *Haplochromis callipterus*) and *Pseudocrenilabris philander* are possible ancestors of the haplochromine fishes in that lake. The suggestion could not be corroborated even if it were expressed as "species resembling or closely related to *A. calliptera* and *P. philander*". Roberts' view that some Malawi haplochromines may be related to the genus *Serranochromis* (an endemic genus of the Zambesi province), seems to be a realistic one.

Despite an amazingly close resemblance, in many features, between certain haplochromine species from Lakes Victoria and Malawi, there appears to be no basis for postulating a close phylogenetic relationship between the taxa concerned. In other words, these are examples of homoplasy rather than of holophyly at a low level of universality.

Preliminary results seem to indicate that the cichlid fauna of Lake Malawi was probably derived from several and not very closely related lineages. The relationship of these lineages to others outside the lake remains obscure (Greenwood, in press).

Although some cichlids in Lake Tanganyika are seemingly related to taxa with Zairean provincial affinities (e.g. *Orthochromis*), others probably are related to fishes now occurring in both the Zaire and East Coast provinces (e.g. *Ctenochromis*). Still others belong to a widespread lineage (the genus *Astatotilapia*) whose affinities with the other genera is presently unresolved (Greenwood, 1979).

The non-cichlid fishes in both Lakes Malawi and Tanganyika also present similar problems. Many are endemic to one lake, others are wider ranging taxa, some associated with the Zambesi province (Malawi), others with the Zaire province (Tanganyika), and a few (found in both lakes) with the Nilo-Sudanian region (Roberts, 1975).

Particular interest attaches to a group of catfishes represented in both lakes but nowhere else in Africa. A single species occurs in Tanganyika, and there is a flock of about 12 benthic species in Malawi. Originally considered to be represented by an endemic genus in Lake Tanganyika (*Dinotopterus*) and a second endemic genus in Malawi (*Bathyclarias*), the species were later united in the single genus *Dinotopterus* (Greenwood, 1961). The union has by no means been accepted, although it must be noted that the objectors, bar one, were not taxonomists familiar with the group. Various reasons for keeping the Malawi and Tanganyika species in separate genera have been advanced (see Roberts, 1975:313), but I find none convincing, or at least no more convincing than my original reasons for uniting the genera. Here, typically we have one of those impasses that is unlikely to be cleared until all the related "genera" have been reconsidered from a phylogenetic viewpoint. That will involve testing the validity of current "generic" characters as indicators of phyletic relationship and not merely as, at present, indicators of morphological gaps.

So far I have considered mainly those taxa with a relatively or actually restricted geographical distribution. Some thought must also be given to the not inconsiderable number of taxa which are widely distributed, often over the greater part of tropical Africa and even beyond the tropics.

Species in that category encompass or almost encompass most faunal provinces, from the Nilo-Sudanian in the North to the Zambesi province in the south; many have equally wide distributions from east to west. As examples one may cite the catfishes *Malapterurus electricus*, *Schilbe mystus* and *Heterobranchus longifilis*, the characins *Hydrocynus vittatus*, *Alestes imberi*, *Alestes macrolepidotus* and *Hepsetus odoe*, and the cyprinid *Barbus paludinosus*. In addition there are many genera, most of which seem to be truly holophyletic taxa, with a similarly wide distribution. A few so-called genera with a wide distribution are, it is thought, non-holophyletic (e.g. *Barbus*, *Labeo*, *Varicorhinus*). Thus although, in their entirety, the components of the lineage currently recognized as a genus may have a wide distribution, its truly holophyletic components will have much narrower ranges.

This almost pan-African distribution of certain taxa seems strongly to suggest that at some stage in the past the waterways of Africa were, from the fishes' point of view, accessibly interconnected. Such a situation would permit the wide dispersal of the fauna ancestral to that which we study today (with, I think it is fair to say, unflagging pleasure tinged with occasional bouts of irritation born of frustration).

The break-up of that early hydrographic pattern, the consequent isolation of biotas, and their differentiation (taxonomically and phylogenetically) in that isolation, are all elements of a classical vicariance pattern in historical biogeography.

The finer details of that pattern will only begin to emerge when we know more about the phyletic relationships of the present-day fauna. Until we have this information available on a much broader scale than at present, I do not believe we can advance beyond the level of bio-accountancy into the realms of true biogeographical research.

ACKNOWLEDGEMENTS

As ever, my colleagues, Keith Banister and Gordon Howes, in the Freshwater Fish Section of the British Museum (Natural History) have provided assistance and support in many ways, and I am most grateful to them. The maps were prepared by Gordon Howes, for which task I owe him especial thanks.

REFERENCES

Adams, G.C. (1981). An outline of Tertiary palaeogeography. *In* "Chance, Change and Challenge: The evolving earth" (P.H. Greenwood and L.R.M. Cocks, eds), pp. 221–235. British Museum (Natural History) and Cambridge University Press, Cambridge.

Banister, K.E. and Bailey, R.G. (1979). Fishes collected by the Zaire River Expedition. *Zool. J. Linn. Soc.* **66**, 205–249.

Boulenger, G.A. (1905). The distribution of African fresh water fishes. *Rept. Meet. Br. Assoc. Adv. Sci. (S. Afr.)* **75**, 412–432.

Briggs, J.C. (1979). Ostariophysan zoogeography: an alternative hypothesis. *Copeia* **1979** (1), 111–118.

Darlington, P.J. (1948). The geographical distribution of cold-blooded vertebrates. *Q. Rev. Biol.* **23**, 1–26, 105–23.

Darlington, P.J. (1957). "Zoogeography: the geographical distribution of animals". John Wiley, New York.

Fink, S.V. and Fink, W.L. (1981). Interrelationships of the ostariophysan fishes (Teleostei). *Zool. J. Linn. Soc.*, **72**, 297–353.

Gosline, W.A. (1975). A reexamination of the similarities between the freshwater fishes of Africa and South America. *Mem. Mus. natn Hist. nat., Paris*, Sér. A (Zool.) **88**, 146–155.

Greenwood, P.H. (1959). Quaternary fish fossils. *Explor. Parc. natn Albert, Miss. J. de Heinzelin de Braucourt* **4** (1):1–80.

Greenwood, P.H. (1961). A revision of the genus *Dinotopterus* Blgr. (Pisces, Clariidae) with notes on the comparative anatomy of the suprabranchial organs in the Clariidae. *Bull. Br. Mus. nat. Hist., Zool.* **7**, 215–241.

Greenwood, P.H. (1970) On the genus *Lycoptera* and its relationship with the family Hiodontidae (Pisces, Osteoglossomorpha). *Bull. Br. Mus. nat. Hist., Zool.* **19**, 257–285.

Greenwood, P.H. (1973a). The interrelationships of the Osteoglossomorpha. *Zool. J. Linn. Soc.* **53** (Supplement **1**), 307–332.

Greenwood, P.H. (1973b). Fish fossils from the late Miocene of Tunisia. *Notes Serv. géol Tunis* **37**, 41–72.

Greenwood, P.H. (1976). Fish fauna of the Nile. *In* "The Nile, biology of an ancient river" (J. Rzoska, ed.) [Monographiae Biologicae **29**], pp. 127–139.

Greenwood, P.H. (1979). Towards a phyletic classification of the 'genus' *Haplochromis* (Pisces, Cichlidae) and related taxa. Part I. *Bull. Br. Mus. nat. Hist., Zool.* **35**, 265–322.

Greenwood, P.H. (1980). Towards a phyletic classification of the "genus" *Haplochromis* and related taxa. Part II. *Bull. Br. Mus. nat. Hist., Zool.* **39**, 1–101.

Greenwood, P.H. (1981). Species-flocks and explosive evolution. *In* "Chance, Change and Challenge: The evolving biosphere" (P.H. Greenwood and P.L. Forey, eds), pp. 61–74. British Museum (Natural History) and Cambridge University Press, Cambridge.

Greenwood, P.H. (1983a). The *Ophthalmotilapia* assemblage of cichlid fishes reconsidered. *Bull. Br. Mus. nat. Hist., Zool.* **44**, 249–290.

Greenwood, P.H. (1983b). On *Macropleurodus, Chilotilaphia* (Teleostei, Cichlidae), and the interrelationships of African cichlid species flocks. *Bull. Br. Mus. nat. Hist., Zool.* **45** (4) (in press).

Greenwood, P.H. and Jubb, R.A. (1967). The generic identity of *Labeo quathlambae* Barnard (Pisces-Cyprinidae). *Ann. Cape Prov. Mus. (nat. Hist.)* **6**, 17–37.

Greenwood, P.H. and Patterson, C. (1967). A fossil osteoglossoid fish from Tanzania (East Africa). *J. Linn. Soc. (Zool.)* **47**, 211–223.

Günther, A.G.L.C. (1880). "An Introduction to the Study of Fishes". A. and C. Black, Edinburgh.

Howell, F.C. and Boulière, F. (1963), "African Ecology and Human Evolution". Aldine, Chicago.

Howes, G.J. (1980). The anatomy, phylogeny and classification of bariliine cyprinid fishes. *Bull. Br. Mus. nat. Hist., Zool.* **37**, 129–198.

Liem, K.F. (1963). The comparative osteology and phylogeny of the Anabantoidei. *Illinois biol. Monogr.* **30**, 1–149.

Liem, K.F. and Greenwood, P.H. (1981). A functional approach to the phylogeny of the pharyngognath teleosts. *Am. Zool.* **21**, 81–101.

Matthes, H. (1964). Les poissons du lac Tumba et de la region d'Ikela. *Annls. Mus. r. Afr. cent.*, sér. 8vo (*Zool.*) **126**, 1–206.

Menon, A.G.K. (1951). Distribution of clariid fishes and its significance in zoogeographical studies. *Proc. nat. Inst. Sci. India* **17**, 291–299.

Menon, A.G.K. (1964). Monograph of the cyprinid fishes of the genus *Garra* Hamilton. *Mem. Indian Mus.* **14**, 173–260.

Nelson, G.J. (1969). Infraorbital bones and their bearing on the phylogeny and geography of osteoglossomorph fishes. *Am. Mus. Novit.* **2394**, 1–37.

Nelson, J.S. (1976). "Fishes of the world". John Wiley, New York.

Novacek, M.J. and Marshall, L.G. (1976). Early biogeographic history of ostariophysan fishes. *Copeia* **1976** (1), 1–12.

Patterson, C. (1975). The distribution of Mesozoic freshwater fishes. *Mém. Mus. natn Hist. nat., Paris*, sér. A, *Zool.* **88**, 156–173.

Patterson, C. (1981). The development of the North American fish fauna—a problem of historical biogeography. *In* "Chance, Change and Challenge: The evolving biosphere" (P.H. Greenwood and P.L. Forey, eds), pp. 265–281. British Museum (Natural History) and Cambridge University Press, Cambridge.

Poll, M. (1957). Les genres des poissons d'eau douce de l'Afrique. *Annls. Mus. r. Congo Belge*, sér. 8vo (*Zool.*) **54**, 1–191.

Poll, M. (1963) Zoogéographie ichthyologique du cours supérieur du Lualaba. *Publications de l'Université d'Elisabethville* **6**, 95–106.

Poll, M. (1973). Nombre et distribution géographique des poissons d'eau douce africains. *Bull. Mus. natn Hist. nat. Paris*, sér. 3 **150**, 113–128.

Poll, M. (1976). Poissons. *Explor. Parc. natn. de l'Upemba, Miss. G.F. de Witte, 1946–49*, **73**, 1–127.

Regan, C.T. (1922). The distribution of the fishes of the order Ostariophysi. *Bijdr. Dierk.* **22**, 203–207.

Roberts, T.R. (1975). Geographical distribution of African freshwater fishes. *Zool. J. Linn. Soc.* **57**, 249–319.

Rosen, D.E. and Greenwood, P.H. (1970). Origin of the Weberian apparatus and relationships of the ostariophysan and gonorynchiform fishes. *Am. Mus. Novit.* **2468**, 1–49.

Rosen, D.E. and Greenwood, P.H. (1976). A fourth neotropical species of synbranchid eel and the phylogeny and systematics of synbranchiform fishes. *Bull. Am. Mus. nat. Hist.* **157**, 1–70.

Skelton, P.H. (1976). Preliminary observations on the relationships of *Barbus* species from Cape coastal rivers, South Africa (Cypriniformes: Cyprinidae). *Zoologica Africana* **11**, 399–411.

Stiassny, M.L.J. (1982). The relationships of the neotropical genus *Cichla* (Perciformes, Cichlidae): a phyletic analysis including some functional considerations. *J. Zool. Lond.* **197**, 427–453.

Vari, R.P. (1979). Anatomy, relationships and classification of the families Citharinidae and Distichodontidae (Pisces, Characoidea). *Bull. Br. Mus. nat. Hist., Zool.* **36**, 261–344.

Whitehead, P.J.P. (1962). A new species of *Synodontis* (Pisces Mochocidae) and notes on a mormyrid fish from the eastern rivers of Kenya. *Revue Zool. Bot. afr.* **65**, 97–120.

Wiley, E.O. (1976). The phylogeny and biogeography of fossil and Recent gars (Actinopterygii: Lepisosteidae). *Misc. Publs Mus. nat. Hist. Univ. Kans.* **64**, 1–111.

8 Speciation in the Shallow Sea: General Patterns and Biogeographic Controls

J.W. VALENTINE

Department of Geological Sciences, University of California, Santa Barbara, California 93106, USA

and

D. JABLONSKI

Department of Ecology and Evolutionary Biology, University of Arizona, Tucson, Arizona 85721, USA

Abstract: A wide variety of animal speciation processes has been described, almost exclusively from terrestrial examples. The marine environment differs importantly from the terrestrial with regard to pattern of dispersal and dispersion of genotypes, and these differences will affect speciation modes and patterns. The presence of a pelagic larval stage in the life cycle of many shallow water benthic marine species imparts a potential level of gene flow far greater than in the terrestrial taxa that form the basis for current models of speciation. This effect is particularly marked in species with planktotrophic larvae (feeding larvae with high dispersal capability), and less so in nonplanktotrophic larvae (feeding larvae with low dispersal capability). Configurations of geographic ranges of parent populations (along one-dimensional, linear shelves or within two-dimensional, broad seas) will be a second important factor in determining mechanisms of speciation. Since the breakup of Pangaea, vicariance and clinal modes that result in adaptive divergences (*sensu* Templeton, 1980a) may have predominated on the common linear continental shelves, while founder (or

Systematics Association Special Volume No. 23, "Evolution, Time and Space: The Emergence of the Biosphere", edited by R.W. Sims, J.H. Price and P.E.S. Whalley, 1983, pp. 201–226. Academic Press, London and New York.

small isolate) modes have predominated in two-dimension provinces such as the present Indo-Pacific or on the occasional broad shelves. The latter mode may promote adaptive divergences, or, more rarely, transiliences in which intermediate stages are absent or highly unstable (Templeton, 1980a). During the Paleozoic, two-dimensional provinces were common within epicontinental seas and founder modes may have predominated throughout the sublittoral environment. The rapid rise of evolutionary novelty, believed to be associated chiefly with founding sorts of events that involve transiliences, would have been particularly favoured during late Precambrian and early Phanerozoic times.

INTRODUCTION

Renewed paleontological interest in problems of the tempo, mode, pattern and frequency of speciation has been stimulated by the punctuational model of Eldredge and Gould (1972). Initially, their model was essentially an extension of Mayr's (1963, 1970) model of allopatric speciation into the time scales presented by the fossil record. Many speciation events do appear to be allopatric or parapatric and to involve biogeographic factors such as dispersion patterns, dispersal abilities and the nature of environmental barriers to gene flow. However, none of the important contributions to the speciation literature has dealt specifically with marine animals, among which patterns of dispersal and of gene flow commonly differ significantly from terrestrial forms. In this paper, we make an attempt to modify the appropriate aspects of speciation models for marine conditions in order to make a first-order assessment of their relative importance for marine species during the Phanerozoic.

MODES OF SPECIATION

1. General

The literature on speciation is so extensive that only recent papers of special interest to the present topic can be mentioned. Just since the publication of Eldredge and Gould (1972), large numbers of works on animal speciation have appeared which have greatly broadened our understanding of the multiplicity of speciation processes and have also amended some earlier speciation models (for example, Croizat *et al.*, 1974; Bush, 1975; Carson, 1975; Endler, 1977; White, 1978; Rosenzweig, 1978; Pimm, 1979; Lande, 1980; Templeton, 1980a,b; and references in those works).

Templeton's (1980a) approach to the classification of speciation modes is particularly useful for our purposes because it relies upon mechanisms of speciation rather than upon geographic or other descriptive criteria. As indicated in Fig. 1, the major subdivision he recognized is between *divergent speciation*, in which isolating barriers evolve continuously and are largely mediated by natural selection, though perhaps helped along by founder effects and drift, and *transilitences*, wherein the isolating barriers involve a genetic discontinuity, with extreme instability or absence of intermediate stages. Divergent speciation includes allopatric modes, i.e. the classic founder model involving small isolates plus models of more equally-subdivided populations such as may commonly result from vicariance events, as well as parapatric

	TYPE OF SPECIATION			
POPULATION SPLIT ?	DIVERGENCE		TRANSILENCE	
	ALLOPATRIC	CLINAL	GENETIC	CHROMOSOMAL
●•	++	−	+	+
 ●●	++	−	−	+
⬤	−	++	−	+

Fig. 1. Classification of speciation by nature of the relationship between parental and daughter populations, and by genetic mechanisms involved, after Templeton (1980a). Only those speciation modes believed to be important in marine animals are included. In the "Population Split?" column, upper box depicts a small founder drawn from a parental population, middle box depicts subequal daughter populations separated by an externally imposed barrier, and lower box represents a no-split situation. For the sea, two crosses indicate a common mode, one cross a rarer mode, and a dash means the mode cannot exist. For further explanations, see text.

speciation such as inferred for clines. Other divergent modes occur, for example sympatric speciation modes, but for reasons discussed below, it appears likely that the allopatric divergences, and to a lesser degree clinal divergence modes, are most important in marine speciation and they will be emphasized here.

Transilience speciations, like divergences, may arise allopatrically. Indeed, *genetic transiliences* are defined as resulting from strong epistasis among a few major genes in small or inbreeding founder populations (Templeton, 1980a,b). Other types of transilience, such as through the appearance and fixation of a chromosomal mutation, occur chiefly or entirely in founder populations or in small inbred demes as visualized by Wright (1951, 1977). Speciation via hybridization may also involve a transilience, and large numbers of hybrid zones have been described in terrestrial settings. There are so few data on marine invertebrates, however, that we omit further discussion of this sort of speciation. In general, the following account of speciation events also omits modes that are thought to have been relatively rare during the history of marine animals, although in some instances those modes might have led to major innovations. We do not consider patterns of chronospeciation.

2. *Small Founder Populations and Similar Isolates*

(a) Allopatric divergence. One formulation of allopatric speciation involves a large founder effect. A large parental population is regarded as essentially panmictic and highly co-adapted genetically, with gene flow suppressing local genotypes even though they may be favoured locally by selection (Mayr, 1954, 1963). From this parental source, a small population is split off to found a new population. This founder samples only a little of the genetic variability of the large parental population from which it has been drawn (Mayr, 1942), and is subjected to novel environmental conditions. In this situation some, perhaps many, genes acquire new selective values. It is inferred that the isolate, representing a small founding sample of the parental gene pool and subject to drift, is unlikely to retain co-adaptation of the parental gene pool, which breaks down. A genetic revolution then sweeps the isolate, affecting a great many loci, until a new state of co-adaptation to newly valuable genes is re-established (Mayr, 1954, 1970). The transformation of the gene pool can be rather rapid, and morphological novelty and reproductive isolation from the parental species may result. Thus, a

morphologically distinct species may arise in a small population, usually marginal to the parental population, in what is geologically a very short time.

This scenario has undergone considerable revision since its original inception; here we rely chiefly on the recent contributions of Lande (summarized in Lande, 1980) and Templeton (1980a,b) who lean in turn upon the works of Sewall Wright and others whom they cite. These workers believe that the founder effect is overrated in Mayr's scenario, because a small founding population can sample a large proportion of the gene pool from which it is derived. An effective founding size of only 50 individuals will actually sample about 99% of the variability in a panmictic parental population. Moreover, the variation in many phenotypic characters is considered to be polygenic, with each gene contributing only a relatively small effect. Thus with a reasonably large sample of the parental gene pool a founding population will usually differ little from the parental one.

On the other hand, mutation rates are high for small effects (Lande cites a number of studies suggesting rates on the order of 10^{-2} mutations per gamete per character per generation). When such rates of change are combined with genetic drift, they can shift the morphology of small isolated populations (perhaps a few hundred individuals at most) through distances equivalent to those between sister morphospecies. This may occur during a few hundred to a few thousand generations, even though the morphological pathway may lead through states that are relatively poorly adapted. Once through an adaptive valley, directional selection will, of course, speed the process of change within the isolate. Thus, the creation of a small marginal isolate can, indeed, lead to a new morphospecies in a niche somewhat disjunct from the parental niche and at a geologically rapid rate even in the absence of the process of genetic revolution that has received such emphasis in the past (see Lande, 1980).

The phenotypic homogeneity of large populations is considered by Lande (1980) to be a plausible consequence of stabilizing selection, which is a powerful force in maintaining constant phenotypes; processes of developmental homeostasis or of extensive genetic co-adaptation, maintained by gene flow against some external selective pressures, may operate but are not necessarily required. If stabilizing selection plays the main role in enforcing phenotypic constancy, then populations with uniform morphologies occupy adaptive peaks (Wright, 1931) or plateaux with rather steep sides; morphological variants will be strongly selected against. The adaptive topography of a given region is obviously very different for different phenotypes, since some species vary over regions where others are morphologically uniform.

b. Transilience

Some genes certainly do have significant epistatic effects, and in certain instances a founding event may set the stage for their expression as genetic transiliences (Templeton 1980a,b). These are most likely in small founding populations drawn from large panmictic parental populations. Changes appear to be restricted to a polygenic system with a few to several major genes, which may have effects on a few minor genes (Templeton, 1980b). The bulk of the genome, however, is neutral with respect to the changes involved in the transilience. The major genes have strong epistatic interactions that create adaptive peaks; the valleys between peaks cannot be crossed by selection alone, and thus represent unstable intermediate states. Templeton (1980b) has concluded that genetic transilience is a relatively rare speciation mode.

Chromosomal transiliences are most likely to be effective in speciation when they occur in small populations (Lewis, 1966; Wright, 1941; Stebbins, 1971; Wilson *et al.*, 1975; White, 1978; Templeton, 1980a) which may be newly founded or may be isolates of long standing. These transiliences depend upon the fixation of chromosomal mutation, which usually involves passage through a heterozygous state of lowered fitness and adaptedness—the unstable intermediate. It is this passage that must usually require a small or locally inbred population. It is suspected that chromosomal mutations leading to transiliences often affect gene regulation, and for that matter the major genes involved in genetic transiliences may usually have regulatory effects.

As a special case, chromosomal mutations have been postulated to create new species in a single generation (Lewis, 1966). This might occur when a mutation arises in a selfing plant, which may not be able to back-cross with members of its parental population but which can propagate itself if sufficiently well adapted. Suspected cases of such speciation events have proven difficult to confirm, and no such "instant species" have been documented in animals, although it would seem to be within the realm of possibility for hermaphroditic forms that have some asexual generations.

3. Large Vicariant Populations

Under this heading we consider populations that are split into two (or more) large daughter populations through the imposition of an external barrier. Inter-population gene flow is terminated or so attenuated as not to affect significantly the future of the daughter populations and, consequently, adaptive divergence ensues (Mayr, 1942; Dobzhansky, 1951). This is the sort of speciation labelled as "vicariant" by Hennig (1966) and discussed at length by Croizat *et al.* (1974; see also Nelson and Platnick, 1981). For such speciation events, all daughter populations are "large"—usually much larger than the "small" isolates (of perhaps 100 individuals or less) although there would obviously be an intergradation between these classes. Theoreticians and experimentalists are in general agreement that large vicariant populations will diverge only slowly if they are panmictic (Wright, 1941; Wilson *et al.*, 1975; White, 1978; Templeton, 1980a).

4. Clines

The efficacy of selection to maintain locally advantageous genotypes even in continuous populations has been stressed by White (1978) and by Endler (1977) in his work on speciation in clines. Endler in particular makes the point that dispersion and gene flow as reported in natural terrestrial populations is generally low. The usual dispersion pattern is leptokurtic with most individuals moving minute distances compared with the geographic distribution of their population (or in absolute terms for that matter). A few individuals do disperse over large distances but Endler argues that this does not usually result in much gene flow. The frequencies of the long-distance travellers are low, and random mating encounters greatly favour nearby mates; furthermore the "alien" genes imported by the long-distance traveller face attenuation in frequency for ethological, ecological and physiological reasons; and, finally, the chance loss of any alien genes that *are* introduced will be high because they are rare.

When populations extend over large geographic distances, differential selection pressures arising among different localities can lead to distinctive gene frequency differences which gene flow will not offset. Gradual trends in selection pressure may give rise to clines in gene frequencies (and in morphology). Under some circumstances, narrow zones of rapid gene frequency change may come to separate broader regions of relatively low change in gene frequencies even though trends of selection are gradual, and it is postulated that parapatric speciation may ensue (Endler, 1977). The restricted gene flow discussed by Endler is cited by Lande (1980) in support of his contention that selection and not gene flow is chiefly responsible for the phenotypic homogeneity found in many large, central populations.

DISTINCTIVE CONDITIONS THAT AFFECT SPECIATION IN THE SHALLOW MARINE BENTHOS

Most of the speciation models that have been suggested are based on terrestrial examples (as are all of those discussed above). While it is quite possible that marine equivalents of all terrestrial speciation modes can be found, the distinctive environmental conditions in the sea should result in important differences in the frequencies at which the modes occur and perhaps in some unique details of the modal styles. The main differences between terrestrial and marine speciation are expected to arise from differences in patterns of dispersal and dispersion. Restriction of gene flow is fundamental to speciation models, and both dispersal capacities and geographic range patterns may affect gene flow significantly.

1. Dispersal Patterns

The density of the water medium, in which organisms may live in suspension, creates a major difference between the marine and terrestrial ecosystems. One important consequence is that suspended propagules of benthic species—eggs or larvae—may be planktonic and widely dispersed by currents, maintaining gene flow between populations whose adult members are otherwise quite isolated. Furthermore, minute food items are suspended in the water, making it possible for larvae to prolong their planktonic period by feeding, although not all do so. The reproductive and developmental types of benthic invertebrates have been reviewed and interpreted by a number of workers (for example Thorson, 1950, 1965; Ocklemann, 1965; Scheltema, 1971, 1972, 1977; Mileikovsky, 1971; Vance, 1973; Crisp, 1976; Strathmann, 1978, 1980; Christiansen and Fenchel, 1979; and Jablonski and Lutz, 1980).

For our purposes, the most important distinction among larval types is between *planktotrophs*, forms that have pelagic feeding larvae and that may live in the plankton for weeks to months; and *nonplanktotrophs* that do not. Nonplanktotrophs include *lecithotrophs*, species with planktonic larvae that do not feed, and so-called *direct developers* that hatch directly into benthic juveniles and lack a free larval stage altogether. Lecithotrophic larvae may spend hours to days in the plankton, but often are brooded within parental body spaces or at least remain close to the parent. Since the nonplanktotrophs do not feed they are provided with nourishment (yolk) in the egg, which imposes on the mother a larger energetic cost per egg than the non-yolky eggs of planktotrophs. Fecundity of nonplanktotrophs is, therefore, significantly lower than for planktotrophs, and obviously their dispersal powers are much lower as well. Therefore, we will regard the dispersal of benthic invertebrates to be markedly bimodal, for although there is an area of overlap, the average dispersal capabilities of planktotrophs are clearly greater than those of nonplanktotrophs.

The proportions of developmental types vary in distinctive ways among the major marine environments within the benthic fauna as a whole, although there are many exceptions within individual clades. In shallow waters of the continental shelves, species are overwhelmingly planktotrophic in low latitudes. Among the advantages that have been suggested for this strategy of high fecundity and long larval life are enhanced dispersal of larvae from parents or from siblings, increased gene flow, increased colonizing ability, recruitment success, and greater opportunity to select favourable habitats (see papers cited above). Observed trends are for planktotrophy to decrease in frequency with increasing latitude and increasing water depth. In very high latitudes and in the deep sea, planktotrophs are few, and many of the species are nonplanktotrophs that either brood or are direct developers.

A further trend in marine invertebrate reproductive patterns relates body size to developmental type: small-bodied species generally have non-planktotrophic strategies (which correlates well with the trend towards non-planktotrophy in waters of increasing depth). A common explanation is that all these trends represent trade-offs between fecundity and larval mortality (see especially Vance, 1973; Christiansen and Fenchel, 1979; Valentine and Jablonski, in press). In regions with adequate resources for planktotrophic larvae, enough young may be produced to more than offset the high mortality that accompanies long larval lives. In regions where productivity is seasonal and inclement conditions recur regularly or, worse, irregularly but frequently, fecundity alone may be inadequate to offset mortality so that the best developmental strategy is to invest more energy in each offspring; yolky eggs and, often, brooding behaviours result, and although these strategies lower fecundity, mortality is reduced enough that survivorship is increased. The trend towards nonplanktotrophy in small animals can be explained by assuming that they cannot usually produce enough planktotrophic eggs to opt for a strategy that relies on fecundity, so they enhance survival of fewer young instead (Jablonski and Lutz, 1980 and references therein). Whatever their explanations, the trends are empirically established and must have important consequences for marine speciation patterns.

2. Dispersion Patterns

The continental shelves of the world, together with other marine environments of comparable depths (as around islands), contain a well-defined division of the earth's biota. The shelf fauna is certainly quite different from the terrestrial one in general composition although numbers of marine families possess species that have penetrated fresh water and a few terrestrial forms have become marine. Similarly, although moderate numbers of marine species range across the boundary between the shelf and the deep sea, most shallow-water groups differ at least at the generic and commonly at the familial level from their nearest deep sea allies (Ekman, 1953; Briggs, 1974), and the major radiations in the deep sea have come from a relatively small number of lineages that have penetrated the deep sea environment (e.g. Allen, 1979; Hessler *et al.*, 1979). It is clear that shallow-water marine habitats are distinctive and represent important loci of speciation, inasmuch as more than 90% of marine benthic animal species inhabit such settings today (Valentine, 1973).

Today there are two major patterns of dispersion of benthic shallow-water species. One is associated with the relatively narrow shelves along the present continental margins, bordered by land on one side and deep sea on the other; by virtue of their linear nature, we can regard these as one-dimensional line segments (Fig. 2). The other pattern is associated with broad two-dimensional provinces, as in the Indo-Pacific region where island chains and continental shelf segments form a wide

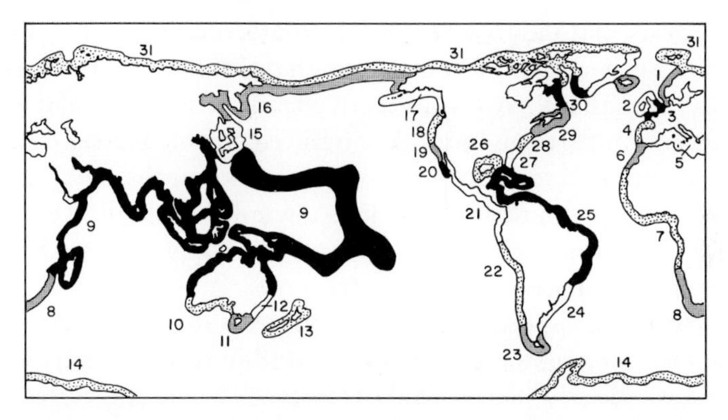

Fig. 2. The major shelf and shelf-depth provinces in the present ocean. All are markedly linear except the Indo-Pacific province (no. 9) and the small Mediterranean Province (no. 5), which are somewhat two-dimensional. After Valentine (1973).

scattering of disjunct but rather similar shallow habitats. It is expected that gene flow patterns would differ between these two types, populations being more continuous along linear shelves but subjected to regional or clinal selection pressures, while patterns of selection and gene migration among scattered two-dimensional arrays of habitats should be more discontinuous and perhaps patchy. Therefore, we treat these patterns separately.

(a) Linear shelves and shelf-depth habitats. There are some data available on patterns of species ranges and compositions on linear shelves. On the extra-tropical northeastern Pacific shelf, the average latitudinal range of shelled molluscan species is about 900 km (Valentine, 1966). For a variety of gastropod families with very different life habits on the shelf (to 200 m) of this same region, the average depth range is 56·7 m, and the median is 40 m (n = 202 species; families included are: Terebridae, Conidae, Olividiae, Littorinidae, Fissurellidae, Haliotidae, Trochidae, Turbinidae, Phasiannelidae, Acmaeidae, Muricidae, Pyrenidae, Nassariidae, and Epitonidae; Jablonski and Valentine, 1981). For the same sample, 93 (46%) range below 50 m and of these, 51 (25%) range below 100 m, but only 17 (8·4%) range into water deeper than 200 m. These figures will certainly vary somewhat from group to group and province to province, but inspection of records of bivalves in the Pacific and of molluscs and other invertebrates in the Atlantic (e.g. Franz and Merrill,

1980) suggests that they are reasonably representative. Thus, there is a clear compositional change in species with depth, but it is not clear from the data at hand that there are distinctive faunal depth boundaries within the shelf of more than local significance. Infaunal assemblages are commonly reported to be the most species-rich near the outer shelf edge (e.g. Carey, 1972), but for the shelf as a whole, the shallower portions are more environmentally heterogenous and support more species of molluscs, many of which are epifaunal.

To summarize, species on the shelf tend to have rather linear distributions, ranging long distances but within relatively narrow depth limits. Most species are in shallow water, where a greater habitat patchiness occurs and where more epifaunal opportunities occur in particular. However, within-habitat species richness may increase in an offshore direction.

Longshore range endpoints of species are not randomly distributed along linear shelves but tend to be clustered at biodistributional boundaries that separate well-known biotic provinces. Environmental discontinuities that produce the boundaries can usually be associated with the thermal regime. At some boundaries different water masses meet over the shelf; at others, currents carrying water from different sources meet over the shelf; while at still others, hydrographic complications within a water mass locally steepen the latitudinal thermal gradient (Valentine, 1966, 1973). Many species end their ranges at points other than the provincial boundaries. Nevertheless, when the molluscan faunas from contiguous provinces are contrasted they usually differ by considerably more than 50% of their species. These provincial boundaries run right across the shelf, evidently terminating only at the shelf-edge thermocline where deep-sea distribution patterns begin (Jablonski and Valentine, 1981).

(b) Two-dimensional shelf-depth provinces. The Indo-Pacific is the outstanding present example of a physically two-dimensional province (Fig. 2). More narrowly ranging species are generally concentrated in the central region and, as one moves towards the provincial margin, these drop out, and one finds progressively fewer but longer-ranging species; some species range from east Africa to the Tuamotu and Marquesa Islands. Diversity tends, therefore, to be highest in the central region and lower towards the provincial margin. There is a low-diversity band in the eastern Indian Ocean that evidently delineates a

west Indian Ocean subprovince (Rosen, 1971; Foin, 1976). Furthermore, at scattered localities around provincial margins are a number of centres of high endemism, as in the Red Sea in the northwest part of the province, and the Hawaiian Islands in the northeast. There are few data on the depth-related changes in diversity and habitat patterns within the province; presumably the trends found on linear shelves would apply.

BENTHIC SPECIATION IN THE SHALLOW SEA

Once the shallow marine biota has achieved a level of species richness appropriate to ambient environmental conditions, speciation should be most likely to occur when environmental changes create new opportunities, either through the extinction of lineages or the expansion of environmental diversity. There are three common sorts of environmental change that seem likely to lead to major speciation opportunities in the shallow sea, either by extinction or by faunal expansion. These are (1) sea level changes that affect the formation and maintenance of barriers; (2) large-scale climatic changes that affect patterns of endemism; and (3) geographic changes that affect patterns of endemism. We now examine the opportunities for different speciation modes provided by such changes to planktotrophs and nonplanktotrophs.

1. Planktotrophic Lineages

In planktotrophs we are faced with species whose powers of dispersal are far greater than in the usual case, by an order of magnitude or more, so that gene flow via pelagic larval dispersal can be maintained over great distances relative to the movement of individual benthic adults. There is substantial morphologic and electrophoretic evidence for the magnitude and homogenizing effect of gene flow in planktotrophic molluscs (Jablonski and Lutz, 1980; Ward and Warwick, 1980). Therefore, speciation that arises from isolation will ordinarily occur only over a broader geographic distance or across stronger ecologic barriers than on land. Speciation models that depend upon differential selection pressures combined with low gene flow will require greater differentials, longer distances, and/or the aid of some partial ecological barriers to gene flow to be as effective as in terrestrial settings.

(a) Linear shelves. The high rate of gene flow among planktotrophs

should provide little opportunity for allopatric or transilient speciation on linear shelves in the absence of some sort of environmental change. Speciation via dispersing founders would ordinarily be restricted to the extreme ends of the elongate ranges. Even there an unusual concomitance of circumstance would be required to isolate a founding population of fewer than a few hundred individuals rather than larger segments. On the other hand, the elongate distributions of species usually entail selection gradients that can shape post-settlement populations (especially today when so many shelves trend north–south and climatic gradients are steep). Morphoclines are certainly common among shelf species, as are clines in electrophoretic gene frequencies (Jablonski and Lutz, 1980; Koehn *et al.*, 1980). Thus, genetic gradients probably exist within many elongate populations so that if gene flow were ever lowered, clinal speciation might occur. Furthermore, if such populations were to be broken into independent segments, there would be an initial genetic difference and divergence could occur at a faster rate than if populations had been identical.

The lowering of gene flow between, or the fragmentation of, linear shelf populations is easily achieved by geographic change. The clearest examples involve tectonic events that create deep-water or land barriers. Deep-water barriers most often arise via continental drift movements that separate formerly contiguous shelves (Valentine, 1971, 1973; Hallam, 1980). Continental rifting occurs at low rates and as it progresses, is accompanied by the appearance of narrow marine arms along the rift zone. Environmental conditions on opposite sides of such a zone would be rather similar and most of the sublittoral populations that migrate into the area would live along both; they would be separated by a widening barrier as rifting continues. Eventually gene flow should be severed, generally first among the populations that lack planktonic larvae, then among planktonic lecithotrophs, and finally among planktotrophs. A few planktotrophs have such long larval lives and are so fecund that it is inferred that they maintain gene flow across the tropical Atlantic even today (Scheltema, 1971, 1977), but tens of thousands of species do not. This isolation by rifting of habitats is in the vicariance mode; Bretsky (1975, 1979) has made this point explicitly.

The most frequently cited example of the creation of a land barrier to marine dispersal is the rise of the Isthmus of Panama between about 2 and 5 million years ago (Woodring, 1966). This vicariant event subdivided many species into Atlantic and Pacific populations, some pairs of which are even now, though morphologically distinguishable, generally quite similar (geminate or twin species; Ekman, 1953; Croizat *et al.*, 1974; Vermeij, 1978).

Climatic change may also lower gene flow or subdivide species ranges and may be the single most important underlying cause of speciation events along linear shelves with north–south trends. Cooling or warming episodes in the sea must usually involve a steepening or lowering of the latitudinal thermal gradient. When the global thermal gradient steepens, there is a sharpening of the thermal change across water mass contacts that create provincial boundaries—both cooler and warmer waters are now available, and the velocity of circulation, horizontal and vertical, increases (Valentine, 1973). However, the geographic location of the provincial boundary, governed by parameters of planetary and atmospheric shape, size and motion, does not change significantly. With an increasing gradient, then, an increasing discontinuity appears at a boundary or incipient boundary. The potential for speciation at such boundaries has been noted by Levinton and Simon (1980).

For example, the Cenozoic has been characterized by an increase in the latitudinal thermal gradient. Temperature distribution on shelves is patchy; bays and shallow areas upcurrent from points or headlands are often cooler owing to upwelling. Thus, as species ranges contracted in the face of a steepening thermal gradient, they commonly left behind somewhat disjunct populations in refuges, in warm spots poleward or cool spots equatorward of the main population. This is particularly noticeable in the fossil record of the marine Pleistocene, wherein range change may be traced because the record has a broad geographic extent and because fossil and living representatives of species commonly have different geographic ranges (for an example in the northeastern Pacific see Valentine, 1961 and references therein). When the disjunct population is separated from the main population by a provincial boundary, the intensification of the boundary discontinuity might well reduce or sever larval dispersal and gene flow and thus promote clinal or vicariant speciation (populations cut off in this manner would nearly always be genetically large, though they might be much smaller than the main population).

This sort of speciation should accompany steepening of the latitudinal thermal gradient as in the Cenozoic example. Not only are boundaries sharpened during such events, but the shortening of the species ranges, as their limiting temperatures are crowded into narrower zones, creates opportunities for speciation. During phases of diminishing thermal gradients, discontinuities become less sharp and the capacity of the shelf to contain species diminishes (Valentine, 1967, 1973). It has been suggested that the provincialization of the world's shelves since the beginning of the Mesozoic and especially during the Cenozoic, both by geographic and climatic changes, has increased the number of invertebrate lineages by over an order of magnitude (Valentine, 1969); most of this speciation, we postulate, would have occurred in clinal or vicariance modes mediated by these geographic and climatic changes. Founding events are not ruled out, of course, and might occur across sharpening provincial boundaries, for example, but we suggest that they play only a minor role on linear shelves.

Sea-level changes are invoked in the classic scenario of marine speciation (for example, Moore, 1954). Falling sea levels have been postulated to reduce habitat diversity and area in epicontinental seas and, therefore, increase extinctions. Rising sea levels permit the appearance or reappearance of new or broad shelf habitats and an accompanying biotic rediversification. The actual speciation events associated with such rising sea level diversifications have not been documented in enough detail to determine the frequencies of various speciation modes. In tropical and subtropical settings at least, the bulk of shallow-water species are planktotrophic and have dispersal powers that on average far surpass those of terrestrial forms. On linear shelves one would expect areal effects to be minimal, since by definition the shelf width is small even before the regressions. Indeed, the Pleistocene invertebrates along the linear eastern Pacific shelves have suffered relatively little extinction, and so far as can be told have undergone little speciation, despite repeated sea-level changes.

(b) Two-dimensional shelf-depth provinces. The geographic pattern of the present Indo-Pacific province, in which habitats are patchily distributed in two dimensions, suggests that for planktotrophs the chief speciation pattern is via founder events to create isolates. The studies of speciation in Hawaiian drosophilids by Carson and his colleagues (review in Carson *et al.*, 1970), while not precisely analogous, provide an example on which to base notions of speciation in this setting. In the case of drosophilids, some speciations have resulted from the chance dispersal of one or few founding individuals from a parental island. On

theoretical grounds this is a likely interpretation of the speciation mode among many of the Hawaiian drosophilids (see Templeton, 1980a). A similar process may also operate among marine planktotrophs in two-dimensional shelf provinces. Chance events permitting rare dispersals into new areas (e.g. variations in current patterns) should often involve large numbers of planktotrophic larvae: founding populations could thus be quite large, at least at settling. However, because high mortalities are associated with settling, metamorphosis and early post-larval ontogeny (e.g. Thorson 1950; other references in Jablonski and Lutz, 1980), large founding larval populations would often be reduced by several orders of magnitude to quite modest founding adult populations. The generally large gene flow assumed among the scattered but relatively densely distributed islands of the central Indo-Pacific region would provide a fairly panmictic gene pool from which small founding adult populations would be drawn. These are ideal conditions for transiliences. It seems reasonable to ascribe the great species richness of many Indo-Pacific clades to founder speciations (sometimes with accompanying transiliences), with secondary sympatry achieved commonly enough that large numbers of sympatric congeners accumulate in particularly successful groups.

Climatic and geographic changes would be a potential source of vicariant speciation opportunities in the Indo-Pacific styles of two-dimensional provinces. Falling current velocities associated with lowered climatic gradients could sever gene flow to outlying populations, as could switches in current directions associated with changing geographies. Plate movements may shift the geographic positions of islands, severing old routes of gene flow and giving rise to new ones (Rotondo *et al.*, 1981). Subsidence of island tracts could create gaps between shallow habitats and permit vicariance. Rising sea levels would change available habitat areas and could also create gaps through drowning of island tracts. Oceanic island provinces that span the margins of oceanic boundary currents are liable to dissection.

Another sort of two-dimensional province that has been common in the past is produced by marine transgressions over broad expanses of the continental platforms. Such epicontinental seas are environmentally patchy, because relief on the platforms creates islands, basins, straits and embayments, but the patches are coarser and fewer than for the oceanic Indo-Pacific province. Speciation potential in this setting would probably be intermediate between oceanic two-dimensional provinces and narrow linear shelves. It is on these broad epicontinental shelves that opportunities for speciation are most likely to accompany sea level changes (Fig. 3). Broad shelf seas should not usually support geostrophic currents, so gyral current systems are not expected. Instead, distinctive water types should be associated with specific embayments and "arms of the sea", areas somewhat cut off from others. These embayments have served as centres of endemism; whether they were usually as faunally distinctive as the present shelf provinces is uncertain. Examples include the distinctive eastern and western regions of the North American craton during early and middle Devonian time (Johnson, 1971), and the relatively shallow series of embayments of the Atlantic and Gulf coastal plains of North America during the late

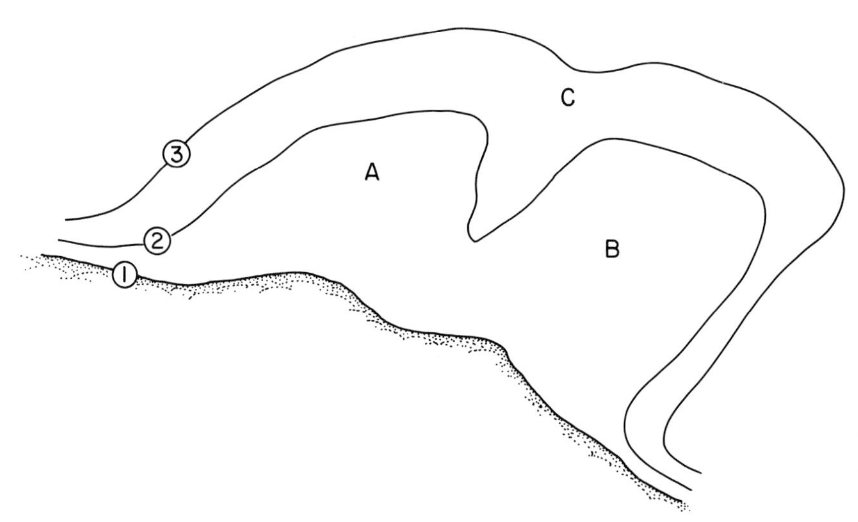

Fig. 3. Hypothetical topography to show possible effects of changing sea levels on provinciality in an epicontinental sea. Coastline no. 1, a low sea level, is fairly straight and embraces a single province (visualized as perhaps 1000 to 2000 km across). Coastline no. 2, resulting from transgression, is broken by an elongate high into two deep embayments A and B, which may contain enough endemic species to qualify as separate provinces. Coastline no. 3 results from further transgression which drowns the high and causes a reduction from two provinces to a single province in the large embayment, C. Regression from this position may recreate two provinces if there is a sea stand near coastline 2, but regression from coastline 2 to coastline 1 will reduce provinciality. Thus either transgression or regression may provincialize, depending upon conditions.

Cretaceous (Jablonski, 1980). We would expect much of the original endemism in these examples to be due to lowered gene flow related to the distinctive embayments created during the sea-level change, and thus, to be clinal or vicariant. Either rising or falling sea levels can promote isolation and endemism, depending upon the topographic situation (Fig. 3). Increased endemism is reported to be associated with both transgressions (Kauffman, 1973) and regressions (Hallam, 1977) in different circumstances. If broader shelves can accommodate more species than narrow ones as suggested by island biogeographic studies (MacArthur and Wilson, 1967), then transgressions should create opportunities for further species enrichment within embayments. These would ordinarily be exploited by founders from adjoining provinces or endemic centres.

2. Nonplanktotrophic Species

We shall consider nonplanktotrophs to be rather poor dispersers in order to contrast them with planktotrophs, although some are probably fair dispersers. For poor dispersers, one expects a low ability to found populations, but a great likelihood of founded populations being isolates. It does not seem possible to estimate from first principles the evolutionary outcome of these two tendencies, but available empirical evidence indicates that nonplanktotrophs display more rapid speciation rates—and extinction rates—than planktotrophs (Shuto, 1974; Scheltema, 1977, 1978; Hansen, 1978, 1980; Jablonski, 1980). Vicariant and clinal speciation should be more common among the poorly dispersing nonplanktotrophs, however, and it is possible that the available data merely reflect this tendency rather than ease of isolation of founder populations. Morphologic and electrophoretic evidence does indicate more heterogenous population structures in nonplanktotrophic molluscs (reference in Jablonski and Lutz, 1980), and these questions are certainly testable in the fossil record as well.

(a) Linear shelves. Nonplanktotrophic populations should be relatively easily segmented along linear shelves when appropriate environmental discontinuities arise. This is most likely in nearshore settings where habitat patchiness is greatest. When linear shelves are not interrupted by major habitat discontinuities, even nonplanktotrophic dispersal may occur over long distances rendering founder events relatively unlikely,

but greatly increasing the possibilities of clinal speciation. Therefore, we expect vicariant and clinal speciation to be dominant among nonplanktotrophs as it is among planktotrophs in linear shelf settings.

(b) Two-dimensional shelf-depth provinces. Since these provinces are characterized by moderately highly discontinuous habitats, poor dispersers will tend to be highly isolated, especially in oceanic provinces, and founding events, uncommon as they might be, would usually result in new isolates. The creation of externally-imposed barriers would result in widespread isolation among nonplanktotrophs. Whether speciation in such provinces is more common via vicariance or founders would depend upon local history. In epicontinental two-dimensional provinces we expect founding events to be proportionately more common among nonplanktotrophs than planktotrophs, because nonplanktotrophs will perceive this environment as much patchier with more barriers, while opportunities for isolation by vicariant events or for clinal speciation may be nearly the same for both types. Founding events, though probably *more* common in nonplanktotrophs would be less likely to be accompanied by genetic transiliences, because founders would probably not be drawn from a panmictic population, as they would for planktotrophs. Chromosomal transiliences, on the other hand, are considered to be more likely when the founder is drawn from an already fragmented parental gene pool (Templeton, 1980a).

GENERAL TRENDS OF PHANEROZOIC SPECIATION IN THE SEA

Opportunities for speciation have clearly varied in number over geologic time, and it seems likely that there has been variation in the favoured mechanisms of speciation as well, resulting in changing frequencies of speciation modes. We will try to sketch only the broadest of outlines here.

During the Paleozoic, major opportunities for vicariant speciation seem to have been relatively few. The general pattern of continental movement was convergent, and any speciation from this pattern should have been via founders that first crossed closing seaways. Shelves were often broad epicontinental seas and seaways. Provinciality was low, and endemic centres were distributed in subcontinental embayments and arms more than in latitudinal provinces, though some isolated continents certainly supported large endemic faunas. The most opportunity for clinal and vicariant speciation was probably associated with the subdivision of epicontinental seas by changing sea levels. Many Paleozoic benthic communities were dominated by articulate brachiopods among the shelled forms; today at least articulates are entirely nonplanktotrophic (Strathmann, 1978). If there was a large nonplanktotrophic element in Paleozoic shelf sea faunas, the epicontinental provinces would have been perceived as highly patchy and isolation of the nonplanktotrophic populations relatively easily achieved.

By contrast, the late Mesozoic and Cenozoic have been characterized by numerous opportunities for clinal and vicariant speciation. Pangaea has been fragmented and scattered and the latitudinal thermal gradient has increased, provincializing the marine biosphere. Seas have usually been restricted to rather narrow shelves, although the great Cretaceous transgression provides an important exception. We therefore suggest that while founding events dominate or at least loom large during the Paleozoic, vicariant and clinal speciation have been dominant during at least late Mesozoic and Cenozoic times.

Vicariant and clinal speciation would ordinarily involve rather large daughter populations at least when compared to founder populations, and should not provide appropriate conditions for transiliences. Indeed, the adaptive divergence of vicariant sister species is rather slow, from both theoretical considerations and observational data, unless they are highly subdivided into small semi-isolated populations as discussed by Wright (1941). Vicariance events are, therefore, unlikely to appear as instantaneous as founder-associated "punctuational" speciations, given a reasonable unbroken fossil record. It would be of considerable interest to attempt to trace the pace of morphological change between lineages that have clearly been separated vicariantly, as across the Atlantic during the Cretaceous and early Cenozoic, and across the Isthmus of Panama during the last 2 to 5 million years. These would be the most likely conditions for gradual patterns; successful detection of such phenomena in these settings would increase our confidence in punctuational patterns when they are found elsewhere in the fossil record.

This first-order gradualistic pattern could be obscured, however, if rapid within-province speciations were frequent during intervals of pervasive between-province vicariance (as suggested, for example, by the Late Cretaceous and Early Tertiary studies of Jablonski, 1980, and Hansen, 1980). In this kind of situation, the ultimate mode of speciation might often appear punctuational, because of the rate differentials of the speciation modes involved. Populations isolated by plate motions could give rise to intraprovincially derived species before those populations had a chance to exhibit gradual morphologic divergence from their extraprovincial sister populations. Consideration of these and other intraprovincial processes will be the subject of another paper (Jablonski and Valentine, in prep.).

Founder-established speciation events will usually appear to be punctuational because they must be completed rapidly, at least during the small-population stage when morphological change is inferred to be fastest. According to our reasoning they would have been proportionately much more common during the Palaeozoic than in the post-Palaeozoic. Also, transiliences seem most likely to occur in founder-like situations and it is here that truly novel morphologies can be developed. Major novelties may well require rare, major or prolonged transilience episodes. Given late Precambrian and early Palaeozoic environments that were relatively empty of higher metazoan adaptive types, and a speciation mode that permits the broaching of unstable intermediates between stable adaptive types, the road to major novelty, though containing many hazards, would seem to have been passable.

ACKNOWLEDGEMENTS

We thank J.A. Endler, C.S. Hickman, S.M. Kidwell, L. Lugar, R.E. Michod, and S.S. Sweet for their critical reviews of the manuscript. J.D. Taylor provided information on Indo-Pacific molluscan patterns. Jablonski gratefully acknowledges support from the Miller Institute for Basic Research in Science, and Valentine acknowledges a Faculty Research Grant from the University of California, Santa Barbara. This research was supported in part by the National Science Foundation, Earth Sciences Section, Grant No. EAR78–15536.

REFERENCES

Allen, J.A. (1979). The adaptations and radiation of deep-sea bivalves. *Sarsia* **64**, 19–27.

Bretsky, P.W., Jr. (1969). Evolution of Paleozoic benthic marine invertebrate communities. *Palaeogeogr., Palaeoclimat., Palaeoecol.* **6**, 45–59.

Bretsky, S.S. (1975). Allopatry and ancestors: A response to Cracraft. *Syst. Zool.* **24**, 113–119.

Bretsky, S.S. (1979). Recognition of ancestor-descendent relationships in invertebrate paleontology. *In* "Phylogenetic analysis and paleontology" (J. Cracraft and N. Eldredge, eds), pp. 113–163. Columbia University Press, New York.

Briggs, J.C. (1974). "Marine Zoogeography". McGraw-Hill, New York.

Bush, G.L. (1975). Modes of animal speciation. *A. Rev. Ecol. Syst.* **6**, 339–364.

Carey, A.G., Jr. (1972). Ecological observations on the benthic invertebrates from the central Oregon continental shelf. *In* "The Columbia River Estuary and Adjacent Waters" (A.T. Pruter and D.L. Alverson, eds), pp. 422–443. University of Washington Press, Seattle.

Carson, H.L. (1975). The genetics of speciation at the diploid level. *Am. Nat.* **109**, 83–92.

Carson, H.L., Hardy, D.E., Spieth, H.T., and Stone, W.S. (1970). The evolutionary biology of the Hawaiian Drosophilidae. *In* "Essays in Evolution and Genetics in Honor of Theodosius Dobzhansky" (M.K. Hecht and W.C. Steere, eds), pp. 437–543. Appleton-Century-Crofts, New York.

Christiansen, F.B. and Fenchel, T.M. (1979). Evolution of marine invertebrate reproductive patterns. *Theor. Pop. Biol.* **16**, 267–282.

Crisp, D.J. (1976). The role of the pelagic larva. *In* "Perspectives in Experimental Zoology" (P. Spencer-Davies, ed.), pp. 145–155. Pergamon, Oxford.

Croizat, L., Nelson, G., and Rosen, D.E. (1974). Centers of origin and related concepts. *Syst. Zool.* **23**, 265–267.

Dobzhansky, T. (1951). "Genetics and the Origin of Species" (3rd edn). Columbia University Press, New York.

Ekman, S. (1953). "Zoogeography of the Sea." Sidgewick and Jackson, London.

Eldredge, N. (1971). The allopatric model and phylogeny in Paleozoic invertebrates. *Evolution* **28**, 479–481.

Eldredge, N. and Gould, S.J. (1972). Punctuated equilibria: An alternative to phyletic gradualism. *In* "Models in Paleobiology" (T.J.M. Schopf, ed.), pp. 82–115. Freeman, Cooper, San Francisco.

Endler, J.A. (1977). "Geographic Variation, Speciation and Clines". Princeton University Press, Princeton, N.J.

Foin, T.C. (1976). Plate tectonics and the biogeography of the Cypraeidae (Mollusca: Gastropoda). *J. Biogeogr.* **3**, 19–34.

Franz, D.R. and Merrill, A.S. (1980). Molluscan distribution patterns on the continental shelf of the Middle Atlantic Bight (northwest Atlantic). *Malacologia* **19**, 209–225.

Futuyma, D.J., and Mayer, G.C. (1980). Non-allopatric speciation in animals. *Syst. Zool.* **29**, 254–271.

Gould, S.J., and Eldredge, N. (1977). Punctuated equilibria: The tempo and mode of evolution reconsidered. *Paleobiology* **3**, 115–151.

Hallam, A. (1977). Jurassic bivalve biogeography. *Paleobiology* **3**, 58–73.

Hallam, A. (1980). Relative importance of plate movements, eustasy, and climate in controlling major biogeographical changes since the early

Mesozoic. *In* "Vicariance Biogeography: A Critique" (G. Nelson and D.E. Rosen, eds), pp. 303–330. Columbia University Press, New York.

Hansen, T.A. (1978). Larval dispersal and species longevity in Lower Tertiary gastropods. *Science, N.Y.* **199**; 885–887.

Hennig, W. (1966). "Phylogenetic Systematics". University Ill. Press, Urbana.

Jablonski, D. (1980). Apparent versus real biotic effects of transgressions and regressions. *Paleobiology* **6**, 397–407.

Jablonski, D. and Lutz, R.A. (1980). Molluscan larval shell morphology: Ecological and paleontological applications. *In* "Skeletal Growth of Aquatic Oraganisms" (D.C. Rhoads and R.A. Lutz, eds), pp. 323–377. Plenum, New York.

Jablonski, D. and Valentine, J.W. (1981). Onshore-offshore gradients in Recent Eastern Pacific shelf faunas and their paleobiogeographic significance. *In* "Evolution Today: Proceedings of the 2nd International Congress of Systematic and Evolutionary Biology" (G.G.E. Scudder and J.L. Reveal, eds), pp. 441–453. Carnegie-Mellon University, Pittsburgh.

Jackson, J.B.C. (1974). Biogeographic consequences of eurytopy and stenotopy among marine bivalves and their evolutionary significance. *Am. Nat.* **108**, 541–560.

Johnson, J.G. (1971). A quantitative approach to faunal province analysis. *Am. J. Sci.* **270**, 257–280.

Kauffman, E.G. (1973). Cretaceous Bivalvia. *In* "Atlas of Paleobiogeography" (A. Hallam, ed.), pp. 353–383. Elsevier, Amsterdam.

Koehn, R.K., Bayne, B.L., Moore, M.N., and Siebenaller, J.F. (1980). Salinity related physiological and genetic differences between populations of Mytilus edulis. *Biol. J. Linn. Soc.* **14**, 319–334.

Lande, R. (1980). Genetic variation and phenotypic evolution during allopatric speciation. *Am. Nat.* **116**, 463–479.

Levinton, J.S. and Simon, C.M. (1980). A critique of the punctuated equilibria model and implications for the detection of speciation in the fossil record. *Syst. Zool.* **29**, 130–142.

Lewis, H. (1966). Speciation in flowering plants. *Science, N.Y.* **152**, 167–172.

MacArthur, R. and Wilson, E.O. (1967). "The Theory of Island Biogeography". Princeton University Press, Princeton, N.J.

Mayr, E. (1942). "Systematics and the Origin of Species". Columbia University Press, New York.

Mayr, E. (1954). Change of genetic environment and evolution. *In* "Evolution as a Process" (J. Huxley, A.C. Hardy and E.B. Ford, eds), pp. 157–180. Allen and Unwin, London.

Mayr, E. (1963). "Animal Species and Evolution". Harvard University Press, Cambridge, Mass.

Mayr, E. (1970). "Populations, Species, and Evolution". Harvard University Press, Cambridge, Mass.

Mileikovsky, S.A. (1971). Types of larval development in marine bottom invertebrates, their distribution and ecological significance: A re-evaluation. *Mar. Biol.* **10**, 193–213.

Moore, R.C. (1954). Evolution of late Paleozoic invertebrates in response to major oscillations of shallow seas. *Bull. Mus. Comp. Zool. Harvard Coll.* **112**, 259–286.

Nelson, G. and Platnick, N. (1981). "Systematics and Biogeography: Cladistics and Vicariance". Columbia University Press, New York.

Ocklemann, W.K. (1965). Developmental types in marine bivalves and their distribution along the Atlantic coast of Europe. *Proc. Eur. Malac. Congr.* **1**, 24–35.

Pimm, S.L. (1979). Sympatric speciation: a simulation model. *Biol. J. Linn. Soc.* **11**, 131–139.

Reaka, M.L. (1980). Geographic range, life history patterns, and body size in a guild of coral-dwelling mantis shrimps. *Evolution, Lancaster, Pa* **34**, 1019–1030.

Rosen, B.R. (1971). The distribution of reef coral genera in the Indian Ocean. *Symp. Zool. Soc. London* **28**, 263–299.

Rotondo, G.M., Springer, V.C., Scott, G.A.J., and Schlanger, S.O. (1981). Plate movement and island integration—a possible mechanism in the formation of endemic biotas, with special reference to the Hawaiian Islands. *Syst. Zool.* **30**, 12–21.

Rosenzweig, M.L. (1978). Competitive speciation. *Biol. J. Linn. Soc.* **10**, 275–289.

Scheltema, R.S. (1971). The dispersal of the larvae of shoal-water benthic invertebrate species over long distances by ocean currents. *In* "Fourth European Marine Biology Symposium" (D.J. Crisp, ed.), pp. 7–28. Cambridge University Press, Cambridge.

Scheltema, R.S. (1972). Reproduction and dispersal of bottom-dwelling deep sea invertebrates: A speculative summary. *In* "Barobiology and the Experimental Biology of the Deep Sea" (R.W. Brauer, ed.), pp. 58–66. North Carolina University Press, Chapel Hill, N.C.

Scheltema, R.S. (1977). Dispersal of marine invertebrate organisms: Paleobiogeographic and biostratigraphic implications. *In* "Concepts and Methods of Biostratigraphy" (E.G. Kauffman and J.E. Hazel, eds), pp. 73–108. Dowden, Hutchinson and Ross, Stroudsburg, Pa.

Scheltema, R.S. (1978). On the relationship between dispersal of pelagic veliger larvae and the evolution of marine prosobranch gastropods. *In* "Marine Organisms: Genetics, Ecology and Evolution" (B. Battaglia and J.A. Beardmore, eds), pp. 302–322. Plenum, New York.

Shuto, T. (1974). Larval ecology of prosobranch gastropods and its bearing on biogeography and paleontology. *Lethaia* **7**, 239–256.

Smith, J.M. (1966). Sympatric speciation. *Am. Nat.* **100**, 637–650.

Stanley, S.M., Addicott, W.O., and Chinzei, K. (1980). Lyellian curves in paleontology: Possibilities and limitations. *Geology* **8**, 422–426.

Stebbins, G.L. (1971). "Chromosomal Evolution in Higher Plants". E. Arnold, London.

Strathmann, R.R. (1978). The evolution and loss of feeding larval stages of marine invertebrates. *Evolution, Lancaster, Pa* **32**, 894–906.

Strathmann, R.R. (1980). Why does a larva swim so long? *Paleobiology* **6**, 373–376.

Templeton, A.R. (1980a). Modes of speciation and inferences based on genetic distances. *Evolution, Lancaster, Pa.* **34**, 719–729.

Templeton, A.R. (1980b). The theory of speciation via the founder principle. *Genetics* **94**, 1011–1038.

Thorson, G. (1950). Reproductive and larval ecology of benthic marine invertebrates. *Biol. Rev.* **25**, 1–45.

Vance, R.R. (1973). On reproductive strategies in marine benthic invertebrates. *Am. Nat.* **107**, 339–352.

Valentine, J.W. (1961). Paleoecologic molluscan geography of the California Pleistocene. *Univ. California Publs. Geol. Sci.* **34**, 309–442.

Valentine, J.W. (1966). Numerical analysis of marine molluscan ranges on the extratropical northeastern Pacific shelf. *Limnol. Oceanogr.* **11**, 198–211.

Valentine, J.W. (1967). Influence of climatic fluctuations on species diversity within the Tethyan provincial system. *In* "Aspects of Tethyan Biogeography" (C.G. Adams and D.V. Ager, eds), pp. 153–166. Systematics Association, London.

Valentine, J.W. (1969). Patterns of taxonomic and ecological structure of the shelf benthos during Phanerozoic time. *Palaeontology* **12**, 684–709.

Valentine, J.W. (1971). Plate tectonics and shallow marine diversity and endemism, an actualistic model. *Syst. Zool.* **20**, 253–264.

Valentine, J.W. (1973). "Evolutionary Paleoecology of the Marine Biosphere". Prentice Hall, Englewood Cliffs, N.J.

Valentine, J.W., Foin, T.C., and Peart, D. (1978). A provincial model of Phanerozoic marine diversity. *Paleobiology* **4**, 55–66.

Valentine, J.W. and Jablonski, D. (in press). Larval strategies and brachiopod diversity patterns. *Evolution, Lancaster, Pa.*

Vermeij, G. (1978). "Biogeography and Adaptation; Patterns of Marine Life". Harvard University Press, Cambridge, Mass.

Ward, R.D. and Warwick, T. (1980). Genetic differentiation in the molluscan species *Littorina rudis* and *Littorina arcana* (Prosobranchia: Littorinidae). *Biol. J. Linn. Soc.* **14**, 417–428.

White, M.J.D. (1978). "Modes of Speciation". Freeman, San Francisco.

Wilson, A.C., Bush, G.L., Case S.M., and King, M.C. (1975). Social structuring of mammalian populations and the rate of chromosomal evolution. *Proc. natn. Acad. Sci. U.S.A.* **72**, 5061–5065.

Woodring, W.P. (1966). The Panama land bridge as a sea barrier. *Proc. Am. Phil. Soc.* **110**, 425–434.

Wright, S. (1931). Evolution in Mendelian populations. *Genetics* **16**, 97–159.

Wright, S. (1941). On the probability of fixation of reciprocal translocations. *Am. Nat.* **75**, 513–522.

Wright, S. (1951). The genetical structure of populations. *Am. Eugen.* **15**, 323–354.

Wright, S. (1977). "Evolution and the Genetics of Populations", vol. 3, "Experimental Results and Evolutionary Deductions". University of Chicago Press, Chicago.

9 The Origin and Biogeography of Malacostracan Crustaceans in the Deep Sea

R.R. HESSLER and *G.D.F. WILSON*

Scripps Institution of Oceanography, La Jolla, California 92093, USA

Abstract: Malacostracans show three deep-sea distribution patterns: poor representation (decapods and stomatopods), diverse representation with high taxonomic endemicity (isopods), and diverse presence with endemicity at only low taxonomic levels (amphipods and cumaceans). These patterns have both historical and ecological causes. Poor representation in decapods and stomatopods reflects inability to cope with deep-sea conditions rather than recency of invasion. Potential immigrants must adjust to low nutrient supply, high pressure, absence of light, muddy substrate and low temperature. Today, high latitudes are the most likely places for deep-sea invasions because the water column is isothermal. However, prior to the Eocene, the oceans were warmer and relatively isothermal worldwide. This, plus the different distribution of land masses, must have resulted in different invasion patterns. The contrast in level of endemicity in deep-sea isopods and amphipod-cumaceans is yet unexplained because the three orders seem equally suited to deep-sea conditions.

INTRODUCTION

As long as man has known about the deep sea, its strangeness has intrigued him. It is an environment of near-freezing temperature and extraordinarily high pressure. Physically derived light is lacking, and the rate of nutrient supply is extremely low. One would expect that if

---

Systematics Association Special Volume No. 23, "Evolution, Time and Space: The Emergence of the Biosphere", edited by R.W. Sims, J.H. Price and P.E.S. Whalley, 1983, pp. 227–254. Academic Press, London and New York.

life could exist at all, it would be different from that of shallow seas. Animals collected from the deep-sea bottom over the past 150 years show that this is indeed the case. Faunal differences occur at all taxonomic levels. They are most pronounced in lower categories, but there are even phyla (Pogonophora) or classes (Monoplacophora) which basically are limited to the deep sea. When a taxon occurs in both shallow and deep water, it often experiences a change in relative abundance. This paper describes some of these patterns in an important marine crustacean group, the Malacostraca, and considers the ways in which these patterns might have come about.

DEPTH PATTERNS OF MALACOSTRACAN DIVERSITY

Some taxa that are important faunal elements in shallow water are poorly represented in the deep sea. Among crustaceans, the Stomatopoda and Decapoda are dramatic examples. Of approximately 300 known stomatopod species, only 14 occur deeper than 300 m (Manning and Struhsaker, 1976). In the western Atlantic 57 species occur in the littoral (<200 m), 15 in the upper bathyal (200–500 m), but only two stomatopod species are found below 500 m (Manning, 1969). They are not known deeper than 1300 m.

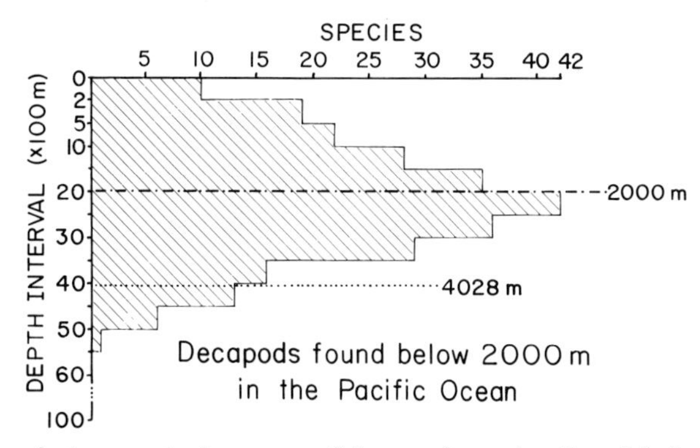

Fig. 1. Cumulative vertical ranges of decapod species found below 2000 m in the Pacific Ocean (data from Zarenkov, 1969). Because many species are eurybathic, a substantial portion of the cumulative ranges lie above 2000 m. The data include 55 species of 25 genera and 13 families; 39 species have ranges greater than 1000 m. The depth at 4028 m marks the mean depth of the Pacific Ocean, including adjacent seas (Sverdrup *et al.*, 1942). This shows that the bulk of decapod diversity lies on bathyal slopes or shallower.

In a summary of all known Pacific records of deep-sea decapods (those whose range includes depths greater than 2000 m), Zarenkov (1969) recorded a precipitous decline in species with depth; only six are known deeper than 4500 m, and only one below 5000 m (Fig. 1). This is a minimum number because lower sampling effort at greater depths introduces a bias, as does the probable lower efficiency of sampling animals capable of net avoidance. This sampling bias is shown by baited camera studies which record natantians at 5500–6000 m in the central North Pacific (Hessler, personal observations). Nevertheless, the general pattern of decrease is probably valid. Extensive trawlings to 4000 m in the Gulf of Mexico (Pequegnat *et al.*, 1971) (Fig. 2) and to 3000 m off New England (Wenner and Boesch, 1979) corroborate the Pacific pattern. Because of the shoaler lower limit of the studies, several species are present in the deepest depth interval. Paguridea were not included in the Gulf of Mexico study.

Not all decapodan taxa are represented equally well at depth (Balss, 1955). Brachyura are restricted almost entirely to the littoral and upper bathyal; of the approximately 220 species of Brachyura occurring in the Gulf of Mexico, only 36 live at or below 180 m (Pequegnat, 1970b). Off eastern Australia, only 30 species (5% of the total Australia crab fauna) have been recorded from the continental slopes (Griffin and Brown, 1975). The Anomura are more diverse in deep water and display a variety of vertical distribution patterns. Paguridea are most diverse in

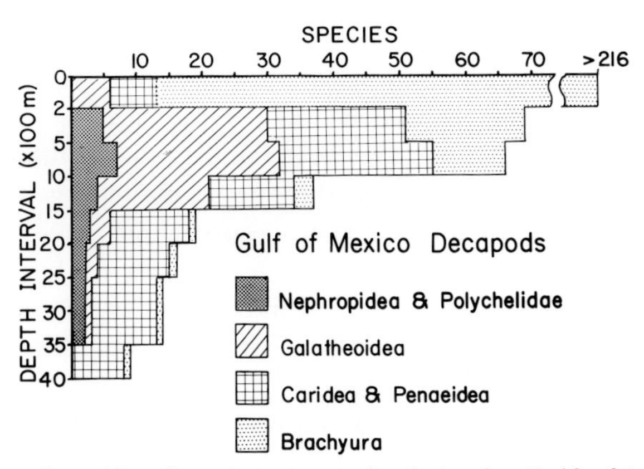

Fig. 2. Decapod species diversity versus depth in the Gulf of Mexico (data summarized in Pequegnat *et al.*, 1971). Caridea and Penaeidea found above 200 m include only the outer shelf species. Paguridea were not included in the study.

littoral regions of the topics, but have important genera in the deep sea (Wolff, 1961). Lithodidae, on the other hand, are found in shallow water only in boreal and arctic seas, but are characteristic in bathyal faunas (Makarov, 1938; Ekman, 1953). The Galatheoidea, whose greatest shallow-water diversity is in the tropics, are the most varied anomurans in the deep sea (Dolflein and Balss, 1913; Tirmizi, 1966; Pequegnat and Pequegnat, 1970; Ambler, 1979). In contrast to groups that are most diverse in shallow water, the Nephropoidea (Holthuis, 1974) and Polychelidae (Firth and Pequegnat, 1971) have centres of diversity in the bathyal, and the polychelid *Willemoesia* is purely abyssal. Natantians are diverse in shallow water and are a common component among decapods in the deep sea (Crosnier and Forest, 1967; Roberts and Pequegnat, 1970; Pequegnat, 1970a).

In contrast to the decapods, peracarids thrive in the deep sea. Here, we will not consider the Mysidacea, which are basically pelagic (Mauchline, 1980). Cumaceans, tanaids, isopods and amphipods are all well represented (Sanders *et al.*, 1965; Hessler and Sanders, 1967). High species diversity in individual samples documents peracarid success (Table I). It far exceeds that which can be found on comparable bottoms in shallow water. From 32 to 51% of the species (all phyla) in these diverse, deep-sea samples are peracarids.

Changes in the diversity of a taxon can also involve changes in the proportions of its subtaxa. Complete replacements of species with depth takes place in all higher taxa, and genera which are endemic to the deep sea or shallow water are common. The distributional patterns of families are more varied.

In the gammaridean amphipods, no benthic families are endemic to the deep sea (Barnard, 1973). Of the 48 families with benthic representatives (Barnard, 1969), 31 have benthic species in the deep sea. Of the 20 families having 10 or more genera, only three (two littoral, one terrestrial) do not have deep-sea representatives (Table II; Barnard, 1969). At the same time, most of these families are dominated by shallow-water genera; only six have over 25% of their genera represented in the deep sea, and only two have over 50%. Thus, gammaridean families are broadly represented in the deep sea, but very few families have special development there (Barnard, 1961, 1962). This is the pattern of a group that invaded the deep sea many times.

Cumaceans display a similar pattern (Jones and Sanders, 1972). Only one (Ceratocumidae) of the eight families is limited to the deep sea, and

Table I. Peracarid diversity and abundance in single epibenthic sled samples from the Gay Head-Bermuda Transect in the northwestern Atlantic. From Hessler and Sanders (1967).

Station	73		62		64		72		70	
Depth (m)	1400		2496		2886		2864		4680	
	spp.	ind.	spp.	ind.	spp.	ind.	spp.	ind.	spp.	ind.
Cumacea	20	794	20	267	36	2178	23	378	13	224
Tanaidacea	23	891	15	128	23	210	1	10	17	40
Isopoda	30	1538	27	262	53	1225	38	410	34	688
Amphipoda	43	1005	35	301	47	621	21	102	17	198
Total	116	4228	97	958	159	4234	83	900	81	1150
Total fauna	365	25,242	257	13,425	310	12,083	208	5897	196	3737
% peracarids	31·8	16·8	37·7	7·1	51·3	35·0	39·9	15·3	41·3	30·8

Table II. Marine benthic gammaridean amphipods: percent generic representation in the deep sea of large (greater than or equal to 10 genera) families. Shallow-water distribution abbreviations:
BP = bipolar; LL = low lattitudes; COS = cosmopolitan; CW = cold water; WW = warm water; data from Barnard (1969).

Family with 10 or more Genera	Marine benthic (N)	*Genera* Purely deep-sea (%)	Littoral to deep-sea (%)	*Shallow-water Distribution*	*Eyes:* with (w); absent (a)
Acanthonotozomatidae	21	24	24	BP, CW	w
Amphilochidae	17	12	12	COS	w
Aoridae	18	22	6	COS	w
Calliopiidae	18	39	33	COS	w
Corophiidae	19	11	32	COS	w
Dexaminidae	10	0	0	COS	w
Eusiridae	24	33	17	BP, CW	w
Gammaridae	37	14	16	COS	w
Haustoriidae	19	16	5	BP, CW	w
Isaeidae	14	14	21	COS	w
Lysianassidae	88	23	27	BP, CW	w
Oedicerotidae	28	11	50	BP, CW	w
Paramphithoidae	10	30	50	BP, CW	w
Pardaliscidae	12	67	17	BP, CW	w
Phliantidae	11	0	0	LL, WW	w
Phoxocephalidae	12	25	50	COS	w
Stegocephalidae	11	45	45	BP, CW	a
Stenothoidae	13	0	31	COS	w
Synopiidae	10	60	30	BP, CW	w
Talitroidea (3 families)	16	0	0	COS	w

it is monogeneric, with but two species (Jones, 1969). From 25% to 70% ($\bar{x} = 42{\cdot}5\%$) of genera in other families have deep-sea representatives, but only 50% or fewer ($\bar{x} = 14{\cdot}0\%$) of genera in each family are limited to the deep sea (Jones, 1969). Currently, the deep-sea cumaceans are being revised (Jones and Sanders, in preparation), and changes in these statistics are likely. Whether the basic pattern will change remains to be seen.

Deep-sea tanaids are too poorly known for meaningful comparisons (Wolff, 1956; Lang, 1968; Gardiner, 1975). The existence of at least one important endemic family (Neotanaidae) hints that tanaids may follow the pattern of isopods (below).

Isopod distributions contrast strongly with those of amphipods and cumaceans. The isopod pattern was discussed in some detail by Hessler and Thistle (1975) and Hessler *et al.* (1979); it will be treated only briefly here. Of the nine suborders, seven have marine representatives in both shallow and deep water, and their dominant development is in shallow water in all but the Asellota. For example, in the Anthuridea which have no deep-sea families, only 17 of the 51 known genera have deep-sea (>250 m) representatives (Kensley, 1982); only three genera are limited to the deep sea.

By contrast, the Asellota are most diverse by far in the deep sea. The pattern is highly structured, showing a clear gradient of evolutionary advancement with depth. The most primitive of the four superfamilies (Aselloidea) is found in fresh water (Table III). Stenetroidea and Gnathostenetroidoidea, which are moderately advanced, are primarily shallow marine (0–500 m). The most advanced superfamily Janiroidea occurs from fresh water to the greatest ocean depths, but is best developed in the deep sea (Table IV). Of the 24 janiroidean families, six are solely or principally distributed in shallow water: Joeropsididae,

Table III. The environmental distribution of asellotan isopod superfamilies. Primary habitat shown by capital letters; secondary habitats are shown by lower case letters.

Aselloidea	F–W	—	—
Stenetrioidea	—	S–W	d–s
Gnathostenetroidoidea	—	S–W	—
Janiroidea	f–w	s–w	D–S

Abbreviations: f–w, fresh water; s–w, shallow marine; d–s, deep sea.

Table IV. Families of janiroidean isopods: percent generic representation in the deep sea.

Family	(N)	*Genera* Purely deep-sea (%)	Littoral to deep-sea (%)	*Shallow-water Distribution*	*Eyes* with (w) absent (a)
Abyssianiridae	2	100	0	None	a
Acanthaspidiidae	1	100	0	None	a
Dendrotionidae	4	75	25	SH, CW	w
Desmosomatidae	19	79	21	BP, CW, MED	a
Echinothambematidae	1	100	0	None	a
Eurycopidae	27	78	22	BP, CW, MED	a
Haplomunnidae	4	75	25	MED	a
Haploniscidae	5	100	0	None	a
Ilyarachnidae	6	67	33	BP, CW, MED	a
Ischnomesidae	12	100	0	None	a
Janirellidae	2	100	100	None	a
Janiridae	30	33	23	COS	w
Joeropsididae	1	0	100	COS	w
Macrostylidae	2	50	50	NH, CW	a
Mesosignidae	1	100	0	None	a
Microparasellidae	7	0	0	COS (interstitial)	a
Mictosomatidae	1	100	0	None	a
Munnidae	6	33	33	COS	w
Munnopsidae	4	75	25	BP, CW	a
Nannoniscidae	12	83	17	BP, CW, MED	a
Paramunnidae	20	50	25	BP, CW, MED	w
Pleurocopidae	5	0	0	LL, WW	w
Pseudomesidae	1	100	0	None	a
Thambematidae	2	100	0	None	a

Shallow-water distribution abbreviations: BP = bipolar; SH = southern hemisphere; NH = northern hemisphere; LL = low latitudes; COS = cosmopolitan; MED = Mediterranean; CW = cold water; WW = warm water. Some of these families need revision, so that the number of genera is an estimate.

Janiridae, Microparasellidae, Munnidae, Paramunnidae and Pleurocopidae. Among these is the most primitive family Janiridae. The remaining 18 families are concentrated in the deep sea, with a great diversity of genera and species and an exuberant array of morphological specializations (Fig. 3).

A major exception to the endemism of the deep-sea janiroidean families is their presence in shallow water at high latitudes, such as the

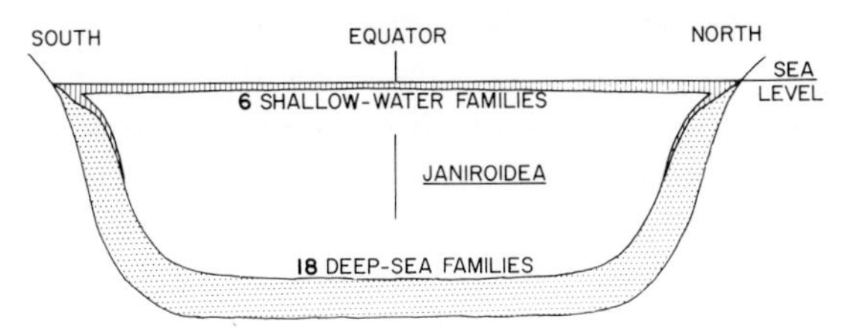

Fig. 3. A diagrammatic depth-latitude distribution of janiroidean isopods, comparing the six shallow-water families with the 18 deep-sea families.

Antarctic, Arctic and Scandinavian fjords. This is the basis of the hypothesis that the deep sea was populated through emigration from high latitude, shallow-water centres of origin (Dahl, 1954; Wolff, 1960; Kussakin, 1973; Menzies *et al.*, 1973). At high latitudes, temperature, which is an important environmental feature affecting species distributions, would be absent as a barrier between shallow and deep water. However, Hessler and Thistle (1975) and Hessler *et al.* (1979) argue that the primarily deep-sea families evolved in deep water and subsequently migrated into shallow water at high latitudes (high-latitude emergence). The lack of a thermal barrier would apply in this case as well.

Several significant lines of evidence are consistent with the high-latitude emergence hypothesis and stand against an origin in high latitude shallow waters.

(1) the centre of generic and specific diversity is in the deep sea, resulting in the "spindle-like" distributions of Kussakin (1973);

(2) in the few primarily deep-sea families where general phylogeny is known, the most primitive genera or species live in deep water;

(3) the deep-sea families lack any vestige of eyes, even their shallow-water species.

The third point requires amplification. Because eyes are plesiomorphic in isopods (and other malacostracans), lack of them implies regression in an environment where they offer no selective advantage. The deep-sea bottom is unquestionably such a place; deep benthic animals are almost always blind. Once lost, eyes as we know them cannot re-evolve. On the other hand, one might argue that blindness can and does evolve in shallow water as well, particularly on mud bottoms. Thus, animals without eyes may be pre-adapted for the deep benthos. The latter scenario is unlikely in the present case. The majority of the deep-sea families showing high-latitude emergence are well adapted for swimming, yet they are blind. In contrast, five of the six principally shallow-water families are ambulatory, yet they have well-developed eyes, even at high latitudes (the sixth, the Microparasellidae, are interstitial and therefore blind). It would be difficult to explain why swimming forms should lose their eyes while ambulatory taxa should keep them. Thus, the condition of eyes is a powerful tool for discerning the history of a taxon.

The isopod genus *Serolis* is an outstanding example (Fig. 4). Its species diversity is highest in shallow water at southern high latitudes and decreases markedly with increasing depth. At the same time visual capacity shifts from 100% presence of well-developed eyes to 100% blindness (Menzies *et al.*, 1968, 1973). Bathyal occurrences are also southern, primarily at high latitudes. At abyssal depths, only a single species is found in the North Atlantic (Bastida and Torti, 1969–70). The janiroidean isopod *Paramunna* (Fig. 5) (Wilson, 1980) displays a similar pattern, except there is some blindness in the littoral zone, and fewer deep-water species are limited to southern latitudes.

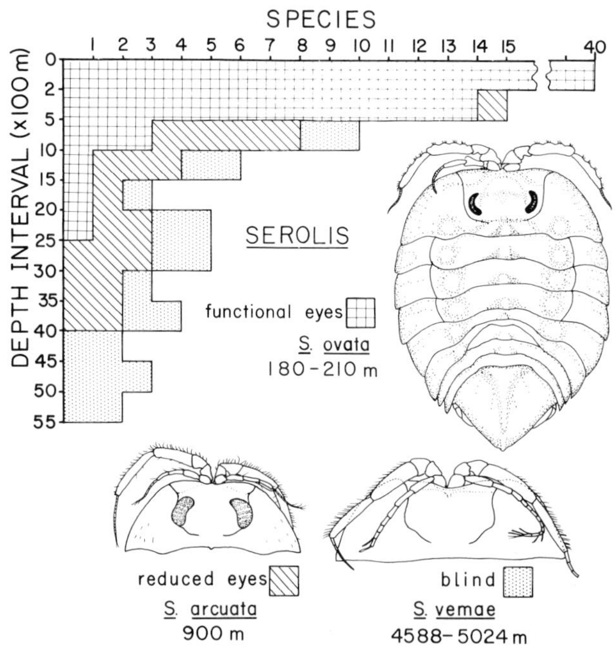

Fig. 4. Species diversity of the isopod genus *Serolis* versus depth, showing the distribution of eye morphology. The illustrations of eye development, *S. ovata*, *S. arcuata*, and *S. vemae*, are after Sheppard (1957), Moriera (1977), and Menzies (1962), respectively.

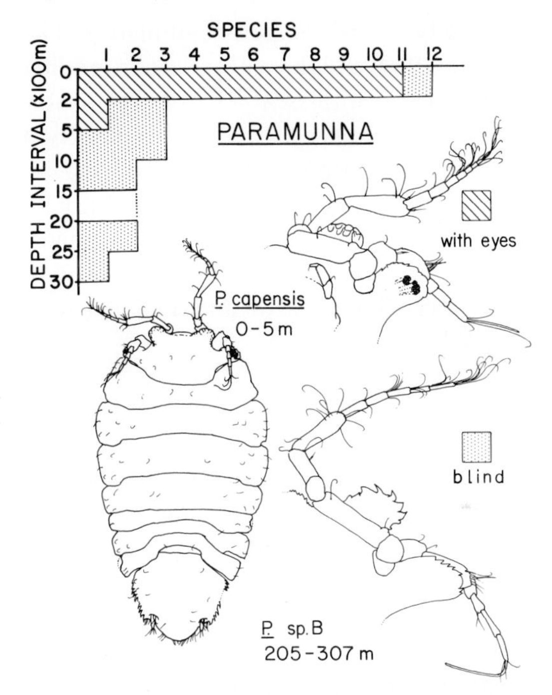

Fig. 5. A cumulative vertical distribution of species of the isopod genus *Paramunna* versus depth, showing the distribution of eye morphology. Two closely related African species *P. capensis* and *P.* sp. B (undescribed) show the loss of eyes with depth.

Many deep-water amphipod genera belong to families whose shallow-water genera have eyes and live primarily at high latitudes (Barnard, 1969) (Table II). All these examples suggest current deep-sea immigration; however, such interpretations must always be made with caution because it is difficult to distinguish between patterns that indicate active immigration from those that reflect a diminishing capacity of shallow-water taxa to cope with conditions at greater depths, e.g. decapods (below). Both patterns display "funnel-shaped" distributions (Kussakin, 1973), and eyes may be present in shallow-water representatives in either case.

The two possibilities may be distinguished using the degree of taxonomic isolation between shallow and deep members of a group. If active immigration is taking place, a taxon will have deep derivatives of many littoral subtaxa, thus showing a low degree of taxonomic isolation, or conversely, a high degree of similarity. The gammaridean amphipods (Barnard, 1969) and the janiroidean isopod family Paramunnidae (Wilson, 1980) are examples. If a taxon has difficulty adapting to deep-sea conditions, few subtaxa will make this transition. The deep-sea members, through specialization or subsequent extinction of shallow-water relatives, will bear little similarity to the littoral members, and thus be taxonomically isolated. Examples of this phenomenon are the brachyuran crabs (Pequegnat, 1970b), the Palinura, which include the wholly deep-sea Polychelidae (Firth and Pequegnat, 1971), and the munnid isopods (Wilson, 1980).

ENVIRONMENTAL FACTORS INFLUENCING DEEP-SEA IMMIGRATION

The patterns described so far suggest that the presence of crustaceans in the deep sea must be the result of varying histories and ecological responses, but all are compatible with the hypothesis that the ultimate origin was in shallow water. The presence of well-developed eyes in all higher taxa is a compelling indication of this. What factors might have influenced immigration into the deep sea? Do these factors document some special geographic area as a centre of origin?

It has been suggested that the shrinking area of continental-shelf (epicontinental) seas resulted in overpopulation, the ultimate effect of which was increased pressure for shallow-water species to invade other habitats, including the deep sea (Beurlin, 1931; Menzies *et al.*, 1973). Kussakin (1973) argued that the sinking of the Antarctic continental shelf stimulated migration of shallow-water species to deeper waters. These hypotheses are difficult to accept. The rate of sea-level change is slow compared to the life span of marine invertebrates, and therefore the only result of a shrinking habitat area would be a gradual decrease of population size, with maintenance of the same average density and habitat limits. The reason for this is simply that whatever environmental factors determined the density and ecological range of a species before will continue to do so. If by sinking, the Antarctic continental shelf experienced environmental alteration, its fauna would be replaced by the existing deeper fauna into whose zone the shelf had descended.

Low level of nutrient input is probably the most restrictive parameter of the deep sea, one likely to require the most extensive modifications, involving body size, ambit, feeding morphology, reproductive strategy, digestive physiology, growth rate, life span and behaviour. Our poor understanding of the natural history of deep-sea organisms allows only the most general glimpses of how an attenuated food supply might operate. The smaller size of deep-sea organisms (Barnard, 1962) and the relative enhancement of meiofaunal taxa (Thiel, 1975) may be attributed to this factor, although mobile scavengers may not be affected in this way (Haedrick and Rowe, 1977). Ambulatory scavengers, such as crabs, that might depend on large organic falls, cannot compete with more quickly ranging, swimming scavengers, such as fish, shrimp or amphipods (Hessler and Jumars, 1974). The low concentration of food in the water above the bottom combines with less active water movement to select against suspension feeders (Hessler and Jumars, 1974). Lower prey densities strongly reduce the opportunities for carnivorous taxa (Pequegnat, 1970b; Rex, 1976). Thus, deposit feeding taxa dominate (Beurlen, 1931; Sanders *et al.*, 1965; Sanders and Hessler, 1969). For these reasons, past changes in oceanic fertility (Thierstein and Berger, 1978; Berger, 1979) should have had a distinct effect on deep-sea life styles and community structure.

The absence of nonbiogenous light would be important to taxa that depend on vision, such as visual predators. The rarity of stomatopods in deep water (Manning and Struhsaker, 1976) may result partly from this, although carnivory is selected against in its own right. Bioluminescence is not a significant environmental factor for deep benthic animals. The rarity of eyes and the absence of bioluminescence in benthonts of the abyss shows this.

High pressure requires metabolic accommodation (Siebenaller and Somero, 1978; George, 1979; Somero *et al.*, in press), but too little is known about the physiologies of benthic taxa to predict the impact of this factor on the success of immigration.

It is difficult to appreciate the importance of substrate type because of severe sampling bias. The vast majority of deep-sea samples come from soft, mud bottoms. Most sampling on hard substrata is unsuccessful; therefore investigators avoid this environment, with the result that the fauna of rocky bottoms is poorly known. One can predict that crustaceans such as decapods, whose success is enhanced by a highly structured environment (Abele, 1974), do poorly on the relatively featureless mud bottoms of the deep sea (Balss, 1955).

Nutrient supply, pressure, light and substrate are neutral in implicating specific shallow-water geographic areas as centres of emigration to the deep sea. Temperature differences, however, offer a clear barrier and are distributed in a way that points to specific geographic regions as likely starting points. The zonal distribution of marine life strongly suggests that temperature is an ultimate determinant of species ranges. Nevertheless, differences in the temperature regime at the northern and southern ends of the range of a shallow-water species show that regional accommodation is possible where the gradient is not steep. These local genetic differences are made possible in part by incomplete gene flow. The thermocline is steeper in going from warm, shallow water at low or middle latitudes to the adjacent cold, upper bathyal waters (Sanders, 1968a), so that appreciably colder temperatures are obtained only a few kilometres away. Here, local genetic accommodation to lower temperatures is more difficult because genetic exchange is better over these short distances. The transition is yet more complicated because to be effective, the adjustment to colder water must include the ability to reproduce, a function usually limited to the warmer part of the year for shallow-water animals.

For these reasons, emigration to the deep sea should be most easily accomplished where the water column is relatively isothermal, that is, where animals are pre-adapted to the deep-sea thermal regime. Thus, at least today, high latitudes offer the most likely source of potential immigrants. Evidence for this comes from distributions of high latitude species and genera, where broad bathymetric ranges with the upper limit extending into littoral depths are more common than elsewhere (George, 1977; Dahl, 1979). In genera, the presence of eyes in at least the shallow-water representatives shows that these cases indeed involve migrations into deep water.

Distribution of isopods in the Mediterranean validates the concept of a isothermal conduit for migration. Here, deep-sea janiroidean families can be found at littoral depths (Lo Bianco, 1903; Schiecke and Fresi, 1969; Schiecke, 1973) where the temperature is approximately 14°C. Unlike the open ocean, deep-water temperatures are only a little lower, yet deep-sea taxa occur there as well (Lo Bianco, 1903; Chardy, 1973, 1974).

Some species are also found outside the Mediterranean in much cooler water. *Munella danteci* (100–200 m in the Mediterranean) is also found at cold, bathyal depths in the Bay of Biscay. Three species from the Bay of Naples have been recorded from cold, littoral-bathyal Scandinavian waters (Hult, 1941). The explanation for these distributions is not clear, but the species in the Mediterranean may be glacial relicts (Ekman,

1953). Thus, deep-sea taxa are not necessarily obligate psychrophiles ("cold lovers"), although their populations are stenothermal (Hessler, 1970). The existence of both cold and warm water stenotherms indicates the capacity for temperature adaptation, raising the important question of which way this adaptation went.

PALEOCEANOGRAPHIC INFLUENCES

It is well known that deep, open-ocean water masses have not always been cold (Berger, 1979; Schopf, 1980). In the Eocene and earlier, bottom temperatures exceeded 10°C (Savin *et al.*, 1975). About 38 million years ago, an abrupt temperature drop reduced bottom temperatures to a little more than 5°C (Kennett and Shackleton, 1976). In the Middle Miocene (13 million years ago), bottom temperature dropped again, and for the first time high and low latitude surface temperatures diverged in a major way. In short the psychrosphere, did not develop until mid-Cenozoic times (Benson, 1975, 1979; Kennett and Shackleton, 1976; Berger *et al.*, 1981), an event with important implications.

First, if there were a deep-water malacostracan fauna at that time (see below), it either adapted or suffered some degree of extinction. The Eocene-Oligocene event is estimated to have occurred in only 75000–100000 years (Kennett and Shackleton, 1976) or 500000 years (Thierstein and Berger, 1978). The record of deep benthic ostracodes (Benson, 1975, 1979) and foraminiferans (Douglas, 1973; Douglas and Woodruff, 1981) displays contemporaneous faunal replacements indicating that the existing fauna could not fully adjust to so rapid a change. Bruun (1956, 1957) and Wolff (1960) postulated that the temperature drop heralding the onset of the Pleistocene glaciations must have resulted in massive extinctions of the stenothermal deep-sea fauna. [We now know the temperature began dropping much earlier (Savin *et al.*, 1975).] Barnard (1961) suggested that the similarity between shallow and deep amphipod faunas reflects their youth in the deep sea and that, in view of the temperature change, the present fauna of bathyal depths would have more representatives of the early Cenozoic abyssal zone because of the similarity in temperature. Impressed by the high diversity of the deep-sea fauna, Hessler and Sanders (1967) doubted that major extinctions accompanied the Late Tertiary-Pleistocene temperature changes; seemingly, enough time had not elapsed to evolve so rich a fauna. Instead, they suggested that the change in temperature was sufficiently slow to allow evolutionary accommodation. The Mediterranean janiroideans show this.

Second, if bottom water was so much warmer before the Middle Tertiary, its high density, the driving force of bottom-water formation, might not have been the result of low temperature, but high salinity (Berger, 1979). If so, the isopycnal conduit between shallow and deep water existed at mid-latitudes (Chamberlin, 1906), with origins in semi-isolated basins (Schopf, 1980) such as the Mediterranean is today (Sverdrup *et al.*, 1942). The deep occurrence of decapods in the Red Sea exemplifies present migration under these circumstances (Balss, 1955). Past possibilities would be the North or South Atlantic basins at intermediate stages of development when they were large but isolated by shallow sills (Sclater *et al.*, 1977; Thierstein and Berger, 1978; Barron *et al.*, 1981). However, during those times, no particular geographic conduit to the deep sea may even have existed. The water column was relatively isothermal world-wide, and depth-related salinity differences would have been modest. In terms of water properties, much of the world's littoral zone would have been equally suitable as a source of immigrants.

At the same time these conditions may have induced bottom water stagnation. If severely reduced oxygen concentrations in deep water were the result (Berger, 1979), it could have had a major effect on the fauna. In the Peru-Chile Trench, the bathyal bottom is essentially azoic where the anoxic, oxygen minimum layer intersects it (Frankenberg and Menzies, 1968), similar to conditions off West Africa (Sanders, 1968b). However, oxygen levels must be extremely low to have an effect; the bathyal San Diego Trough has a very rich fauna in spite of oxygen levels of only 0·6 ml^{-1} litre (Jumars, 1976; Thistle, 1978). Temporal and spatial distribution of low oxygen events are not known well enough to evaluate their importance in the present context.

With these differences, the dissimilar arrangement of past continental shelves (Firstbrook *et al.*, 1980) produced oceanographic systems very unlike those of today. Consequently, much of the deep-sea fauna must have immigrated under conditions little resembling the high-latitude, cold-water submergence which now dominates.

Shallow-water distributions must play a role. The high-latitude concentrations of serolid isopods makes them especially available for the mode of deep-sea immigration occurring today. Many amphipod families (Barnard, 1969) and lithodid crabs (Makarov, 1938; Ekman,

1953) might also follow this pattern. On the other hand, the brachyuran decapods are poorly represented at high latitudes and are probably excluded from immigration now, except in places like the Mediterranean and the Red Sea (Balss, 1955).

EVOLUTIONARY PATTERNS

When did deep-sea invasions take place? Logically, immigration can occur at any time after a taxon has evolved, except when peculiarities of their shallow-water distributions, as just mentioned, cause exclusion. Because they are rarely preserved as fossils, little specific information is available on peracaridan evolution. The earliest record for tanaids is Lower Carboniferous (Schram, 1981). The oldest isopod and cumacean fossils are Upper Carboniferous and Permian, respectively. Amphipods are not seen until the Tertiary. These first fossils are well-developed representatives of their group, indicating that the taxa evolved at an earlier time. The earliest eumalacostracans, having a generalized form, come from the Devonian (Schram *et al.*, 1978). These data suggest that eumalacostracan diversification, including peracarids, was rapid, particularly in the late Devonian-early Carboniferous.

The late appearance of amphipods is strange; no fossils of potential ancestors exist and Recent taxa do not help. They may be a recent innovation, as indicated by the concentration of their high diversity at relatively low taxonomic levels (Bousfield, 1978). If this is so, amphipods are currently undergoing an explosive radiation, and compared to other taxa, entry into the deep sea is recent. Shallow-water biogeography opposes this explanation by suggesting many amphipod taxa must have already existed in the early Mesozoic (ibid.).

Asellotes have been regarded as primitive isopods (Hansen, 1916; Schram, 1977) so that the extraordinary pattern of their presence in the deep sea might reflect an ancient origin (Zenkevitch and Birstein, 1960). However, the primitive condition is not well founded (Kussakin, 1973), and their success in the deep sea is probably due to a detritivorous life style.

Unlike the peracaridan situation, the decapodan fossil record documents important aspects of their entry into the deep sea (Beurlin, 1931; Glaessner, 1969). The Eryonidae and Homolodromiinae are limited to the deep sea today, but the fossil record indicates a diverse shallow-water fauna until the end of the Jurassic, by which time they must have evolved deep-sea representatives. Some Nephropidae may follow this pattern, but other nephropids and members of Galatheidae indicate submergence by the end of the Cretaceous. Thus, the possibility that attenuation of decapodan diversity with depth reflects active immigration is not supported by fossil evidence; their presence in the deep sea is partially relict.

HORIZONTAL DISTRIBUTIONS

The horizontal distribution of taxa in the deep sea offers little help in the inquiry about local shallow-water centres of emigration. Except in the case of extremely recent activity, as reflected by bathyal serolids for example, the distribution of higher taxa is usually homogeneous. Even genera tend to be cosmopolitan in the deep sea. This is best illustrated by asellote isopods, whose genera are well studied. The frames of reference are the Atlantic as a whole, which has been well sampled (Sanders, 1977) and the Gay Head-Bermuda Transect (Sanders *et al.*, 1965) off the northeastern United States, the most thoroughly sampled are in the deep sea. Of the 143 genera known from the deep sea (>200 m), only nine Pacific genera have not been found in the Atlantic (unpubl. data). It is uncertain how widespread or important these nine genera are in the Pacific because of the relatively poor sampling there, but they do not seem to be major, being known from only 19 localities and by 14 species. Of the 134 genera found in the Atlantic, 33 have not been seen on the Gay Head-Bermuda Transect. Of these, only 22 could be considered significant on the basis of the number of stations where they are found, and 16 on the basis of number of individuals by which they are represented. The list of absences from the Gay Head-Bermuda Transect is high because, by deep-sea standards, the Transect's species diversity is surprisingly low. Even so, the comparison shows a remarkable ubiquity of deep-sea genera. Thus, at least in the case of isopods, most genera have been in the deep sea long enough to become evenly distributed.

The data also indicate that, in contrast to the shallow-water or terrestrial situation, environmental conditions are sufficiently homogeneous in the deep sea that life styles represented by levels as low as because, if severe localized conditions such as total oxygen depletion cause extinction of the deep fauna, the only source for subsequent replacement would be local shallow waters. The Mediterranean

(Benson, 1979) and Arctic Ocean-Norwegian Sea (Gurjanova, 1938; Dahl, 1979; Dahl *et al.*, 1976; Just, 1980) are probable examples of this phenomenon, yet endemism is seen only at the species level. This probably reflects the recentness of the critical events. In the case of Arctic Ocean-Norwegian Sea isopods, except for the valviferan *Saduria*, the deep-water fauna is almost exclusively typical deep-sea janiroidean families. This fauna may be a remnant of the original fauna, but it is also a product of the adjacent, high-latitude, shallow-water fauna being dominated by deep-sea families, thus allowing invasion of the basin by shallow-water representatives of deep-sea groups.

For isopods, diversity trends have yet to be analysed within individual lower taxa. The isopods in aggregate show a distinct increase in diversity toward the equator (Fig. 6), probably the result of ecological

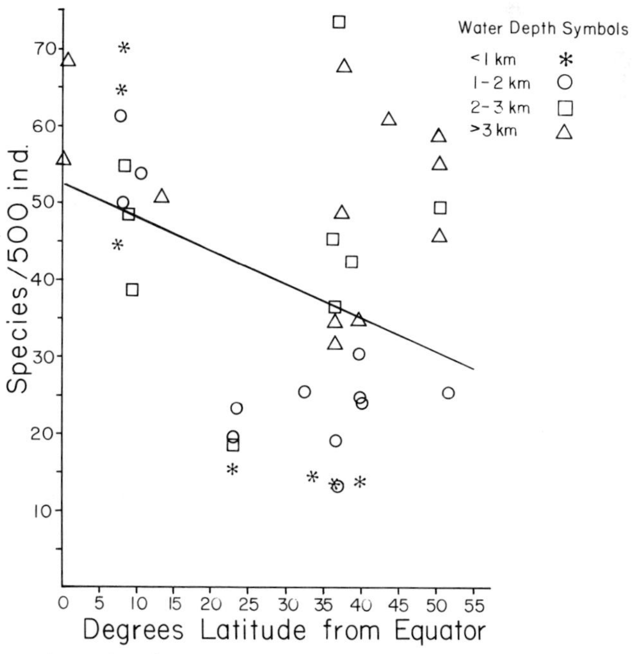

Fig. 6. Isopod species diversity, rarefied to 500 individuals (Hurlbert, 1971), plotted against degrees latitude from equator. The data come from large epibenthic sled samples taken below 500 m in the Atlantic Ocean (both north and south latitudes). The samples are classified into four water depth categories, and the heavy line is a least squares estimate of the relationship. The regression coefficient shows that the inverse relationship between distance from the equator and species diversity is significant.

rather than historical causes. For example, the greater stability of nutrient input at lower latitudes could account for this effect (Sanders, 1968a).

CONCLUSIONS

This paper has documented three kinds of distribution for major malacostracan taxa: dwindling representation with increasing depth, as typified by the Decapoda and Stomatopoda; abundant representation in the deep sea without proliferation of special families and with many genera also occurring in shallow water, as in the Amphipoda and Cumacea; abundant representation in the deep sea, where there are many families and genera which are endemic or but poorly represented in shallow water, as in Isopoda and perhaps Tanaidacea. The decapod-stomatopod pattern reflects decreasing success with increasing depth. The major enigma is the contrast between the amphipod-cumacean and isopod patterns; why do isopods show so much special development in deep water? Perhaps isopods invaded the deep sea much earlier. This might explain the contrast with amphipods, but not with cumaceans. Available evidence does not yield a satisfactory historical explanation.

Ecological considerations are little more revealing. Among isopods, asellotes may be best suited for a deep-sea existence. Flabelliferans, anthurideans, gnathiideans and microcerberideans have substrate or feeding preferences largely different from the soft-bottom detritivorous habits that dominate in the deep sea. Many valviferans are herbivorous and would fare poorly, but others are soft-bottom dwellers that would seem well suited, in spite of their absence. Arcturid valviferans are epizoans that are occasionally abundant in the deep sea, but show no unusual taxonomic development there. While detritivory may explain why asellotes excelled among isopods in the deep sea, it does not account for asellotes having a major radiation there when equally suitable amphipods or cumaceans did not.

In discussing potential causes for differences in distribution of major taxa, both ecological and historical factors have been implicated. In truth, these cannot be isolated from each other. Unique events must have had ecological causes; it is only the juxtaposition of these events that may be historically unique. For example, if brachyurans are not actively immigrating into the deep sea today, part of the reason must be their absence from the Antarctic, the cause of which surely has a strong ecological component. Only rarely can historical factors be isolated on the basis of recent distribution patterns alone, and our understanding of the natural history of deep-sea organisms is so poor that defining the ecological features which yeild their success is still well beyond our grasp.

SUMMARY

1. Different major malacostracan taxa have contrasting distributions with respect to the deep sea.
   (a) Decapods and stomatopods decrease markedly in representation with depth, while benthic peracarids (cumaceans, tanaids, isopods and amphipods) flourish.
   (b) Deep-sea amphipods and cumaceans do not differ greatly at higher taxonomic levels from those in shallow water, whereas isopods contrast strongly in kind and proportion.
   (c) Distribution and morphological evidence suggest the great diversity of janiroidean isopods in the deep sea is the product of *in situ* evolution, with subsequent immigration into shallow water at high latitudes.
   (d) The presence or absence of eyes in both littoral and deep-sea taxa can be used to interpret the activity of migration to and from the deep sea.
   (e) In those groups with funnel-shaped vertical distributions, recent active immigration must be distinguished from diminished capacity to adapt to deep-sea conditions. The degree of taxonomic isolation of deep groups from their shallow-water relatives can be used to make this distinction.
2. Both biological and physical factors influence the potential for immigration to the deep sea.
   (a) Low nutrient concentrations and input rates may require a variety of profound modifications in morphology and life style, and may exclude some types of organisms entirely.
   (b) Among physical factors, light, pressure, substrate and temperature may all play a part, but today, temperature exerts the most obvious influence.
   (c) Because temperature conditions limit species distributions so strongly, the most likely sites of interchange today are where the

water column is isothermal, such as at high latitudes. Some taxa give evidence that this is indeed the case.

(d) The presence of deep-sea isopods in shallow water in the warm, but nearly isothermal Mediterranean shows the importance of its isothermicity as well as illustrating that, while stenothermal, deep-sea taxa can adapt to warmer water.

(e) The sinking of continental shelves or shrinking of epicontinental seas is not likely to have stimulated emigration to the deep sea.

3. Not all the conditions influencing invasion of the deep sea today pertained in the past.

   (a) Mesozoic and early Cenozoic seas were more isothermal, and dense bottom water may well have been the product of high salinities generated at mid-latitudes.

   (b) With these conditions, mid-latitude invasion may have dominated, but it is equally likely that there was no particular geographic centre for emigration.

   (c) Shallow-water distribution are part of the initial boundary conditions that determine the possibility of emigration.

4. The fossil record shows that most higher eumalacostracan taxa had appeared by the end of the Carboniferous.

   (a) Amphipods may be an exception in that they are not seen until the Tertiary.

   (b) Several decapod families that are endemic to the deep sea today have a shallow-water fossil record, indicating that their present distribution is relict.

5. For the most part, distributions in the deep sea do not cast light on shallow-water centres of origin.

   (a) Genera tend to be cosmopolitan, indicating enough time has elapsed for them to have spread widely.

   (b) Species diversities are highest toward the equator, probably for ecological reasons.

ACKNOWLEDGEMENTS

We appreciate the stimulating conversations with Drs W. Berger (Scripps Institution of Oceanography), J.L. Barnard and R.H. Benson (United States National Museum). This work was funded by grant DEB 80–07150 of the National Science Foundation.

REFERENCES

Abele, L.G. (1974). Species diversity of decapod crustaceans in marine habitats. *Ecology* **55**, 156–161.

Ambler, J.W. (1979). Species of *Munidopsis* (Crustacea, Galatheidae) occurring off Oregon and in adjacent waters. *Fish. Bull.* **78**, 13–34.

Balss, H. (1955). Decapoda: Okologie. *H.G. Bronns Klassen u. Ordnungen des Tierreichs*, **5** (1), 7 (10), 1285–1367.

Barnard, J.L. (1961). Gammaridean Amphipoda from depths of 400 to 6000 meters. *Galathea Rept.* **5**, 23–128.

Barnard, J.L. (1962). South Atlantic abyssal amphipods collected by R.V. *Vema*. Abyssal Crustacea. *Vema Res. Ser.* (*Columbia Univ., New York*) **1**, 1–78.

Barnard, J.L. (1969). The families and genera of marine gammaridean Amphipoda. *Bull. U.S. natn. Mus.* **271**, 1–535.

Barnard, J.L. (1973). Deep-sea Amphipoda of the genus *Lepechinella* (Crustacea). *Smithson. Contr. Zool.* **133**, 1–31.

Barron, E.J., Thompson, S.L. & Schneider, S.H. (1981). An ice-free Cretaceous? Results from climate model simulations. *Science, N.Y.* **212**, 505–508.

Bastida, R. & Torti, M.R. (1969–70). Crustaceos isopodos: Serolidae. *Res. sci., campagnes "Calypso"* **9**, 61–105.

Benson, R.H. (1975). The origin of the psychrosphere as recorded in changes of deep-sea ostracode assemblages. *Lethaia* **8**, 69–83.

Benson, R.H. (1979). In search of lost oceans: A paradox in discovery. *In* "Historical Biogeography, Plate Tectonics, and the Changing Environment" (J. Gray and A.J. Boucot, eds). *Proc. Biol. Colloquium, Oregon State Univ.* **37**, 379–389.

Berger, W.H. (1979). Impact of deep-sea drilling on paleoceanography. *In* "Deep Drilling Results in the Atlantic Ocean: Continental Margins and Paleoenvironment" (M. Talwani, W. Hay and W.B.F. Ryan, eds), pp. 297–314. American Geophysical Union, Washington, D.C.

Berger, W.H., Vincent, E. and Thierstein, H.R. (1981). The deep-sea record: Major steps in Cenozoic ocean evolution. *In* "Symposium on Results of Deep-Sea Drilling" (R.G. Douglas and E.L. Winterer, eds). SEPM *Special Publ.* **32**, 489–504.

Beurlen, K. (1931). Die Besiedelung der Tiefsee. *Natur und Museum* **61**, 269–279. U.S. Naval Oceanographic Office (Transl.), Washington, D.C.

Bousfield, E.L. (1978). A revised classification and phylogeny of amphipod crustaceans. *Trans. R. Soc. Canada*, ser. 4, **16**, 343–390.

Bruun, A.F. (1956). The abyssal fauna: Its ecology, distribution and origin. *Nature, Lond.* **177**, 1105–1108.

Bruun, A.F. (1957). Deep sea and abyssal depths. [*In* "Treatise on Marine Ecology and Paleoecology" (J. Hedgpeth, ed.).] *Mem. geol. Soc. Amer.* **67**, 641–672.

Chamberlin T.C. (1906). On a possible reversal of deep-sea circulation and its influence on geological climates. *J. Geol.* **14**, 363–373.

Chardy, P. (1973). Dragages profonds en mer Ionienne-donnee. *Rapp. Comm. int. Mer Méditerr.* **22**, 103–105.

Chardy, P. (1974). Deux espèces nouvelles d'isopodes asellotes recoltées en Méditerranée profonde. *Vie et Milieu* **24**, 409–420.

Crosnier, A. and Forest, J. (1973). Les crevettes profondes de l'Atlantique oriental tropical. *Faune Tropicale* **19**, 1–409.

Dahl, E. (1954). The distribution of deep sea Crustacea. *In* "On the Distribution and Origin of the Deep Sea Bottom Fauna". *Internat. Union Biol. Sci.*, ser. B. **16**, 43–48.

Dahl, E. (1979). Amphipoda Gammaridea from the deep Norwegian Sea. A preliminary report. *Sarsia* **64**, 57–59.

Dahl, E., Laubier, L., Sibuet, M. and Stromberg, J.-O. (1976). Some quantitative results on benthic communities of the deep Norwegian Sea. *Astarte* **9**, 61–79.

Doflein, F. and Balss, H. (1913). Die Galatheiden der Deutschen Tiefsee-Expedition. *Wiss. Ergeb. dt. Tiefsee-Expedition "Valdivia"* **20**, 130–84.

Douglas, R.G. (1973). Evolution and bathymetric distribution of Tertiary deep-sea benthic Foraminifera. *Geol. Soc. Am., Abstracts*, 603 (Annual Meeting, Dallas, Texas).

Douglas, R. and Woodruf, F. (1981). Deep-sea benthic Foraminifera. *In* "The Oceanic Lithosphere, the Sea", vol. 7 (C. Emiliani, ed.), pp. 1233–1238. Wiley, New York.

Ekman, S. (1953). "Zoogeography of the Sea". Sidgwick and Jackson, London.

Firth, R.W. and Pequegnat, W.E. (1971). Deep-sea lobsters of the families Polychelidae and Nephroipidae (Crustacea, Decapoda) in the Gulf of Mexico and Caribbean Sea. *Texas A&M Research Foundation Reference* **71–11T**, 103 pp.

Firstbrook, P.L., Funnell, B.M., Hurley, A.M. and Smith, A.G. (1980). "Paleoceanic Reconstructions 160–0 Ma". [Special Publication of National Science Foundation National Ocean Sediment Program.] Scripps Institution of Oceanography (Deep Sea Drilling Project), La Jolla, California.

Frankenberg, D. and Menzies, R.J. (1968). Some quantitative analyses of deep-sea benthos off Peru. *Deep-Sea Res.* **15**, 623–626.

Gardiner, L.E. (1975). The systematics, postmarsupial development, and ecology of the deep-sea family Neotanaidae (Crustacea: Tanaidacea). *Smithson. Contri. Zool.* **170**, 1–265 pp.

George, R.Y. (1977). Dissimilar and similar trends in Arctic and Antarctic marine benthos. *In* "Polar Oceans" (M.J. Dunbar, ed.), pp. 391–408. Arctic Institute of North America, Calgary.

George, R.Y. (1979). What adaptive strategies promote immigration and speciation in deep-sea environment. *Sarsia* **64**, 61–66.

Glaessner, M.F. (1969). Decapoda. *In* "Treatise on Invertebrate Paleontology, R (Arthropoda 4)" (R.C. Moore, ed.), pp. 339–533. Geological Society of America, Lawrence, Kansas.

Griffin, D.J.G. and Brown, D.E. (1975). Deepwater decapod Crustacea from Eastern Australia: Brachyuran crabs. *Rec. Aust. Mus.* **30**, 248–271.

Guryanova, E. (1938). On the question of the composition and origin of the fauna of the polar basin bassalia. *Cr. (Doklady) Acad. Sci. URSS* **20**, 333–336.

Haedrich, R.L. and Rowe, G.T. (1977). Megafaunal biomass in the deep sea. *Nature, Lond.* **269**, 141–142.

Hansen, H.J. (1916). Crustacea Malacostraca III. *Danish Ingolf Expedition* **3**, 1–262.

Hessler, R.R. (1970). The Desmosomatidae (Isopoda, Asellota) of the Gay Head-Bermuda transect. *Bull. Scripps Inst. Oceanogr.* **15**, 1–185.

Hessler, R.R. and Jumars, P.A. (1974). Abyssal community analysis from replicate box cores in the central North Pacific. *Deep-Sea Res.* **21**, 185–209.

Hessler, R.R. and Sanders, H.L. (1967). Faunal diversity in the deep sea. *Deep-Sea Res.* **14**, 65–78.

Hessler, R.R. and Thistle, D. (1975). On the place of origin of deep-sea isopods. *Mar. Biol.* **32**, 155–165.

Hessler, R.R., Wilson, G. and Thistle, D. (1979). The deep-sea isopods: a biogeographic and phylogenetic overview. *Sarsia* **64**, 67–76.

Holthuis, L.B. (1974). The lobsters of the superfamily Nephropidea of the Atlantic Ocean (Crustacea: Decapoda). *Bull. Mar. Sci.* **24**, 723–884.

Hult, J. (1941). On the soft-bottom isopods of the Skager Rak. *Zool. Bidr. Uppsala* **21**, 1–234.

Hurlbert, S.N. (1971). The nonconcept of species diversity: a critique and alternative parameters. *Ecology* **52**, 577–586.

Jones, N.S. (1969). The systematics and distribution of Cumacea from depths exceeding 200 meters. *Galathea Rept.* **10**, 99–180.

Jones, N.S. and Sanders, H.L. (1972). Distribution of Cumacea in the deep Atlantic. *Deep-Sea Res.* **19**, 737–745.

Jumars, P.A. (1976). Deep-sea species diversity: Does it have a characteristic scale? *J. Mar. Res.* **34**, 217–246.

Just, J. (1980). Abyssal and deep bathyal Malacostraca (Crustacea) from the Polar Sea. *Vidensk. Meddr. dansk naturh. Foren.* **142**, 161–177.

Kennett, J.P. and Shackleton, N.J. (1976). Oxygen isotope evidence for the development of the psychrosphere 38 m.y. ago. *Nature, Lond.* **260**, 513–515.

Kensley, B.F. (1982). Deep water Atlantic Anthuridea (Crustacea: Isopoda). *Smithson. Contr. Biol.* **346**, 1–60.

Kussakin, O.G. (1973). Peculiarities of the geographical and vertical distribution of marine isopods and the problem of deep-sea fauna origin. *Mar. Biol.* **23**, 19–34.

Lang, K. (1968). Deep-sea Tanaidacea. *Galathea Rept.* **9**, 23–209.

Lo Bianco, S. (1903). Le pesche abisseli esequite da F.A. Krupp col Yacht Puritan nelle adiacenze de Capri ed in altre localita del Mediterraneo. *Mitt. Zool. Stat. Neapel* **16**, 109–278.

Makarov, V.V. (1938). [Fauna of USSR. Crustacea Anomura.] In ["Fauna of USSR"], vol. 10, no. 3. [Translated by Israel Program for Scientific Translations, Jerusalem, 1962.]

Manning, R.B. (1969). "Stomatopod Crustacea of the western Atlantic". [*Studies in Tropical Oceanography* no. 8.] University of Miami Press, Coral Gables, Fla.

Manning, R.B. and Struhsaker, P. (1976). Occurrence of the Caribbean stomatopod, *Bathysquilla microps*, off Hawaii, with additional records for *B. microps* and *B. crassispinosa*. *Proc. Biol. Soc. Wash.* **89**, 439–450.

Mauchline, J. (1980). "The Biology of Mysids and Euphausiids". Academic Press, London and New York.

Menzies, R.J. (1962). The isopods of abyssal depths in the Atlantic Ocean. Abyssal Crustacea. *Vema Res. Ser.* (*Columbia, New York*) **1**, 79–206.

Menzies, R.J., George, R.Y. and Rowe, G.T. (1968). Vision index for isopod Crustacea related to latitude and depth. *Nature, Lond.* **217**, 93–95.

Menzies, R.H., George, R.Y. and Rowe, G.T. (1973). "Abyssal Environment and Ecology of the World Oceans". John Wiley, New York.

Moreira, P.S. (1977). New bathyal species of *Serolis* (Isopoda, Flabellifera) from the western South Atlantic Ocean. *Crustaceana* **33**, 133–148.

Pequegnat, L.H. (1970a). Deep-sea caridean shrimps with descriptions of six new species. *In* "Contributions on the Biology of the Gulf of Mexico", [Texas A&M Univ. Oceanogr. Studies, vol. 1.] (F.A. Chace, ed.), pp. 59–124. Gulf Publishing, Houston.

Pequegnat, W.E. (1970b). Deep-water brachyuran crabs. *In* "Contributions on the Biology of the Gulf of Mexico" [Texas A&M Univ. Oceanogr. Studes, vol. 1.] (F.A. Chace, ed.), pp. 171–205. Gulf Publishing, Houston.

Pequegnat, W.E. and Pequegnat, L.H. (1970). Deep-sea anomurans of superfamily Galatheoidea with descriptions of two new species. *In* "Contributions on the Biology of the Gulf of Mexico" [Texas A&M Univ. Oceanogr. Studies, vol. 1.] (F.A. Chace, ed.), pp. 125–170. Gulf Publishing, Houston.

Pequegnat, W.E., Pequegnat, L.H., Firth, R.W., James, B.M. and T.W. Roberts (1971). Gulf of Mexico deep sea fauna. Decapoda and Euphausiacea. *In* "Serial Atlas of the Marine Environment" (W. Webster, ed.), pp. 1–12. American Geographical Society, New York.

Rex, M. (1976). Biological accommodation in the deep-sea benthos: Comparative evidence on the importance of predation and productivity. *Deep-Sea Res.* **23**, 975–987.

Roberts, T.W. and Pequegnat, W.E. (1970). Deep-Water decapod shrimps of the family Penaeidae. *In* "Contribution on the Biology of the Gulf of Mexico" [Texas A&M Univ. Oceanogr. Studies, vol. 1.] (F.A. Chace, ed.), pp. 21–58. Gulf Publishing, Houston.

Sanders, H.L. (1968a). Marine benthic diversity: A comparative study. *Am. Nat.* **102**, 243–282.

Sanders, H.L. (1968b). Benthic marine diversity and the stability-time hypothesis. *Brookhaven Symp. Biol.* **22**, 71–81.

Sanders, H.L. (1977). Evolutionary ecology and the deep-sea benthos. *Acad. Nat. Sci. Spec. Publ.* **12**, 223–243.

Sanders, H.L. and Hessler, R.R. (1969). Ecology of the deep-sea benthos. *Science, N.Y.* **163**, 1419–1424.

Sanders, H.L., Hessler, R.R. and Hampson, G.R. (1965). An introduction to the study of deep-sea benthic faunal assemblages along the Gay Head-Bermuda Transect. *Deep-Sea Res.* **12**, 845–867.

Savin, S.M., Douglas, G. and Stehli, F.G. (1975). Tertiary marine paleotemperatures. *Bull. Geol. Soc. Am.* **86**, 1499–1510.

Schiecke, U. (1973). "Ein Beitrag zur Kenntnis der Systematik, Biologie und Autokologie mariner Peracarida (Amphipoda, Isopoda, Tanaidacea) des Golfes von Neapel". Dissertation, Math.-Naturwiss. Fakultat, Christian Albrechts Universitat (Kiel).

Schiecke, U. and Fresi, E. (1969). Further desmosomids (Isopoda: Asellota) from the Bay of Naples. *Pubbl. Staz. Zool. Napoli* **37**, 156–169.

Schopf, T.J.M. (1980). "Paleoceanography". University Press, Cambridge, Mass.

Schram, F.R. (1977). Paleozoogeography of Late Paleozoic and Triassic Malacostraca. *Syst. Zool.* **26**, 367–379.

Schram, F.R. (1981). On the classification of the Eumalacostraca. *J. Crust. Biol.* **1**, 1–10.

Schram, F.F., Feldmann, R.M. and Copeland, M.J. (1978). The late Devonian Palaeopalemonidae and the earliest decapod crustaceans. *J. Paleontol.* **52**, 1375–1387.

Sclater, J.G., Hellinger, S. and Tapscott, C. (1977). The paleobathymetry of the Atlantic Ocean from the Jurassic to the present. *J. Geol.* **85**, 509–552.

Sheppard, E.M. (1957) Isopod Crustacea. Part II. The Sub-order Valvifera. Families: Idoteidae, Pseudidotheidae and Xenarcturidae fam. n. with a supplement to Isopod Crustacea, Part I. The family Serolidae. *"Discovery" Rep.* **29**, 141–198.

Siebenaller, J.F. and Somero, G.N. (1978). Pressure-adaptive differences in lactate dehydrogenases of congeneric fishes living at different depths. *Science, N.Y.* **201**, 255–257.

Somero, G.N. Siebenaller, J.F. and Hochachka, P.W. (In press). Biochemical and physiological adaptations of deep-sea animals. *In* "The Seas". vol. 7 (G.T. Rowe, ed.).

Sverdrup, H.U., Johnson, M.W. and Fleming, R.H. (1942). "The Oceans, Their Physics, Chemistry and General Biology". Prentice-Hall, Englewood Cliffs, N.J.

Thiel, H. (1975). The size structure of the deep-sea benthos. *Int. Rev. Ges. Hydrobiol.* **60**, 575–606.

Thierstein, H.R. and Berger, W.H. (1978). Injection events in ocean history. *Nature, Lond.* **276**, 461–466.

Thistle, D. (1978). Harpacticoid dispersion patterns: Implications for deep-sea diversity maintenance. *J. Mar. Res.* **36**, 377–397.

Tirmizi, N.M. (1966). Crustacea Galatheidae. *John Murray Expedition, Sci. Rept.* **11**, 169–234.

Wenner, E.L. and Boesch, D.F. (1979). Distribution patterns of epibenthic decapod Crustacea along the shelf-slope coenocline, Middle Atlantic Bight, USA. *Bull. Biol. Soc. Washington* **3**, 106–133.

Wilson, G.D. (1980). New insights into the colonization of the deep sea: Systematics and zoogeography of the Munnidae and the Pleurogoniidae *comb. nov.* (Isopoda; Janiroidea). *J. Nat. Hist.* **14**, 215–236.

Wolff, T. (1956). Crustacea Tanaidacea from depths exeeding 6000 meters. *Galathea Rept.* **2**, 187–241.

Wolff, T. (1960). The hadal community, an introduction. *Deep-Sea Res.* **6**, 95–124.

Wolff, T. (1961). Description of a remarkable deep-sea hermit crab with notes on the evolution of the Paguridea. *Galathea Rept.* **4**, 11–32.

Zarenkov, N.A. (1969). Decapoda. *In* "Biology of the Pacific Ocean. Part II. The Deep-Sea Bottom Fauna" (L.A. Zenkevich, ed.), vol. 7, pp. 79–82. US Naval Oceanographic Office (transl.), Washington, D.C.

Zenkevitch, L.A. and Birstein, J.A. (1960). On the problem of the antiquity of the deep-sea fauna. *Deep-Sea Res.* **7**, 10–23.

10 Speciation, Phylogenesis, Tectonism, Climate and Eustasy: Factors in the Evolution of Cenozoic Larger Foraminiferal Bioprovinces

C.G. ADAMS

Department of Palaeontology, British Museum (Natural History), London SW7 5BD, UK

Abstract: The roles of speciation, phylogenesis, tectonism, eustasy, and climate are examined in relation to the development, through dispersal/migration, of the three main Cenozoic faunal provinces (Central American, Mediterranean, and Indo-West Pacific). Species delineated on two distinct types of criteria are illustrated with reference to lineages drawn from the Miliolacea, Miogypsinidae, and Lepidocyclinidae. In the context of phylogeny, these differently discriminated species seem to demonstrate the reality of the evolutionary processes known as punctuated equilibra and phyletic gradualism, and it is argued that these are but two manifestations of a single process and should not be regarded as mutually exclusive. The reality of both recapitulation and orthogenesis is reiterated, the latter being regarded as probably responsible for the parallel development shown by some foraminiferal groups in different faunal provinces. The direct effect of tectonism (mountain building and plate tectonics), although regionally important, is considered to have been globally insignificant in Tertiary times. Climatic changes are likewise seen as having had little effect upon provinces beyond causing them to expand and contract latitudinally in response to warming and cooling phases. On the other hand, tectonically induced eustatic changes, although in no way responsible for provincialism, are thought to have been the most likely cause of some important changes in the composition of larger foraminiferal faunas during the Cenozoic.

Systematics Association Special Volume No. 23, "Evolution, Time and Space: The Emergence of the Biosphere", edited by R.W. Sims, J.H. Price and P.E.S. Whalley, 1983. pp. 255–289. Academic Press, London and New York.

INTRODUCTION

This paper examines the importance of five factors in the development of the three main Tertiary bioprovinces currently recognized on the basis of larger foraminifera. Since the generic composition of the relevant faunas has already been described (Adams, 1967, 1973) and their development outlined in relation to the role of migration (dispersal), this ground is not covered again here. At first sight it may seem that only migration, climatic change, and tectonic activity are of prime importance in the development of faunal provinces, but this is not so. Evolution (speciation and phylogenesis) is a principal and easily overlooked factor, which, together with migration or dispersal, largely determines the eventual composition of the faunas in a particular region. The examples discussed below show that two distinct types of speciation can be recognized, and that both phyletic gradualism and punctuated equilibria may be recognized within a single family of the Foraminiferida.

The larger foraminifera constitute a group of protozoans whose main similarity is that they are all readily visible to the naked eye. Their principal morphological characters can, however, be seen only with the aid of a microscope. Both living and fossil larger foraminifera have been studied since the early part of the nineteenth century, but more especially after about 1920, since when they have been systematically and widely applied to oil exploration. A rare combination of structural beauty, academic interest, and commercial importance ensured that the Foraminiferida became one of the most intensively studied of all groups of fossils, and it is not therefore surprising that a great deal is now known about their evolutionary history, and geographical and stratigraphical distributions. Much of this knowledge was collated or referred to in *Evolutionary Trends in Foraminifera* (Koenigswald *et al.*, 1963), but the specialized articles in that volume were neither related to one another, nor reviewed in the context of geological events or evolutionary theory.

All foraminifera in which the reproductive cycle is known show an alternation of asexual and sexual generations which, in the adult stage, can look very different—especially in size (Pl. 1, nos 1 and 2). The asexually produced, haploid generation usually begins with a relatively large (megalospheric) initial chamber (proloculus or protoconch), whereas the sexually produced diploid generation possesses a much smaller (microspheric) initial chamber. These are often called the A and B generations respectively. This alternation of generations is not necessarily regular, and in most living populations and fossil assemblages there are many more megalospheric than microspheric individuals. This is explained by the occurrence of apogamic schizogony (schizogony without meiosis) resulting in the paratrimorphic life cycles demonstrated in a number of larger foraminifera (for summary see Leutenegger, 1977).

Laboratory experiments (Bradshaw, 1961) have shown that although individual foraminifera will tolerate a variety of adverse physico-chemical conditions, they normally reproduce only within a narrow environmental range. For this reason, the early or embryonic part of the test, which in some groups is quite complex, is believed to be the most useful for evolutionary studies. In morphologically simpler forms, e.g. *Nummulites*, the larger microspheric test is more useful for this purpose. Differences in the morphology of the embryonic apparatus (protoconch, deuteroconch (second chamber) and any periembryonic chambers) are more likely to be genetically determined than are minor variations in morphological features such as size, shape, ornament etc. of the adult test, since these may merely reflect environmental conditions. Characters controlled largely by environment are likely to be developed and expressed differently by individuals of the same species living under different ecological conditions. Myers (1943a,b) claimed that operculinids developed a flat shell on soft, muddy substrates and a more inflated shell on firm, sandy bottoms. Recent work (Hottinger and Dreher, 1974; Hottinger, 1977) in the Red Sea has to some extent borne this out. However, these authors also reported that *Operculina ammonoides* (Gronovius) *sic* and *Heterostegina depressa* d'Orbigny were represented by involute, thick forms in shallow water and by evolute, thin forms in deeper environments. A similar trend was reported (Larsen, 1976) for *Amphistegina* in the same area. Depth, and light penetration seem therefore to have more effect than substrate on the growth mechanisms which mainly determine shape. The same factors may also affect the development of surface ornament.

Most larger foraminifera (*Discospirina* is an exception) live only in the warm shallow waters of the tropics and subtropics where the minimum summer surface temperature is not less than 18°C (Murray, 1973). For most species 20–22°C is the probable minimum. They usually enjoy a symbiotic relationship with photosynthesizing unicellular algae incorporated in their cytoplasm. Living forms are normally associated with carbonates and calcareous clays (marls), and occurrences of fossils in coarse terrigenous sands or other clastics representing unstable substrates at the time of deposition are to be regarded with suspicion since empty tests easily suffer post-mortem transport. Carbonates, however, afford a wide range of habitable environments in the tropics. The most important (see Fig. 8) are open shelf, fore-reef, top-reef, and back-reef (lagoonal), with a multitude of subsidiary habitats being provided by channels, light penetration, algal and sea grass growths etc. Most of these environments existed throughout the Tertiary, but were being continuously modified, and their faunas subjected to stress, by sea-level changes and other geological and hydrological events. Stress was probably most marked when eustatic changes periodically affected the photic zone of the sea floor, but must have been almost continuous at high latitudes near provincial boundaries.

Extant larger foraminifera live on or near the sea-floor, and are distributed by currents, either in their very young stages when they are briefly free-drifting, or by rafting on marine vegetation torn up by storms. They therefore find it difficult to cross wide oceanic barriers such as the Atlantic and Eastern Pacific, and are prevented from migrating around them by the cool waters of the temperate and polar zones. These factors are reflected in their geographical distribution.

All that need be said here about the Central American, Mediterranean, and Indo-West Pacific faunal provinces is that they are separated longitudinally rather than latitudinally; that they existed throughout the Cenozoic; that one at least (the Indo-West Pacific) may be susceptible to

---

Plate explanation

Plate 1: **1** and **2** *Nummulites laevigatus* (Bruguière). 1, microspheric form ×5. P.50894. 2, Megalospheric form; a ×5, P.50895; b ×15 (same specimen). **3** = *Cycloclypeus carpenteri* Brady, ×2. Recent, off Borneo. 51.1.20.39. External view showing annular chambers. **4** = *Lepidocyclina* (*Nephrolepidina*) *howchini* Chapman and Crespin. Megalospheric form ×21, Early Miocene, Batesford Limestone, Australia. Note absence of initial coil. P.50896. **5** and **6** = *Lepidocyclina* (*Nephrolepidina*) sp. Microspheric form, to show initial coil. Early Miocene, Borneo. 5, ×20; 6, ×580. P.50897. **7** = *Austrotrillina striata* Todd and Post. Megalospheric form, ×53. Oligocene, Eniwetok drill hole E–1. P.47598. **8** and **9** = *Austrotrillina howchini* (Schlumberger). 8, Centre of megalospheric form (outer chambers removed) ×50, P.47588; 9, Microspheric form, ×30, P.47608, Pata Limestone, Australia.

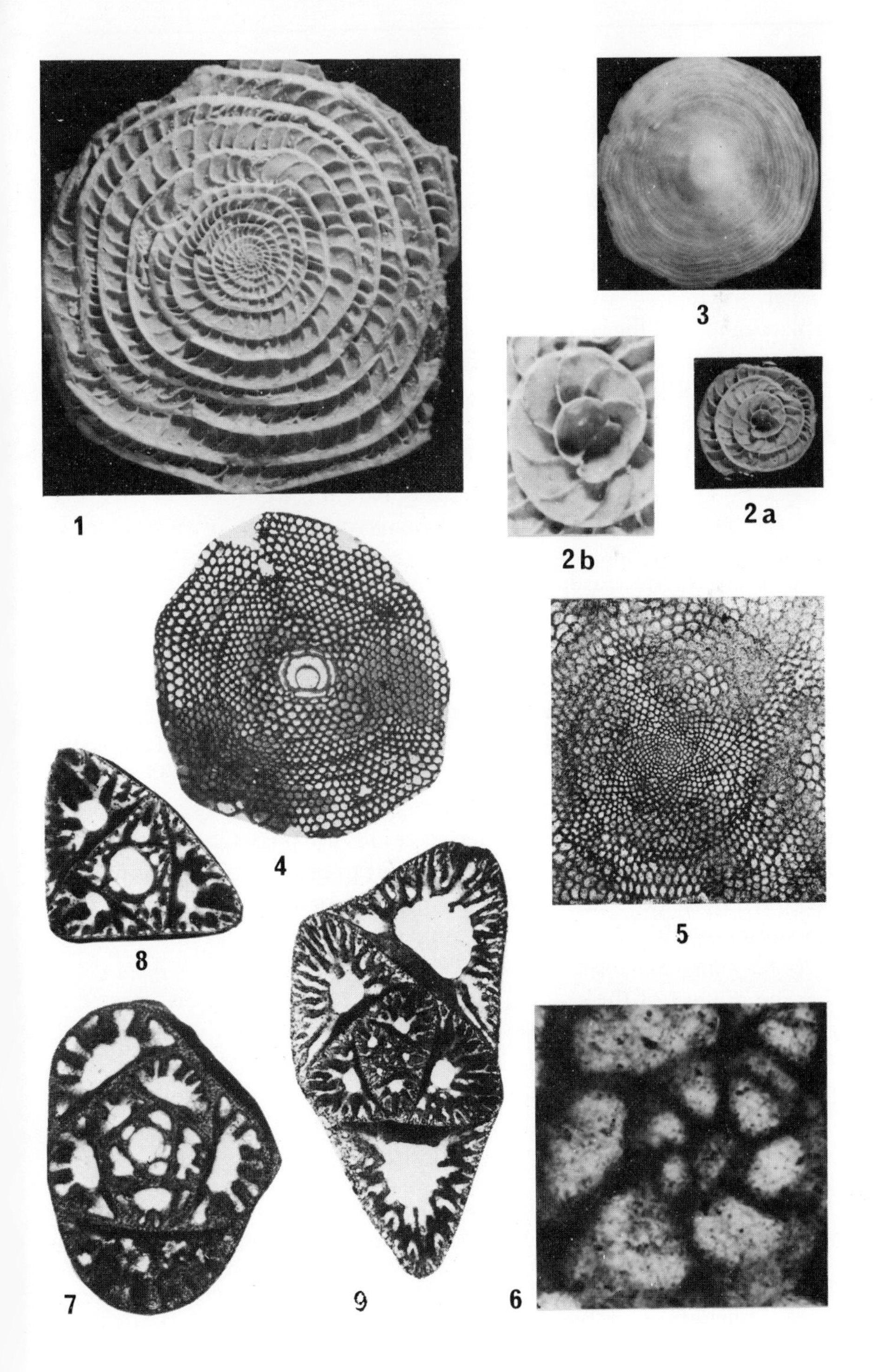
1
2a
2b
3
4
5
6
7
8
9

further division into subprovinces; and that all are currently situated within a broad circumglobal belt, approximately within the range 35°N to 35°S (Fig. 1). During times of climatic warming they extended as far north and south as the prevailing combination of warm and shallow water allowed (up to 60°N and 45°S in the Eocene): during cold periods, although they contracted towards the equator, their boundaries were never situated in latitudes lower than about 30°N and 30°S. From Palaeocene to Oligocene times (approximately 40 my) the Mediterranean and Indo-West Pacific were in direct communication across what is now the Middle East, and throughout all but the last 4 million years the west and east coasts of the Americas were in open marine communication across the present Isthmus of Panama.

SPECIATION AND PHYLOGENY

There are three reasons for considering evolution (speciation and phylogenesis) in relation to provincialism. The first is that our methods of delineating taxa are basic to the recognition of provinces. Dilley (1973) stated that he could not distinguish an American Late Cretaceous province on the basis of larger foraminifera. Yet van Gorsel (1978) using the same organisms, but discriminating differently between taxa, came to the unequivocal conclusion that such a province existed. The second reason is that our view of speciation must determine how we calculate rates of evolution, and very different results are possible using different methods. The third, is that it is important to distinguish between inter-regional gene flow maintained by dispersal or migration, and parallel evolution which may produce similar results without the need for regular gene exchange.

The gravity of the species problem in the Foraminifera is attested to by the vast number of specific names with which the literature is burdened, and by the difficulty specialists have in recognizing species. For most practical purposes an extant species may be defined as an interbreeding community of organisms, or as Hennig (1966) has it, as a "well integrated gene pool" (*vide* Hull, 1979:431). But when fossils are

Fig. 1. Approximate areas occupied by the three main larger foraminiferal provinces during much of the Tertiary. The Mediterranean and Indo-Pacific were connected across the Middle East during the Palaeogene. Mixed American and Mediterranean Eocene faunas sometimes occurred along the N.W. coast of Africa. Based on data in Adams (1967; 1973).

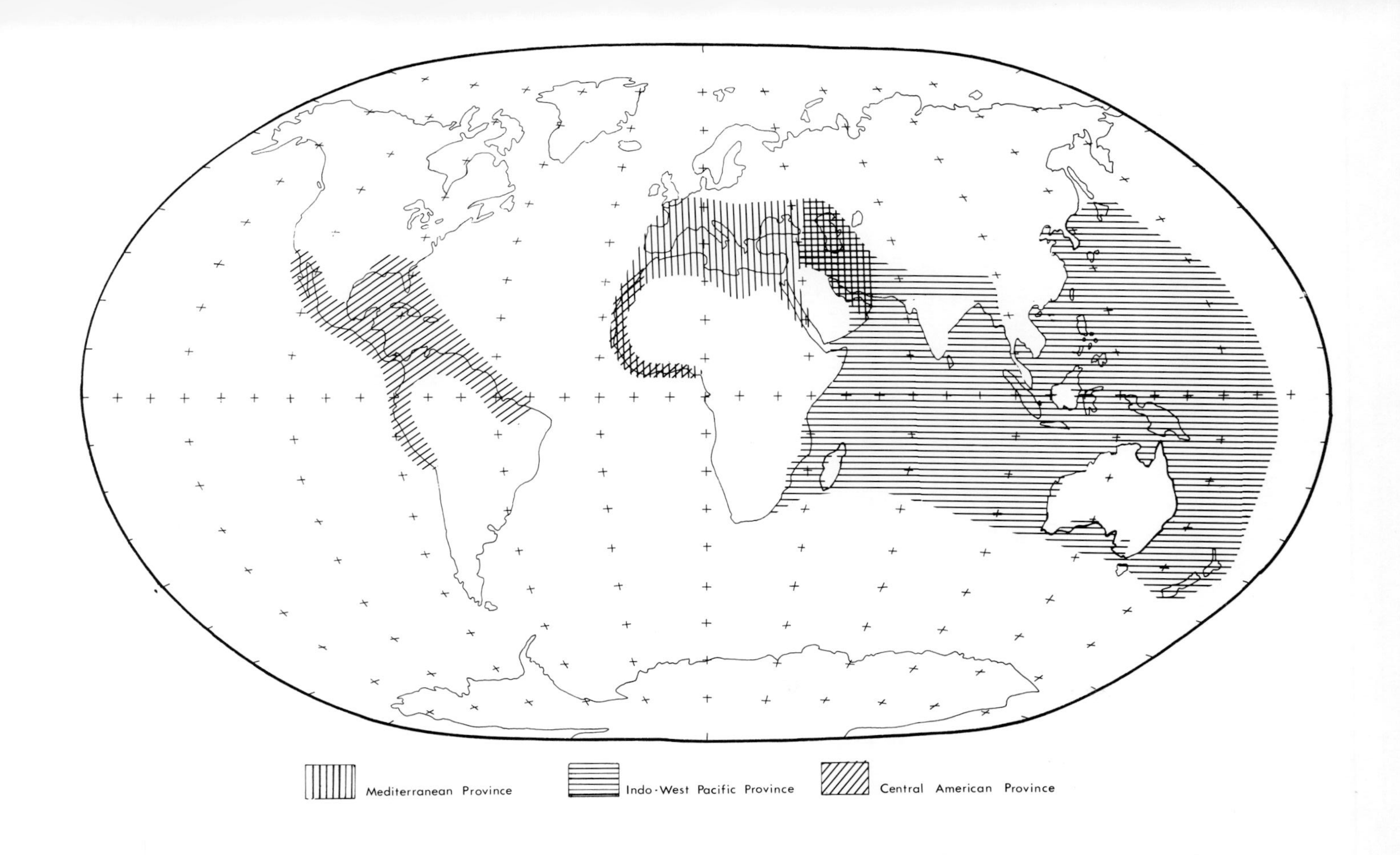
Mediterranean Province
Indo-West Pacific Province
Central American Province

considered, Wiley's (1978) modification of Simpson's earlier definition may be preferred:

> a species is a lineage of ancestral descendant populations which maintained its identity from other such lineages and which has its own evolutionary tendencies and historical fate.

Since fossils cannot be subjected to breeding tests, species delineation must be based on morphological similarities between assemblages. If extreme members of an assemblage seem to be connected by an adequate number of transitional forms they are regarded as belonging to a single species. This method works well enough for large assemblages found at a single geological horizon, but is less effective for species represented only by small numbers of individuals, or for those occurring at different stratigraphical levels or in widely separated areas. The problem of temporal morphological gradation is the root cause of most taxonomic disagreements between palaeontologists and biologists, and renders the formal definition of species difficult.

The most cursory perusal of publications on systematic palaeontology suffices to show that practising systematists rarely pause to consider the nature and relative value of specific characters when erecting new species (such matters presumably being regarded as too theoretical or too difficult), and it is evident that a specific character may be any morphological feature capable of recognition and description. In the Foraminifera these characters range from basic features such as wall structure and chamber arrangement to minor features like test size and ornament. Some are held to be of greater significance than others because they characterize higher taxa (genera, families etc.). Yet all characters are intially only of specific value since they merely distinguish new species from their immediate ancestors. Indeed they must be of less than specific importance otherwise they could not spread throughout a population. All genera are initially monospecific, and some remain so. Mutations that are theoretically of immediate specific significance may sometimes prevent reproduction and therefore prove lethal.

Examination of two or three well-established foraminiferal lineages reveals the consequences of according equal taxonomic status to characters that are not strictly comparable, and of attributing equal systematic status to organisms accorded specific rank on this unsatisfactory basis. *Austrotrillina* is a small, benthic miliolacean that lived in the same carbonate environments as larger foraminifera during Oligocene to mid-Miocene times (i.e. for about 20 m y). It possessed an alveolar wall which became progressively more complicated with time, (cf. *A. striata* Todd & Low and *A. howchini* (Schlumberger) Pl. 1, nos 7–9). The end members of the lineage *A. paucialveolata* to *A. howchini* look very different and would certainly be regarded as distinct species if found on a single time plane since they do not intergrade. Yet all gradations can be found between them in time (Adams, 1968). Differences in wall structure can be seen readily but are not measured easily. They are, therefore, verbalized, and the existence of a lineage is asserted rather than proved. Proving the reality of evolutionary lineages usually depends upon consideration of characters that can be counted or measured; i.e. those found in genera such as *Miogypsina* and *Lepidocyclina*.

Miogypsina (*Miogypsinoides*) and *M.* (*Miogypsina*) differ in that the former appears to have solid lateral walls (although the ventral wall is laterally canaliculate. See Bock, 1976) while the latter possesses lateral chamberlets (Fig. 2). In both subgenera the species are distinguished mainly on the condition of the embryonic apparatus and initial spire or spires (proloculus size, length of spire(s), degree of spiral symmetry etc.) these characters all being measurable and susceptible to statistical expression using both simple and advanced techniques (Drooger, 1963; Raju, 1974). Both subgenera show a gradual shortening of a long initial spire; in *Miogypsinoides* (Mid Oligocene to Early Miocene) a coil of some 20–30 chambers is reduced to one only 5–9 chambers long, while in *Miogypsina* (Late Oligocene-Mid Miocene) a coil some 17 chambers in length is reduced to 2–4 very short spirals each with only a few chambers (Fig. 2).

In the lineages so far mentioned (*Austrotrillina*, *Miogypsinoides*, and *Miogypsina*), the individual species are based upon the gradual modification of a character or group of characters. It does not matter whether their differences are verbalized or expressed statistically, they demonstrate that the slow modification of existing characters can produce distinctly different end members within a lineage (phyletic gradualism). Gould and Eldredge, the prime advocates of the evolutionary process known as "punctuated equilibria" were obliged to admit (1977) that they could find no fault with the case made out by Ozawa (1975) for phyletic gradualism in the Permian larger foraminifer *Lepidolina multiseptata* (Deprat). The changes in proloculus size and test shape demonstrated by Ozawa (Fig. 3) parallel those seen in numerous lineages of Palaeozoic, Mesozoic and Tertiary larger foraminifera. However, Ozawa has certainly provided some of the best stratigraphical and statistical documentation for these changes. In all such lineages the species are distinguished, verbally or statistically, on the basis of stratigraphy since they grade into one another through time.

But what of the first species in a lineage? How does that differ from its immediate ancestor? The principle of phyletic gradualism demands that any difference should be gradational. However, *A. paucialveolata* Grimsdale, the earliest and most primitive species of the *Austrotrillina* lineage, presumably differs from its immediate ancestor (genus and species unknown but a quinqueloculine or triloculine miliolacean of some kind) in possessing alveoli in the wall. *Miogypsinoides* differs from the ancestral rotaliid (perhaps *Pararotalia* as suggested by several authors) in possessing equatorial chambers (Fig. 4); *Miogypsina* differs from *Miogypsinoides* in possessing lateral chamberlets. These differences are not gradational but result from the appearance of new characters (novelties) which individuals either do or do not possess. True gradation from ancestor to descendant is not seen, although one individual may have more or less of a particular character than another. It seems, therefore, that we are using two kinds of criteria to delineate species, and can recognize both phyletic gradualism and punctuated equilibria in a single family. Novelty produces the appearance of punctuated equilibria; modification of existing characters, that of phyletic gradualism.

It can of course be argued that failure to trace species from one lineage into another using biometrical methods is simply a reflection of the limits of accuracy of stratigraphical zonation. The finest recognizable division of geological time (using foraminifera) is between half and one million years—but this fails to explain why evolution within lineages proceeded more slowly than between lineages. Hottinger (1981) regarded genera as possessing two types of characters, (a) qualitative (unchanging) characters such as the arrangement of septulae

---

Fig. 2. Stratigraphical distribution of some miogypsinids. Data mainly from Drooger (1963) and Raju (1974). Drawings show embryonic apparatus only. Walls of protoconch and deuteroconch clear; walls of spiral chambers with single apertures, black: walls of primary and secondary spiral chambers with retrovert appertures, stippled. Chamber walls of 3rd and 4th spirals, clear. Closed circles mark relative positions of *Miogypsinoides* species; open circles the relative positions of Miogypsina species. A and B, vertical sections through *Miogypsinoides* and *Miogypsina* to show the lateral walls.

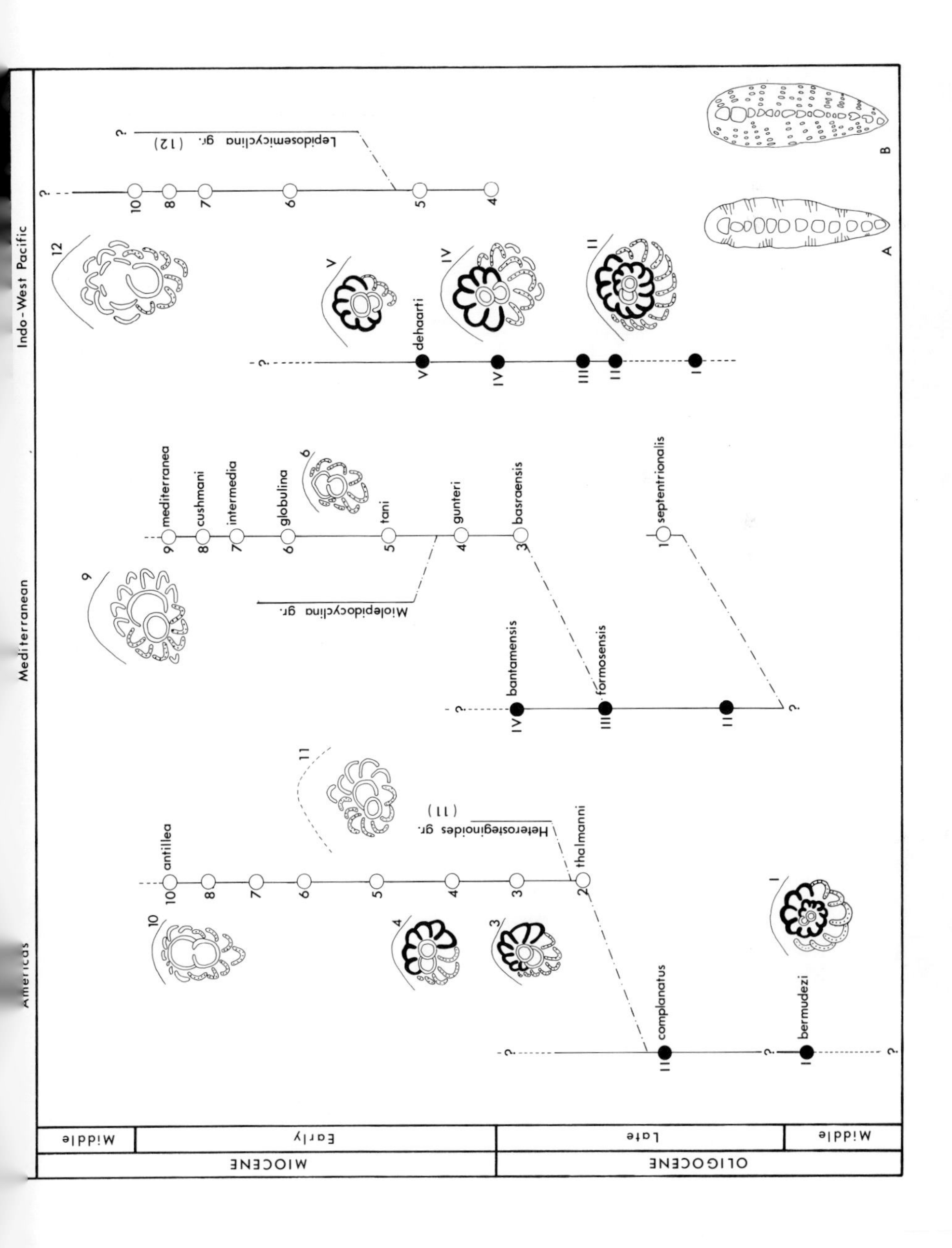
Indo-West Pacific
Mediterranean
Americas
Lepidosemicyclina gr. (12)
dehaarti
Miolepidocyclina gr.
mediterranea
cushmani
intermedia
globulina
tani
gunteri
basraensis
septentrionalis
bantamensis
formosensis
Heterosteginoides gr. (11)
antillea
thalmanni
complanatus
bermudezi
A
B
Middle
Early
Late
Middle
MIOCENE
OLIGOCENE

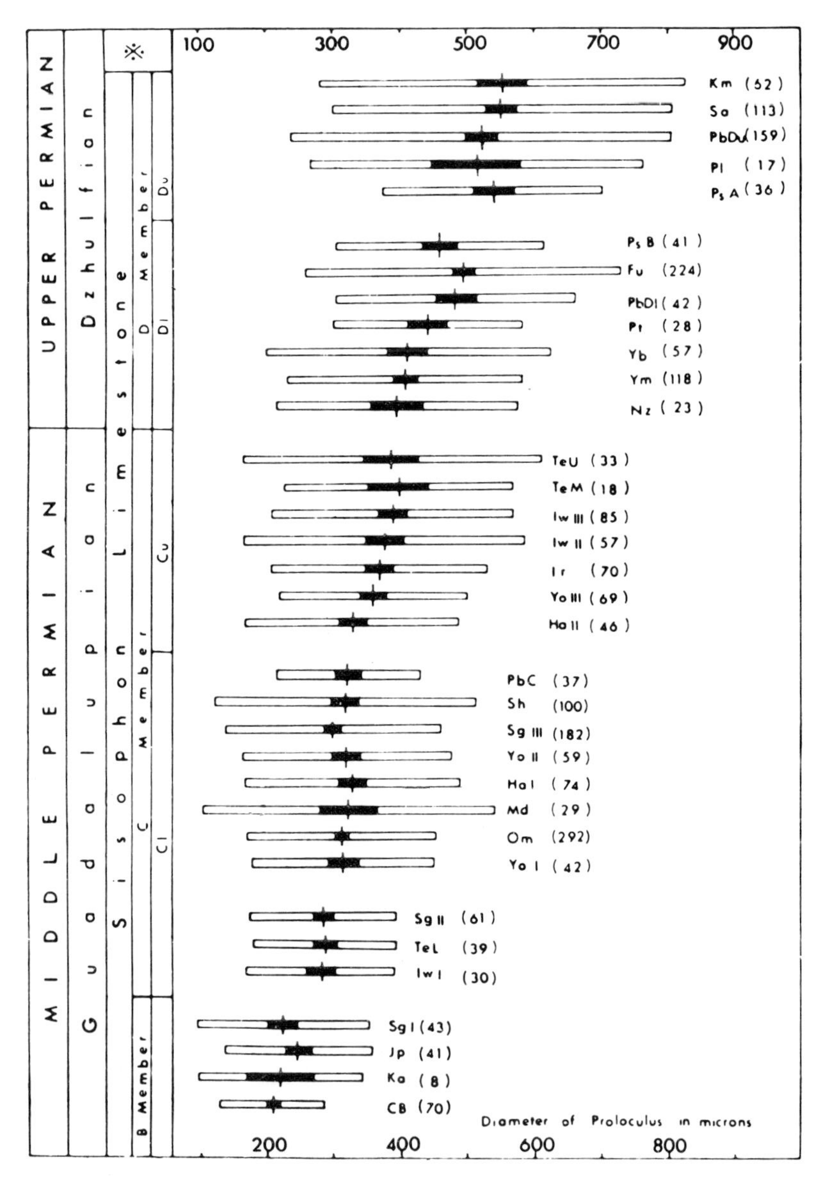

Fig. 3. Geographical and temporal variation in prolocular diameter for the Permian foraminifer *Lepidolina multiseptata* in East Asia. Black rectangles are 95% confidence limits for the mean; white rectangles span two standard deviations. From Ozawa (1975).

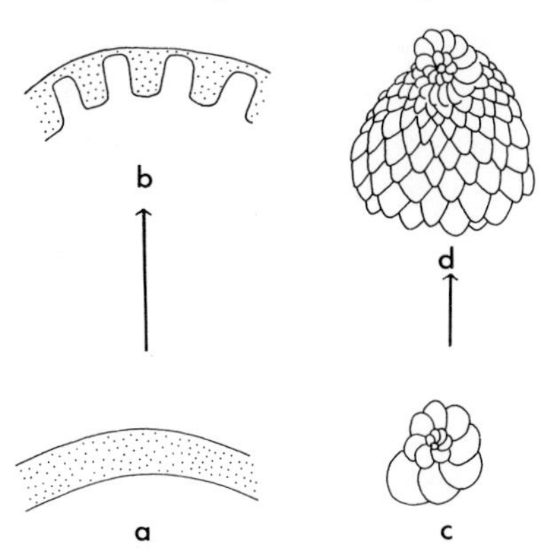

Fig. 4. The production of novelty in *Austrotrillina* and *Miogypsinoides*. The ancestor (a) of *Austrotrillina* (b) had no alveoli in the chamber walls. The ancestor (c) of *Miogypsinoides* (d) possessed no equatorial chambers.

and the double rows of alternating foramina in *Alveolina*, and (b) quantitative (changing) characters such as megalosphere size and the index of elongation in the same genus. The quantitative characters show unidirectional evolution.

Changing quantitative characters of a minor nature are the basis of all biometric studies of evolution and are of great importance in biostratigraphy. They often obviate the necessity to name species (see p. 271) thus rendering a great service to biostratigraphers, but they reflect the detail of evolution rather than the major processes. Drooger (1980) unfairly mocked other biostratigraphers by quoting J.C. Brouwer's unpublished remark, "quantitative methods are complex whereas most biostratigraphers are simple-minded". But the methods he advocated have not solved any major problems of evolution even in the families to which they have been most intensively applied. Indeed, Barker and Grimsdale (1936) and Caudri (1975) said more about the origin of the Lepidocyclinidae in descriptive terms than has the sum total of statistical work—valuable though this has been—about their subsequent evolution.

The fundamental difference between the conventional descriptive approach to lineage construction and the modern morphometric method is that the former depends largely on the subjective appreciation of a large number of variable characters, whereas the latter is based upon the mathematical expression of a small group of characters or even of a single character. The classical method allows us to see similarities between the species in one lineage and those in another; the morphometric method convincingly demonstrates unidirectional changes in the characters measured, but tends to obscure other similarities. These methods have not, however, replaced one another (see Scott 1974, for discussion). The earlier descriptive method tends to be applied where biometry is inappropriate, i.e. to genera within a family or families within a class. Biometry, on the other hand, is considered to be especially valuable in relation to species within a genus, since it is believed to measure phyletic speciation.

Cladists would argue that each of the lineages so far mentioned represents only one species, however different in appearance the end members may be. They could be right. Pheneticists, and these would necessarily include most biostratigraphers, would argue that since the end members do not intergrade morphologically they must be regarded as distinct species. They too could be right. It is therefore clear that we must be prepared to use whatever classificatory method is appropriate to the solution of a particular problem. A new classification of taxa on a single time plane (e.g. Recent) will be most informative if it is cladistic, but the dating and correlation of sedimentary successions will be better accomplished if phenetic classifications are used. However, any "good" classification will automatically reflect evolutionary history. If it does not, the classification is probably faulty.

Having asserted that the first species in some lineages are based upon novelty, it is necessary to ask whether this is true of all lineages. Unfortunately, it is not. *Nummulites* and *Assilina* are important members of the same family. Species of *Nummulites* (*sensu* Cushman, 1948) are nearly all involute; those of *Assilina* are all evolute. But the condition of the spire, however stable it may be in some groups of species, cannot be construed as constituting novelty. The same is true of many other genera. It is worth noting that the concept of novelty, if applied to the classification of planktonic foraminifera or to some groups of benthic foraminifera (e.g. the Mesozoic Nodosariidae) would have devastating consequences.

It follows from the foregoing remarks that before trying to calculate rates of evolution it is first necessary to decide whether the taxa concerned are based upon new characters (novelties), modified characters, or new combinations of characters. True novelties do not seem to have developed gradually, but to have arisen rapidly by mutation and to have spread quickly through populations. There is clearly little chance of finding the earliest representatives of taxa based on novelty in the fossil record. New combinations of characters, such as those defining the genus *Assilina*, also appear suddenly but can hardly be said to have evolved in the same way as novelties. The Family Miogypsinidae includes at least 15 generally recognized Indo-West Pacific species and 28 species world-wide (see Drooger, 1963; fig. 25; Raju, 1974: fig. 34). Based on the production of true novelty (median layer and lateral chamberlets), and allowing 15 my for the range of the family, the speciation rate could be regarded as one per 7·5 my (a ludicrously low and totally unacceptable estimate to systematists and stratigraphers alike) to one per 0·94 my (one per 0·53 my world-wide). If novelty is defined to cover particular arrangements of apertures, then intermediate rates of speciation are obtained. The six species of *Miogypsinoides* listed by Raju (1974) represent a speciation rate of about one per 1·66 my.

If the nature of speciation determines whether "punctuated equilibria" or "phyletic gradualism" is observed in the fossil record; that of phylogenesis determines whether or not inter-regional gene-flow or parallel evolution seems to occur. The nature of phylogenesis can be illustrated by reference to two well-known genera, *Cycloclypeus* and *Lepidocyclina*.

Cycloclypeus appeared during the Early Oligocene, some 35 million years ago, and lives on in the Indo-West Pacific where it is represented today by one species (Adams and Frame, 1979). Three stages of growth can be recognized in the megalospheric test (Fig. 5). An embryonic stage consisting of the first two chambers, a nepionic stage comprising one (rarely more) undivided "operculine" chambers and up to 38 divided "heterostegine" chambers, and a neanic stage represented by numerous annular (cyclic) chambers forming most of the test. Tan (1932) demonstrated that specimens collected at successively younger levels in the Oligocene and Miocene show a progressive shortening of the initial coil, a process referred to as *nepionic acceleration* (a form of tachygenesis). All gradations can be found between primitive and advanced forms, and statistical methods are usually employed to distinguish between species. The entire test may be said to demonstrate Haeckel's Principle of Recapitulation. The first two growth stages, "operculine" and "heterostegine" are believed to represent the ancestral genera *Operculina* and *Heterostegina*, and it is evident that as new characters have been added the earlier growth stages have been reduced. Nepionic acceleration can, of course, be recognized only when new morphological characters are introduced at a late stage of development. The introduction of new

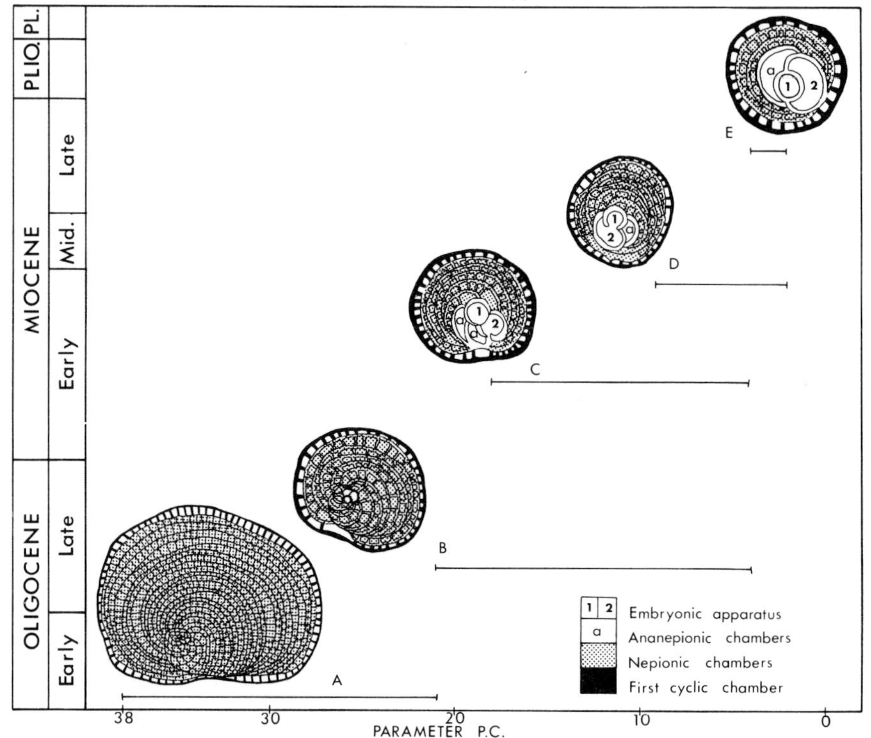

Fig. 5. Evolution of the nepionic stage of *Cycloclypeus*. The long "heterostegine" initial coil (stippled) of the primitive Oligocene species is reduced (nepionic acceleration) through time thus hastening the onset of cyclic growth. The direction, but not the rate, of evolution was constant. Only the first annular chamber is shown although these comprise the greater part of the adult test. A–D, based on figures by Tan (1932); E, a Recent specimen from Funafuti Atoll figured by Adams and Frame (1979). P.C. = number of nepionic chambers; horizontal lines indicate range of parameter p.c. at various levels (data from Tan, 1932). A × 3; B–D × 10; E × 20.

characters into the early stage of growth would either be lethal or would produce an adult so different in appearance from its ancestor that the two might not seem to be closely related. This would make the erection of a classification based on clades extremely difficult if not impossible.

Recapitulation has been the basis of many described evolutionary lineages including those proposed by Barker and Grimsdale (1936) and

Caudri (1975) for the early lepidocyclinids, and by Tan (1932) for the cycloclypeids. It also formed the basis of Cushman's (1928–48) classification of the Foraminifera.

One of the more structurally complex foraminifera, *Lepidocyclina* (*Nephrolepidina*), appeared in the Mediterranean and Indo-West Pacific regions in Mid-Oligocene times having evolved earlier in the Americas. From then on it occurred in all three provinces and underwent broadly similar evolutionary changes in each. These seem mainly to have affected the embryonic apparatus and equatorial chambers (Fig. 6). Between Mid-Oligocene and Mid-Miocene times (about 17 my) the deuteroconch came to embrace the protoconch to a progressively greater degree. The number of adauxiliary chambers increased, and the equatorial chambers, in one lineage at least, changed from rhombic to hexagonal in shape. At all stratigraphical levels there is a wide range of variation in the embryonic apparatus, but as van der Vlerk (1959, 1974) and van der Vlerk and Postuma (1967) have shown, by expressing the grade of enclosure (Factor A) as a percentage, it is possible to determine the relative age of an assemblage without necessarily having to name the species present.

The reduction of the initial spire in *Miogypsinoides*, *Miogypsina*, and *Cycloclypeus*, and the progressive changes seen in the embryonic apparatus of *Lepidocyclina* (*Nephrolepidina*) can be regarded as examples of orthogenesis. The term is used here descriptively in the sense of MacGillavry (1968), and is not intended to denote a mechanism or directing force. Unidirectional evolution is, perhaps, a better expression. Some specialists have argued that changes of this kind are adaptive, and have implied that their purpose was to shorten the nepionic stage, thus benefiting the organism by hastening the onset of the adult mode of growth. But end members having attained this condition, were not always more successful (in the sense of having longer geological ranges) than their ancestors—indeed, many soon became extinct, e.g. *C.* (*Katacycloclypeus*) *annulatus* Martin, (see Adams and Frame, 1979). It is possible, as Hottinger (1981) has suggested, that the benefit was functional, although it remains difficult to understand why the changes should have been so slow when the initial variability of populations was high.

If the reality of recapitulation and orthogenesis (unidirectional evolution) is acknowledged, it is easy to understand why some genera are represented by similar species in different provinces while others are not. Dispersal is difficult for all species of larger foraminifera, but especially for those spending no part of their life cycle on weeds since they cannot be drifted on floating vegetation. However, it is unnecessary to postulate repeated trans-oceanic crossings if the passage of just one species can produce the same evolutionary result in each region. Unidirectional evolution could ensure the production of similar end members in lineages on either side of major barriers provided they shared a common ancestor. This seems to have occurred both with *Miogypsinoides* and *Lepidocyclina*, although Freudenthal (1972) has postulated a separate origin for Tethyan and American species of *Lepidocyclina*. The occurrence of *Miogypsinoides complanatus* (Schlumberger) and *Miogypsina gunteri* Cole (Fig. 2) in all three provinces could have ensured the subsequent appearance of similar descendants in each region provided local extinction did not supervene. Regional side branches could, of course, be expected and were indeed produced (the *Heterosteginoides* group in the Americas; *Lepidosemicyclina excentrica* group in the Indo-Pacific; and the *Miolepidocyclina* group in the Mediterranean). Representatives of these sub-groups do not seem to have gained access to other provinces, thus further indicating that dispersal was difficult. Drooger (1979) has argued the case for parallel evolution in the Mediterranean and Indo-Pacific regions.

Extinctions can occur at any time, and lineages once initiated do not necessarily proceed to completion in all areas. *Cycloclypeus* did not persist beyond the *C. eidae* stage in the Mediterranean; *Miogypsinoides* did not continue beyond the level of *M. complanatus* in the Americas or of *M. bantamensis* Tan in the Mediterranean (records of *M. dehaarti* van der Vlerk in this region require confirmation), while *Nephrolepidina* only attained the acme of its development in the Indo-Pacific. Local

---

Fig. 6. Evolution of the embryonic apparatus in *Lepidocyclina* (*Nephrolepidina*) in the Indo-Pacific region. The sub-equal protoconch and deuteroconch of the primitive Oligocene species gradually increased in size and the protoconch became progressively more highly embraced by the deuteroconch. The number of adauxiliary chambers also increased. Factor A measures the degree to which the protoconch is embraced by the deuteroconch and is expressed as a mean for each assemblage. The trend is unidirectional. Illustrations based on figures in van Vessem (1978). Abbreviations: l.c.w., length common wall; i.c.p., internal circumference of protoconch. See Pl. 1, nos 4–6 for equatorial sections of megalospheric and microspheric forms.

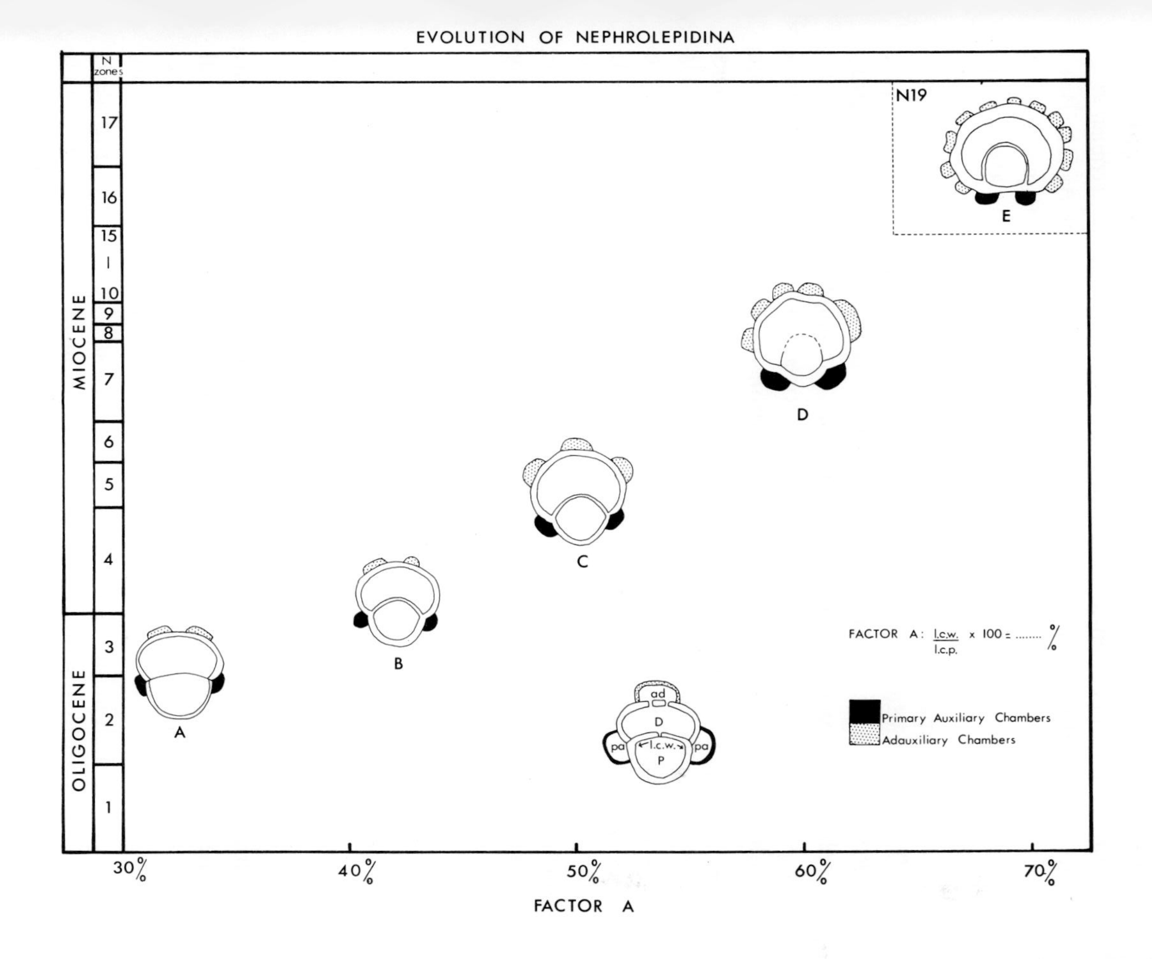

EVOLUTION OF NEPHROLEPIDINA
N zones
17
16
15
10
9
8
7
6
5
4
3
2
1
MIOCENE
OLIGOCENE
N19
E
D
C
B
A
ad
D
pa
l.c.w.
P
pa
FACTOR A : l.c.w. / l.c.p. x 100 = ... %
Primary Auxiliary Chambers
Adauxiliary Chambers
30%
40%
50%
60%
70%
FACTOR A

retardation in the development of characters can sometimes be recognized between taxa in a region or between those in one region and another, e.g. between *M'oides complanatus* and *M'oides* cf. *bermudezi* in India, and between *Lepidocyclina* and *Miogypsina* in West Africa (Brun and Wong, 1974). This creates both local and inter-regional correlation problems and indicates that the first records of a species in an area should be treated with caution until their ages have been confirmed by reference to other groups.

The distribution tables produced by some authors (e.g. Drooger, 1963) imply that primitive forms disappear completely as more highly evolved species appear. Hence, *Miogypsinoides complanatus* should not be found in strata younger than Late Oligocene. However, there are a number of records of primitive species from much higher stratigraphic levels than would be expected if ancestors became extinct as new species arose. Adams and Belford (1974) found *M. complanatus* in Miocene beds on Christmas Island. Glaessner (pers. comm.) has recorded the same species from beds of similar age in Papua New Guinea, and Raju (1974) found *M.* cf. *bermudezi* in beds younger than those with *M. complanatus* in Cutch.

It is possible that the length of the initial spire may be controlled by water depth as has been suggested by Freudenthal (pers. comm.), although this does not seem to be a likely explanation for the occurrences of *M. complanatus* in India and Christmas Island since the sedimentary sequences in these two areas are fairly uniform.

TECTONISM

Three aspects of tectonism require consideration in the present context. The first is concerned with mountain building and uplift on a regional scale: so-called orogenic and epeirogenic movements. These can transform marine into continental environments thus destroying carbonate and other shelf habitats, and causing the seaward (downslope) migration of faunas and subsequent intense competition for space near the shelf margin. The second concerns the movement of large segments (plates) of the Earth's crust relative to one another. This may involve the movement of continents and associated shallow water areas from one latitude and temperature zone to another, and may bring continental areas into collison. Once this happens, associated orogenic movements can produce land barriers to impede further marine migration or dispersal. The third aspect concerns the vertical or lateral crustal movements of oceanic ridges or trenches. By altering the volumetric capacity of the world's ocean basins these may cause desiccation or flooding of the continental margins on a global scale. The consequences of such movements are dealt with below under eustasy.

The Cenozoic Era saw two orogenic events of potential regional biogeographic importance. One was the creation of the Panama Isthmus about 4 my ago; the other was the occlusion of the ancient Tethys Sea by the formation of a land bridge across the area of the present Persian Gulf during the Late Oligocene to Early Miocene.

Little is known about the larger foraminiferal faunas of the west coast of the Americas after Early Miocene times, possibly because the sediments of the region are mainly developed in unsuitable facies. However, the creation of the Panama Isthmus brought about the final separation of the Atlantic and Pacific faunas and probably contributed to the disappearance of the remaining post-Early Miocene larger foraminifera, such as *Amphistegina*, on the Pacific coasts. This event is still too recent to be of interest in the present context.

Of much greater palaeobiogeographic importance was the final disconnection of the proto-Mediterranean from the Indian Ocean during the Early Miocene. Distribution maps of genera such as *Miogypsina*, *Spiroclypeus*, *Borelis*, and *Lepidocyclina* in the Middle East show that they are all confined to a narrow zone between Arabia and central Iran. By the Late Oligocene this was all that remained of the much broader seaway which existed in this region during early Tertiary and pre-Tertiary times. The distributions of *Borelis melo curdica* Reichel and *Flosculinella bontangensis* (Rutten) in late Burdigalian* times indicate that a land bridge had by then appeared across the central part of the Persian Gulf (Fig. 7). This prevented the migration of *B. melo curdica* into the Indian Ocean and of *F. bontangensis* into the Mediterranean. Thereafter, new taxa (e.g. *Discospirina italica* and *Alveolinella quoyi*) arising in the Mediterranean and Indo-Pacific respectively, were each excluded from the other area. It is relatively easy to demonstrate how tectonism modifies faunas but the distribution of *B. melo curdica* shows that faunas can sometimes be used to indicate both the timing and provenance of important geological events.

Although orogenesis and epeirogenesis can be important regional factors in biogeography, in order to observe the effects of tectonism on a

* *See* postscript, p. 289.

large scale it is necessary to look to plate tectonics, and for this purpose the 65 million years of the Cenozoic Era is hardly long enough. The Central American Province already existed during the Cretaceous (van Gorsel, 1978), having been created by the westward drift of the Americas from Europe and Africa, and further movement during the Cenozoic merely maintained the differences between these provinces.

The only Cenozoic plate movement of major palaeobiogeographical significance in the present context was that of the Indian plate (including Australia and New Guinea) northward from Antarctica. It is generally accepted that Australia was attached to Antarctica throughout most of the Palaeogene and that the two continents were seprated by a narrow, relatively shallow, seaway during the early part of this period. Final separation, marked by the appearance of a deep trough did not occur until Mid to Late Oligocene times (Kennett *et al.*, 1975)—Chattian according to Jenkins (1978)—at which time the Circum-Antarctic Current is believed to have come into existence. This being so, western Australia lay farther south in Eocene times than it does today. It then supported a fauna of larger foraminifera which included *Discocyclina* and *Pellatispira* in the Carnarvon Basin (Chapman and Crespin, 1935). No larger foraminifera of this age have been reported from southern Australia, but *Asterocyclina* occurs in Middle Eocene sediments of New Zealand and the Chatham Islands (Cole, 1962). In this connexion, it is interesting to note that *Discocyclina* and its close relative *Asterocyclina*, have the widest latitudinal distribution (Alaska to New Zealand) of all Tertiary larger foraminiferal genera, and it is therefore possible that they could tolerate slightly cooler conditions than other forms. Well diversified larger foraminiferal assemblages had a somewhat narrower latitudinal range in Eocene times.

No larger foraminifera are known from southern Australia until the Middle Miocene, by which time this area lay much farther north than during the Palaeocene. It had, in fact, almost reached its present position. Chaproniere (1980) has attributed the known distribution of Tertiary larger foraminifera in Australasia directly to the northward

---

Fig. 7. Map showing the known distribution of *Borelis melo curdica* and *Flosculinella bontangensis* in the later part of the Early Miocene. Southern shoreline of the Mediterranean shown by a solid black line. Northern shoreline of the Indian Ocean by a broken line. Taxa common to both regions not shown although their presence helps to delimit the shorelines.

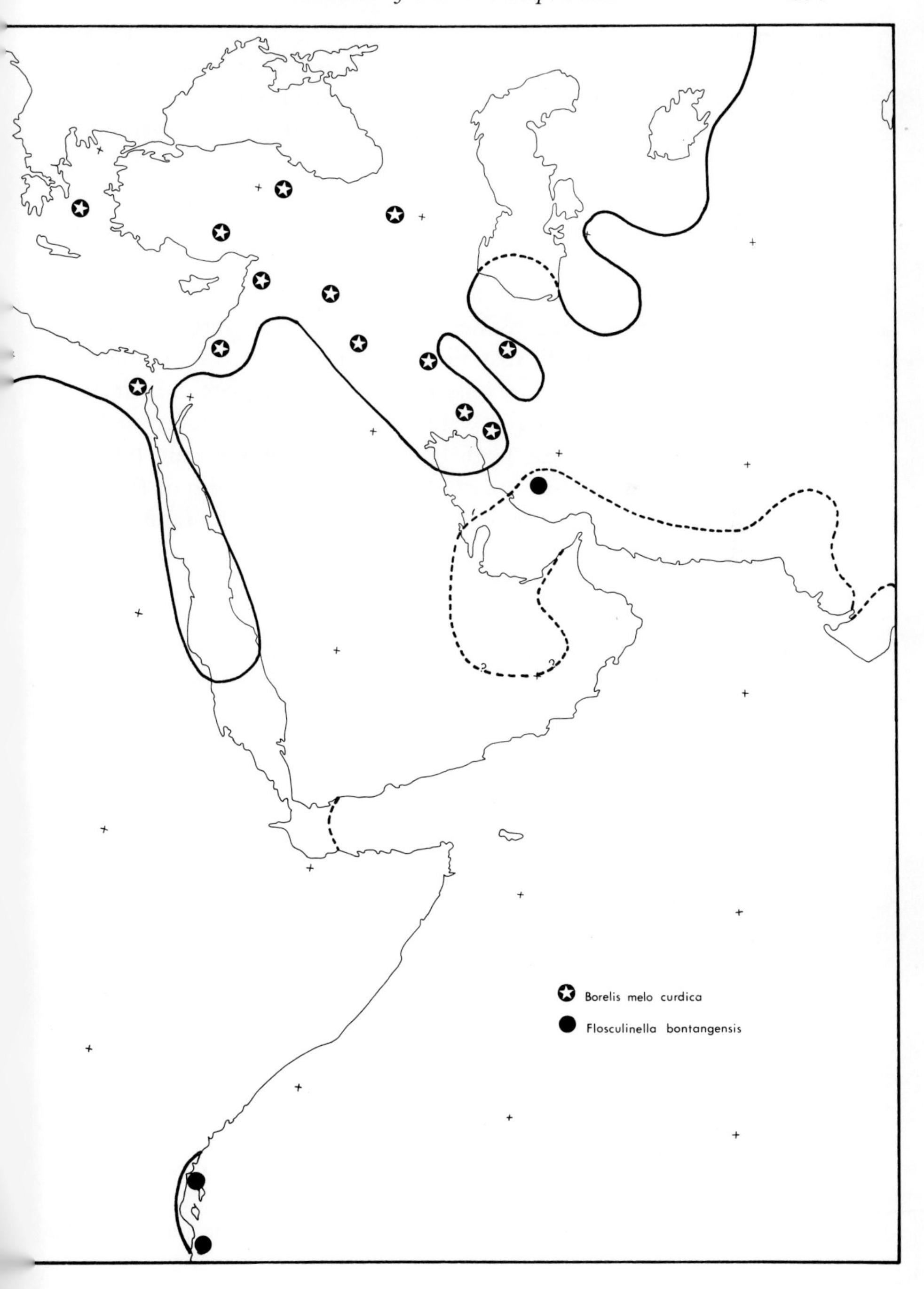
Borelis melo curdica
Flosculinella bontangensis

movement of the Indo-Australian plate, drawing special attention to the distribution of *Cycloclypeus*, *Lepidocyclina*, and *Miogypsina*, which he considers to have arrived in Australia when the distance between New Guinea and the Malay Archipelago had shortened sufficiently to permit crossings. However, this explanation does not take account of the fact that other foraminiferal genera (e.g. *Pseudorbitoides*, *Nummulites*, *Discocyclina* etc.) all arrived in New Guinea during the Cretaceous and Eocene when this area lay far to the south of its present position (see below), nor does it allow for the distribution of *Asterocyclina* already mentioned. Chaproniere was nevertheless right to suppose that the northward movement of Australia had some effect on foraminiferal distribution. It is certainly the only known Tertiary plate movement to have effectively diminished a water barrier to migration, partly by moving shallow shelf areas from high to low latitudes (and therefore into warmer conditions) and partly by reducing the width of the water crossing. Other Tertiary plate movements acted to hinder rather than to facilitate marine dispersals.

There are no known records of Palaeocene to Mid-Eocene larger foraminifera south of the latitude of New Guinea which, in Palaeocene times is believed to have lain at about 35°S (Smith and Briden, 1977), but this does not mean that they could not have penetrated to higher latitudes. The extreme limits (Alaska to New Zealand) of larger foraminifera during the Eocene was 60°N to 45°S on present-day latitudes, but in Mid-Eocene times the latitude of New Zealand may have been somewhat different (Stevens, 1980). The absence of larger foraminifera from the west coast of Australia is probably attributable to the northward movement of cool water from the Antarctic, while on the eastern side of the continent, the lack of suitable facies in the few Tertiary rocks so far discovered may well be the explanation.

CLIMATE

Climate is an important factor in maintaining provincialism amongst larger foraminifera, since cool surface waters at high latitudes currently prevent migration along the shelves bordering the ocean basins. This seems to have been the case throughout the Cenozoic except, perhaps, briefly during Middle Eocene times. Most present day larger foraminifera live successfully between about 40°N and 40°S, but during periods of climatic warming (e.g. Early-Mid Eocene) their ranges were extended. Indeed, the only direct effect of climatic change on larger foraminifera during Cenozoic times seems to have been to cause the boundaries of the three provinces to move away from or contract towards the equator during warming and cooling phases.

Climatic conditions are known to have fluctuated considerably during the Cenozoic. According to Savin *et al.* (1975), after a temperature drop near the Cretaceous/Tertiary boundary, conditions improved rapidly and the climate remained warm and relatively constant from the Palaeocene to the Middle Eocene. This allowed larger foraminifera to attain a wide latitudinal distribution (see Adams, 1973). The Late Eocene saw a sharp drop in temperatures after which conditions stabilized temporarily before a further deterioration led to the Tertiary minimum being reached during the Oligocene, about 30 my BP (see Fig. 9a). Temperatures then rose again, the high point for the Cenozoic being attained in Mid-Miocene times: thereafter, surface temperatures declined rapidly at high latitudes while remaining steady in the tropics.

The widest latitudinal distribution of larger foraminifera seems to have occurred in Mid-Eocene rather than Mid-Miocene times when surface temperatures were at their highest. This apparent anomaly may reflect the fact that the Eocene and Miocene taxa were not the same and could therefore have had slightly different temperature tolerances, but it may also have been related to the relative dispositions of land and sea. In Mid-Miocene times, Indo-Pacific larger foraminifera no longer had access to the Mediterranean and Europe across the Middle East, while in high latitudes generally, sedimentary facies do not seem to have favoured the spread of these organisms. The only significant conclusions to be drawn from a comparison of foraminiferal distributions and surface temperatures, are that at no time in the Tertiary were larger foraminifera ever in danger of extinction through climatic cooling alone, and that they never occupied a latitudinal zone less than 60° wide. Except along the northern and southern boundaries of the three faunal provinces they could, therefore, never have undergone severe stress as a direct consequence of climatic change.

EUSTASY

An indirect effect of climatic cooling and warming is the lowering and raising of sea level through the formation and melting of polar ice. The resulting glaciations and deglaciations could, it has been said, account for sea level changes of the order of 120 m (180 m less about 60 m attributed to isostic compensation: John, 1979). But crustal movements in the ocean basins have been held to be responsible for even greater sea level changes.

Vail *et al.* (1977, Table 1) claim to have recognized five major lowstands (Mid-Palaeocene; Early–Mid Eocene; Middle late Oligocene; Late Miocene and Late Pliocene–Early Pleistocene) and four highstands during the Cenozoic alone. Lowstands are said to have occurred when sea level fell below the edge of the continental shelf in most regions (*op. cit.*:92); highstands when the sea covered or rose above the shelves—sometimes by as much as 300 m or more. Hancock and Kaufmann (1979) postulated a maximum Late Cretaceous highstand of up to 650 m above the present level, so the more modest estimates of Vail and collaborators for the Cenozoic are by no means extraordinary. Changes of these magnitudes must necessarily have had dramatic effects on the shelf faunas of the world, and should be strongly reflected by larger foraminifera living in shallow water carbonate environments. According to Vail *et al.*, all lowstands developed rapidly and were of short duration, while highstands developed slowly and tended to be of much longer duration. Climatic fluctuations (glaciations and deglaciations) were invoked to explain some second and third order changes, but geotectonic movements (expansion and contraction of ocean ridges, and subsidence and elevation of basins) were postulated to account for all first order movements. Donovan and Jones (1979) doubted this particular conclusion but speculated that rapid subsidence of continental margins could provide an alternative explanation. Subsequent sea level recoveries were, however, left unexplained.

The effect of marked sea-level changes on the faunas of shallow water carbonate environments can be deduced from Fig. 8. Any major fall in sea level would force the entire fauna to migrate down slope or die out. Individual foraminifera would not, of course, migrate, but successive generations would colonize the downslope side of the habitat, and those on the upslope side would disappear. Eventually, there would be great competition for space in the narrower euphotic zone towards the shelf margin, and many species would either become extinct or suffer a great diminution in their numbers. During this time, the newly exposed limestones on the upper part of the shelves would be subjected to erosion, and when sea level eventually rose again, newly deposited sediments would be laid down disconformably over a weathered surface. The faunas in the overlying sediments would certainly differ in

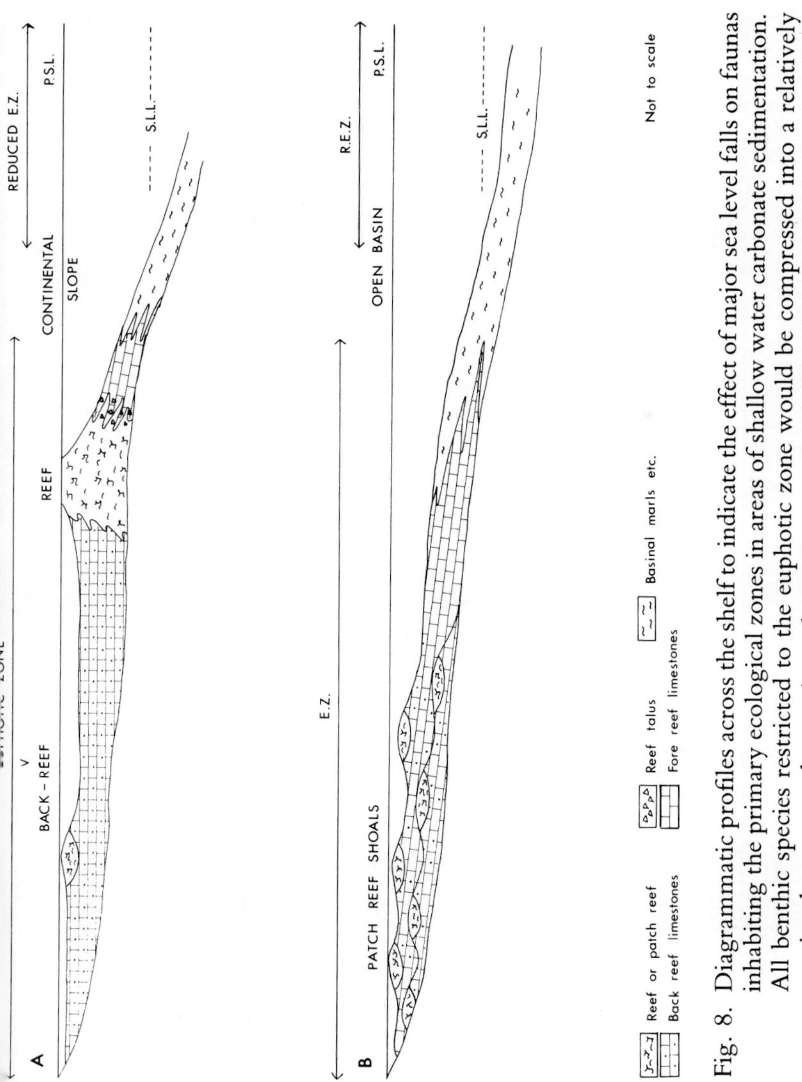

Fig. 8. Diagrammatic profiles across the shelf to indicate the effect of major sea level falls on faunas inhabiting the primary ecological zones in areas of shallow water carbonate sedimentation. All benthic species restricted to the euphotic zone would be compressed into a relatively restricted area near the continental margin as sea level fell. Modified after Henson (1950). P.S.L. = Present sea level. S.S.L. = Sea Level Lowstand. E.Z. = Euphotic zone. R.E.Z. = Reduced euphotic zone.

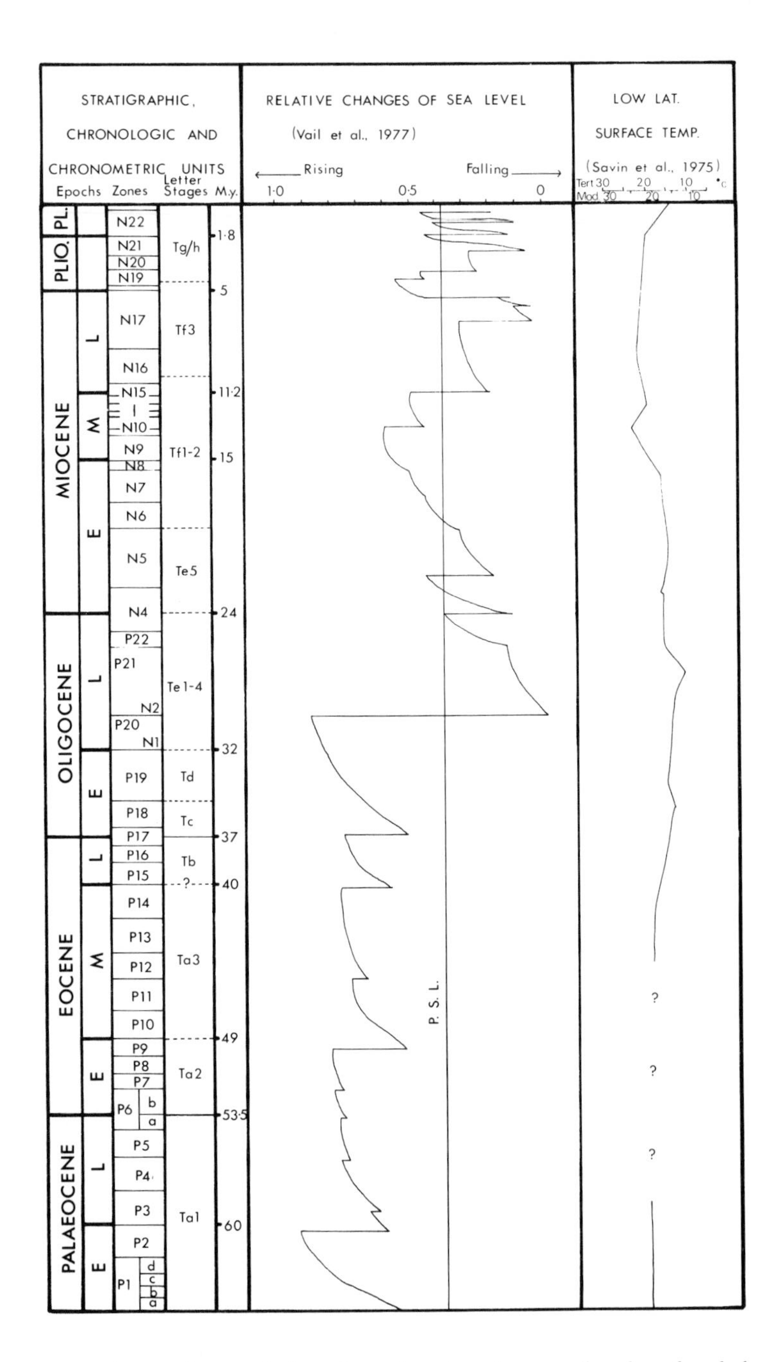

Fig. 9a&b. Cenozoic correlation chart showing the postulated sea level changes of Vail *et al.* (1977), and temperature changes at Site 167 in the tropical Pacific by Savin *et al.* (1975), plotted against three important carbonate sequences; the Asmari Limestone, Iran, based on data from Thomas (1950a,b); the Melinau Limestone, Sarawak (Adams, 1965, 1970); and the Eniwetok drill holes (Cole, 1958). There is no obvious correlation between cooling phases and

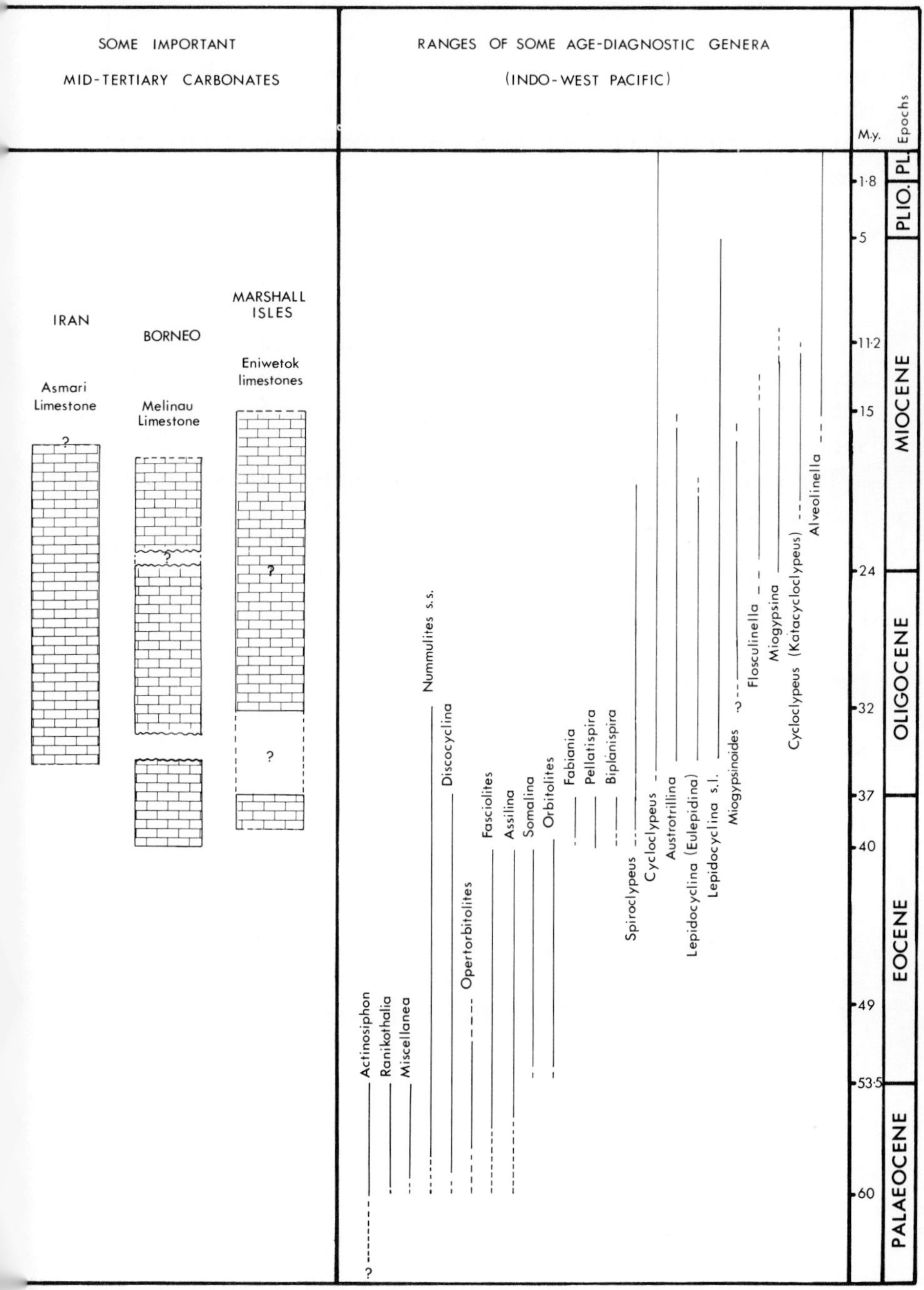

SOME IMPORTANT MID-TERTIARY CARBONATES

RANGES OF SOME AGE-DIAGNOSTIC GENERA (INDO-WEST PACIFIC)

M.y.

Epochs

IRAN

Asmari Limestone

BORNEO

Melinau Limestone

MARSHALL ISLES

Eniwetok limestones

Actinosiphon
Ranikothalia
Miscellanea
Nummulites s.s.
Discocyclina
Opertorbitolites
Fasciolites
Assilina
Somalina
Orbitolites
Fabiania
Pellatispira
Biplanispira
Spiroclypeus
Cycloclypeus
Austrotrillina
Lepidocyclina (Eulepidina)
Lepidocyclina s.l.
Miogypsinoides
Flosculinella
Miogypsina
Cycloclypeus (Katacycloclypeus)
Alveolinella

1·8
5
11·2
15
24
32
37
40
49
53·5
60

PL.
PLIO.
MIOCENE
OLIGOCENE
EOCENE
PALAEOCENE

extinctions or between sea level lowstands (except at the E/M Eocene and Eocene/Oligocene boundaries) and extinctions. Correlation of the planktonic zonation and the chronometric scale after Berggren and Van Couvering (1974) and Hardenbol and Berggren (1978). Letter "Stage" boundaries currently under revision (Adams, in prep.); the Ta_3/Tb boundary is almost certainly too high.

many respects from those below. Not only should there be evidence of shallowing below the disconformity and of deepening above, but the faunas of similar depth zones should show differences. Those above the disconformity would be impoverished relative to those below or would contain new species produced during the period of high stress on the shelf edge. The reality of first order sea-level changes such as those postulated by Vail *et al.* (*op. cit.*) should therefore be testable by reference to major shallow water limestone sequences described from different parts of the world. A global fall in sea level should affect carbonate deposition everywhere, and although other factors, e.g. rapid local subsidence, might mask the eustatic changes in a few localities, they should certainly show up clearly in most areas.

Three important mid-Tertiary carbonate units are shown in Fig. 9a, b together with the ranges of larger foraminifera in the Indo-Pacific region and the sea level changes postulated by Vail *et al.* (1977). Unfortunately, there are no described continuous sequences of shallow water limestones straddling the Cretaceous/Tertiary boundary or these too would have been included. The marked change shown by larger foraminifera between the Cretaceous and Palaeocene is certainly consistent with a major sea level fall, as is the general absence of continuously deposited shallow-water carbonates across this boundary. If falls in sea level were partly responsible for the terminal Cretaceous extinctions and for the clear faunal changes at the Eocene/Oligocene boundary, it would be reasonable to expect the supposedly greater Late Oligocene eustatic event to have had at least an equal effect. Apparently it did not. Although there were some important faunal changes in the Oligocene; for example, the extinction of *Nummulites fichteli* Michelotti and *N. vascus* Joly and Leymerie, and the first appearances of *Lepidocyclina* and *Miogypsina* in the Indo-West Pacific region, these did not occur simultaneously (Fig. 9b). Neither the Eniwetok sequence nor the Melinau Limestone seems to include an hiatus within Tertiary lower *e* (= Chattian), and although such a break could occur in the Asmari Limestone, it is unclear. There is evidence for a faunal change throughout the region at the lower/upper *e* boundary, but this is too young to be coeval with the Chattian lowstand. The apparent rarity of described shallow-water carbonates straddling the Middle/Upper Oligocene (Td/lower *e*) boundary and continuing into the upper part of the Chattian, nevertheless suggests that a drawdown may have occurred at about this time. This matter certainly merits further investigation.

SUMMARY AND CONCLUSIONS

The foraminiferal lineages discussed in this paper seem to have developed through an evolutionary process indistinguishable from phyletic gradualism as usually defined. Although end members look very different and would be regarded as distinct species if found at the same horizon, the species within each lineage can be delineated and recognized only by arbitrary statistical or verbal methods. The individual trends are apparently orthogenetic (unidirectional). The first species in some lineages differ from their ancestors in possessing new characters (novelties) and the existence of such species demonstrates that punctuated equilibria is also a feature of the evolutionary history of the Foraminiferida. The time required to produce a stem species based on novelty is short compared with that needed to produce a new "species" by modification of existing characters. Speciation based on the production of novelty can, in fact, be regarded as a geologically instantaneous event which is rarely recorded by fossils.

Specific characters are therefore of two or more types and cannot be regarded as equal in rank. In order to calculate rates of evolution it is first necessary to discriminate between species based on different kinds of criteria.

Tectonism has had little effect on foraminiferal distribution, except locally when barriers betwen regions (e.g. the Panama Isthmus) have been erected. Plate tectonics has, on the other hand, been responsible for creating and maintaining the three main faunal provinces recognized during the Cenozoic. The principal effects of this process during Tertiary times were to widen the Atlantic, occlude the Tethys, and to carry Australia northwards into the subtropics. The movement of India, although geographically important, had no significant effect on the regional distributions of marine faunas.

Climatic change has been relatively unimportant in determining the distribution of Cenozoic larger foraminifera. It has, however, limited their latitudinal range and caused their boundaries to advance and retreat in accordance with temperature fluctuations. There is no evidence that climatic change has had any significant effect on diversity or has brought about extinctions. World-wide faunal changes have, however, probably been caused by major eustatic movements resulting from tectonic activity in ocean basins, although the three carbonate successions mentioned in this paper do not appear to support previous suggestions of a major eustatic fall in the Late Oligocene.

ACKNOWLEDGEMENTS

Thanks are due to Professor Lukas Hottinger (University of Basel) for his constructive criticism of the text; to my colleague, Dr J.E. Whittaker, for reading and correcting the typescript, and to Miss Lois Cody for drafting the figures.

REFERENCES

Adams, C.G. (1965). The Foraminifera and stratigraphy of the Melinau Limestone, Sarawak, and its importance in Tertiary correlation. *Q. Jl. geol. Soc. Lond.* **121**, 283–338.

Adams, C.G. (1967). Tertiary Foraminifera in the Tethyan, American, and Indo-Pacific provinces. *In* "Aspects of Tethyan Biogeography" (C.G. Adams, and D.V. Ager, eds), pp. 195–217. Systematics Association, London.

Adams, C.G. (1968). A revision of the foraminiferal genus *Austrotrillina* Parr. *Bull. Br. Mus. nat. Hist., Geol.* **16**, 73–97.

Adams, C.G. (1970). A reconsideration of the East Indian Letter Classification of the Tertiary. *Bull. Br. Mus. nat. Hist., Geol.* **19**, 87–137.

Adams, C.G. (1973). Some Tertiary Foraminifera. *In* "Atlas of Palaeobiogeography" (A. Hallam, ed.), pp. 453–468.

Adams, C.G. and Belford, D.J. (1974). Foraminiferal biostratigraphy of the Oligocene-Miocene limestones of Christmas Island (Indian Ocean). *Palaeontology* **17**, 475–506.

Adams, C.G. and Frame, P. (1979). Observations on *Cycloclypeus* (*Cycloclypeus*) Carpenter and *Cycloclypeus* (*Katacycloclypeus*) Tan (Foraminiferida). *Bull. Br. Mus. nat. Hist., Geol.* **32**, 3–17.

Barker, R.W. and Grimsdale, T.F. (1936). A contribution to the phylogeny of the orbitoidal foraminifera, with descriptions of new forms from the Eocene of Mexico. *Journ. Paleont.* **10**, 231–247.

Berggren, W.A. and van Couvering, J.A. (1974). Biostratigraphy, geochronology and paleoclimatology of the last 15 million years in marine and continental sequences. *Palaeogeog., Palaeoclim. Palaeoecol.* **16**, 1–216.

Bock, J.F. de (1976). Studies on some *Miogypsinoides—Miogypsina s.s.* associations with special reference to morphological features. *Scripta Geol.* **36**, 1–137.

Bradshaw, J.S. (1961). Laboratory experiments on the ecology of foraminifera. *Contr. Cushman Fdn. foramin. Res.* **12** (3), 87–106.

Brun, L. and Wong, T.E. (1974). Larger foraminiferal assemblages from Nigeria. *Proc. Kon. ned. Akad. Weten.*, B **77**, 418–431.

Caudri, C.M.B. (1975). Geology and paleontology of Soldado Rock, Trinidad (West Indies). Pt. 2. The larger foraminifera. *Eclog. geol. Helv.* **68**, 533–589.

Chapman, F. and Crespin, I. (1935). Foraminiferal Limestones of Eocene age from North-West Division, Western Australia. *Proc. R. Soc. Vict., n.s.* **48**, 55–62.

Chaproniere, G.C.H. (1980). Influence of plate tectonics on the distribution of Late Paleogene to Early Neogene larger foraminiferids in the Australasian region. *Palaeogeog., Palaeoclim. Palaeoecol.* **31**, 299–317.

Cole, W.S. (1958). Larger Foraminifera from Eniwetok Atoll Drill Holes. *Prof. pap. U.S. geol. Surv.* **260–V**, 741–784.

Cole, W.S. (1962). Asterocyclina from New Zealand and the Chatham Islands. *Bull. Am. Paleont.* **44** (no. 203), 343–357.

Cushman, J.A. (1928). Foraminifera. Their Classification and Economic Use. Cushman Laboratory of Foraminifera Research, Sharon, Mass.

Cushman, J.A. (1948). Foraminifera. Their Classification and Economic Use (4th edn.). Cushman Laboratory of Foraminifera Research, Sharon, Mass.

Dilley, F. (1973). Cretaceous Larger Foraminifera. *In* "Atlas of Palaeobiogeography" (A. Hallam, ed.), pp. 403–419. Elsevier, Amsterdam.

Donovan, D.T. and Jones, E.J.W. (1979). Causes of world-wide changes in sea level. *Jl. geol. Soc. Lond.* **136**, 187–192.

Drooger, C.W. (1963). Evolutionary trends in the Miogypsinidae. *In* "Evolutionary Trends in Foraminifera" (G.A.R. von Koenigswald, *et al.*, eds), pp. 314–349. Elsevier, Amsterdam, London, and New York.

Drooger, C.W. (1979). Marine connections of the Neogene Mediterranean, deduced from the evolution and distribution of larger foraminifera. *Ann. geol. Pays Hellen.* [VIIth Int. Cong. Med. Neogene, Athens.] **1**, 361–369.

Drooger, C.W. (1980). Progress of IGCP Projects. No. 1 Accuracy in time. Geological Correlation. *Rept. Int. Geol. Correlation Programme* (IGCP) **8**, 22–32.

Freudenthal, T. (1972). On some larger orbitoidal foraminifera in the Tertiary of Senegal and Portugese Guinea (Western Africa). *Colloquium African Micropal.* **4**, 144–164.

Gorsel, J.T. van (1978). Late Cretaceous orbitoidal foraminifera. *In* "Foraminifera" (R.H. Hedley and C.G. Adams, eds) **3**, 1–120. Academic Press, London and New York.

Gould, S.J. and Eldredge, N. (1977). Punctuated equilibria: the tempo and mode of evolution reconsidered. *Paleobiology* **3**, 115–151.

Hancock, J.M. and Kaufmann, E.G. (1979). The great transgression of the Late Cretaceous. *Jl. geol. Soc. Lond.* **136**, 175–186.

Hardenbol, J. and Berggren, W.A. (1978). A New Paleogene Numerical Time Scale. *Am. Assoc. Pet. Geol., Studies in Geol.* **6**:213–234.

Hennig, W. (1966). "Phylogenetic Systematics" [translated by D.D. Davis and R. Zangerl). University of Illinois Press, Urbana, Chicago and London.

Henson, F.R.S. (1950). "Middle Eastern Tertiary Peneroplidae (Foraminifera), with remarks on the phylogeny and taxonomy of the family. Wakefield.

Hottinger, L. (1977). Distribution of larger Peneroplidae, Borelis and Nummulitidae in the Gulf of Elat, Red Sea. *Utrecht Micropal. Bull.* **15**, 35–109.

Hottinger, L. (1981). The resolution power of the biostratigraphic clock based on evolution and its limits. *Int. Symp. Concpt. Meth. Paleo. Barcelona*, pp. 233–242.

Hottinger, L. and Dreher, D. (1974). Differentiation of protoplasm in nummulitidae (foraminifera) from Elat, Red Sea. *Mar. biol.* **25**, 41–61.

Hull, D.L. (1979). The limits of cladism. *Syst. Zool.* **28**, 416–440.

Jenkins, D.G. (1978). *Guembelitria* aff. *stavensis* Bandy, a paleooceanographic marker of the initiation of the Circum-Antarctic Current and the opening of the Drake Passage. *In* "Initial Repts Deep Sea Drilling Project XL" (H.M. Bolli, *et al.*), 687–693.

John, B.S. (1979). "The Winters of the World". David & Charles, London.

Kennett, J.P., Houtz, R.E. *et al.* (1975). Cenozoic paleooceanography in the Southwest Pacific Ocean, Antarctic glaciation, and the development of Circum-Antarctic current. DSDP Leg 29. *In* "Initial Reports of the Deep Sea Drilling Project" (J.P. Kennett, R.E. Houtz *et al.*) **29**, 1155–1169.

Koenigswald, G.H.R. von, Emeis, J.D., Buning, W.L. and Wagner, C.W. (eds) (1963). "Evolutionary Trends in Foraminifera". Elsevier, Amsterdam, London, New York.

Larsen, A.R. (1976). Studies of Recent *Amphistegina*, Taxonomy and some Ecological Aspects. *Israel Inst. Earth-Sciences* **25**, 1–26.

Leutenegger, S. (1977). Reproductive cycles of larger foraminifera and depth distribution of generations. *Utrecht Micropal. Bull.* **15**, 26–34.

MacGillavry, H.J. (1968). Modes of evolution mainly among marine invertebrates. *Bijdragen tot de Diekunde* **38**, 69–74.

Murray, J.W. (1973). "Distribution and Ecology of Living Benthic Foraminiferids". Heinemann, London.

Myers, E.H. (1943a). Ecologic relationships of larger Foraminifera. *Rep. Comm. Marine Ecol. rel. to Pal.*, **3**, 26–31.

Myers, E.H. (1943b). Ecologic relationships of some Recent and fossil Foraminifera. *Rep. Comm. Marine Ecol. rel. to Pal.* **3**, 31–36.

Ozawa, T. (1975). Evolution of *Lepidolina multiseptata* (Permian Foraminifer) in East Asia. *Mem. Fac. Sci. Kyushu Univ.*, D (*Geol.*) **23**, 117–164.

Raju, D.S.N. 1974. Study of Indian Miogypsinidae. *Utrecht Micropal. Bulls.* **9**, 148 pp.

Savin, S.M. Douglas, R.G. and Stehli, F.G. (1975). Tertiary marine paleotemperatures. *Bull. geol. Soc. Am.* **86**, 1499–1510.

Scott, G.H. (1974). Biometry of the foraminiferal shell. *In* "Foraminifera" (R.H. Hedley, and C.G. Adams, eds) **1**, 56–151. Academic Press, London and New York.

Smith, A.G. and Briden, J.C. (1977). "Mesozoic and Cenozoic Palaeocontinental Maps". Cambridge University Press, Cambridge.

Stevens, G.R. (1980). Southwest Pacific faunal palaeobiogeography in Mesozoic and Cenozoic times: a review. *Palaeogeog, Palaeoclim., Palaeoecol.* **31**, 153–196.

Thomas, A.N. (1950a). The Asmari Limestone of South-West Iran. *Rpt. 18th Session Int. Geol. Cong.* 1948, **6** (E), 35–44.

Thomas, A.N. (1950b). Facies variation in the Asmari Limestone. Ibid. "Pt. X, Sect. J, 74–81.

Tan, S.H. (1932). On the genus *Cycloclypeus* Carpenter, Pt. I and an appendix on the Heterostegines of Tjimanggoe, S. Bantam, Java. *Dienst. Mijnb. Wetensch. Neded. Batavia* **19**, 1–194.

Vail, P.R., Mitchum, R. M. jr[2] and Thompson, S[3]. (1977). Seismic stratigraphy and global changes of sea level, Pt.4: Global cycles of relative changes of sea level. *Mem. Am. Assoc. Petrol. Geol.* **26**, 83–97.

Vessem, E.J. van (1978). Study of Lepidocyclinidae from South East Asia, particularly from Java and Borneo. *Bull. Utrecht. Micropal.* **19**, 163 pp.

Vlerk, I.M. van der (1959). Problems and Principles of Tertiary and Quaternary Stratigraphy. *Q. Jl. geol. Soc. Lond.* **115**, 49–64.

Vlerk, I.M. van der (1974). Nomenclature and Numerical Taxonomy (Name and Number). *Verhandl. Naturf. Ges. Basel*, **84**, 245–255.

Vlerk, I.M. van der and Postuma, J.A. (1967). Oligo-Miocene lepidocyclines and planktonic foraminifera from East Java and Madura, Indonesia. *Proc. Kon. Ned. Akad. Wet.*, B, **70**, 391–398.

Wiley, E.O. (1978). The evolutionary species concept reconsidered. *Syst. Zool.* **27**, 17–26.

POSTSCRIPT

Recent studies have shown that most records of *B. melo curdica* are probably of Langhian age, and that in Burdigalian times an arm of the Indian Ocean extended northwards as far as the Kirkuk area (Adams, in prep.). It now seems likely that the faunal sequence in the Asmari Limestone may be too incomplete to provide a useful test of the supposed late Oligocene sea level lowstand.

11 Patterns in Plankton Distribution and the Relation to Speciation: The Dawn of Pelagic Biogeography

S. VAN DER SPOEL

Institute for Taxonomic Zoology, University of Amsterdam, Postbus 20125, 1000 HC Amsterdam, the Netherlands

Abstract: Distribution of recent neritic and oceanic plankton in the three-dimensional biotope of the oceans reflects the influences of recent conditions and historical events of the last 100 million years. Inter-ocean connections of the present and the past strongly influenced speciation and dispersal. However, plankton transport over continents occurred relatively recently, as demonstrated by the amphipods in crossing Brazil from the Pacific to the Atlantic as a result of flow reversal of the Pre-Amazonas. For planktonic taxa one has to accept a westward dispersal before and an eastward one after the Miocene; this Miocene reversal may be an overall phenomenon for the oceanic biota. Faunal centres for planktonic taxa are difficult to discover, but it is possible to prove that one centre has been found off Dakar in the N. Atlantic. Central-Water ranges seem to be the oldest (of Eocene age) of present distributions, while belt-shaped patterns developed in several periods. North–south and east–west discontinuities and connections are important features for belt-shaped patterns, but also for many others, as they are directly linked with variation and speciation. Evolution in plankton, though in small part due to allopatric developments, is largely dominated by mechanisms of parasympatric speciation, as is demonstrated by comparing different forms of clinal variation and geographic diversity.

INTRODUCTION

To restrict this paper to marine plankton means discussing the biogeography of 70% of the earth surface and less than 5% of biogeographical literature (and probably also of the knowledge on

Systematics Association Special Volume No. 23, "Evolution, Time and Space: The Emergence of the Biosphere", edited by R.W. Sims, J.H. Price and P.E.S. Whalley, 1983, pp. 291–334. Academic Press, London and New York biogeography). These are percentages which conflict strongly, due to the facts that the sea is far less studied than the continents and that it is an environment in which geographers have nothing to go by. In order to postulate useful theories on biogeography one needs isolated areas, with developing or reducing barriers affecting active and passive dispersal and speciation. In the open ocean, all this is absent; theoretically, any organism can move from any point in the sea to any other locality. During such a journey the organism does not have to cross abrupt environmental changes greater, or of another type, than those it undergoes moving daily upwards and downwards in its local area, for conditions in the ocean change 10^5 times more slowly in the horizontal than in the vertical direction.

The open sea lacks barriers; physical and chemical properties change gradually and so do the ecological conditions in this huge three-dimensional environment. Plankton, moreover, is that passively floating cloud of organisms unable to cross barriers by active waves of migration as they lack the required swimming ability; that is, if there were any barriers to cross.

Are the species in the plankton, a concept defined by Hensen as early as 1887, therefore all cosmopolitan? Sixty years ago biogeography of the pelagic realm consisted of nothing more than distinguishing cold- from warm-water taxa. Twenty years ago Fraser (1962) stated "... it is only to be expected that there will be geographic and seasonal variation". At present, we know that cosmopolitan species are far from common and in a multitude of taxa the traces of recent speciation can be perceived. To trace these processes exactly, however, is difficult as most planktonic groups do not leave fossil records, while faunal centres, refugia and faunal elements are terms not applicable to the study of plankton. The published distributions of species which could be effective tools for biogeography are usually pictures of oceanographic expeditions, while species borders reflect preferences of scientists for visiting specific areas in specific seasons. The borders for warm- and cold-water faunas, as given by Orthmann (1896), Meisenheimer (1905) and Dahl (1923) for example, show differences identical with those shown by isotherms in the different seasons.

The only possibility left to the pelagic biogeographer is to study recent patterns of the few taxonomically well-studied lower taxa and to explain these as the result of chiefly recent dispersal and speciation. Higher planktonic taxa are usually too old to provide indications as to the course of evolution and dispersal, as the conditions under which they developed disappeared long ago. As most Recent species, according to fossil records of some groups, date back to the Miocene, or exceptionally to the Middle-Cretaceous, and as the present configuration of oceans became established some 100 million years ago, one should explain distributions by way of historical events of the last 100 million years.

Plankton groups known from sediments show periods of strong diversification, followed by massive extinction. The latest period of extinction is found in the Late-Cretaceous (Herman, 1979; Pierrot-Bults and van der Spoel, 1979), so we may accept that the planktonic community as a whole chiefly bears the traces of Tertiary and Quaternary events. The latest movements of continental drift, the Alpine folding, the Ice Ages, and the hydrographical phenomena of the Holocene, supply the scenery for recent pelagic biogeography, and therefore for the few examples discussed in this paper.

THE RANGE

In planktonic study, one has to look for patterns rather than for barriers; the concept "range" should be well defined. All localities where a species can be found do not necessarily belong to the reproductive range of that entity. As, by definition, plankton is dependent for its movements upon water circulation, currents easily transport specimens out of their range. This expatriation may ultimately bring specimens into areas where they fail to reproduce, and this transport may be both horizontal and vertical. The total distribution of a taxon thus consists of a reproductive range, surrounded by a non-sterile and a sterile expatriation area. In the non-sterile expatriation area, specimens reproduce but cannot support a population without additional specimen influx from the reproductive area. In the sterile expatriation area, the specimens live without sexual activity and this usually results in prolonged growth.

For taxonomy and biogeography only the reproductive area (and to a lesser degree the non-sterile area) are of importance, although the sterile expatriation area can be of interest when genetic contacts between disjunct populations have to be explained.

Besides passive expatriation, many species show a vertical, active migration which may influence the distribution of a taxon. This migration is frequently of an ontogenetic nature, so that the exact place where reproduction is found can be much smaller than the total area where mature specimens occur.

The passive movements of populations and the three-dimensional character of the range make interpretations of distributional records rather difficult (Fig. 1). With the sampling techniques and ecological knowledge available today, it is very difficult to discriminate between the different sections of a range. The large distances over which expatriation can sometimes be effective also make it difficult to incorporate surface sediment records into studies of recent biogeography (Fig. 1). This is very well illustrated by *Globorotalia hirsuta*. This foraminiferan lives in the N. Atlantic and the waters around S. Africa and Madagascar, but it is recorded from sediments over the whole Atlantic Ocean and eastwards beyond New Zealand.

The differences between distributions in living plankton and in surface sediments may result from transport but also from fluctuations

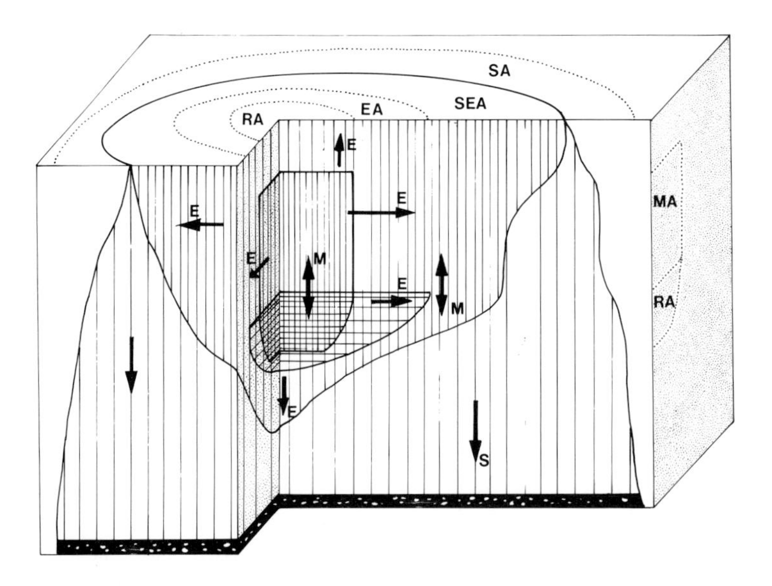

Fig. 1. Block diagram of the components of the range of planktonic taxa. E = directions of expatriation; EA = non-sterile expatriation area; M = migratory movements; MA = migratory area; RA = reproductive area; S = sedimentation; SA = sediment range; SEA = sterile expatriation area.

of the range in the near past. Dramatic changes of distribution are known from the last 9000 years, as shown for example by *Coccolithus pelagicus*, a bianti-tropical species to conclude from surface sediments, but which today is only found in the plankton of the Subarctic waters. To explain this difference, McIntyre *et al.* (1970) postulated a recent extinction of the southern hemisphere population during a temperature drop at about 8000 BC.

However, when other methods cannot explain a recent range, sediment distribution may sometimes provide a tool to identify the distance and direction of transport. *Euphausia pseudogibba*, for example, occurs between 40°N and 20°S in all oceans, but it does not show provincialism. From the recent plankton pattern, it cannot be explained how the Atlantic population maintains genetic contact with the Indo-Pacific one. *Pulleniatina obliquiloculata*, however, shows the same plankton pattern, but the sediment records of this foraminiferan show a continuous range also south of Africa and through the Indo-Malayan archipelago. For the Recent species with this distribution one thus may accept, based on the sediment records, a transportation of living specimens from one ocean to the other. This is an example of an expatriation area that connects the components of a disjunct range.

Palaeontologists frequently present distributions with areas of high and low abundance. Though this may seem to represent respectively the reproduction and expatriation areas, there is no proof of such. In *Limacina lesueuri*, the areas of high abundance coincide with central water masses, but normal reproduction of this species is also recorded outside this area. Probably the areas of high abundance provide an indication of the nature of the species, so that *L. lesueuri* is to be considered a Central-Water species.

TYPES OF PATTERNS

Plankton can be divided into neritic plankton, epi-, meso-, and bathyplankton. It is questionable, however, whether one should consider the epi- and meso-planktonic distributions separately, as they are influenced by approximately the same phenomena and as the ecological characteristics of these groups are closely linked. The bathyplanktonic species occupy a separate position, with their usually wide distributions. These bathyplanktonic taxa are rarely collected, so that they are not discussed in this paper.

Coastal geomorphological and ecological conditions influenced by the continents chiefly determine the distribution of neritic and inlet-water taxa, so that they too fall outside the scope of this paper which will focus on distant- or open-water-neritic taxa and oceanic taxa of the upper layers.

1. Neritic Patterns

The coast-bound, extreme, neritic and inlet-neritic taxa are very recent in origin (Tokioka, 1979), which makes them less interesting to discuss, but the distant-neritic forms seem evolutionarily older. There are even some families and genera, like *Pterotrachea*, which are restricted to the distant-neritic realm. The total area covered by all warm-water distant-neritic species is given in Fig. 2. This area is almost identical with the distribution of *Pterotrachea*. Some authors, like Beklemishev *et al.* (1977), included species with much greater penetration into the oceanic biotope, such as *Electrona rissoi*, but these distributions are more influenced by oceanic than by neritic conditions.

The range occupied by *Pterotrachea* and most other distant-neritic taxa shows three discontinuities, found also for the separate species of *Pterotrachea*, *viz.* the central Pacific, the central Atlantic and the Panama discontinuities. In some species, a discontinuity is found west of South Africa. This distribution can have developed out of the Atlantic or Indo-Pacific Oceans before or after the Panama closing. When these three discontinuities or barriers separate different taxa, e.g. the representatives of the genus *Acetes*, the patterns are more interesting and easier to explain. This genus is slightly more neritic than *Pterotrachea*, and it is absent from the eastern Atlantic Ocean (Fig. 3). The phylogeny of *Acetes* (*cf.* Omori, 1975) shows this taxon to be of Indo-Pacific origin. It shows its greatest diversity along the coasts of East Asia, Indonesia and the western and eastern Indian Ocean. In the eastern Pacific, however, one species (*A. binghami*) is found and from the western Atlantic two species are known, one brackish and one marine (Fig. 4). A dispersal in a westward direction towards the western Atlantic Ocean is not expected for two reasons; an eastern Atlantic population is lacking, and the Atlantic species are closely related to Pacific populations. The marine, western Atlantic, form *Acetes americanus* is most closely related to the eastern Pacific *A. binghami*, so the crossing of the Panama Passage by *Acetes* in Pre-Miocene or earlier periods must be accepted. The brackish water species *A. marinus* is, however, not related to *A. binghami*.

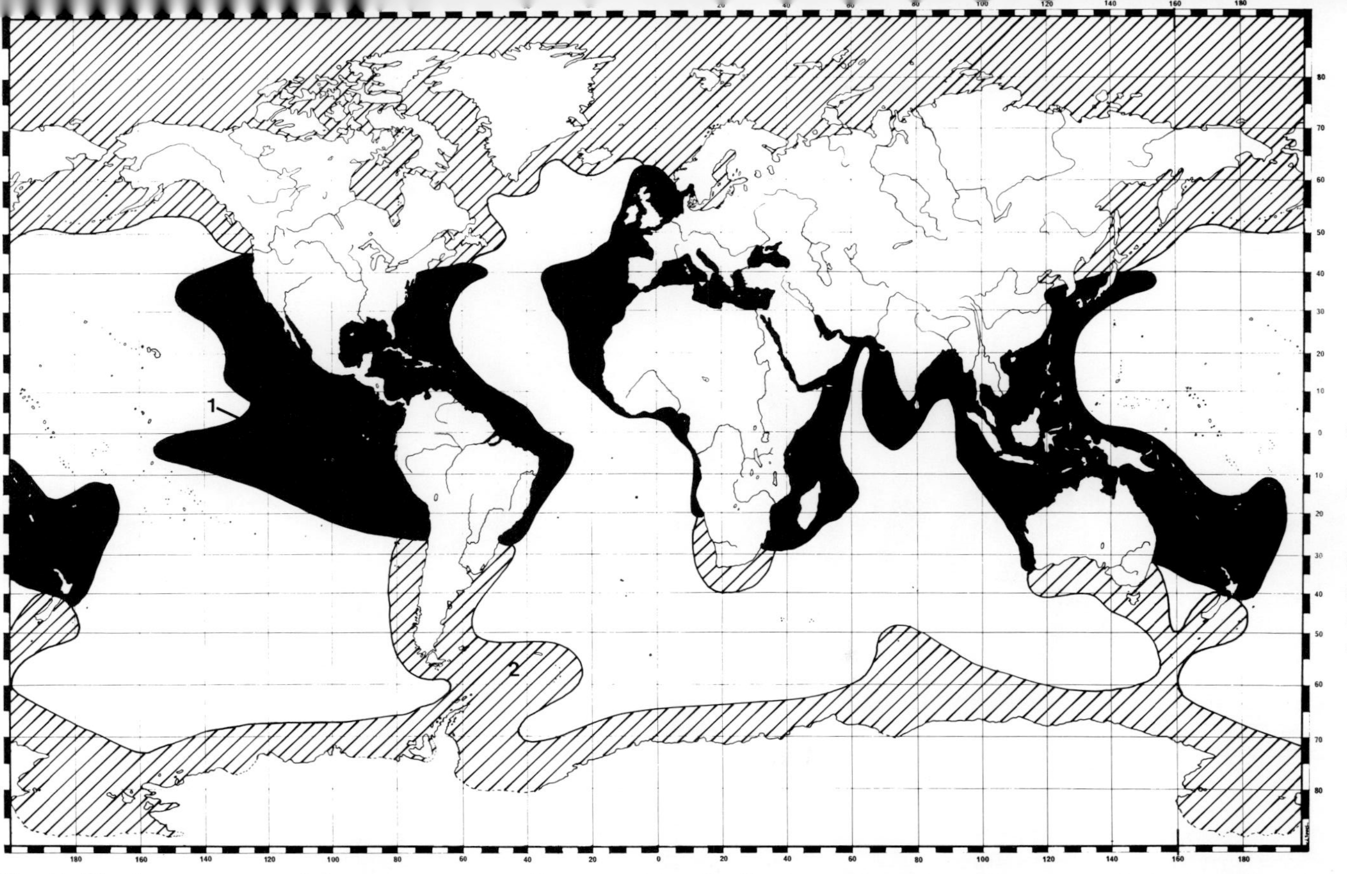

Fig. 2. Total area covered by warm-water distant-neritic distributions (1) and cold-water and ice-bound-neritic distributions (2).

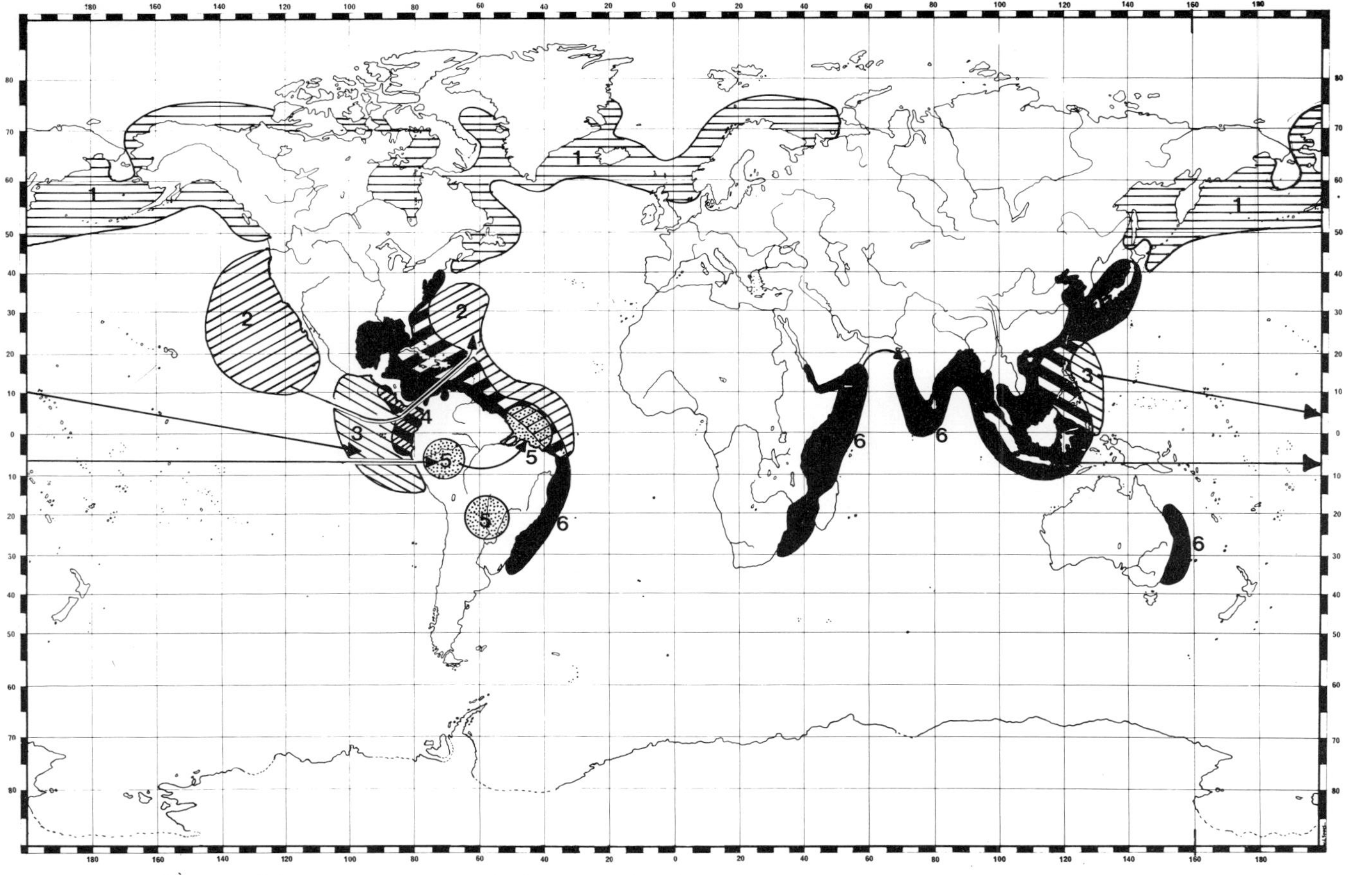

Fig. 3. Distribution of the genus *Acetes*, and of *Diacria schmidti*, *Corolla spectabilis* and *Thysanoessa raschi*; the arrows indicate migration in the geological past. 1 = *Thysanoessa raschi*; 2 = *Corolla spectabilis*; 3 = *Diacria schmidti*; 4 = *Acetes binghami*;

(a) Amazonas inter-ocean connection. *Acetes marinus* is most closely related to *A. paraguayensis* found in the Rio Paraguay, Rio Nanay, Rio Paraná and Rio Amazonas (Fig. 4); this latter fresh-water species is in its turn most closely related to western Pacific and Indian Ocean representatives of the genus (Omori, 1975). This *Acetes* is not the only descendant from marine forms in the fresh water of the Upper Amazonas and Paraguay river system. Dolphins, *Inia* and marine fishes like the sardines are also recorded from these waters (Dr P.J.H. van Bree, *pers. comm.*). There must have been a period favourable to these taxa to penetrate the Pre-Amazonas. The Pre-Amazonas drained into the Pacific Ocean in Pre-Tertiary times. The Brazilian plate showed a geosyncline downfold where the Andes began to fold up, and at this place an inland sea was separated off from the Pacific. In the Middle-Miocene, a land-locked sea received the Pre-Amazonas outflow. In the Pleistocene, the flow direction reversed and, at first near the present city Abuña, water from

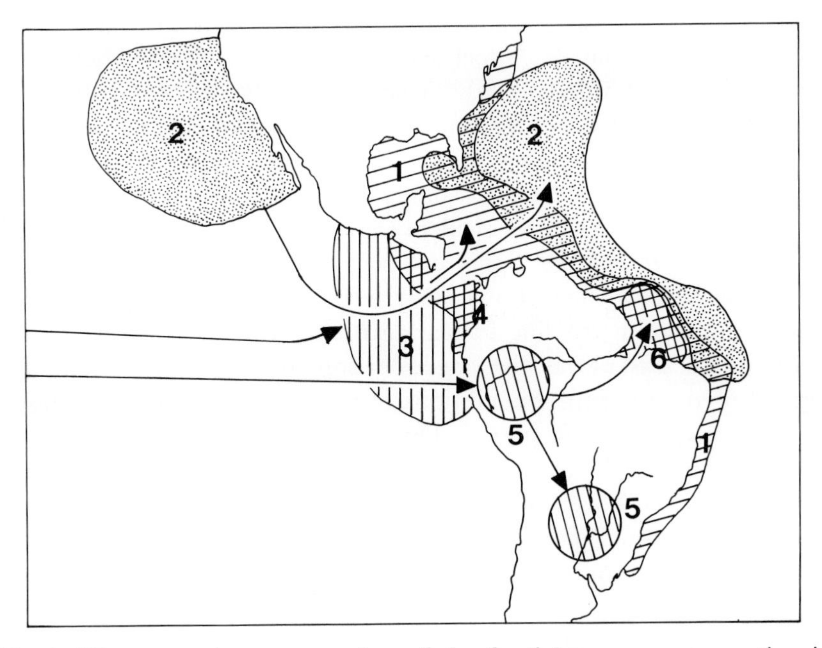

Fig. 4. Diagrammatic representation of the fossil inter-ocean connection in Middle America with the distributions of: 1 = *Acetes americanus*; 2 = *Corolla spectabilis*; 3 = *Diacria schmidti*; 4 = *Acetes binghami*; 5 = *A. paraguayensis*; 6 = *A. marinus*.

the land-locked sea started to flow eastwards to the Atlantic (Damuth and Kumar, 1975; Grabert, 1971a,b).

The hypothesis is that a population of the Indo-West Pacific *Acetes erythraeus* group penetrated from the Pacific into the inland sea of Brazil and adapted to fresh-water conditions. *A. paraguayensis* demonstrates this phenomenon. During eastward draining of the Rio Amazonas, *A. marinus* developed along the Atlantic coast of S. America. Thus there has been a Pacific-Atlantic inter-ocean connection via Brazil, only effective for eastward dispersal of taxa which could temporarily adapt to the lacustrine conditions. Comparable "overland" connections for marine organisms (such as *Pallasea quadrispinosa*, *Mysis relicta*, *Limnocalanus grimaldii*) are stated to have existed in northern Europe and North America during and after the Pleistocene Ice Ages (Sergerstråle, 1962, 1971), so that the Brazilian inter-ocean connection is not unique.

(b) Eastern Pacific fauna. The ancestors of *Acetes binghami* and *A. paraguayensis* are supposed to have lived in the western Pacific but there is at present an ecological and distance barrier between the eastern and western Pacific (van der Spoel and Pierrot-Bults, 1979). The eastern Pacific, roughly bordered by the Americas, 160°W, 40°N and 40°S, is at present characterized by a special fauna composed of oceanic and neritic taxa, among which are numerous endemic ones. The eastern Pacific has its characteristic temperature, salinity, oxygen concentration and currents (cf. Muromtsev, 1958; Beklemishev, 1961, 1971; McGowan, 1971; Reid *et al.*, 1978; van der Spoel and Pierrot-Bults, 1979). In Fig. 4, the tropical eastern Pacific is shown with the most characteristic distribution patterns and with the line indicating the 2 ml $O_2 l^{-1}$ concentration at 100 m depth. East of this line, concentrations are lower throughout. The typical physico-chemical conditions of this area have not been present for a geologically long period. Due previously to the absence of the Panama isthmus, the blocking of the Indo-Malayan inter-ocean connection and an even lower mean global temperature, the eastern Pacific can only have acquired its special conditions and separate position during the Quaternary. Another argument for the rather recent development of the eastern Pacific fauna is the fact that taxa endemic to this area are also phylogenetically young.

As this fauna evidently developed after the closure of the Panama isthmus, it cannot be expected to display many Atlantic relations or influences. It is composed of relicts from the pre-closing period, with or without relatives in the Atlantic, and of taxa, originating from the western Pacific, which adapted to the special conditions.

There are two types of amphi-American taxa, those which are almost identical on both sides and those which developed into distinctly different taxa in the Atlantic and Pacific Oceans. An example of an amphi-American taxon lacking differentiation is *Corolla spectabilis* (Fig. 4). Species of this type show a tendency to more northern Pacific distribution, as this section (north of 20°N) changed far less in conditions in the past than did the central and southern parts of the eastern Pacific. Amphi-American taxa with a distinct degree of speciation occur more frequently south of 20°N in the Pacific. The species *Acetes binghami* and *A. americanus* form a typical example of the latter type of distribution.

Another possibility is that the eastern Pacific representative of the amphi-American population disappeared under the strongly changing conditions. This has to be supposed for the ancestral form of *A. paraguayensis*, related to the western Pacific *A. erythraeus* group. There are also eastern Pacific taxa whose western Pacific counterparts disappeared. *Diacria schmidti* (Fig. 3) lives in the tropical eastern Pacific but is found as subfossils in the surface sediments around the Philippines (van Leyen and van der Spoel, 1982). The subfossil population differs slightly from the recent eastern Pacific population. In all probability, a distant-neritic *Diacria* species lived in both sides of the Pacific before the Ice Ages, and in recent times the western Pacific population became extinct while the eastern population adapted to the special conditions.

Oceanic taxa of the eastern Pacific show relations with the Indo-western Pacific, rather than with the Atlantic Ocean. Epiplanktonic tropical and equatorial taxa, some mesoplanktonic and bathyplanktonic taxa, and even abyssal taxa, developed separate eastern Pacific forms. The cosmopolitan epiplanktonic *Phronima colleti* and *P. stubbingi* are represented by taxonomically different forms in the eastern Pacific, while the northern tropical *Sergestes consobrinus* developed, near California, an eastern Pacific form; similarly, the southern tropical *S. gibbilobata* developed the eastern Pacific species *S. gemisus* (Fig. 5). The mesopelagic fish *Stomias* is represented by an endemic species *S. atriventrix*, and the abyssal molluscs *Abra profundorum* and *Poromya tornata* developed into the endemic *A. californicus* and *P. pula* in the eastern Pacific. Even pelagic birds and dolphins are represented by some endemic taxa in this area. All these examples point to a dispersal from

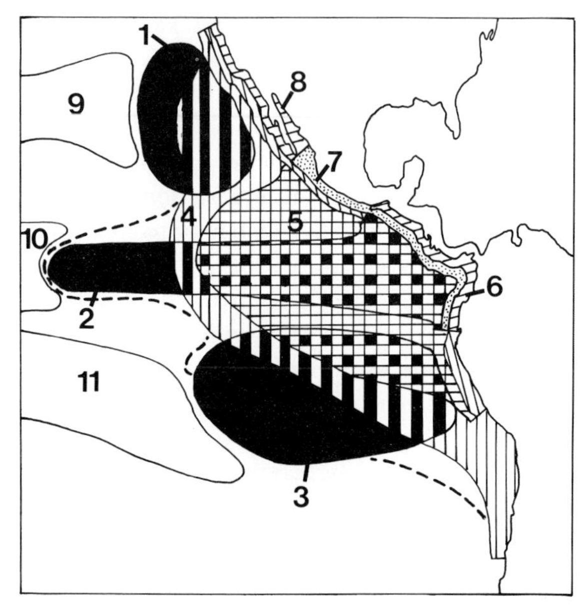

Fig. 5. Diagrammatic representation of E. Pacific fauna elements: 1 = *Desmopterus pacificus*, *Sergestes consobrinus* Californian form; 2 = *Eucalanus inermis*; 3 = *Sergestes gemisus*; 4 = *Sagitta bierii*; 5 = *Pontellina sobrina*; 6 = *Lobidocera lubbocki*; 7 = *Temora discaudata*; 8 = *Lobidocera jollae*; 9 = *Sergestes consobrinus* s. str.; 10 = *S. tantillus*; 11 = *S. gibbilobata*.

the western to the eastern Pacific, and not from the Atlantic to the Pacific Ocean.

Eastern Pacific elements of higher latitudes are not usually endemic and have their counterparts in the western Pacific, like *Idiacanthus astrostomus* which occurs at about 40°N at both sides of the Pacific. The taxa of Subantarctic and Antarctic waters are identical in all the oceans, but if there is dispersal it has to be an eastward one, due to the West Wind Drift. Only in the north, for Sub-arctic and Arctic neritic taxa, which show ranges covering the North Pacific and North Atlantic, does eastward and westward dispersal seem possible, although currents, as in the geological past, favour transportation from the Pacific to the Atlantic (Dunbar, 1979). The major trend of dispersal in distant-neritic taxa thus demonstrates eastward direction.

The Indo-West Pacific coasts form an area of prominent diversity of neritic species (Tokioka, 1979); this is chiefly due to the present geomorphological conditions of the coast line. From the south-east

African coast to northern Japan, a continuous series of biotopes suitable for neritic species is found. This makes it difficult to prove the direction of dispersal in this area. The biogeography of the Atlantic and Indian Oceans gives more information on this point.

(c) Atlantic and Indian Ocean neritic patterns. Contact between Atlantic and Indian Ocean distributions is frequently lost near the tip of South Africa (see also p. 296). West and East Indian Ocean faunas are closely related. Frequently, the neritic taxa of the east coast of Africa and the coast of India and Burma are identical; even when they are not present in a continuous range, they are separated by only a small gap in the north of the Arabian Sea. It is usually accepted that the Indo-Malayan archipelago is a centre of speciation for the neritic realm, so that expansion of taxa out of this centre resulted in the utilization of all biotopes along the tropical shores of the Indian Ocean. This expansion did not usually result in populating the eastern Atlantic waters.

The Pacific-Atlantic distributions of *Sagitta friderici* and *S. tenuis* represent a pattern more common than Indo-Atlantic distributions. *S. friderici* lives along the west coasts of America and Africa, in warm water. *S. tenuis* is nothing but the inlet-water form of *S. friderici* and it occurs in the same areas as *S. friderici*, as well as in the E. Mediterranean and along the east coast of America (Tokioka, 1979). As these species lack relatives in the Indian Ocean, it seems most appropriate to assume a dispersal through the Panama passage, before the Miocene, from the eastern Pacific to the Atlantic. The presence of *S. tenuis* in the Caribbean Sea also supports this hypothesis. The pattern of *S. friderici* is not at all unique, as shown e.g. by *S. bierii*, which even has no representatives in the western Atlantic.

The general absence of neritic species between 10°S and 30°S along East African shores makes it probable that most tropical Atlantic neritic taxa originate from Pacific ancestors.

Cold-water neritic species are frequently found around the southern tips of the African, American and Australian continents, but they do not explain inter-ocean contacts of warm-water faunas. Studying warm-water faunas with a great tolerance of cooler conditions, one discovers other neritic Atlantic faunal contacts, but these again are contacts with the Pacific, as represented in patterns like that of *Centropagus brachiatus*. This species occurs around South America from 30°S in the Atlantic to 5°S in the Pacific. A dispersal in westward direction towards the Pacific seems impossible. Transport around the southern continents is always eastward by the West Wind Drift. The major trend in dispersal of all neritic taxa is thus from west to east.

(d) Cold-water neritic patterns. The terms *distant-neritic* or *open-water-neritic* include all ice-covered areas, most cold water masses, and large transitional areas characterized by strong fluctuations in temperature, salinity and other ecological conditions. Most Arctic and Antarctic taxa may therefore be considered to be of the distant neritic type.

The large cold-water neritic fauna thus occurs in two separate belts, one around the north pole and one around the south pole. In the southern seas, separation of neritic waters is not by continents but by oceanic regions. In the Arctic, the land masses are barriers to Subarctic taxa (Fig. 2).

The distributions of the species groups *Calanus finmarchicus* and *C. helgolandicus*, together with those of *Pontegenia inermis* and *Euphausia crystallorophus*, give an almost complete picture of the overall cold-water neritic distribution. The most striking fact is that there is a strong affinity between (Sub)Arctic and (Sub)Antarctic faunas as represented by, e.g. closely related species in the genus *Calanus* or the subspecies and forms in the genera *Limacina* and *Clione*; this is the well-known bipolarity. When one realizes that in many plankton species with geographic discontinuities of Miocene or older age no diversity developed, as for example in *Sagitta friderici* and *Corolla spectabilis*, it may seem correct to date as long ago the separation of the Arctic and Antarctic faunas. To conclude from their fossil records, many species (like those in *Limacina*) are not much older than the Oligocene so that long-lasting disjunction cannot have there imposed its influence. It is more appropriate in these species and genera to accept that radiation was the mechanism causing their evolution. Warm-water species developed cold-water descendants, with parallel developments in both hemispheres. The bipolarity of these groups is also discussed, on page 330, in the light of north–south contacts.

Antarctic neritic taxa like *Euphausia crystallorophus* are usually monotypic and circum-polar; for Subantarctic taxa, the same pattern is found (cf. *Calanus australis*) but the range is frequently discontinuous with populations around the continents and gaps in the three ocean centres. Only west of South America are endemic taxa found, such as *Calanus chilensis*. This type of endemism is also found in more oceanic plankton like *Salpa gerlachei*; it is probably due to the fact that westward dispersal is strongly reduced at present by the narrow Drake passage and was more so in the past by possible absence of this passage.

The Subarctic fauna is more complex. The North Atlantic and North Pacific Oceans usually show taxonomic differences within a single taxon, and there are also differences found between the East and West Pacific in higher latitudes. When there is biogeographic connection between Pacific and Atlantic in the north, dispersal is eastward (Dunbar, 1979) and it is chiefly the eastern Pacific faunal elements that reached the Atlantic, since westward dispersal, in a direction opposite to currents, was difficult. The climatologically- and ecologically-defined Subarctic Zone is discontinuous, with a gap of 40° longitude north of Canada and of 120° north of Siberia, so that it is much easier to migrate from the Pacific to the Atlantic than the reverse.

Eastern Pacific taxa may be endemic, like *Calanus marshallae*, as they frequently cannot manage to disperse against the current direction. Western Pacific taxa, like *C. sinicus*, should be easily transportable from the eastern shore to the western shore of the ocean. Western ranges like that of *C. sinicus*, however, are found more to the south than eastern ranges with the same conditions, and an eastward transportation will bring these taxa into the transitional area instead of into a Sub-arctic area. A real pathway does not exist between ranges of eastern and western cold-water taxa.

2. Oceanic Patterns

In the oceanic patterns, especially, the three-dimensional nature of the range plays a role. Frequently, it is not the sea-surface conditions that determine distribution and usually expatriation is more extensive at some depth levels than at others; sometimes species occur at different depths in different areas.

In this paper, attention will only be paid to some of the major trends in oceanic biogeography. The Central-Water distribution, belt-shaped patterns, and dispersal around faunal centres will receive special attention.

(a) Central-Water patterns. The distribution of Central-Water taxa represents a phenomenon unique in plankton. Russian scientists, focusing their attention on the Pacific Ocean, provided the concept with

a firm basis (Vinogradov, 1970; Beklemishev, 1971; Voronina, 1978). The distributions concerned cover, in discontinuous or continuous patterns, some or all of the five Central-Water masses in the Atlantic, Indian and Pacific Oceans. The most typical feature of these distributions is that they are not the result of recent dispersal but of disruption of an original area, combined with extinction and adaptations without expansion, i.e. vicariance. In Fig. 6 the Central-Water patterns are given and they are listed in Table I.

Besides the seven combinations of the disjunct areas found (Table I: 1,2,3,5,6,8,9) two combinations (4,7) are given which have not been found in plankton distribution so far, but which are very likely to exist. The many other combinations which can be made are considered not to occur as they do not fit in with the ideas given below.

The most typical Central-Water pattern is shown by *Euphausia brevis* and theoretically this species may have lived in these Central-Water masses almost since the Early-Cretaceous, the period in which Central-Waters first developed in the Proto-Atlantic. Until the Tertiary these, and especially the Central-Waters of the northern Proto-Atlantic, continually received water, with its fauna, from the Tethys Sea. The waters of the Tethys showed, over large areas, the character of Central-Water masses (Fig. 7). The Tethys formed an open connection for Central-Water forms between Atlantic and Pacific (*cf.* Berggren and Hollister, 1974). In the Eocene, real Central-Water circulation also developed in the Proto-Indian Ocean (Fig. 7), in open connection with the southern Central-Waters of the Pacific Ocean. At the end of the Eocene the Tethys ceased to form an open connection. The present discontinuous pattern of, e.g., *Euphausia brevis*, is thus not older than the Eocene. At the end of the Miocene, average seawater temperatures began to drop and simultaneously the sizes of the Central-Water gyres decreased so that their longitidunal connections were cut off in the south by the continents. Patterns exactly like that of *E. brevis* (Table I:6) can have existed since the Miocene or Pliocene period. One may assume that expatriation made possible some gene flow between the disjunct populations so that they remained identical.

After the Pliocene, periods of strong cooling affected such patterns. The first result of lower temperatures was the complete isolation of the Atlantic Ocean. In this isolated Atlantic, the Central-Water species could adapt to equatorial water conditions which showed manifestly lower temperatures than the equatorial waters of the Indo-Pacific

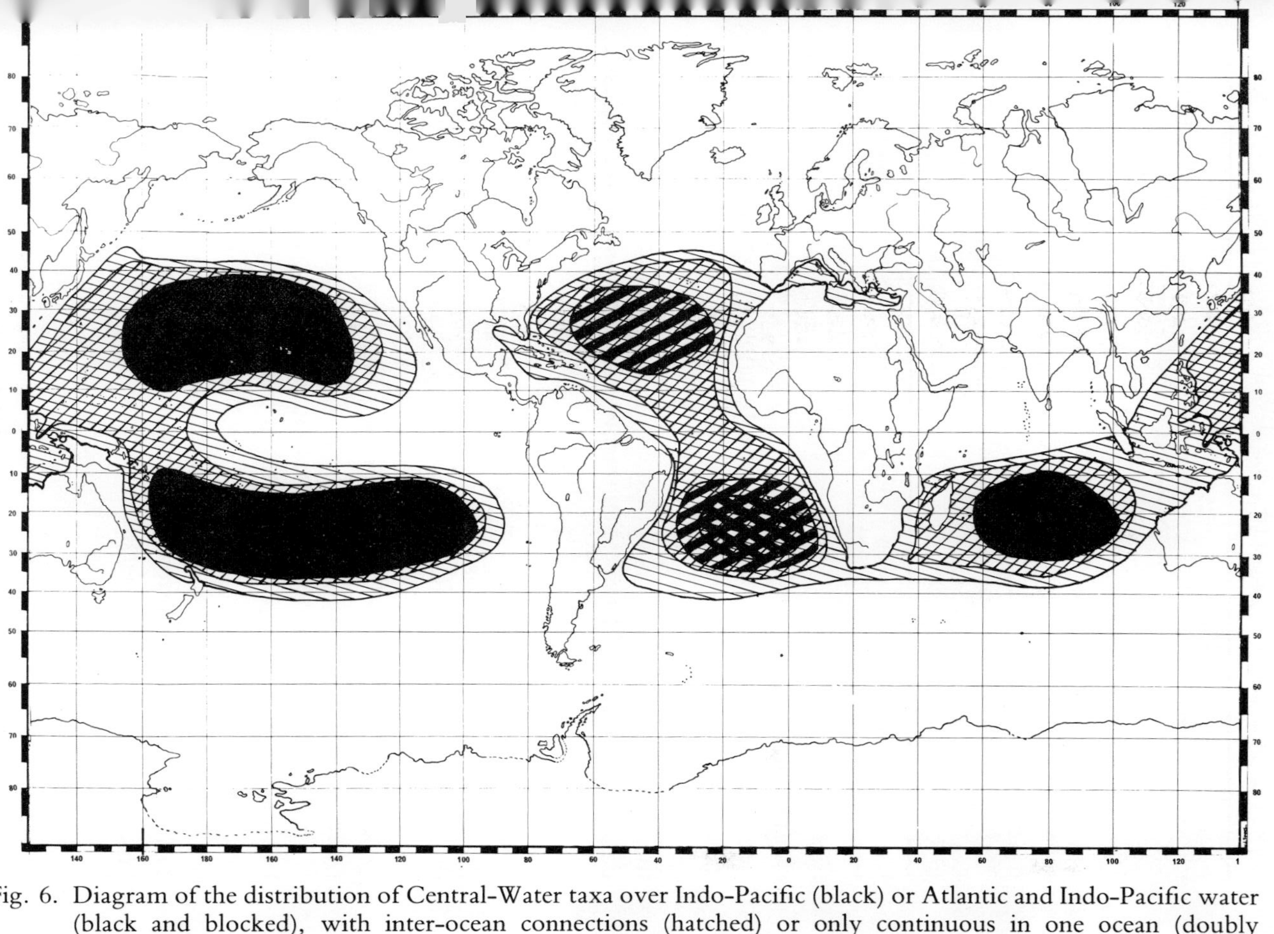

Fig. 6. Diagram of the distribution of Central-Water taxa over Indo-Pacific (black) or Atlantic and Indo-Pacific water (black and blocked), with inter-ocean connections (hatched) or only continuous in one ocean (doubly hatched).

Table I. Patterns of Central-Water distributions.

Pacific	Indian	Atlantic	Taxon
1 SC—TW—NC—T—SC—SC—NC			*Diacria trispinosa*
2 SC—TW—NC---T—SC ---SC—NC			*Thysanopoda obtusifrons*
3 SC—TW—NC---T—SC ---SC			*Stylocheiron robustum*
4 SC—TW—NC---T—SC		absent	?
5 SC ---------NC --------SC ---SC—NC			*Stylocheiron suhmii*
6 SC ---------NC --------SC ---SC ---NC			*Euphausia brevis*
7 SC ---------NC --------SC ---SC			?
8 SC ---------NC --------SC		absent	*Stylocheiron affines* "Central form"
9/SC/ ---------NC --------SC ---SC ---NC			*Euphausia gibba—E. hemigibba*

C = Central Water; N = north; S = south; T = tropical; W = west; / / = represented by a related taxon, —— = continuous distribution; --- = discontinuous distribution.

Oceans. Subsequent rises of temperature resulted in a reoccupation of all Central-Waters, but a continuous Atlantic population was maintained as demonstrated in *Stylocheiron suhmii* (Table I:5). During periods of low temperature, more flexible species also adapted to equatorial waters in the Indo-Pacific Oceans, as illustrated by *Thysanopoda obtusifrons* (Table I:2). Sometimes the latter phenomenon is synchronous with greater effects in the Atlantic Ocean, where the cooling was always stronger; sometimes the North Atlantic Central-Water populations became extinct; sometimes both the northern and southern populations disappeared. *Stylocheiron robustum* (Table I:3) shows the first type of distribution. An example of the complete absence from the Atlantic (Table I:4) could not be found, so that either cooling was not profound enough to give this result, or science has thus far failed to discover the representatives of this pattern.

During warmer periods a pattern as given for *Diacria trispinosa* (Table I:1) cannot develop, as the equatorial waters are then unfavourable; neither can a cold-tolerant species show this pattern. Only species which are less stenothermal than normal Central-Water species can disperse in this way. The eurythermal character of such species may have developed during the temperature fluctuations of the Ice Ages. If these ideas are correct, it is also only logical that a continuous Indo-Pacific and a discontinuous Atlantic distribution are not found to exist simultaneously in one taxon.

Central-Water taxa restricted to the Indo-Pacific, like *Stylocheiron affines* "Central-Water form" (Table I:8), seem to provide an example of

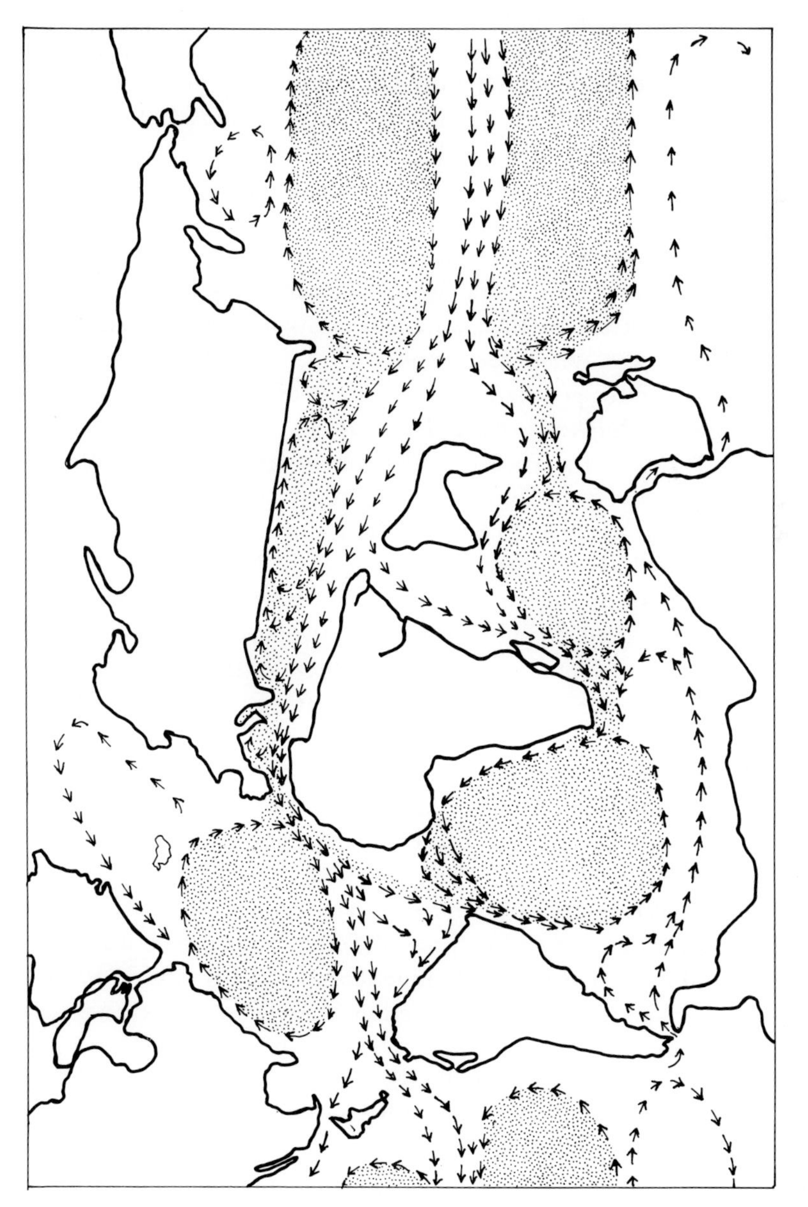

Fig. 7. Hypothetical diagram of Eocene dispersal of Central-Waters with continents and currents.

complete extinction in the Atlantic but this contradicts the above hypothesis as the pattern shows no influence of the colder periods in the Indo-Pacific, neither do the species penetrate equatorial waters at any place. *S. affines* "Central-Water form" thus probably developed in the Indo-Pacific more recently and failed to penetrate the Atlantic basin. If this hypothesis is correct, it is also logical that the pattern given in Table I:7 was never found. Yet this pattern is incorporated in the list of possible ranges, as expatriation even on a small scale may give, or may already have given, rise to a South Atlantic Central-Water population.

That contact between Central-Water masses is restricted when the taxa did not adapt to warmer waters is shown by speciation within Central-Water taxa. When speciation occurs, it is always the South Pacific Central-Water population which splits off as a separate taxon, a pattern shown by *Euphausia gibba–E. hemigibba.* The northern Central-Water population in the Pacific was never isolated; in the geological past, it connected with the Tethys Sea and at present it is connected with the Indian Ocean via the Indo-Malayan archipelago (Brinton, 1975). This lack of isolation prevented speciation. In the North Atlantic, long-lasting isolation seems to have been present but no clear speciation is found. This can be explained by the fact that the North Atlantic Central-Water populations adapted to equatorial waters in their early development and thus broke down the isolation from the South Atlantic populations.

The Tethys Sea is here mentioned as being of importance to the distribution of the Central-Water fauna, but probably this is the only pattern in plankton related to the Tethys Sea. The patterns of some rare and not easily identifiable species, such as *Fowlerina punctata*, *Pneumoderma meisenheimeri* and *Pneumodermopsis paucidens*, covering Atlantic and Indo-Malayan waters, resemble those of Tethys Sea relicts but one may seriously doubt whether the total range of these taxa is thus far known. Moreover, in plankton geography, Tethys Sea relict patterns are usually difficult to distinguish from patterns caused by other influences. A discontinuous Atlantic/Indo-Pacific distribution may indicate a Tethys relict, but it can also have developed by dispersal through inter-ocean connections during warmer periods.

(b) Indo-Malayan inter-ocean connection. The Indo-Malayan archipelago forms a longitudinal connection between the Pacific and Indian Oceans for warm-water species distributed over Central-Waters and in belt-

shaped patterns (*cf*, p. 315). Epipelagic species and mesopelagic species occurring in the upper 200 m utilize the Molucca-Banda-Timor Sea passage (Brinton, 1975). Indian Ocean, and from the other side, Pacific Ocean taxa penetrate into the archipelago up to the Ceram-North Borneo line (Fig. 8). A group of species is continuous in this area; others are clearly discontinuous. Concerning this problem, Brinton (1975) stated "Rare or inconsistent occurrence in these areas of certain species ... indicates ... transient or 'puddle jumping' east–west communication of genetic material". The prevailing currents, with east–west direction, suggest that only a westward dispersal is possible.

Some subtropical species of the Indo-Pacific are continuous south of Australia (Fig. 8). Species with this pattern are "cold-tolerant", but it is clear that with a small increase in temperature an effective Indo-Pacific pathway south of Australia is created. This will be a west to east connection as the West Wind Drift circulation is always dominant in these areas.

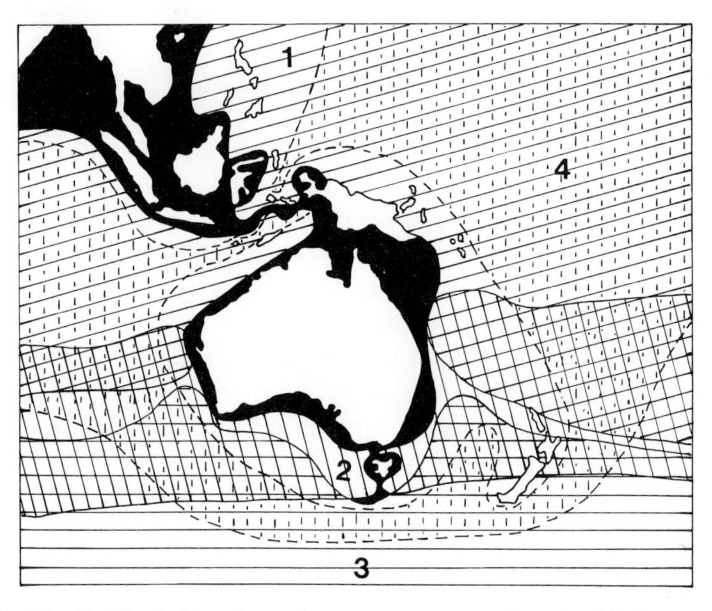

Fig. 8. The Pacific-Indian Ocean inter-ocean connection, the black area represents the barrier function of the area. 1 = warm-water species identical in both oceans but not really continuous near Timor; 2 = cold-tolerant species continuous, or almost, south of Australia; 3 = circumpolar cold-water species; 4 = meso- and bathypelagic species.

Distributions of the epiplanktonic genera *Phronima* and *Desmopterus*, however, seem only amenable to explanation by a passage through the Indo-Malayan archipelago itself in an eastward direction. The same conclusion may at first seem correct for the dispersal of the mesoplanktonic genus *Peraclis* (*cf.* p. 313). A possibility for eastward dispersal of epiplanktonic species such as *Phronima* and *Desmopterus* can be found in the current reversal during the monsoon periods in the North Indian Ocean. During the NE monsoon, all currents which may contribute to inter-ocean transport run in a westward direction. During the SW monsoon, from June to September, a few current branches have a direction favourable to transport towards the Pacific but, also in this period, they form only a minority of the total current system in the passage area. The dispersal of mesoplankton like *Peraclis* is, however, unaccountable regardless of what kind of present-day conditions are considered. There are two possible explanations for the *Peraclis* patterns, and they may then also be applied to epiplankton; these are, (1) dispersal south of Australia with the characteristics of an Indo-Malayan dispersal; and (2) dispersal before the present land mass configuration established itself, therefore before the Miocene.

In both cases, one has to explain why the present patterns have their greatest extension in the North Pacific; why the connections in the south, near New Zealand, ceased; why there seems to be a contact in the Banda-Molucca-Timor Sea. It is indeed possible to explain these. The surface waters directly east of the Indo-Malayan archipelago show conditions rather similar to those in the warm Indian Ocean, while those south of about 10°N in the Pacific Ocean are at present very different. Species that once penetrated the Pacific thus became extinct in the southern parts, when they maintained their ecological adaptation to Indian-Ocean-like conditions. The deep-water elements of the Pacific have also shifted to the north after their penetration via the South Pacific, concluding from the present conditions in the deeper South-West Pacific, which differ from those in the deep North Pacific and deep Indian Ocean. Moreover, the distribution of the Pacific deep-water taxa, like that of the surface forms, approaches the Ceram-North Borneo line. Thus, the area of the Indian Ocean populations is reached, and this may give the impression of either a direct contact here or dispersal from the Indian to the Pacific Ocean.

For mesoplanktonic forms, the Banda-Molucca-Timor passage is rather shallow (200–2000 m) but not insurmountable. During some periods of the Pleistocene Ice Ages, sea level was approximately 100 m higher than at present; this may have provided more opportunities to cross the passage near Timor. The deep-water currents have a tendency to flow eastwards to south-eastwards at these latitudes, hence transport in that direction was possible especially in the Miocene or earlier periods, when the northward shift of the Australian continent had not yet rendered the passage narrow.

Stylocheiron robustum seems to represent a dispersal through the Indo-Malayan archipelago. It is a Central-Water species and the dispersal may be considered of greater age since the pattern developed before the Miocene (*cf.* p. 306). This taxon is thus an example of the second possiblity mentioned for Indian to Pacific Ocean dispersal. That for some species an Indian Ocean–Pacific Ocean penetration was possible south of Australia is demonstrated by, e.g. *Limacina helicoides* in the deep sea, and by *Clausocalanus arcuicornis* in the epiplankton. Species passing along this way are usually of Atlantic origin or are cold-tolerant warm-water species with circumglobal distributions.

(c) Faunal centres. For *Peraclis* and *Phronima*, it was accepted above that they show an eastward dispersal between the Indian and Pacific Oceans. This hypothesis is based on the fact that both genera are faunal elements of the Dakar Centre. When considering the ranges of species of the genus *Peraclis*, one discovers that they all become sympatric in an area off Daker at approximately 20°N 20°W (Fig. 9). A large number of other species of the Order Pseudothecosomata and other groups (amongst which is also *Phronima*) show this same phenomenon. It does not seem likely that the Dakar area is the only faunal centre for plankton, but good indications of others are not available. The distribution of many species around the Indo-Malayan archipelago and around the North Indian Ocean suggests that these two areas are also faunal centres, but it is not possible to prove this. The eastern Pacific may also seem to be a faunal centre, but it certainly is not. There is no phylogenetic relation between the eastern Pacific taxa and it is not a centre with expansive characters. One major reason for the lack of evidence on other faunal centres is the fact that they are geographically mobile. With each temperature fluctuation, for example, the position of an original centre shifts, and it is clear that during the Ice Ages centres only kept their positions in tropical areas. As this geographical shift is not only shown by the centre but also by its elements, all indications on which centres

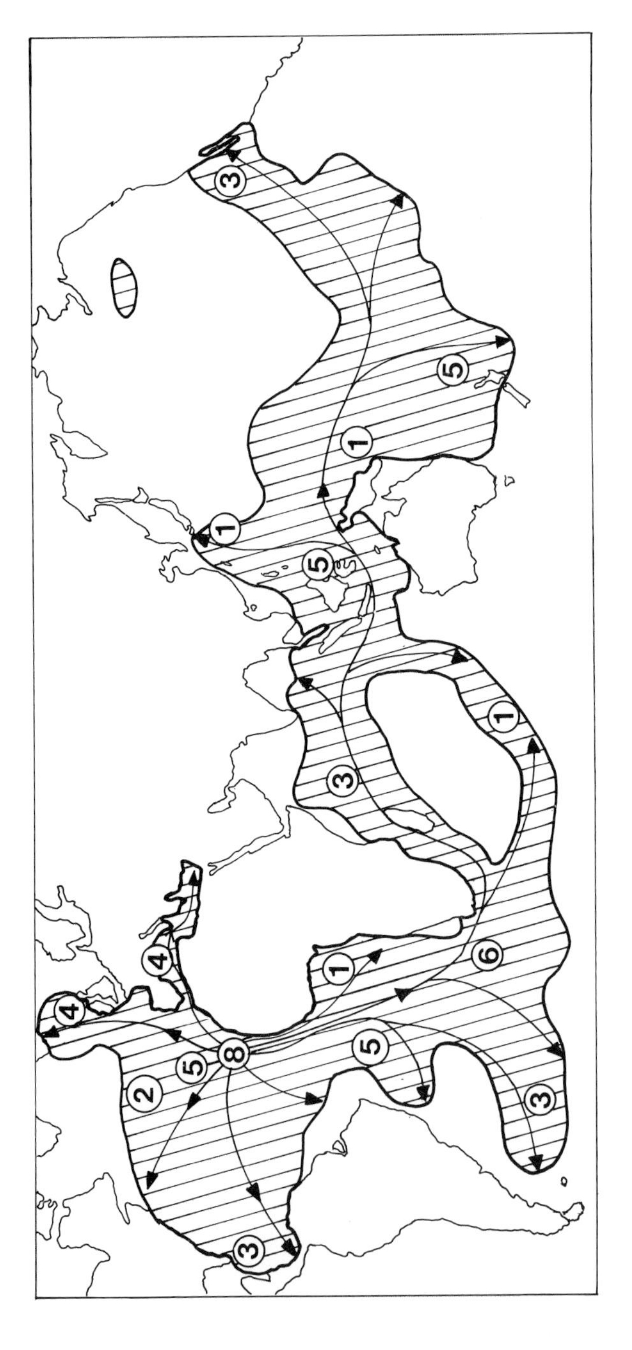

Fig. 9. Distribution of the genus *Peraclis* around the faunal centre off Dakar, with total number of species in the different areas given in circles and the probable migration indicated by arrows.

can be reconstructed have by now faded. Another factor making it difficult to locate the faunal centres is the fact that most centres are not geographical "points", as in terrestrial faunas, but are usually large belts or water masses dominated by a large current system and climate regime.

Examining the pelagic fauna on a large scale, it is possible to consider the warm-water belt to be the area from which radiation gave rise to the cold-water faunas. These cold-water faunas either developed out of the low latitudes or they spread out after periods of cooling. The deep-water faunas chiefly developed from the cold-water faunas. Most planktonic groups have different degrees of species diversity in the different oceans. For some groups (like the Pteropoda), this diversity is greatest in the Atlantic Ocean; for others (e.g. Chaetognatha), it is greatest in the Pacific Ocean; for most neritic groups, the Indian Ocean shows the highest diversity. These differences point to the fact that for some groups, usually the older ones, the Pacific Ocean is a centre of speciation and dispersal, while other groups seem to have originated in the Atlantic Ocean.

(d) Belt-shaped patterns. A large number of planktonic taxa are circum-globally distributed over one or more latitudinal belts. Antarctic, Subantarctic, and bisubtropical species are well-known examples of such belt-shaped patterns. They can be subdivided into two types: the north-south continuous patterns and the biantitropical patterns. The term biantitropical here covers the concepts bipolar, bisubpolar, bisub-tropical and so on, but the Central-Water taxa are not included as they represent an essentially different type; they are non-expansive in character, and they stay within the central gyres. The belt-shaped patterns are, however, occupied by expansive taxa, which is proved by the complexity of patterns and a diversity of adaptations. On the other hand, it is evident that any expansive oceanic species will ultimately acquire a belt-shaped pattern.

Few comments can be made concerning the age of belt-shaped distributions, as species of very different age may have developed such distributions without leaving any trace of their origin or original distribution. At present, the belt-shaped patterns are chiefly determined by climatological conditions and to a lesser degree by currents and vertical migration. That the climate has an important influence, compared with currents, is concluded from the fact that most belt-shaped patterns are only slightly, or not at all, affected by boundary currents. Poleward transport at the west-side, and equator-ward transport at the east-side, of oceans is a common effect of the boundary-current system, but it is not so frequently observed in belt-shaped patterns. The westward currents at low latitudes and the eastward currents at high latitudes are, of course, responsible for the transport and circumglobal distribution. Connection of horizontally separated biantitropical taxa by deep living populations, the tropical submergence phenomenon, shows the impact of vertical distribution on belt-shaped patterns. Vertical migration and dispersal are related to the eurythermal nature of a taxon; the greater the vertical penetration the more eurythermal is the taxon and the wider usually is the belt in which it can live. In Table II, the different types of belt-shaped patterns are given; in Figs 10 and 11, they are given diagrammatically.

Fleminger and Hulsemann (1973) suggested two

> generalities of biogeographical significance to the epiplankton of the Indian Ocean: a) warm-water species that breed regularly up to mid-latitudes tend to be circumglobal in distribution and probably maintain geneflow around South Africa; b) warm-water species that breed regularly only in low latitudes are provincial and may have one or more tropical cognates elsewhere in the other oceans.

These two rules also hold good for neritic and mesoplanktonic species, and they are especially demonstrable in belt-shaped patterns. Only the cosmopolitan taxa (Fig. 10:1 and 2) are continuous in all the oceans but may be discontinuous off South Africa, off South Australia and in the Indo-Malayan archipelago. Isolation in the three oceans is, however, not strong enough to create more diversity than clinal east-west variation. Tropical and equatorial taxa (Fig. 10:5 and 6) are always discontinuous as the African and American continents (and for some taxa, the Indo-Malayan archipelago) form an interruption. Still there are some tropical species, like *Euphausia pseudogibba*, which are discontinuous but monotypic. Incidental dispersal from ocean to ocean seems to connect these ranges genetically. The equatorial species are usually restricted to the Indo-Pacific or to the Atlantic Ocean. The Central-Water species usually just reach the mid-latitudes and thus remain monotypic.

For biantitropical species, the continents of the northern hemisphere are barriers between the northern populations, so that here the similarity between North Pacific and North Atlantic Oceans requires explanation. The similarity of populations of the northern and southern belts of

Table II. Belt-shaped patterns. A subdivision of the most common belt-shaped patterns with references to Figs 10 and 11 and examples of the most typical representatives.

Pattern	Range	Example	Fig:
1. Continuous over N. and S. hemispheres:			
a. Cosmopolitan	75°N–75°S	*Rhyzosolenia alata*	10:1
b. Restricted cosmopolitan	70°N–60°S	*Stylocheiron maximum*	10:2
c. Like b with tropical submergence	60°N–60°S	*Sagitta planctonis*	
d. Warm water	50°N–45°S	*Clausocalanus paupulus*	10:3
e. Tropical-subtropical	45°N–35°S	*Euphausia tenera*	10:4
f. Tropical	30°N–20°S	*Euphausia pseudogibba*	10:5
g. Equatorial	10°N–10°S	*Clio convexa*	10:6
2. Continuous in one hemisphere:			
a. Polar	50–90°N	*Calanus glacialis*	
b. Subpolar	50–65°S	*Euphausia triacantha*	
c. Transitional	40–50°S	*Euphausia lucens*	
d. Temperate (Cold-)	30–45°S	*Limacina helicina rangi*	
3. Biantitropical:			
a. Bisubpolar	above 40°N&S	*Globorotalia pachyderma*	
b. Bitransitional	20–50°N&S	*Thysanoessa gregaroides*	11:2
c. Bitransitional with absence in one of the northern oceans	absent N. Atlantic	*Euphausia recurva*	11:3
4. Patterns with latitudinal diversity:			
a. Like 1a with latitudinal belts of different infraspecific taxa		*Clio pyramidata*	10:1
b. Like 1c with latitudinal belts of different infraspecific taxa		*Salpa fusiformis*	10:2
5. Pattern with longitudinal diversity:			
a. Like 1d with different infra-specific taxa in the 3 oceans		*Eucalanus subtenuis*	10:3

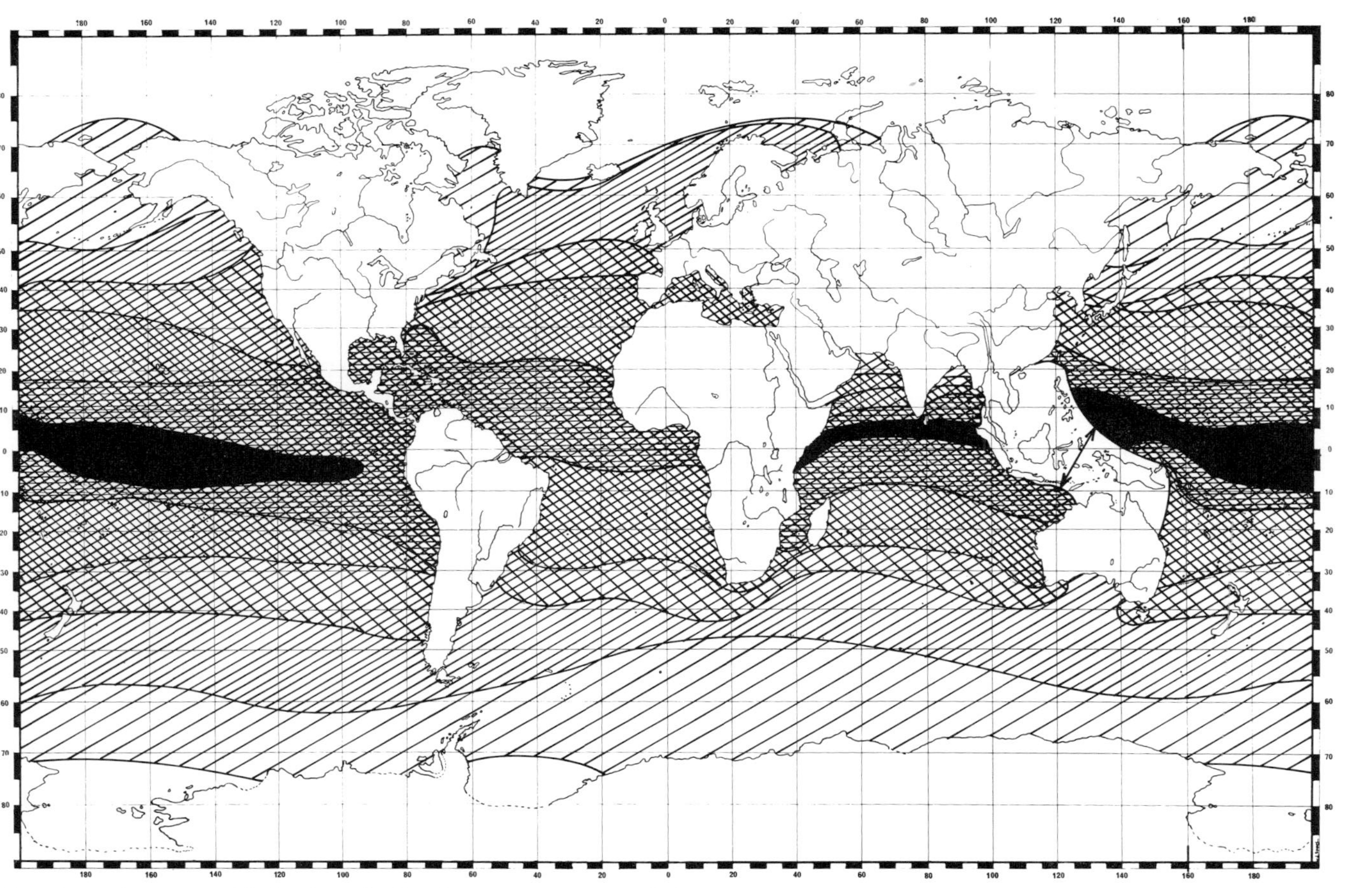

Fig. 10. Diagram of belt-shaped distribution patterns of the north–south continuous type. The belts represent the

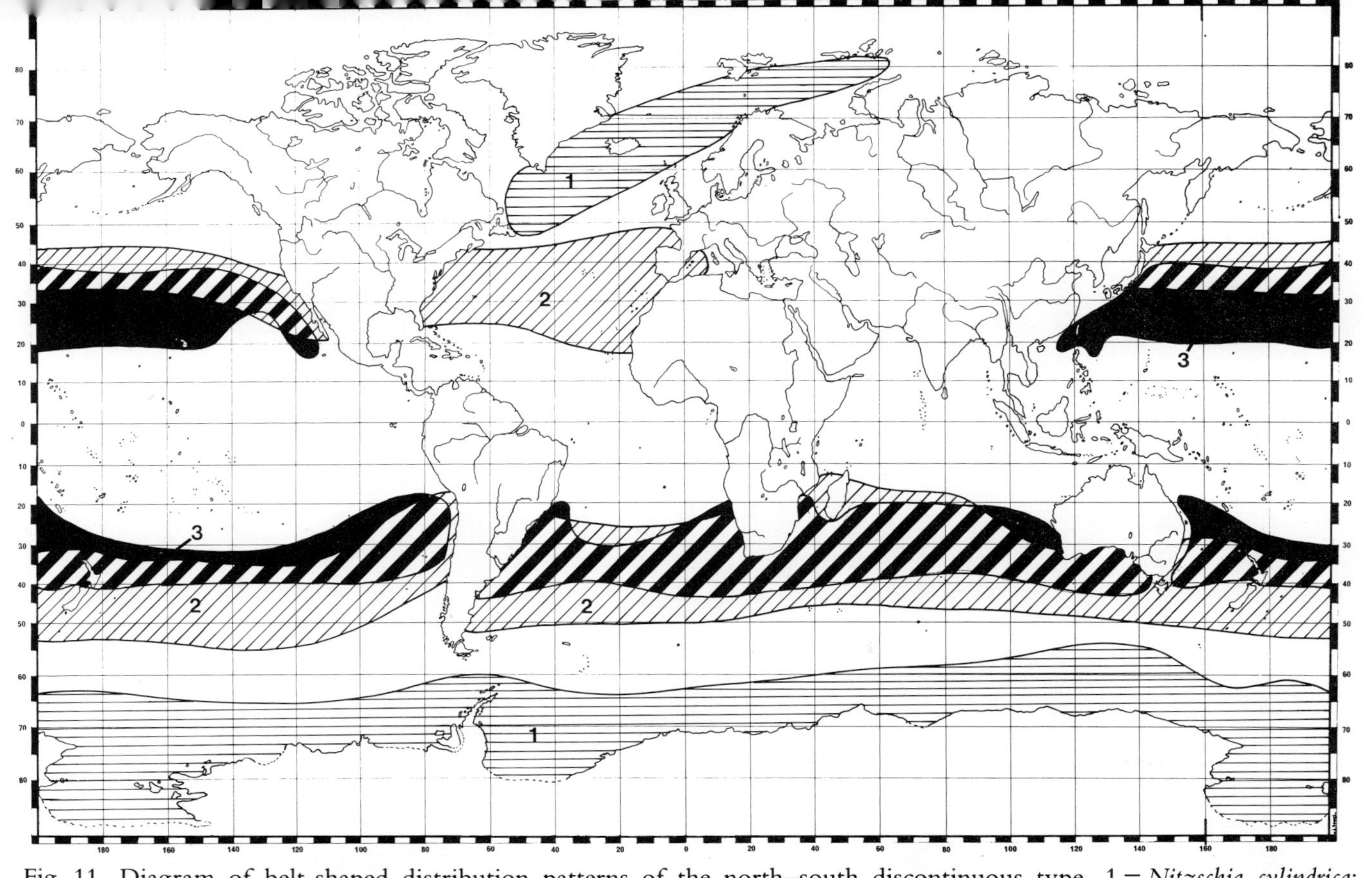

Fig. 11. Diagram of belt-shaped distribution patterns of the north–south discontinuous type. 1 = *Nitzschia cylindrica*; 2 = *Thysanoessa gregaria*; 3 = *Euphausia recurva*.

biantitropical taxa poses the same question, which also has to be explained. The problem is: wherein consist the north-south contacts between the belts. There are four types of north-south contacts:

(1) contact through deep-water layers by tropical submergence;

(2) contact through surface layers by incidental transport (expatriation) bringing specimens from one belt to the other;

(3) contact in the geological past, during cooler periods, when the two belts were more or less unified for restricted periods;

(4) phylogenetic contact due to the fact that northern and southern populations developed from the same warm-water taxon.

Sagitta planctonis is an example of the first type: it lives in higher latitudes and is replaced at lower latitudes by *S. zetesios* (for discussion on speciation, see p. 322), but in the deep sea it is connected by means of submergence. The populations in north and south above 50°S are taxonomically identical (Pierrot-Bults, 1976). *Sagitta tasmanica* may be an example of the second possibility. The North Atlantic population and the circumantarctic population come very close to each other near North-West Africa and incidental exchange of specimens may provide a genetic contact (Pierrot-Bults, 1976). The third possibility can be illustrated with the distribution of *Globorotalia inflata*. This species is bitransitional, but in the sediment records it shows a continuous north-south range. There is evidence that in colder periods the transitional belts were forced towards the equator, where they fused. The last possibility is shown by, e.g. *Salpa fusiformis* and *Limacina helicina*. In *Salpa fusiformis*, almost identical forms developed in the northern and southern colder waters in parasympatry with the warm-water population (*cf.* p. 324). *Limacina helicina* belongs to a chiefly warm-water genus, and one may suppose that from an extinct warm-water species different subpolar taxa developed, *L. helicina helicina* in the north and *L. helicina antarctica* in the south (van der Spoel, 1967).

A large number of species show an incomplete belt-shaped pattern: either the North Atlantic population (Fig. 11:3), or the North Pacific population, or the southern belt, is absent. The absence of the southern belt can be explained by extinction of populations during warmer periods in the geological past. *Coccolithus pelagicus*, for example, is in principle a "bipolar" species, although at present it is found only in the northern hemisphere; however, sediments from the southern hemisphere of *c.* 12000–8000 years ago contain representatives of this species. *C. pelagicus* has a temperature range of 6°–14°C. In the Antarctic Ocean, the belt of 6°–14°C is very narrow and even small postglacial periods of warming up, like the one of 8000 BC, could have disturbed the southern populations (McIntyre *et al.*, 1970).

(e) North Atlantic and North Pacific Oceans. Absence from either the North Pacific or the North Atlantic shows that the north–south contact between latitudinal belts and the east–west contact both failed to function. Subarctic, and more especially Northern Transitional and Northern Temperate, taxa make no contact and do not migrate from Pacific to Atlantic through the northern passage, so that for them only the north–south contact remains. Contact and migration by tropical submergence are identical in both oceans, so that species demonstrating these phenomena will show no differences in the two northern basins.

Contact and repopulation by migration through surface layers is, however, much easier in the Atlantic than in the Pacific due to different configurations of the boundary current systems. The occupation of the North Atlantic by migration waves coming from the southern belts is much easier than in the Pacific. Temperature fluctuations in the past were simultaneous in Atlantic and Pacific, but in the Atlantic they were greater so that less cold-tolerant taxa were more likely to survive in the North Pacific than in the North Atlantic. On the other hand, cooling will have favoured more north–south contacts in the Atlantic than in the Pacific (see also p. 330).

It is easy to understand how the isolated populations in the two basins developed differently in cases in which the populations or taxa from the northern belt developed from a warm-water species.

Other differences between the taxa in the two northern oceans are the results of the different hydrological conditions north of 40°N. In the Atlantic, the North Atlantic Current and the Gulf Stream cross the ocean obliquely; this gives rise to a dispersal of warm-water taxa to much higher latitudes in the Atlantic than in the Pacific, and these currents also disrupt the transitional (and to a lesser degree the subarctic and temperate) belts of the Atlantic. In the Pacific, by contrast, a well-developed transitional belt is found probably even with an internal gyral flow (McGowan, 1971). The most spectacular differences in biogeography resulting from these dissimilarities are the greater radiation and adaptation of warm-water species in the North Atlantic than in the Pacific, and the larger number of Transitional-Water species in the North Pacific which have no counterparts in the Atlantic. That only present-day hydrographical conditions are responsible, not the geological history of the areas, is proved by species which can still develop adapted populations in the two oceans but which only develop these populations in the Atlantic. *Clio pyramidata* and *Salpa fusiformis* are examples of such species. *Clio pyramidata* forma *lanceolata* lives in warm water of all the oceans and in both Pacific and Atlantic it occurs to approx. 40°N. In the Atlantic, *Clio* only developed the forma *C. pyramidata* forma *pyramidata* at latitudes north of 40°N (van der Spoel, 1967). More striking is the absence of *Salpa fusiformis* north of 40°N in the Pacific while it penetrates to 70°N in the Atlantic. This species adapts to colder water by an increase in the number of fibres in the body muscles. In the tropics the number, for aggregates, is ±30; at 40°N the number is ±55 in the Pacific as well as in the Atlantic; in the Atlantic the number increases to 65 near 70°N (van Soest, 1975). Therefore, a mere restriction of northward transport, and not an inability of the species to adapt, causes this species not to penetrate further into the North Pacific. The hydrographically well-developed transitional zone seems an obstruction to southern North Pacific influences. The transitional zone and its fauna in the North Pacific are well-documented by McGowan (1971). A remarkable phenomenon is the relation of North Pacific transitional elements and eastern Pacific faunal elements, and for some species this relation extends to the warm waters of the Atlantic Ocean.

SPECIATION

To this point, some of the most common plankton patterns have been discussed and little attention has been given to speciation. The history of most ranges cannot be traced back far enough to elucidate the evolution of the taxa found in these ranges. One can also state that the species evolved too slowly to throw any light on speciation by way of their traceable history. The few indications for speciation found, in the eastern Pacific and amphi-American faunas, all concern speciation by isolation. However, to explain the development of so many taxa in an almost continuous environment one should look not only for allopatric but also for sympatric or parasympatric speciation. For a few species, it seems evident that they developed in parasympatry, but most examples suggested for parasympatric speciation processes consist of developments below the species level. Special attention is therefore also given to variation, as it is a phenomenon usually preceding speciation and frequently present where isolation or other factors have not yet resulted in speciation. In the open ocean where evolution proceeds so slowly, variation is even more to be expected than completed speciation.

The ecological relations between species, especially those concerned with feeding, predation and reproduction, make their own important contribution to speciation. The population dynamics and structures of plankton communities are only vaguely known and they are of major importance to understanding this speciation mechanism; in this paper, therefore, no attention is given to "ecological speciation". In *Zoogeography and diversity of plankton* (van der Spoel and Pierrot-Bults, 1979) full attention was paid to the various mechanisms of speciation and variation; here only *some* of the biogeographically interesting models of speciation are given as illustrations.

1. Isolation

The real isolating barriers are the continents, recently completed when the Panama isthmus and the South Asian landmasses closed the Panama passage and the Tethys Sea respectively. Incomplete barriers are found in the Indo-Malayan archipelago, the Straits of Gibraltar, those of Bab el Mandeb, and the Drake Passage. The influence of these barriers on dispersal and speciation was discussed above; the Red Sea and the Mediterranean Sea alone were not mentioned as isolated basins. The Red Sea fauna is essentially an impoverished Indo-Pacific fauna and only a few endemic forms are found (Rao, 1979). Likewise, the Mediterranean fauna is an impoverished Atlantic fauna, with only a few forms restricted to this basin (Furnestin, 1979). Moreover, endemic taxa in both enclosed seas are for the most part at the infraspecific level, which shows that the isolation by the straits and sills was either not strong enough or not of long enough duration to cause diversification to the species level.

Distance barriers are less effective mechanisms for isolation of plankton. At present, the dispersal of plankton goes chiefly in an eastward direction; in the geological past, it probably went in a westward direction. Distance barriers running north–south are in general therefore not effective. The only one known to have any effect is the mid-Pacific distance barrier. For the Pacific, the effects on neritic taxa were discussed above but oceanic plankton is not expected to show the influences of this barrier. The distance separating taxa restricted to certain latitudes of the northern and of the southern hemisperes is sometimes an isolating mechanism resulting in speciation, as shown for *Euphausia gibba* and *E. hemigibba* (*cf.* p. 310). It has been mentioned already, however, that climatic changes in the geological past, and either continuous or incidental tropical submergence, are responsible for a breakdown of this barrier type.

To summarize the conclusions for oceanic plankton:

(1) only a few Atlantic species developed by isolation due to continents;

(2) a few Indian, Atlantic and Pacific Ocean equatorial species developed by isolation of continents;

(3) very few species developed by isolation in the Red Sea and in the Mediterranean;

(4) some species developed by distance isolation in separated latitudinal belts and probably in the eastern Pacific Ocean.

Not explained is how the diversity developed in continuous series of populations and taxa. Such series are usually composed of forms, subspecies or species which succeed each other latitudinally, longitudinally or bathymetrically. The development of cold-water forms from warm-water ancestors was mentioned, but not yet explained. These problems can be solved by studying present-day variation.

2. Clinal Variation and Parasympatric Speciation

Ecological, climatological and physico-chemical conditions change gradually in the ocean. Gradients in the conditions usually run from the equator to the poles, from the surface to the ocean floor or, though less pronounced, from east to west. Under the influence of these changing conditions, the taxa develop variation in which similarly three gradients can be found, the north–south, east–west, and bathymetrical. When this variation is found in continuous ranges, it usually becomes a clinal variation, a type of variation which seems to dominate in oceanic plankton. As these clines are usually not smooth curves but of the stepped type (Fig. 12), we can distinguish a variety of types of clinal variation and the more typical ones are discussed in the following examples.

(a) Salpa fusiformis *and* Eucalanus subtenuis *variation.* Van Soest (1972, 1974, 1975) made a thorough investigation of the variation in the species

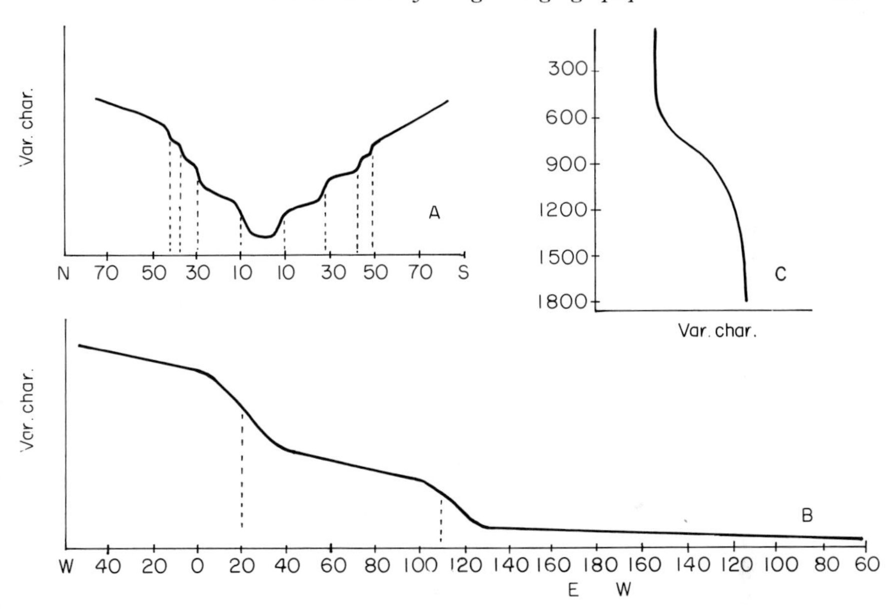

Fig. 12. Clinal variation curves. A. north–south variation (x-axis), variable on y-axis; B. east–west variation (x-axis), variable on y-axis; C. Bathymetric variation (y-axis), variable on x-axis.

group of *Salpa fusiformis*. Of this group, *S. fusiformis* occurs between 70°N and 50°S, *S. thompsoni* between 40°S and 70°S, and *S. gerlachei* only in the Pacific between 60°S and 75°S. The number of muscle fibres of the body muscles varies strongly in this group and seems to be related to temperature and water density. The North Atlantic population (70°N) shows a muscle-fibre number, in the aggregates, of 65, in the south (50°S) comparable numbers are found; in the Antarctic species the number varies between 112 and 228, while in the tropics the number drops to 30. However, no absolutely linear correlation was found between the numbers and water temperature or density. Firstly, there is a large increase in the number between *S. fusiformis* and the two Antarctic species near 50°S, the area where the Antarctic Convergence occurs. North of the convergence, the number of fibres is maximally 65 and south of it the minimum is 110. Near 40°–35°N and near 30°–40°S, the gradient of variation in *S. fusiformis* shows an abrupt change, not coinciding with changes in temperature or density. The two Antarctic species differ only biometrically and the difference in muscle fibres, 112–142 in *S. gerlachei* and 137–228 in *S. thompsoni*, may very well be correlated with water density; if so, the continued separation at species level of the latter two taxa becomes dubious, the more so when other characteristics of the 'species' are considered (van Soest, 1975).

The clinal variation described by Van Soest for the muscle fibres is the result of selection and gene flow. It was clearly shown that this variable character, which contributes to the taxonomic differences between geographically separated taxa (*S. fusiformis*, *S. thompsoni* and *S. gerlachei*), may establish clinal variation within these taxa and thus contribute to taxonomic differences between groups of geographically continuous populations. Near the convergence, the variation is discontinuous due to the fact that *S. fusiformis* and *S. thompsoni* probably do not there interbreed. The phenomena near 40°–35°N and 30°–40°S similarly cannot be considered mere ecophenotypic variation. In these areas, the northern and southern boundaries of many plankton species occur and many other clines show a step in the gradient of variation at these latitudes. Geneflow and selection seem not to be in equilibrium, so that variation is greater over a shorter north–south distance than is normally the case. The factor responsible for this "disturbance" has not yet been fully elucidated, but the Subtropical Convergence is likely to have some influence. For some taxa, the Subtropical Convergence area forms a limit to their dispersal; for others, it generates speciation to the species level; for yet others it only results in a step in their cline. The Antarctic Convergence is a much greater hydrographic discontinuity and it usually either forms a barrier to dispersal or leads to speciation. *Salpa fusiformis* and *S. gerlachei* are isolated by this convergence. Some species, however, are continuous over the convergence region, but they usually show adaptations to the water masses north and south, which merge near the convergence, as demonstrated by *Sagitta gazellae* and its variation (David, 1963).

The variation in *Eucalanus subtenuis* is of the same type as in *Salpa fusiformis*, but the variation gradient runs east–west instead of north–south. *E. subtenuis* lives between 30°N and 30°S; it is continuous south of Africa and in the Indo-Malayan area. The total number of perforations of the integument in adult females of this species has been studied by Fleminger (1973). The number shows a clinal variation induced by different selection in the different oceans and by restricted gene flow between oceans. The within-ocean variation is consequently smaller than the inter-ocean variation. In *Salpa fusiformis*, initiation of speciation is induced by the Subtropical Convergence or by other phenomena in this area; in *Eucalanus subtenuis*, initiation of speciation is effected by restriction of gene flow in the inter-ocean connections.

(b) Clio pyramidata *variation.* The species group of *Clio pyramidata* shows a variation pattern comparable with that of *Salpa fusiformis*, but the differentiation has proceeded much further. In the equatorial waters of the Indo-Pacific and in the Red Sea, *Clio convexa*, an extreme warm-water species formerly considered a forma, is found; this is closely related to *Clio pyramidata* forma *lanceolata*. This latter taxon is a warm-water form which, together with *C.p.* forma *pyramidata* and *C.p.* forma *antarctica*, covers the same area as *Salpa fusiformis*. *C.p.* forma *sulcata* is sympatric with *S. thompsoni* and *S. gerlachei* (van der Spoel, 1967). In the species group of *Clio pyramidata*, the shell shape, shell size, and volume of embryonic shell show clinal north–south variation, so that there are great resemblances between *S. fusiformis* and *C. pyramidata*. Differences are found, however, in the degree of speciation. In the North Atlantic where *S. fusiformis* developed higher numbers of muscle fibres, *Clio pyramidata* developed an almost isolated form *Clio pyramidata* forma *pyramidata*. In the Southern Ocean, the *Salpa fusiformis* representatives with highest muscle fibre number are sympatric with *C.p.* forma *antarctica*. In *Clio*, the clinal variation of the characters mentioned shows much steeper steps near the Subtropical Convergence than does *Salpa*; also in *Clio*, interbreeding between cold-water populations (formae *pyramidata*, *antarctica* and *sulcata*) and warm-water populations (forma *lanceolata*) is restricted by differences in the reproductive biology of the populations (van der Spoel, 1981). *C.p.* forma *sulcata* remains at infraspecific level, and is therefore not as far evolved as *S. thompsoni*. The populations with very low muscle fibre numbers in *Salpa* are taxonomically identical with other populations; in *Clio*, however, the populations of equatorial waters have evolved to the species level, giving rise to *Clio convexa*.

From the above, one may conclude that the same hydrological, ecological and climatological influences have had a different impact on two groups of animals. Variation found in one group may be considered the start of the real speciation found in other groups. There are no essential differences between the two phenomena.

For a great number of cases, the evolution of Arctic and Antarctic species can be explained by processes described above. A species group like *C. pyramidata* is eurythermal, but the populations composing such a group may be stenothermal. A period of strong global cooling will in this case destroy the warm-water populations, which will result in a biantitropical species group of the type of *Limacina helicina*. When thermal energy flow in the ocean from the equator to the poles decreases, the latitudinal discontinuities increase and a north–south series of infraspecific taxa then easily develops into a series of valid species. The present-day north–south variation can thus be considered a phenomenon leading to the development of polar and subpolar faunas under conditions as known from the geological past.

(c) Sagitta planctonis *and tropical submergence.* When horizontal hydrological, climatological, and ecological differences occur they may cause clinal variation, also a vertical cline may be expected to exist, parallel with the bathymetrical variation in such conditions. In the tropics, the vertical differences are large, whilst in the high latitudes they become negligible. Between 30°N and 30°S at a depth of about 1000 m are found the greatest differences in conditions over the smallest distances. Real clinal bathymetrical variation has not yet been found, as the vertical ecocline is never smooth except probably for the pressure factor, but Pierrot-Bults (1970, 1975) found infraspecific diversity in *Sagitta planctonis* which nicely illustrates bathymetric adaptation resulting in a certain degree of speciation. The *Sagitta planctonis* group consists of *S. marri* in the Antarctic waters; *S. planctonis* in the upper 700 m between 40°N and 40°S; *S. zetesios* between 60°N and 45°S (at the surface in higher latitudes but below *S. planctonis* in warmer waters); and also forms intermediate between *S. planctonis* and *S. zetesios* in the area of contact. *S. marri* may be a taxon at infraspecific level, or a recently developed species which is isolated by the Antarctic Convergence. *S. planctonis* and *S. zetesios* are of infraspecific level and consequently will be treated as such from now on. *S. zetesios* forma *zetesios* and *S.z.* forma *planctonis* occur over different temperature ranges, the former seeming to be especially adapted to low temperatures and not specifically to greater depth (Pierrot-Bults, 1975). Consequently, it cannot be proved that this variation developed primarily along a bathymetrical gradient, as a north–south gradient may have generated the same variation. It is, however, evident that at present two closely related forms, with intermediates, remain separated by a bathymetrical gradient in conditions in the tropical latitudes. As it was accepted that such gradients may generate speciation in horizontal directions, one should also accept that, along the vertical, speciation can

occur by comparable influences. Deep-sea species can therefore have developed either as high latitude species which afterwards descended to greater depth, or as species which differentiated to species level only after descent; *S.z.* forma *planctonis* may show a phase in this process. Finally, it cannot be excluded that deep-sea taxa may have developed from tropical species by vertical variation and speciation.

3. Species groups

Species groups in warm-water planktonic animals like the *Pontellina plumata* group (*P. plumata*, *P. platychela*, *P. sobrina*, *P. morii*), the *Eucalanus hyalinus* group (*E. hyalinus*, *E. elongatus*, *E. bungii*, *E. inermis*), the *Eucalanus attenuatus* group (*E. attenuatus*, *E. parki*, *E. sewelli*, *E. langae*), the *Diacria trispinosa* group (*D. trispinosa*, *D. rampali*, *D. major*), the *Sagitta serratodentata* group (*S. serratodentata*, *S. pseudoserratodentata*, *S. bierii*, *S. pacifica*, *S. tasmanica*) and the species groups in the genus *Stylocheiron*, all seem to show the following biogeographical regularity. Groups of closely-related oceanic warm-water taxa roughly at the species level are composed of, (a) an Indo-Pacific warm-water taxon between 30°N and 30°S or in a narrower belt, which may have split into a Pacific and Indian Ocean taxon; (b) a circumglobal taxon between 40°N and 40°S or wider, which may have split into an Atlantic and Indo-Pacific taxon; (c) a group of Central-Water forms between 30°N and 20°N and between 20°S and 30°S; (d) a group of Transitional-Water forms between 50° and 30°N and between 30° and 50°S.

When two taxa of the same species group both show the distribution-type mentioned under (a) or (b), one of the two taxa is less closely related to the others. In some of the species groups, taxa with restricted distributions developed, for example separate forms in both the eastern Pacific and North Indian Oceans.

The biogeographical hypothesis here proposed, postulates that these species groups developed out of an ancestral Central-Water form. Biogeographic processes as described for the Central-Water forms caused the isolation of the Atlantic populations during cold periods in the Pleistocene. Enclosed in the Atlantic, a group adapted to a much wider range of conditions was able to spread out over the whole circumglobal belt during warmer periods. Depending on the degree of speciation, the warm-water component again mixed with the Central-Water component, or it succeeded in living in sympatry. From the warm-water form living between approximately 40°N and 40°S, the equatorial and Transitional-Water forms could develop, as described for *Clio pyramidata*, but it cannot be excluded that these possibly developed out of the Central-Water form. As in *Eucalanus subtenuis*, the warm-water taxon may finally split up into an Atlantic, an Indian, and a Pacific Ocean or Indo-Pacific Ocean form. Some provincialism may also develop in the equatorial forms, and at a later period the eastern Pacific warm-water and equatorial forms may split off, while diversity can also develop in the Central-Water forms, especially in the South Pacific. Central-Water ranges may start to fuse when, instead of diversification, adaptation to equatorial water is achieved.

Species groups of the higher latitudes like the *Calanus finmarchicus* group, the *Calanus helgolandicus* group, and the *Limacina helicina* group had a less complex development. Usually there exist two North Atlantic, two North Pacific and two Southern Ocean taxa, and in groups with an extreme cold-tolerance (in the high polar groups) one of the two Atlantic and one of the two Pacific taxa may be replaced by one and the same Arctic taxon, as in *Calanus glacialoides*-like groups. The factors determining distribution in these groups were discussed above and speciation is considered to have taken place as radiation, in which warm-water taxa developed forms in the north and in the south (p. 327). If this type of speciation did indeed exist, it is puzzling as to how these populations can remain rather similar and how they can have been in contact ever since they developed. In these species groups, there is no indication of deep-sea contacts or of transportation by boundary currents. One period in the geological past, however, provided an opportunity of north–south exchange for distribution of these types when the Tethys Sea closed and the Panama passage was still open. During that short period, all equatorial water from the Atlantic constantly drained to the Pacific, while there was no supply from the Tethys Sea. This situation made an extra flow of water necessary from the south and from the north. The returning water of the equatorial current flowed through the Pacific and re-entered via the Arctic and Antarctic regions. This configuration brought the northern and southern faunas into close contact along the eastern shores of the Atlantic. Probably this short period can be taken to be responsible for the transfer of high-latitude populations, and even of species, from one hemisphere to the other. It is, however, questionable if such a single event could maintain a great resemblance and degree of genetic similarity between northern and southern populations. As there are presently no other explanations, one has to accept that speciation is so slow in these groups that they did not diverge much after the Miocene. That real "monotypic" bipolar and bisubpolar taxa are absent, or very rare, proves that the north–south contact is indeed much smaller than for the transitional and warm-water taxa.

The two belts in the northern and southern hemispheres are the result of parasympatric speciation along the polar fronts. The influence of the Antarctic Convergence on the variation and speciation in *Clio pyramidata* and *Salpa fusiformis* has been discussed and the development of the two taxa in these species groups in the north and in the south may be explained by the mechanism described.

DISCUSSION

"Pelagic biogeography is to-day constrained to mere description of distributions. Higher taxa tell very little about centres of speciation, or dispersal as they are usually cosmopolitan" (Frost, 1980). Is this statement completely correct? There are distributions which can be explained; there are essential differences between the various types of dispersal; sometimes a faunal centre can be distinguished and dispersal itself can be explained in some cases.

It is probably the first step in real biogeography of plankton to accept the thesis of the Miocene reversal of dispersal. This is taken to mean that plankton dispersal, especially in the lower latitudes, was westwards in pre-Miocene periods and eastwards in the Miocene and later periods.

A second hypothesis is that the Central-Water patterns form a special category, and that they are the oldest patterns present in plankton and had already developed in the pre-Miocene.

Recognition of the belt-shaped patterns is a third working hypothesis which can help detect regularities in circumglobal distributions.

The generalities postulated by Fleminger and Hulsemann (1973) to explain provincialism and monotypy in circumglobal taxa, of the lower and of the higher latitudes respectively, are clearly linked with the circumglobal distributions mentioned.

The theories presented by Brinton (1975) on inter-ocean connections explained the effects of the passages south of Australia, Africa and America, and in the Indo-Malayan archipelago (the remarks on the

Panama and Amazonas inter-ocean connections given above are complementary to Brinton's theories).

The presence of well-defined plankton provinces and the recognition of the special characters of some of these, like the North Atlantic, the North Pacific and the eastern Pacific, make possible a more convincing interpretation of faunal differences.

The hypothesis as to the special composition of species groups provides another tool to make plankton zoogeography a more promising study.

The faunal centre described for the genera *Peraclis* and *Phronima* is another indication that, in the future, plankton biogeography will develop along lines of the biogeography of other organisms.

The generalities postulated for the clinal variation in plankton are probably very special to marine life, but they form a firm basis to study the variation and the first steps in speciation of plankton.

All the postulates concern biological phenomena, therefore exceptions to these rules are to be expected. Nevertheless, the hypotheses or the attempts to arrive at some theses are necessary to lift plankton biogeography out of the purely descriptive level of the moment, although an enormous task remains for the descriptive zoogeographers of plankton.

REFERENCES

Beklemishev, C.W. (1961). [On the spatial structure of plankton communities in dependence of type of ocean circulation. Boundaries of ranges of oceanic plankton animals in the North Pacific.] *Okeanologia* **6**, 1059–1072. [in Russian].

Beklemishev, C.W. (1971). Distribution of plankton as related to micropalaeontology. *In* "The Micropalaeontology of Oceans" (B.M. Funnel and W.R. Riedel, eds), pp. 75–87. Cambridge University Press, London.

Beklemishev, C.W., Parin, N.B. and Semina, G.N. (1977). [IV. Oceanic biogeography. *In* "Biogeographic structures of the Ocean". "Ocean biogeography I"] (M.E. Vinogradov, ed.), pp. 219–261. Publ. Nauk, Moscow [in Russian].

Berggren, W.A. and Hollister, C.D. (1974). Paleogeography, paleobiogeography and the history of circulation in the Atlantic Ocean. *In* "Studies in Paleo-oceangraphy" (W.W. Hay ed.), pp. 126–186. Soc. Econ. Paleontol. Mineralog. Spec. Publ. 20.

Brinton, E. (1975). Euphausiids of Southeast Asian Waters. *Naga Rep.* **4** (5), 1–287.

Dahl, F. (1923). "Okologische Tiergeographie II". G. Fischer, Jena.

Damuth, J.E. and Kumar, J. (1975). Amazone Cone: Morphology, sediments, age, and growth pattern. *Bull. geol. Soc. Am.* **86**, 863–878.

David, P.M. (1963). Some aspects of speciation in Chaetognatha – Speciation in the sea. *Publ. Syst. Ass.* 129–143.

Dunbar, M.J. (1979). The relation between oceans. *In* "Zoogeography and diversity of plankton" (S. van der Spoel and A.C. Pierrot-Bults, eds), pp. 112–125. Bunge Scientific Publishing, Utrecht.

Fleminger, A. (1973). Pattern, number, variability, and taxonomic significance of integumental organs (sensilla and glandular pores) in the genus Eucalanus (Copepoda, Calanoida). *Fish. Bull.* **71** (4), 965–1010.

Fleminger, A. and Hulsemann, K. (1973). Relationship of Indian Ocean epiplanktonic Calanoids to the World Oceans. *In* "The biology of the Indian Ocean" (B. Zeitschel and A. Gerlach, eds), pp. 339–348. Springer, Berlin, Heidelberg, New York.

Fraser, J. (1962). "Nature Adrift, the Story of Marine Plankton". G.F. Foulis, London.

Frost, B.W. (1980). Problems in marine biogeography (bookreview). *Science, N.Y.* **209**, 1112.

Grabert, H. (1971a). Die Prä-andine drainage des Amazones-Stromsystems. *Munster Forsch. Geol. Paläont.* **20–21**, 51–61.

Grabert, H. (1971b). Die Wasserfallstrecke des Rio Madeira (Territorium Randônia, Brasilien) als Rest einer vortertiären Wasserscheide zwischen dem Atlantik und dem Pazifik. *Die Erde* **20**, 53–62.

Herman, Y. (1979). Plankton distribution in the past. *In* "Zoogeography and diversity of plankton" (S. van der Spoel and A.C. Pierrot-Bults, eds), pp. 29–49. Bunge Scientific Publishing, Utrecht.

McGowan, J.A. (1971). Oceanic biogeography of the Pacific. *In* "The Micropalaeontology of Oceans (B.M. Funnel and W.R. Riedel, eds), pp. 3–74. Cambridge University Press, London.

McIntyre, A., Bé, A.W.H. and Roche, M.B. (1970). Modern Pacific Coccolithophoridae, a palaeontological thermometer. *Trans. New York Acad. Sci.* II, **32** (6), 720–731.

Meisenheimer, J. (1905). Pteropoda. *Wiss. Ergebn. Deutsch. Tiefsee Exp. auf dem Dampfer "Valdivia" 1898–1899* **9** (1), 1–314.

Muromtsev, A.M. (1958). "The principal hydrological features of the Pacific Ocean". Gimiz, Leningrad. [Israeli Programme of Scientific Translations, Jerusalem.]

Omori, M. (1975). The systematics, biogeography and fishery of epipelagic shrimps of the genus Acetes (Crustacea, Decapoda, Sergestidae). *Bull. Ocean. Res. Inst. Univ. Tokyo* **7**, 1–91.

Orthmann, A. (1896). Grundzüge der marinen Tiergeographie. G. Fischer Verl., Jena.

Pierrot-Bults, A.C. (1976). Zoogeographic patterns in Chaetognaths and some other planktonic organisms. *Bull. zool. Mus. Univ. Amsterdam* **5** (8), 59–72.

Pierrot-Bults, A.C. and van der Spoel, S. (1979). Speciation in macrozooplankton. *In* "Zoogeography and diversity of plankton" (S. van der Spoel and

A.C. Pierrot-Bults, eds), pp. 144–167. Bunge Scientific Publishing, Utrecht.

Reid, J.L., Brinton, E., Fleminger, A., Venrick, E.L. and McGowan, J.A. (1978). Ocean circulation and marine line. *In* "Advances in Oceanography" (H. Charnock and G. Deacon, eds), pp. 65–130. Plenum, New York.

Segerstråle, S.G. (1962). The immigration and prehistory of the glacial relicts of Eurasia and North America. A survey and discussion of modern views. *Int. Revue ges. Hydrobiol.* **47** (1), 1–25.

Segerstråle, S.G. (1971). The zoogeographic problem involved in the presence of the glacial relict Pontoporeia affinis (Crustacea, Amphipoda) in Lake Washington U.S.A. *Journ. Fish. Resch. Bd. Canada* **23** (9), 1331–1334.

Tokioka, T. (1979). Neritic and oceanic plankton. *In* "Zoogeography and diversity of plankton" (S. van der Spoel and A.C. Pierrot-Bults, eds), pp. 126–143. Bunge Scientific Publishing, Utrecht.

van der Spoel, S. (1967). "Euthecosomata, a group with remarkable developmental stages (Gastropoda, Pteropoda)". J. Noorduijn, Gorinchem.

van der Spoel, S. (1976). "Pseudothecosomata, Gymnosomata and Heteropoda (Gastropoda)." Bohn, Scheltema & Holkema, Utrecht.

van der Spoel, S. (1981). "Gelijk en Ongelijk". Bunge Scientific Publishing, Utrecht.

van der Spoel, S. and Pierrot-Bults, A.C. (eds). (1979). "Zoogeography and Diversity of Plankton." Bunge Scientific Publishing, Utrecht.

van der Spoel, S. and Pierrot-Bults, A.C. (1979). Zoogeography of the Pacific Ocean. *In* "Zoogeography and diversity of plankton" (S. van der Spoel and A.C. Pierrot-Bults, eds), pp. 293–327. Bunge Scientific Publishing, Utrecht.

van Leyen, A. and van der Spoel, S. (1982). A new taxonomic and zoogeographic interpretation of the *Diacria quadridentata* group (Mollusca, Pteropoda). *Bull. zool. Mus. Univ. Amsterdam* **8** (13), 101–118.

van Soest, R.W.M. (1975). Zoogeography and speciation in the Salpidae (Tunicata, Thaliacea). *Beaufortia* **23**, 181–215.

Vinogradov, M.E. (1970). "Vertical Migration of the Oceanic Plankton." Israeli Programme of Scientific Translations, Jerusalem.

Voronina, N.M. (1978). Variability of ecosystems. *In* "Advances in Oceanography" (H. Charnock and G.E.R. Deacon, eds), pp. 221–243. Plenum Press, New York and London.

12 | Biogeographical Explanations and the Southern Beeches

C.J. HUMPHRIES

Department of Botany, British Museum (Natural History), London SW7 5BD, UK

Abstract: Following J.D. Hooker's classical nineteenth century essays on the biogeography of disjunct Southern Hemisphere plants, the southern beeches (*Nothofagus*) have figured as a key taxon in many different biogeographical studies. A review of *Nothofagus* taxonomy and biogeography is presented in considering the theoretical developments in biogeography over the last 130 years. Systematic theory seems to have undergone two major philosophical shifts from a pre-Darwinian empirical phase of species description, through a post-Darwinian era of narrative evolutionary taxonomy, to a new, recent analytical and empirical phase. Within the narrative phase, explanations for disjunctions in *Nothofagus* distribution involve evidence on the taxonomy of modern species in relation to fossils, being influenced by fashionable geological theories, interpretations of places of origin and methods of dispersal. Within an analytical framework, the distribution patterns of *Nothofagus* are considered in terms of explicit cladistic reconstructions in relation to cladograms of other groups occupying similar areas. These are compared with geological area cladograms in an effort to understand area patterns. Despite using the same raw data, the results of the eleven different accounts of *Nothofagus* biogeography considered here are so different from one another that methodological differences can be highlighted.

INTRODUCTION

The Key to the history of terrestrial life in the far south may be *Nothofagus*
P.J. Darlington (1965).

Systematics Association Special Volume No. 23, "Evolution, Time and Space: The Emergence of the Biosphere", edited by R.W. Sims, J.H. Price and P.E.S. Whalley, 1983, pp. 335–365. Academic Press, London and New York.

Nothofagus is uninformative on the interrelationships of southern hemisphere areas.

C. Patterson (1981).

In the introductory essay to his *Flora Novae Zelandiae* (1853), Hooker attempted for the first time to explain the problem of why so many different groups of unrelated organisms should show similar, widely disjunct distribution patterns in the southern hemisphere areas of southern South America, Tasmania, Australia and New Zealand. In an effort to find a general explanation that could account for geographic disjunctions in one hundred or more plant genera, Hooker provided the novel theory that all of the different groups originated and subsequently dispersed in the Southern Hemisphere on a continuous tract of land. Although this was the simplest possible explanation for a major biological problem, Hooker's theory was unacceptable to many of his contemporaries because fashionable geological theories of the day advocated that continents and intervening oceans are and always had been stable. Darwin, for example, could not agree with Hooker on how different groups of plants spread over the globe. Darwin was influenced by stable globe theories and his explanations involved multiple origins in the Northern Hemisphere with individual long-distance dispersal events for repeated disjunctions in the austral biota.

Rather surprisingly similar views still prevail. It is impossible comprehensively to review all the literature, but certain groups of organisms such as the marsupials and the southern beeches (*Nothofagus*; Fagaceae) have frequently figured as important examples. This paper reviews the history of theory in phytogeography, since the middle of the nineteenth century, by examining different explanations for disjunct distributions of *Nothofagus* (see Humphries, 1981). This latter genus has been selected for three reasons. Firstly, it has attracted the attention of many biogeographers working from different theoretical frameworks. The quotation (above) from Darlington is one of many to be found in the writings of classical biogeographers who use theories on earth history, traditional classifications, fossils and dispersal hypotheses as the bases for explaining disjunct distribution patterns. By contrast, the Patterson quotation represents the views of a smaller group of biogeographers who compare cladograms and distribution patterns to geological area cladograms in an effort to find general explanations for disjunctions. Secondly, the interrelationships of *Nothofagus* species and their relationships with other members of the beech family are comparatively well documented. Thirdly, despite the similarity of the supporting evidence for each biogeographical explanation, the hypotheses differ widely one from another.

TAXONOMY AND BIOGEOGRAPHY

Historical biogeography is the study of distribution patterns as connections between areas in terms of time and space (Rosen, 1978). But biogeography is dependent on the theoretical foundations of systematics since it can only be as good as the taxonomy it uses to establish historical distribution. Biogeography must therefore be regarded as an integral part of systematics; indeed, changes in biogeographical explanations follow changes in taxonomic theory.

The developments of taxonomic and biogeographic methods over the last 130 years are analogous. Ball (1976) and Patterson (1980) recognize three sequential phases: *empirical* (or *descriptive*), the *narrative* and the *analytical*. *Empirical* or *α-taxonomy* is data-gathering systematics, seeking to distinguish and describe species. *Narrative taxonomy*, by contrast, attempts to explain evolutionary relationships of different organisms by using so-called traditional methods of analysis for assessing overall similarities/differences. The degree and rate of morphological divergence in contemporary organisms are estimated by comparison with fossils or by considering that some extant organisms are more primitive than others and therefore ancestral. *Analytical taxonomy* describes methods originally developed by, for example, Hennig (1965, 1966); these have formalized concepts and procedures for estimating cladistic relationships.

Narrative biogeography is the practice of using known historical or geological events, with various biological assumptions, as a basis for explaining distribution patterns. *Analytical biogeography* compares cladistic patterns of different groups of endemic organisms occupying similar areas, in attempting to derive from a combined geological and biological point of view general biogeographical explanations.

In narrative approaches, the historical distribution of contemporary species is expressed in terms of taxonomy of modern species as compared with distribution and taxonomy of fossils. Narrative historical biogeography is mostly concerned to provide explanations in terms of six variables: mobile or fixed continents, dispersal or vicariance, and climate or ecology. Usually, the current geological or climatological theory determines the biogeographical explanation. Thus, in a world of fixed continents, intercontinental disjunctions must have resulted from transoceanic migration or long-distance dispersal. With mobile continents, intercontinental disjunctions derive from a dispersal event prior to continental break-up or a vicariance event, e.g. division of a formerly continuous community by continental rafting. Dispersal explanations interpret disjunctions in terms of one group of plants living in one area giving rise to the plants occupying another area. Disjunctions between one or more areas are thus caused (in non-biparental species) by at least one sexually mature individual crossing from one area over a pre-existing barrier and establishing itself in a second area. In other words, the first area is considered to be the "centre of origin", a feature which can be suggested by different criteria (Table I). Sometimes the reverse has been thought true, when the "centre of origin" is that area where the greatest variety of taxa or the most derived taxa are formed within "a centre of diversity". The ancestral or primitive groups are assumed to have migrated to peripheral areas (Table I, Aii) (see Matthew, 1915; Oliver, 1925). The introduction of fossils presents a different picture and they are of cardinal importance to dispersal biogeographers. The mere presence of fossils in one area with modern species in two or more areas can suggest the centre of origin (Table I, Bi). When there is a tolerably good fossil record, i.e. occurring in more than one area, those areas containing the oldest fossils are said to represent the centre of origin (Table I, Bii). In groups with restricted modern distributions, fossils

Table I. Reasons for dispersal in narrative hypotheses (see text for explanation).

		Area 1	*Dispersal Route*	*Area 2*
A	i	Primitive species	⟶	Advanced species
	ii	Advanced species	⟶	Primitive species
B	i	Modern species and fossils	⟶	Modern species
	ii	Modern species and old fossils	⟶	Modern species and young fossils
	iii	Fossils	⟶	Modern species
	iv	Modern species and old fossils	⟶	Young fossils

play a very important role for determining dispersal routes on the basis of relative age—in other words the oldest taxa suggest the centre of origin (Table I, Biii, iv).

The alternative to dispersal explanation of disjunct distribution patterns is vicariance. Wulff (1943) defined vicarious taxa as those that occupy mutually exclusive areas but share a relatively recent common ancestor. The genesis of vicariants or vicariads was seen by Croizat (1952) to be the historical separation of a formerly continuous population, biota or community by the development of a natural barrier. He went on to state that if a natural barrier such as a climatic change or continental drift separated a formerly continuous biota or community into two different groups, the organisms would show the same track of relationship across the barrier. In other words, correspondence between several tracks in both fossil and modern groups would give a "generalized track" linking the vicariant products of a formerly continuous biota or community (see Croizat *et al.*, 1974). Deriving from the vicariance view of disjunction is the idea that historical biogeography is more about the biotic and geological history of areas than of particular groups or organisms. In realizing this, and that dispersal is a random, untestable phenomenon, Platnick and Nelson (1978) proposed a method of historical biogeography, the cladistic-vicariance method, which was put into practice by Rosen (1978). The method is one which combines the cladistic method with the vicariance approach, exploiting to the full the analogy between systematics and biogeography (Patterson, 1980). It applies cladistics to biogeography by producing cladograms or trees of areas; these are then compared with cladograms of different groups of organisms occupying those areas.

THEORIES OF *NOTHOFAGUS* BIOGEOGRAPHY

1. Narrative Biogeographical Theories

(a) Explanations involving stable continents and dispersal. In contrast to Hooker's theory of southern origins and geological continuity during the Mesozoic, Darwin (1859) preferred to consider that wholly southern genera such as *Nothofagus* originated in the northern hemisphere and migrated by dispersal to the Southern Hemisphere. This general view

was rapidly adopted by zoologists and was particularly well articulated by Wallace (1876):

> . . . The north south division of the modern biota truly represents the fact that the great northern continents are the seat and birthplace of all the higher forms of life, while the southern continents have derived the greater part, if not the whole of their vertebrate fauna from the north . . .

This was an interesting comment with startling conclusions since, despite the striking austral affinities in many different taxonomic groups, Wallace went on to say: ". . . but it implies the erroneous conclusion that the chief southern lands—Australia and South America—are more closely related to each other than to the northern continent . . .". Even though there seemed to be little evidence for north–south migration routes, the views of Darwin and Wallace were soon adopted by phytogeographers. For example, Schröter's (1913) "monoboreal relict hypothesis" was an early attempt to justify Darwin's original idea. This hypothesis stated that isolated austral groups were invariably primitive forms, driven southwards from the northern hemisphere through the development of new groups in the north. The evidence came from the distribution of such gymnosperm genera as *Podocarpus* and *Araucaria*, with modern distributions in the Southern Hemisphere and Mesozoic fossil relatives in the Northern Hemisphere. For the Hooker adherent, like Engler (1882), groups with northern and southern complements could best be assumed simply to have been descendants of an earlier, world-wide group. However, Thistelton-Dyer (1909) stated that: ". . . the extraordinary congestion in the peninsulas of the Old World points to the long continued action of a migration southwards". By his adding "the theory of southwards migration is the key to the interpretation of the geographical distribution of plants", dispersal explanations became dogma in phytogeography. Indeed, Oliver (1925) provided a more detailed exposition of Schröter's view for *Nothofagus*: "It is probable that *Fagus* and *Nothofagus* originated in North America and spread thence east, south and west. The western moiety passed, *via* Japan, round the Pacific reaching Australia and New Zealand". He went on to say:

> If the characters of which *Nothofagus* is separated from *Fagus* be considered primitive, then these two genera exemplify the principle enunciated by Matthew, which states that a group should be most advanced at its point

> of original dispersal, the most conservative stages being farthest from it . . .

(see Table I, Aii).

The 36 species of the southern beech genus *Nothofagus* are the only members of the Fagaceae to occur in the Southern Hemisphere, south and east of New Guinea, and New Britain (Table II). The closest relative is the northern beech genus, *Fagus*, widely distributed in Europe and North America; the remaining Fagaceae genera are important trees of the north temperate and southeast Asian to Papuasian forests. The species of *Nothofagus* are almost always tall trees, dwarfed only at the latitudinal and altitudinal limits of tree growth where excessively severe weather conditions prevail. They are mainly evergreen, with several deciduous species. They have small seeds, which do not travel very far and do not survive in sea water. All the species are poorly fitted for dispersal and simply for this reason have always raised intriguing questions regarding their wide distribution in the Southern Hemisphere. They form extensive forests in three principal areas of the southern cold-temperate zone—southern South America, Australia (including Tasmania) and New Zealand; they also occur in the cooler tropical highlands of New Guinea, New Britain and New Caledonia (see van Steenis, 1953, 1971; Soepadmo, 1972).

There are ten species in southern South America, forming almost continuous forest on the tops and along the western slopes of the Andes, variously distributed from latitudes 56°S to 33°S. According to McQueen (1976), the southern limit is due to the lack of suitable land and the northern limit to the aridity which accompanies the Mediterranean-type climate of central Chile. Three species occur in Tasmania and Australia; the deciduous *N. gunnii* is endemic to the mountains of Tasmania, whilst *N. moorei* occurs in New South Wales (the McPherson-Maclay overlap of Burbridge, 1960) and *N. cunninghamii* in Victoria and Tasmania.

The five New Zealand species occur in rather isolated patches, from the lowlands to the timberline on both main islands, particularly on the western side. Some 18 tropical evergreen species are known in the mountains of New Guinea, New Britain and New Caledonia. Here, they are important co-dominants of the rain forest, forming extensive but patchy stands, especially in the lower montane areas between about 2300 and 2800 m.

Most phytogeographical discussion of *Nothofagus* revolves around

Table II. The modern species of *Nothofagus*, based on van Steenis (1953, 1971), Soepadmo (1972), Hanks and Fairbrothers (1976), McQueen (1976) and Allan (1961).

Name	*Distribution*	
N. alpina (Poepp. & Engl.) Oerst.	South America	(S Am)
N. procera (Poepp. & Endl.) Oerst.	South America	(S Am)
N. obliqua (Mirb.) Oerst.	South America	(S Am)
N. antarctica (Forst Oerst.	South America	(S Am)
N. pumilio (Poepp. & Endl.) Krasser	South America	(S Am)
N. betuloides (Mirb.) Oerst.	South America	(S Am)
N. nitida (Phil.) Krasser	South America	(S Am)
N. dombeyi (Mirb.) Oerst.	South America	(S Am)
N. glauca (Phil.) Krass.	South America	(S Am)
N. alexandri Espinosa	South America	(S Am)
N. cunninghamii (Hook f.) Oerst.	Australia	(Aus)
N. moorei (F.r. Muell.) Krasser	Australia	(Aus)
N. gunnii (Hook. f.) Oerst.	Tasmania	(Tas)
N. cliffortioides (Hook. f.) Oerst	New Zealand	(NZ)
N. solandri (Hook f.) Oerst.	New Zealand	(NZ)
N. menziesii (Hook. f.) Oerst.	New Zealand	(NZ)
N. fusca (Hook. f.) Oerst.	New Zealand	(NZ)
N. truncata (colenso) Cockayne	New Zealand	(NZ)
N. aequilateralis (Baum.-Bod.) Steen.	New Caledonia	(NC)
N. codonandra (Baill.) Steen.	New Caledonia	(NC)
N. balansae (Baill) Steen.	New Caledonia	(NC)
N. baumannida (Baum.-Bod.) Steen.	New Caledonia	(NC)
N. discoidea (Baum.-Bod.) Steen.	New Caledonia	(NC)
N. resinosa Steen.	New Guinea	(NG)
N. flaviramea Steen.	New Guinea	(NG)
N. carrii Steen.	New Guinea	(NG)
N. wormersleyi Steen.	New Guinea	(NG)
N. pullei Steen.	New Guinea	(NG)
N. pseudoresinosa Steen.	New Guinea	(NG)
N. crenata Steen.	New Guinea	(NG)
N. grandis Steen.	New Guinea	(NG)
N. perryi Steen.	New Guinea	(NG)
N. suda Steen.	New Guinea	(NG)
N. rubra Steen.	New Guinea	(NG)
N. brassii Steen.	New Guinea	(NG)
N. starkenborghii Steen.	New Guinea	(NG)

establishing the "centre of origin" and then deducing, on the basis of the taxonomic relationships, how the modern species attained their present distributions. Craw (1978) noted that there are generally two types of dispersal hypothesis, depending on the acceptance of either stable or mobile continents. In stable continent theories, disjunctions are seen to have originated by trans-oceanic long-distance dispersal. Acceptance of mobile continents, however, provides pre-drift overland dispersal routes, except where continental drift is believed to have occurred too early to account for the disjunction.

Croizat (1952), originally working within a stabilist framework, suggested that there were several centres for the origin of the angiosperms. The Fagaceae, as with several other "primitive" families, was said to have originated from a New Caledonian centre. Although no precise dispersal routes were postulated, *Nothofagus* was considered to have moved along a circum-antarctic "track", the other members of the family moving through south-east Asia to Europe and North America.

Since it is generally believed that fossils are amongst the best indicators of evolutionary direction, the particularly good record and distinctive fossil pollen of *Nothofagus* have been used extensively in phytogeographical discussion. There are three recognizable pollen morphs named after species—"*brassii*", "*fusca*" and "*menziesii*"; these are considered typical of the respective morphs (see Hanks and Fairbrothers, 1976). The *brassii* morph correlates with the present-day tropical, leathery-leaved species of New Guinea and New Caledonia, whilst the *menziesii* and *fusca* morphs are found in both evergreen and deciduous species of the cold-temperate zone. Despite reports of *Nothofagus* pollen fossils in the Northern Hemisphere (Cranwell, 1963), reliable records are to be found in areas of contemporary forest with appreciable extensions of range only into western Australia, Patagonia and Antarctica (Table III). Acceptance or rejection of northern fossil records greatly influences narrative interpretations. Amongst the earliest fossil records (Table III), the *brassii* and *fusca* morphs occur throughout the Upper Cretaceous of New Zealand. The earliest record for the *menziesii* morph is from the Eocene of Australia and South America. It appears, from fossil evidence, that *Nothofagus* was reasonably widespread in the Cretaceous and became even more widespread during the Tertiary. Today, the genus is absent from Antarctica; the *brassii* morph is restricted to New Guinea and New Caledonia and only the *fusca* and *menziesii* morphs occur in the cooler temperate floras of Australia, New Zealand and South America.

Darlington's (1965) interpretation of the history of *Nothofagus* combined fossil pollen evidence with stable continent geology. Although he acknowledged continental drift as a means of causing disjunction, his

Table III. Stratigraphic record for *Nothofagus* pollen. Based on van Steenis (1971), Hanks and Fairbrothers (1976).

		Upper Cretaceous	*Palaeocene*	*Eocene*	*Oligocene*	*Lower Miocene*	*Upper Miocene*	*Pliocene*	*Recent and Pleistocene*
Australia	b	+	+	+	+	+	+	+	
	f			+	+	+	+	+	+
	m			+	+	+	+	+	+
New Zealand	b	+	+	+	+	+	+	+	
	f	+	+	+	+	+	+	+	+
	m				+	+	+	+	+
Fuegia and Patagonia	b	+	+	+	+				
	f			+	+				+
	m			+	+				+
New Guinea	b						+	+	+
	f								
	m								
New Caledonia	b								+
	f								
	m								
Seymour Island	b		+	+					
	f		+	+					
	m								
McMurdo Sound	b			+					
	f			+					
	m			+					

b = *brassii* type pollen, f = *fusca* type pollen, m = *menziesii*-type pollen.

dismissal of it as a basis rested on the then contemporary belief that it occurred too early to affect the angiosperms. Despite the reliable microfossil record only in the austral continents, Darlington said that *Nothofagus* could not have originated there because its nearest relative, *Fagus*, lives in the Northern Hemisphere. Since he placed the "centre of diversity" for the Fagaceae in south-east Asia, Darlington considered:

> *Nothofagus* may have originated in Asia primarily in sub-tropical parts of South-east Asia during the Cretaceous. It probably was never widespread or dominant in the northern hemisphere. It probably crossed the tropics once, by way of the Indo-Australian archipelago, in the Cretaceous. The *Nothofagus* of the '*brassii*' group now on New Guinea may be descendants of the original tropics crossers . . . (Darlington, 1965:145).

He went on to say that an origin in Asia followed by southwards dispersal across the Indo-Australian archipelago would bring *Nothofagus* into the southern hemisphere to Australia, New Zealand, or both. Considering species of the *brassii* group as ancestral, Darlington then went on to postulate three separate late Cretaceous long-distance dispersal events around the austral land-masses, to account for the modern distributions.

(b) Explanations invoking stable continents and land bridges. The "land-bridge" theory in botany, as elaborated by van Steenis (1962), represents an intermediate stage between the stable and mobile continent theories. Although van Steenis believed in a steady-state world, he equally realized that the major plant disjunctions in both tropical and temperate floras could not be accounted for by trans-oceanic dispersal. In rejecting long-distance dispersal and cumbersome polyphyletic origins for similar floras on different continents, he developed the land-bridge theory. To account for the relationships of the austral biota, land bridges were postulated connecting Australia, Tasmania, New Zealand, Antarctica and South America; the presence of submerged continental shelves was used as evidence for a former Mesozoic highway. His classification of *Nothofagus* (van Steenis, 1953; revised, 1971, and modified in Soepadmo, 1972) based on variations in the leaves, the female flower and the number of florets (Fig. 1) is that most generally accepted today.

According to van Steenis, all Fagaceae originated in Indo-Malesia in an area bounded roughly by Yunnan (China) and Queensland (Australia). A land-bridge migration from south-east Asia into the temperate south was postulated to account for present-day distribution patterns as,

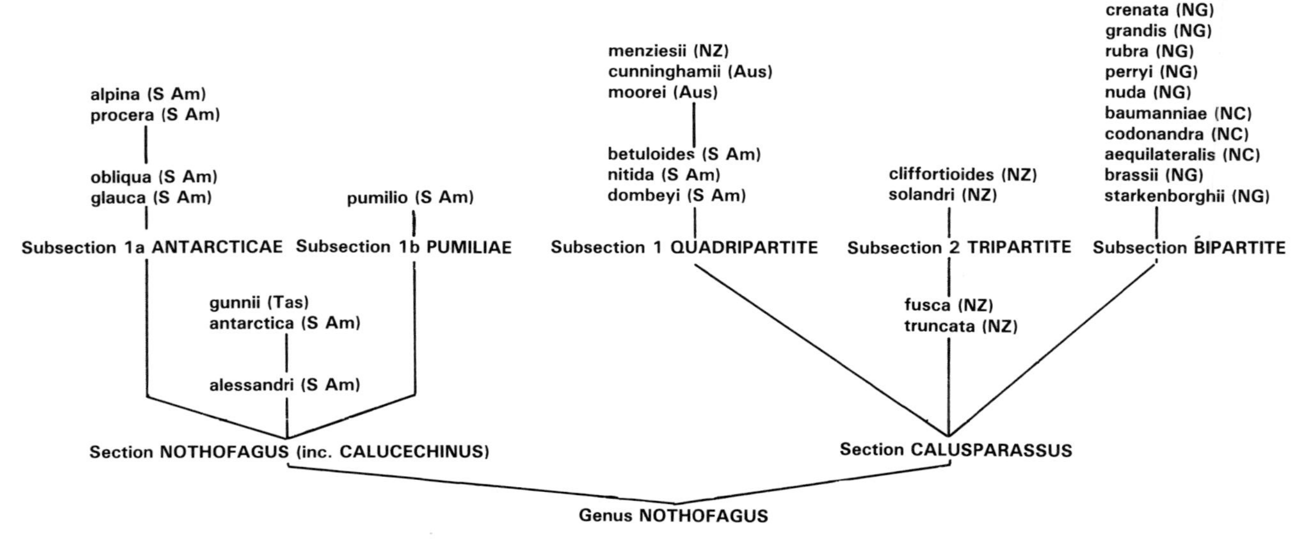

Fig. 1. Scheme of species interrelationships in *Nothofagus* (re-drawn from van Steenis 1953, and modified from van Steenis 1971 and in Soepadmo, 1972).

(a) *Nothofagus* is the only austral genus of Fagaceae; (b) the Papuasian species form a distinct group with *brassii* pollen; (c) the *brassii* pollen is considered primitive. Since the fossil record (Table III) shows the *brassii* pollen morph to have become extinct during the Pliocene in the austral areas, van Steenis considered New Guinea and New Caledonian species as relict survivors of an older, more widespread group.

(c) Explanations invoking mobile continents Other biologists were not wholly convinced that the world was in a steady state or that continental drift had occurred too early. Cranwell (1963) commented the "Flotsam of sea and sky, including the gravid female . . . carried a heavy burden in pan-biogeographic argument". By initiating a fresh look at fossil pollens of *Nothofagus*, she expanded the ideas of Florin (1940) and Couper (1960) that the areas with the oldest fossils are the most likely sites for the origin of particular groups. For these workers, therefore, *Nothofagus* had a wholly southern origin and must have migrated to the other continents by overland trails.

Literature of the 1970s largely reflects a general acceptance of plate tectonic theory; the idea of mobile continents as a causal agent of disjunction, however, in biographical terms simply provides an alternative basis (overland routes) for dispersal. Whereas stable continents would have posed dire transport problems for such poor dispersers as *Nothofagus*, former overland dispersal events are more plausible. Most of the major movements of continents deemed to have affected the distribution of modern biota occurred some 120 million years ago. The main effect of continental rafting has been to shift the emphasis of timing to earlier dispersal events when land masses were closer together. Nevertheless, the accounts remain narrative explanations with the same subjective elements as occur in stabilist theories. Hence, there is considerable variety between explanations, and they are not directly comparable since they are dependent on initial assumptions.

For *Nothofagus*, there are two types of post-tectonic narrative explanation—those that accept northern origins and others favouring southern origins.

Raven and Axelrod (1972:1382; also in 1974) used evidence of relatives of *Nothofagus* occurring in the northern hemisphere with data of various present-day austral gymnosperm groups having Mesozoic and Cainozoic fossil relatives in Europe to suggest that *Nothofagus* ". . . probably passed between the Northern and Southern Hemispheres by way of Africa and Europe since land connections were absent in Middle America". In the Southern Hemisphere, *Northofagus* migrated from Africa into South America and eventually to Australasia. This therefore is a "bandwagon" theory, fitting Schröter's monoboreal relict hypothesis into a pre-continental drift narrative based on present geological theory. Schuster (1976) similarly accepted a northern origin for *Nothofagus* but suggested an alternative migration route because there were land connections between North and South America until the Tertiary. Considering that "centres of diversity" are equivalent to "centres of origin", Schuster adopted a North American origin for this well-travelled genus since the Fagaceae (except for *Nothofagus*) are Eurasian. He affirmed that the ancestor of *Nothofagus* arrived in Gondwanaland by overland transport from North to South America, along the so-called Marsupial route, possibly by mid-Cretaceous times. This far-fetched explanation, completely without evidence, continued: "By Upper Cretaceous times, *Nothofagus* had diffused not only to Antarctica but, more than eighty million years ago, across to Australia and New Zealand." Entry into New Caledonia and New Zealand "... must have occurred well before the end of the Cretaceous because of their separation at this time but did not reach New Guinea until well into the Tertiary" (Schuster, 1976:120). Both the Raven and Axelrod and the Schuster hypotheses accepted the validity of Northern Hemisphere fossils, despite their general rejection (e.g. in Cranwell, 1963). The main alternative is a southern origin and Moore (1971) completed the narrative picture by presenting the post-tectonic version of Hooker's original observations. Moore suggested (1971:131) a starting date for the origin of *Nothofagus* by saying that "... at least some of the Palaeo-austral distributions date from the Cretaceous". Since the oldest fossils occur in New Zealand and Antarctica, and because of poor dispersal qualities of *Nothofagus* seed,

> ... one is forced to the conclusion that *Nothofagus* achieved migration between New Zealand and Antarctica before their late Cretaceous separation, at a time when its major groups were already differentiated, crossing to Australia and South America while connections were still available during the late Cretaceous–early Tertiary and subsequently moving northwards in the east and west.

In a fresh analysis of pollen data Hanks and Fairbrothers (1976:7), who accepted northern fossil pollen evidence, supported Moore's view and extended it by saying that *Nothofagus* "... could have developed in a region between New Zealand, Antarctica, and Australia then migrated through Antarctica into South America; and north through Tasmania, Australia and New Zealand and eventually arrive in Europe". (Presumably it then became extinct in the Northern Hemisphere.)

Summarizing the narrative approach, it can only be said that there are many possible centres of origin for *Nothofagus*, involving almost every major continent on the globe; the main possibilities are North America, Europe, South-east Asia, New Caledonia, somewhere between Yunnan and Queensland, Antarctica and New Zealand. The present-day distribution of *Nothofagus* has occurred by dispersal along a variety of trans-oceanic or overland routes ranging from South America to Japan; choice depends on the acceptance or rejection of continental drift. There is in no case an attempt to use a refined taxonomy demonstrating species level relationships and only vague notions are given as to what are "centres of origin". Furthermore, there is little appreciation that biogeography is more about areas of the globe than about a particular group of organisms.

2. Analytical Theories

Brundin (1966), in his classical investigation of the transantarctic relationships of chironomid midges, states that until there are explicit phylogenetic hypotheses for groups such as *Nothofagus* it will be impossible even to start reconstructing precise distributional patterns. That is, until there are cladistic schemes which account for *all* characters in modern species and fossils, it will be possible neither to provide historical details for particular genera nor to make comparisons between unrelated groups which occupy similar areas.

Melville's (1973) attempt was perhaps the first to reconstruct a fully resolved model of species relationships in *Nothofagus*. His cladogram, based on van Steenis's (1953) classification, is shown in Fig. 2. The relationships are based on the three evolutionary trends of deciduous to evergreen leaves, a gradual reduction in the parts of the female flower, and an elaboration of scales on the lobes of the female cupule. Character polarities are based on Forman's (1966) thesis that *Trigonobalanus*, an unusual tree (from Celebes, Malaya, Sarawak, Sabah, Thailand and, recently discovered (Lozano-C. *et al.*, 1980) the neotropics, in Colombia), has the most primitive female infructescence structure in the Fagaceae.

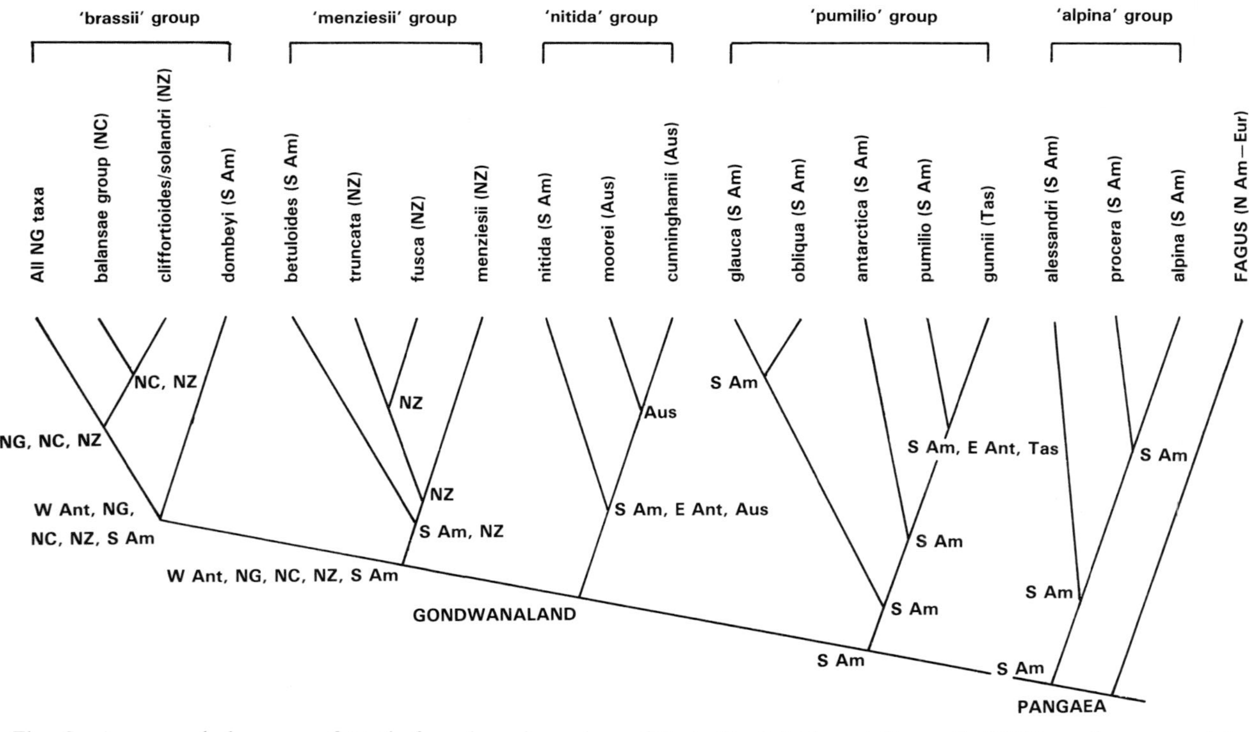

Fig. 2. An area cladogram of *Nothofagus* based on that of Melville (1973) and Cracraft (1975); and corrected for modern species from van Steenis (1971), van Steenis (*in* Soepadmo, 1972), Hanks and Fairbrothers (1976), McQueen (1976) and Allan (1961).

Cracraft (1975) used Melville's cladogram to reconstruct the historical biogeography of modern species of *Nothofagus*. In Fig. 2, Cracraft's selected species names are given to the four trans-antarctic groups: the *brassii* group includes the evergreen *N. dombeyi* from South America and its sister group in New Zealand, New Caledonia and New Guinea; the *menziesii* group comprises *N. betuloides* from South America and its sister species in New Zealand; the *nitida* group comprises *N. nitida* in South America and its sister group of two species in Australia; finally, the *pumilio* group occurs in South America. The reconstruction was an attempt to determine the centre of origin by the Hennigian dispersal method. In this, the descendant (contemporary) distributions are combined at each branch point and the ancestral "centre of origin" established by eliminating unique or unshared elements. The result here was that ancestors of the *brassii* group must have been distributed throughout a land-mass comprising South America, western Antarctica, New Zealand, New Caledonia and New Guinea.

Continental drift is used as an explanation for the present disjunct pattern. Separation of the New Zealand and South American continental blocks could account for the isolation of *N. dombeyi* from its sister group. A different, later event isolated the ancestor of *N. flaviramea* and *N. brassii* in New Guinea, and the ancestor of *N. codonandra* and *N. solandri* in New Caledonia and New Zealand. The ancestor of the *menziesii* group was spread across South America, western Antarctica and New Zealand. Continental break-up could also account for vicariance speciation isolating *N. betuloides* in South America and leaving the ancestor of *N. menziesii* and its relatives in New Zealand. A similar pattern can be reconstructed for the *nitida* group, but over eastern rather than western Antarctica. However, because in the *pumilio* group the stem-group species *N. glauca*, *N. obliqua* and *N. antarctica* all occur in South America and the crown-group species *N. pumilio* and *N. gunnii* in South America and Tasmania, Cracraft suggested that the most parsimonious explanation for the crown-group is a dispersal event from South America to Tasmania. The *alpina* group has a wholly South American origin; since the taxa in New Zealand, New Guinea, New Caledonia and Australia figure as unique events, the obvious deduction for the origin of the whole group is in pre-drift west Gondwanaland, with subsequent dispersals into other areas. The presence of at least four species groups in pre-drift Gondwanaland means that dispersal to east and west Antarctica and into Australia, New Guinea and New Zealand occurred by at least the Turonian, in the upper Cretaceous some 90 my ago (Figs 3–6). According to the model, there are no sister-group relationships between New Zealand and Australia but, across Antarctica to South America, there is a pattern commensurate with several extant groups occurring at the time of continental break-up.

Although Cracraft's analysis is perhaps the most explicit to date, its validity is difficult to assess because it is based on dubious taxonomy and

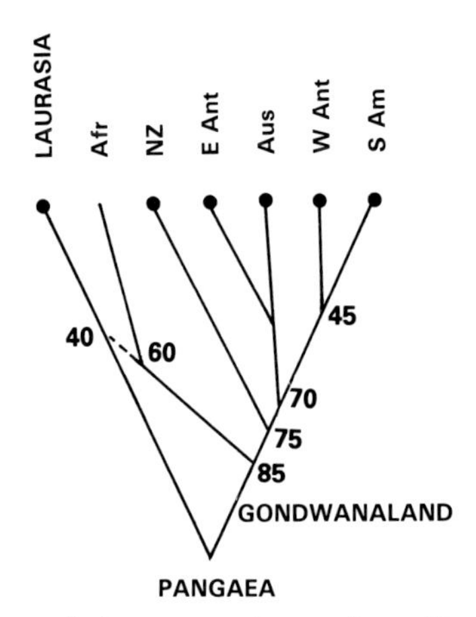

Fig. 3. A geological area cladogram redrawn from Rosen (1978) based on the maps of Dietz and Sproll (1970), Rich (1975) and Ballance (1976). (•) shows areas in which endemic *Nothofagus* and *Fagus* species occur.

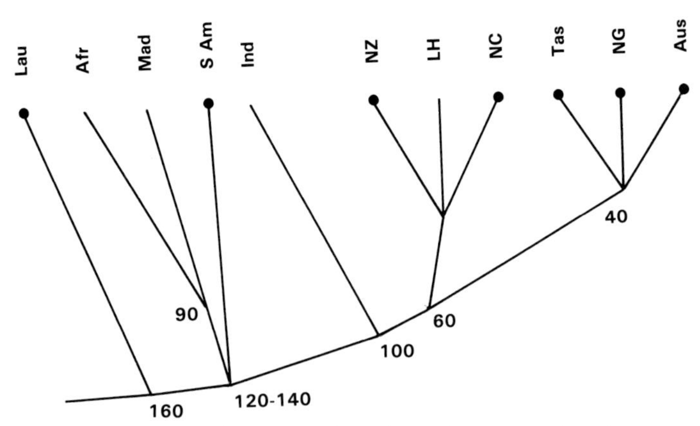

Fig. 4. A geological area cladogram based on the maps of Smith, Briden and Drewry (1973). (•) shows areas in which endemic *Nothofagus* and *Fagus* species occur.

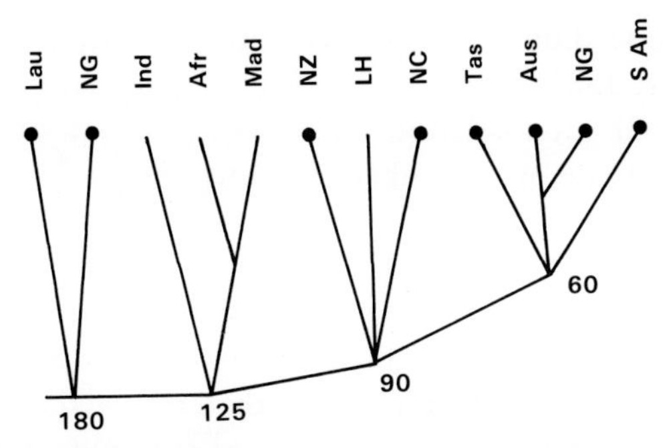

Fig. 5. A geological area cladogram (excluding Antarctica) based on the maps of Owen (1976). (●) shows areas in which endemic *Nothofagus* and *Fagus* species occur.

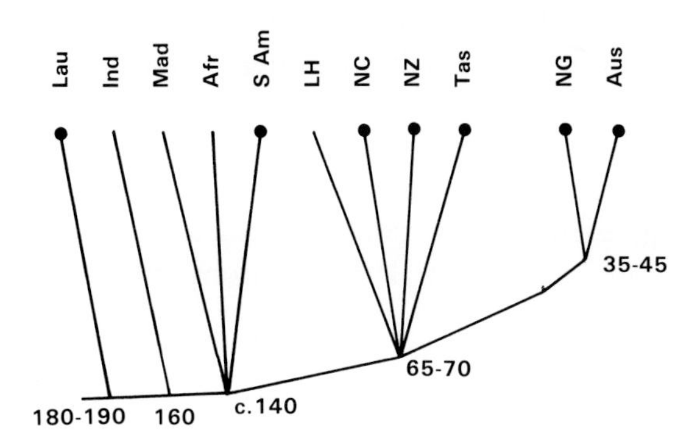

Fig. 6. A geological area cladogram based on the paper of Shields (1979). (●) shows areas in which endemic *Nothofagus* and *Fagus* species occur.

incorporates dispersal explanations for some of the ancestral distribution patterns. Moreover, the model is based on general interpretations of advancement in some morphological characters, rather than on synapomorphies, for establishing monophyletic groups, hence there are several cumbrous and unlikely interpretations in the scheme. For example, the South American species *N. dombeyi* is, on the basis of cupule and floret morphology, one of the most primitive species in the genus and yet is here placed as sister species to some, but not all, of the taxa with derived cupules. Furthermore, there are in the diagram dichotomies which have no evidence at all for their inclusion. For these reasons, and because various other sources of valuable data on wood anatomy and the male inflorescence have been disregarded, another cladogram produced on strictly cladistic lines is here given as Fig. 8 (see also Humphries, 1981).

The scheme of inter-relationships favoured here differs from that proposed by Melville on a number of points. All cool temperate species show a "specialized" wood anatomy and are more closely related to one another than they are to New Caledonian and New Guinean species. Within the two temperate groups, there is insufficient evidence to establish relative order of some main branching points; the divisions are therefore given as polychotomies rather than resolved dichotomies. The net result from rearrangement is that there are only two transantarctic relationships at species level, rather than four as predicted in the Melville diagram, and only one (*gunnii-pumilio*) is identical to both schemes.

To determine if dichotomies in the rearranged *Nothofagus* phylogeny can be accounted for by known events in earth history, the vicariance-cladistic method can be applied; this differs from the Hennig dispersal method in that, instead of eliminating unshared or unique elements, all elements are added together. The additive area sequence for the present *Nothofagus* cladogram appears in Fig. 8. From such cladograms, Nelson (1975) has suggested that, within a vicariance framework, unshared elements indicate not ancestral distributions but exact positions of barriers which caused divergence.

By combining the northern and southern beeches and substituting taxa for areas in a reduced area cladogram, *Fagus* and *Nothofagus* are generally found to form an eight-area pattern of distribution (Fig. 10). Two of the areas represent North America and Europe in Laurasia, the other six the principal land masses of Gondwanaland. This cladogram can now be compared with cladograms which represent patterns of earth history—geological cladograms.

Recent renewed interest in continental drift has led to several historical mapping schemes. Geological cladograms (Figs 3–6) show four different schemes illustrating the break-up sequences postulated by different authors: two fixed earth models (Dietz and Sproll, 1970; Smith *et al.*, 1973) and two expanding earth models (Owen, 1976; Shields, 1979). It is interesting to compare the different schemes, since New Zealand, India and South America give three different sequences on the cladograms (Fig. 7). As Rosen (1978) pointed out, geological data arranged in nested cladograms should reflect the history of areas under study. Within an historical context, related geological areas are two or more such that arose by disruption or fragmentation of an ancestral

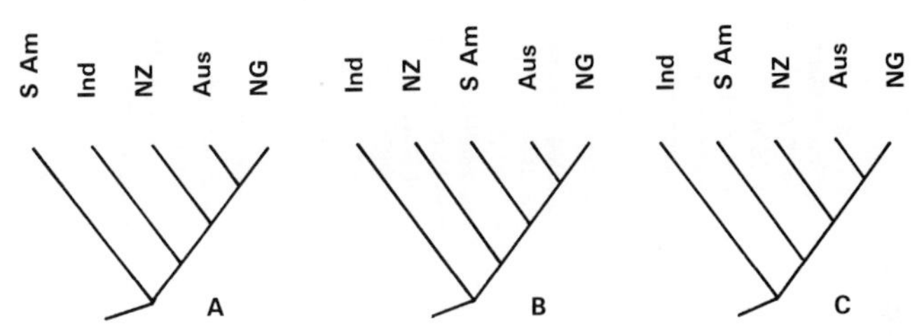

Fig. 7. Reduced geological area cladograms to show the differences in sequence for South America, New Zealand and India in the different mapping schemes. A, based on Smith *et al.* (1973); see Fig. 4; B, based on Rosen (1978) and Owen (1976); see Figs 3 and 5; C, based on Shields (1978); see Fig. 6. See text for explanation.

whole. If, therefore, geological and biological cladograms are constructed on special similarities which have arisen for the same historical reasons, the cladograms should be congruent. Deletion of the unique components in cladograms to be compared should lead to their being sequentially congruent for the areas. In addition to the possibility of thus checking congruence between different endemic groups occurring in similar areas, congruence between biological and geological cladograms should give the most economical break-up sequence for Gondwanaland.

Since three groups emerge from Fig. 8 with different disjunction patterns for South America-Australia, South America-Tasmania and New Zealand-Australia, they have been treated in Fig. 9 as branching from one point because such an incongruent pattern could not have occurred through sequential continental breakup. It is possible to resolve the next branching point because all species of cool temperate austral areas are sister to the New Guinea–New Caledonian group. Although this cladogram gives only low resolution for *Nothofagus*, a similar area diagram for the Melville/Cracraft model (Fig. 10) gives even less resolution, since the South America–New Guinea/New Caledonia disjunction in the *brassii* group represents yet another pattern.

Do fossils improve the resolution of this phylogeny? The only fossils that can be determined and identified with particular modern monophyletic groups are the micro-fossil pollen grains of the *brassii* morph. Adding these fossils to the area diagram does not extend the areas occupied by *Nothofagus*, but does extend the tropical *brassii* group into Australia and New Zealand. The net effect is to produce a reduced area cladogram identical to a reduced area cladogram from the Melville/

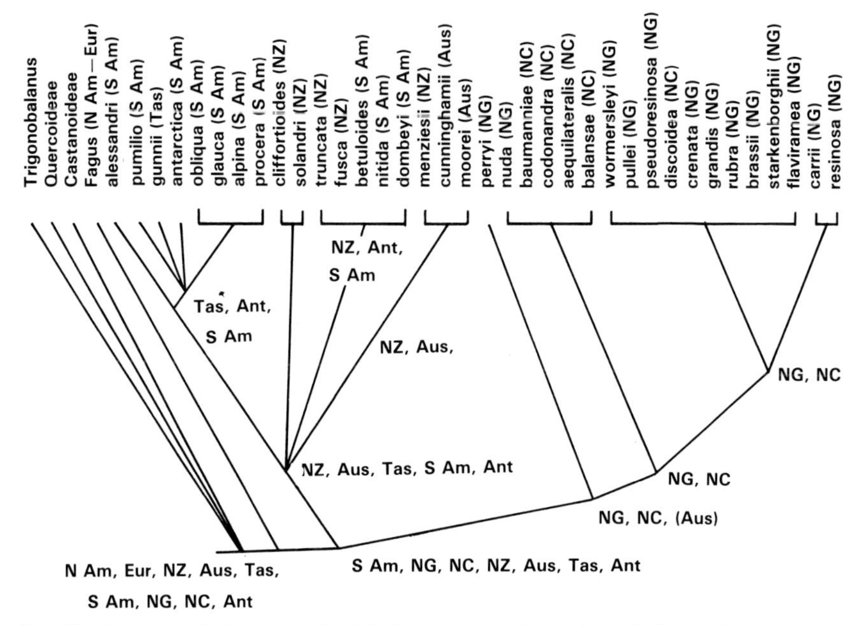

Fig. 8. An area cladogram for *Nothofagus* based on shared derived characters.

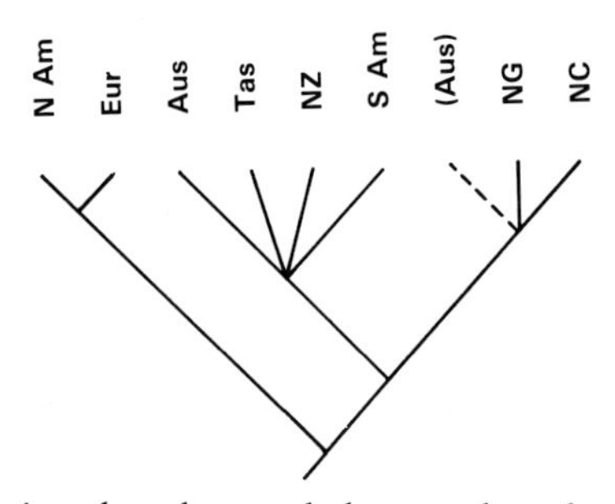

Fig. 9. A reduced area cladogram based on Fig. 8.

Cracraft phylogeny (Fig. 10). Nevertheless, fossils do give the minimum age for some of the area relationships. The Australia–South America–New Zealand pattern is at least as early as the late Cretaceous and, if the new cladogram (Fig. 8) is correct, that pattern must be predated by the New Guinea/New Caledonia–Australia/South America/New Zealand/Tasmania/Antarctica dichotomy. The Northern Hemisphere–Gondwanaland dichotomy for the *Fagus-Nothofagus* split has no fossil record, but it must predate the Australia/South America/New Zealand split. In other words, the beeches relate northern and

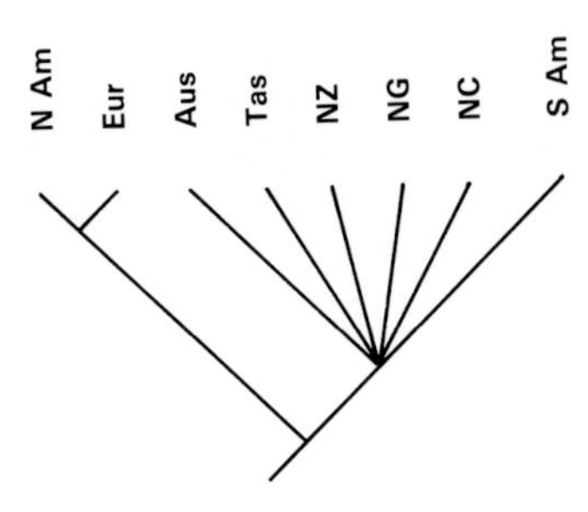

Fig. 10. A reduced area cladogram based on Melville's and Cracraft's cladogram (from Fig. 2).

southern areas and, within the vicariance framework, *Fagus* and *Nothofagus* probably diverged at the break-up of Pangaea. This means that Patterson's comment (P. [MS4]) is nearer the truth; because of the age of endemic taxa in contemporary *Nothofagus* this genus, as a whole, is uninformative on the history of the breakup of Gondwanaland.

It was mentioned earlier that historical biogeography is more concerned with the relationships of areas than with the history of individual taxonomic groups. By using area cladograms selected at random from other taxonomic groups (Fig. 11) with endemic representatives in at least three of the areas occupied by *Nothofagus*, it should be possible to provide a more fully resolved hypothesis of the general southern area pattern. For areas indicated in Figs 8, 9, 10, 11 and 12, the cladograms in Fig. 11 (L, M, N and O) give five 5-area patterns. Figure 11 (A, E, G, H, I, J, K, P) also gives nine 4-area patterns and (B, D) four 3-area patterns of the unresolved *Fagus/Nothofagus* pattern. Fig. 11 (C, F), although resolved for four areas, remains uninformative for the *Fagus/Nothofagus* pattern since the genera occur only in two areas covered. From these cladograms, it does seem certain that area relationships are generally similar in all but a few situations regarding the areas of interest. Exceptions include, for example, the last dichotomy in the genus *Leptocarpus* (Restionaceae, Fig. 11C). Since this genus is the only one to occur in Malaysia, a parsimonious conclusion demands that, because all other groups show austral connections, the crossing of Wallace's line must be a unique, perhaps stochastic, phenomenon (such as a dispersal event) rather than a case of subsequent extinction in Malaysia for *Nothofagus* and the twenty other groups. A somewhat similar situation is found in Fig. 11 A, presenting a reduced area cladogram for *Drapetes* in the southern hemisphere. The genus

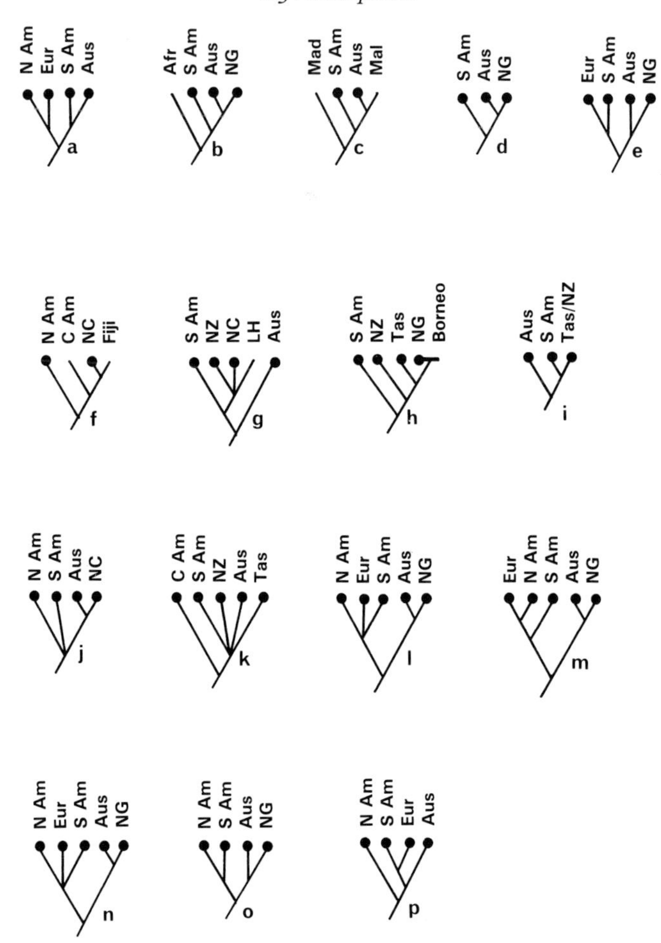

Fig. 11. Reduced area cladograms for randomly selected groups. (●) Shows areas in which endemic *Nothofagus* and *Fagus* species occur. (a) Chironomid midges (based on Brundin, 1966); (b) (i) Proteaceae tribe Gevuineae, (ii) Proteaceae tribe Macadamiae. ((i) and (ii) based on Johnson and Briggs, 1975 and redrawn from Patterson, 1981); (c) *Leptocarpus* in Malaysia, Australia and South America and *Calopsis* in Madagascar (after Cutler, 1972). (d) Osteoglossine fishes (*Osteoglossum* and *Scleropagus*) omitting south-east Asia, the sister area of Australia and New Guinea (after Gaffney, 1977 and redrawn from Patterson, 1981); (e) Higher ratite birds omitting Africa as the sister area of Europe (from Cracraft, 1972, 1973 and redrawn from Patterson, 1981); (f) *Lindenia* (Rubiaceae) (after Darwin, 1976 and Seeman, 1862); (g) a combined cladogram for *Negria*, *Depanthus*, *Rhabobthamnus*, *Corathera* and *Fielda* (Gesneriaceae); (h) *Drapetes* (Thumeleaceae) (based

consists of four specis, with *D. ericoides* restricted to New Guinea and Borneo. It is more reasonable to suggest as a parsimonious solution a dispersal event from New Guinea to Borneo rather than extinction of all the other groups.

The cladograms (except Fig. 12C) all have in common the Australia–New Guinea connection. Although New Guinea is largely of very recent origin, the southern part of the island has had two historical connections with Australia during its geological history—one in the Pleistocene and the other Jurassic. From the *Nothofagus* cladogram (Fig. 8) and fossil evidence it would appear that the austral groups are at least as old as the Cretaceous and that the New Guinea connection therefore belongs to the first of these. On the other hand, since the Australia–New Guinea dichotomies are the most derived in the cladograms of Fig. 11, it is likely in these groups to represent the more recent connection.

The only cladograms to show a pattern similar to the *Fagus/Nothofagus* division are those of the chironimid midges and the Ratite birds (Fig. 11A,E). The three 4-area cladograms for the two tribes of the Proteaceae and *Leptocarpus/Calopsis* (Fig. 11B,C) do show the same general sequence, but with southern African or Madagascan rather than northern hemisphere connections for the first dichotomy; this is a general area pattern which agrees more with the break-up sequence of Gondwanaland.

It becomes more obvious from looking at cladograms in Fig. 11 that the information in them might amount to a more general biogeographical pattern. Humphries (1980) reported that there might be three separate patterns showing serious incongruencies (Fig. 12A,B,C),

---

on data of Moore, 1968 and Ding Hou, 1960); (i) (i) *Phyllacne*, (ii) *Donatia* (Stylidiaceae) (based on van Balgooy, 1975 and Allen, 1961); (j) *Nicotiana* (Solanaceae) (based on the monograph of Goodspeed, 1954); (k) Combined data set for *Lagenifera* and *Solenogyne* (Compositae) (after Drury, 1974); (l) Galliform birds (from Cracraft, 1973 as redrawn from Patterson, 1980); (m) (i) Hylid frogs (redrawn from Patterson, 1980), and (ii) Chaleosyrphus group of xylotine syrphid flies (Syrphidae) (after Hippa, 1978 and redrawn from Patterson, 1981); (n) Recent and fossil groups of Marsupials (from Patterson, 1981); (o) (i) Recent marsupials (from Patterson, 1981), (ii) Hydrobiosine caddisflies (after Ross, 1956 and redrawn from Patterson, 1981), and (iii) *Anisotarsus* (Carabidae) (after Noonan, 1973 and redrawn from Patterson, 1981); (p) *Anisotarsus*.

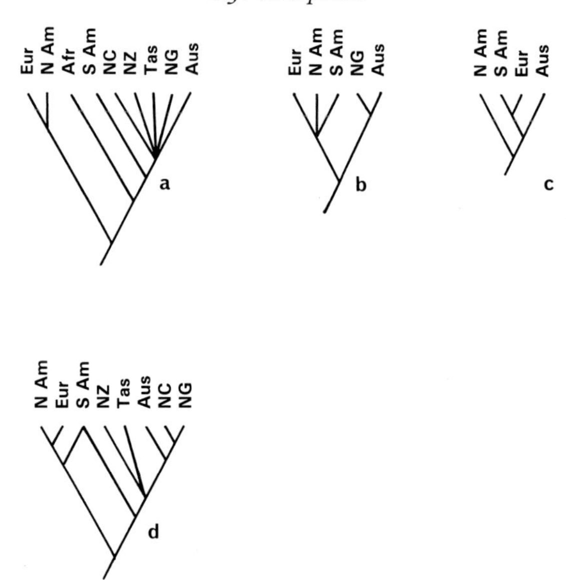

Fig. 12. General patterns of area relationships for the cladograms of Fig. 11. (a) General reduced area cladogram for the taxa included in Fig. 11a,b,c,d,f,g,h and k; (b) General reduced area cladogram for the taxa included in Fig. 11e,j,l,m,n and o; (c) Unique area cladogram for *Anisotarsus* in Fig. 11p; (d) The most informative cladogram for those areas in which *Nothofagus* species occur. See text for explanation.

particularly with respect to the position of South America. This view employs the principle that incongruent patterns represented unresolved sequences. By adopting the alternative method of considering only the resolved patterns, a cladogram of areas inhabited by *Nothofagus* can be drawn (Fig. 12D). Comparison of this cladogram with *Nothofagus* areas in geological cladograms (Figs 3–6) leads to the most parsimonious solution (Figs 4 and 6) demonstrating congruency for all areas except the position of New Caledonia. The cladogram from Rosen (1978) similarly shows incongruency for one area, South America, but must be considered less parsimonious for the whole area since it deals with only 5 of the eight possible areas. The least matching is found in Fig. 5, based on the maps of Owen and showing incongruency for two areas, South America and New Caledonia. The alternative positions in the general cladogram (Fig. 12D) of South America, showing recency of common ancestry with both the southern and northern hemisphere areas, relate to a rather special problem—the difficulty of treating huge continents as comparable areas of endemism. However, as Rosen (1978) pointed out, general patterns (in this case the relationships of continents) require general explanations and the best solution from combining all the data is given in Fig. 12D. The sequence of nested sets shows that Australia is most closely related to New Caledonia and New Guinea, than to New Zealand and Tasmania. North America forms a set with Europe, and South America shows a dual pattern with both northern and southern hemisphere groups.

CONCLUSION

It appears that Patterson's remark, quoted on p. 336, is justified. Previous claims that the distributional history of the southern beeches is in any way special (e.g. is a key to life in the far south) cannot be substantiated. What can be said is that the southern beeches do appear to form a wholly southern group whose precise "centre of origin", if there ever was one, is impossible to locate. Poor resolution in the reduced area cladograms for *Nothofagus* suggests that there were several species groups already extant at the break-up of Gondwanaland, although it does appear that continental drift created barriers in at least two of the modern vicariant species groups. I have tried to demonstrate, by using *Nothofagus* as an example, that narrative theories are dependent on many assumptions, all of which can be legitimately questioned and which must indeed change with time. Earlier narrative theories assumed a world of stable geography, or one in which present land masses were linked by land-bridges. These assumptions are not now fashionable. Later narrative theories involved continental drift, but, like their predecessors, were coupled with the assumption that animals and plants dispersed over large parts of the earth's surface. This leads to deciding from where, by what means, and by what routes, the particular groups of organisms travelled. Narrative theories are often based, as in the case of *Nothofagus*, on imprecise ideas of the interrelationships of organisms; indeed, it is not uncommon for the assumed biogeographic history to determine the scheme of relationships.

On the other hand, the analytical biogeographic approach attempts to determine, as completely as possible, the cladistic inter-relationships of any group of organisms occupying three or more areas. The distributional data in the form of area cladograms are compared with similar data from other groups. If there is a common history, caused in the case of the Southern areas by geological events, then there should be congruence between many different groups of organisms. Congruence between cladograms from different groups of organisms should have a low probability of being due to chance, whether the patterns be due to dispersal or to vicariance. Routes of dispersal and "centres of origin" are never assumptions in analytical methods, and fossils are never given any special importance. This approach is in its infancy and further changes will eventually be required. Cladistic relationships are known in so few groups that analytical historical biogeography thus far lacks precision. *Nothofagus* may not itself be the key to an understanding of the history of southern continents, but any general historical theory of the flora and fauna of the southern hemisphere must explain the present-day distribution of *Nothofagus*. By comparing *Nothofagus* patterns with others, the preliminary results show a more general pattern.

ACKNOWLEDGEMENTS

Sincerest thanks to Professor T.C. Chambers for providing facilities at the Botany School, University of Melbourne, where much of this work was carried out whilst in tenure of a Melbourne University Fellowship. Thanks also to D. Frodin, P. Forey, N.B. K. Robson and A.O. Chater who all gave useful comments on the manuscript, and to R. Press for drawing some of the figures. Finally thanks to Barbara Joyce and Marilyn Humphries for typing various drafts of this paper.

REFERENCES

Allan, H.H. (1961). "Flora of New Zealand", vol. 2. R.T. Owen, Wellington.

Ball, I.R. (1976). Nature and formulation of biogeographical hypotheses *Syst. Zool.* **24**, 407–430.

Ballance, P.F. (1976). Evolution of the Upper Cenozoic Magmatic arc and plate boundary in northern New Zealand. *Earth Plan. Sci. Lett.* **28**, 356–370.

Brundin, L. (1966). Transantarctic relationships and their significance as evidenced by Chironomid midges. *K. svenska VetenskAkad. Handl.*, **11** (1), 1–472.

Burbridge, N.T. (1960). The phytogeography of the Australian region. *Aust. J. Bot.* **8**, 75–209.

Couper, R.A. (1960). Southern hempisphere Mesozoic and Tertiary Podocarpaceae and Fagaceae and their palaegeographic significance. *Proc. R. Soc. B.* **152**, 491–500.

Cracraft, J. (1972). The relationships of the higher taxa of birds: problems in phylogenetic reasoning. *Condor* **74**, 379–392.

Cracraft, J. (1973). Continental drift, paleoclimatology, and the evolution and biogeography of birds. *J. Zool, Lond.* **169**, 455–545.

Cracraft, J. (1975). Historical biogeography and earth history. Perspectives for a future synthesis. *Ann. Mo. bot. Gdn.* **62**, 277–250.

Cranwell, L.M. (1963). *Nothofagus*: living and fossil. *In* "Pacific Basin Biogeography Symposium" (Gressitt, J.K., ed.), pp. 387–400. Bishop Museum Press, Hawaii.

Craw, R.C. (1978). Two biogeographical frameworks: implications for the biogeography of New Zealand, a review. *Tuatara* **23**, 81–114.

Croizat, L. (1952). "Manual of phytogeography: or an account of Plant Dispersal throughout the world". W. Junk, The Hague.

Croizat, L., Nelson, G. and Rosen, D.E. (1974). Centres of origin and related concepts. *Syst. Zool.* **23**, 265–287.

Cutler, D. (1972). Vicarious species of Restionaceae in Africa, Australia and South America. *In* "Taxonomy, Phytogeography and Evolution" (D.H. Valentine, ed.), pp. 73–83. Academic Press, London and New York.

Darlington, P.J. Jr. (1965). "Biogeography of the Southern end of the World". Harvard University Press, Cambridge, Mass.

Darwin, C. (1859). "On the Origin of Species". J. Murray, London.

Darwin, S.P. (1976). The Genus *Lindenia* (Rubiaceae) *J. Arnold Arbor.* **57**, 426–449.

Dietz, R.S. and Sproll, W.P. (1970). Fit between Africa and Antarctica: a continental drift reconstruction. *Science, N.Y.* **167**, 1612–1614.

Ding Hou (1960). Thymelaeaceae. *Drapetes Fl. Males.* **1** (6), 43–44.

Drury, D.G. (1974). A broadly based taxonomy of *Lagenifera* section Lagenifera and Solenogyne (Compositae–Astereae) with an account of the species in New Zealand. *N.Z. Jl. Bot.* **12**, 365–396.

Engler, A. (1879–1882). "Versuch einer Entwicklungsgeschiete der Pflanzenwelt . . .". Engelmann, Leipzig.

Florin, R. (1940). The tertiary fossil conifers of south Chile and their phytogeographical significance. *K. svenska VetenskAkad. Handl.* **14** (2), 1–107.

Forman, L.L. (1966). On the evolution of cupules in the Fagaceae. *Kew Bull.* **18**, 385–419.

Gaffney, E.S. (1977). The side-necked family Chelidae: a theory of relationships using shared derived characters. *Am. Mus. Novit.* **2620**, 1–28.

Goodspeed, T.J. (1954). The genus *Nicotiana Chronica Bot.* **16**, 1–536.

Hanks, S.L. and Fairbrothers, D.T. (1976). Palynotaxonomic investigation of *Fagus* L. & *Nothofagus* Bl.: light microscopy, scanning electron microscopy and computer analyses. *In* "Botanical Systematics" (V.H. Heywood, ed.), vol. 1, pp. 1–141. Academic Press, London and New York.

Hennig, W. (1965). Phylogenetic systematics. *A. Rev. Ent.* **10**, 97–116.

Hennig, W. (1966). "Phylogenetic systematics". University of Illinois Press, Urbana, Chicago and London.

Hippa, H. (1978). Classification of Xylotini (Diptera, Syrphidae). *Acta zool. fenn.* **156**, 1–153.

Hooker, J.D. (1853). Introductory essay. *In* "The Botany of the Antarctic voyage of H.M. Discovery Ships Erebus and Terror in the years 1853–55. II Flora Nova Zelandiae", pp. i–xxxix. Reeve, London.

Humphries, C.J. (1981). Biogeographical methods and the southern beeches (Fagaceae: *Nothofagus*). *In* "Advances in Cladistics: Proceedings of the First Meeting of the Hennigian Society, New York Botanical Garden". (V.A. Funk and D.R. Brooks, eds).

Johnson, L.A.S. and Briggs, B.G. (1975). On the Proteaceae—the evolution and classification of a southern family. *Bot. J. Linn. Soc.* **70**, 83–182.

Lozano-C., G., J.I., Hernandez-C. and J.E. Henao-S. (1980). El genero *Trigonobalanus* Forman, en el Neotropico-I. *Caldasia* **12**, 517–537.

Matthew, W.D. (1915). Climate and evolution. *Ann. N.Y. Acad. Sci.* **24**, 171–318.

McQueen, D.R. (1976). The ecology of *Nothofagus* and associated vegetation in South America. *Tuatara* **22**, 38–68.

Melville, R. (1973). Continental Drift and Plant Distribution. *In* "Implications of Continental Drift to the Earth Sciences" (D.H. Tarling and S. Runcorn, eds), vol. 1, Academic Press, London and New York.

Moore, D.M. (1971). Connections between cool temperate floras with particu- *Antarct.* **60**, 1–202.

Moore, D.M. (1971). Connections between cool temperate floras with particular reference to southern South America. *In* "Taxonomy, Phytogeography and Evolution" (D.H. Valentine, ed.), Academic Press, New York and London.

Nelson, G.J. (1975). (1975). Historical biogeography: an alternative formalization. *Syst. Zool.* **23**, 555–558.

Noonan, G.R. (1973). The anisodactylines (Insecta: Coleoptra: Carabidae: Harpalini): classification, evolution and zoogeography. *Quaest. ent.* **9**, 266–480.

Oliver, W.R.B. (1925). Biogeographical relations of the New Zealand region. *Bot. J. Linn. Soc.* **47**, 99–139.

Owen, H.G. (1976). Continental Displacement and expansion of the earth during the mesozoic and Cenozoic. *Phil. Trans. Roy. Soc.* B. **281**, 223–291.

Patterson, C. (1981). Methods of paleobiogeography. *In* "Vicariance Biogeography: a critique" (D.E. Rosen and G. Nelson, eds), pp. 446–500. Columbia University Press, New York.

Platnick, N.I. and Nelson, G. (1978). A method of analysis for historical biogeography. *Syst. Zool.* **27**, 1–16.

Raven, P.H. and Axelrod, D.I (1972). Plate tectonics and Australasian paleobiogeography. *Science, N.Y.* **176**, 1379–1386.

Raven, P.H. and Axelrod, D.I. (1974). Angiosperm biogeography and past continental movements. *Ann. Mo. bot. Gdn.* **61**, 539–673.

Rich, P.V. (1975). Antarctic dispersal routes, wandering continents and the origin of Australia's non-passerine avifauna. *Mem. Nat. Mus.* **36**, 63–126.

Ross, H.H. (1956). "Evolution and classification of the Mountain Caddisflies". University of Illinois Press, Urbana.

Rosen, D.E. (1978). Vicariant Patterns and Historical Explanation in Biogeography. *Syst. Zool.* **27**, 159–188.

Schröter, C. (1913). Genetishe Pflanzengeographie (Epiontologie). *In* "Handworterbuch der Naturwissenschaften" (E. Korschett *et al.*, eds), pp. 907–942, vol. 4. G. Fisher, Jena.

Schuster, R.M. (1976). Plate tectonics and its bearing on the geographical origin and dispersal of angiosperms. *In* "Origin and early evolution of Angiosperms" (C.B. Beck, ed.), pp. 48–138. Columbia Univ. Press, New York, London.

Seeman, B. (1862). *Lindenia vitiensis. Bonplandia* **10**, 33–34.

Shields, O. (1979). Evidence for initial opening of the Pacific ocean in the Jurassic. *Palaeogeogr., Paleoclimat., Palaeoceol.* **26**, 181–220.

Smith, A.G., Briden, J.C. and Drewry, G.E. (1973). Phanerozoic world maps. *In* "Organisms and continents through time" (N.F. Hughes, ed.), pp. 1–42. Palaeontological Association, London.

Soepadmo, E. (1972). Fagaceae. *Fl. Males.*, ser. 1, **7** (2), 277–294.

Thistelton-Dyer, W. (1909). Geographical Distribution of Plants. *In* "Darwin and Modern Science" (A.C. Seward, ed.), pp. 298–316. University Press, Cambridge.

van Balgooy, M.M.J. (1975). "Pacific Plant Areas" ld. vol. 3. Rijksberbarium, Leiden.

van Steenis, C.G.G.J. (1953). Papuan *Nothofagus. J. Arnold Abor.* **34**, 301–373.

van Steenis, C.G.G.J. (1962). The land-bridge theory in Botany with particular reference to tropical plants. *Blumea* **11**, 235–542.

van Steenis, C.G.G.J. (1971). *Nothofagus*, key genus of plant geography in time and space, living and fossil, ecology and phylogeny. *Blumea* **19**, 65–98.

Wallace, A.R. (1876). "The geographical distribution of animals, with a study of the relations of living and extinct faunas as elucidating the past changes of the earth's surface". 2 vols. Hafner Publishing, New York. [Reprinted 1962.]

Wulff, E.V. (1943). "An introduction to Historical Plant Geography". Chronica Botanica, Waltham, Mass.

13 In the Steps of Alfred Russel Wallace: Biogeography of the Asian–Australian Interchange Zone

J. ALLEN KEAST

Department of Biology, Queen's University, Kingston, Ontario K7L 3N6, Canada

Abstract: Continental drift data explains the striking biotic differences between the Asian and Australian region. Wallace's Line, long regarded as one of the world's most puzzling biogeographic features, can now be seen as an expression of the late arrival of Australia in its present position, the plate tectonics of eastern Indonesia, Pleistocene sea level changes, and climates, past and present. It is, or is not, a very major biological division depending on the ecological attributes of the particular group, particularly capacity for dispersal. An updated review of the avifaunas of the major sub-regions and islands, bird biogeography, and speciation patterns, is provided.

THE MAJOR BIOGEOGRAPHIC SUBREGIONS OF THE AREA

The Australian and Oriental Regions are two of the great biogeographic regions of the globe. Differences in the biotas are profound (Hooker, 1860; Good, 1964; Schuster, 1972, 1976 and others). Australia supports austral gymnosperms (Podocarpaceae, Araucariaceae), *Nothofagus*, Restionaceae, Proteaceae, Myrtaceae, and many others with southern (especially South American) relationships, some 12 endemic plant families including primitive and archaic ones, and a great diversity of endemic genera (540, plus another 100 that probably evolved in Australia and migrated out; see Beadle, 1981; Specht, 1981). The Australian flora was long thought to consist of three elements; southern,

Systematics Association Special Volume No. 23, "Evolution, Time and Space: The Emergence of the Biosphere", edited by R.W. Sims, J.H. Price and P.E.S. Whalley, 1983, pp. 367–407. Academic Press, London and New York.

endemic and Malesian, the last named representing late intruders (mostly rainforest from the northwest). Now most Australian rainforest elements are seen to be locally derived (Specht, 1981), although Malesian elements dominate the New Guinea rainforests (Axelrod and Raven, 1981). Faunal differences between the Australian and Oriental Regions are likewise profound (Darlington, 1980), especially in mammals, Amphibia and freshwater fish. The Asian mammal fauna is composed of "modern" placentals and is highly diversified. In Australia, marsupials and egg-laying monotremes are dominant (Archer, 1981). Differences in the amphibians are as great as in mammals, Australia again sharing relationships with South America (Tyler, 1979). The freshwater fish faunas are very distinct (McDowall, 1981). Here too Asia has a highly diversified modern fauna; that of Australia (except for a couple of archaic forms) is of marine derivation. The reptile and bird faunas of the two regions likewise show fundamental differences (Cogger and Heatwole, 1981, Keast, 1981).

The "new" geological data confirm that some of these basic differences are ancient. Many of them extend back to the time of Pangaea, semi-divided into northern and southern regions and, subsequently, to the Mesozoic division of the world into northern Laurasian and southern Gondwanan supercontinents. Late Paleozoic floras and insect faunas of Australia were moulded by Permo-Carboniferous glaciation and were distinctively Gondwanan. By contrast, as far south as Sumatra, the Asian floras of this age were members of a characteristic tropical Cathaysian assemblage (Jongmans and Gothan, 1935).

A uniquely Australian, as distinct from generalized Gondwanan, identity stems from the progressive breakup of Gondwanaland throughout the Cretaceous and into the Tertiary. The latest dates for the phases of this breakup (summary in Keast, 1981, drawn mainly from data summarized in Hallam, 1981) are (all in millions of years B.P.): isolation of India and the temperate regions of southern South America and Africa, 126; isolation of Laurasia from Gondwana and Africa from Antarctica—end of the early Cretaceous, 156; isolation of New Zealand, 78; final isolation of South America and west Africa, 78–65; separation of Australia from Antarctica, about 50; separation of South America from Antarctica, 22·5–50·0; collision between India and Asia, 49; approach by Australia to its present position adjacent to Asia, 10. Thus, the distinctness of the Australian biota and the basic differences from that of Asia stem from the relatively late approach of the Australian segment of the Indian–Australian plate. Some important aspects of Cretaceous–Early Tertiary biogeography remain to be elucidated. These include the position of India relative to Australia within the Gondwanaland mass; the position and early history of the Ninetyeast Ridge in the Indian Ocean (with its Paleogene pollen of typically Australian plants; Kemp and Harris, 1975); the nature of the assumed austral elements transported north by India (Raven and Axelrod, 1974). One of the most intriguing questions is the entry pathway of distinctive Australian subtropical rainforest plant elements. These would seem to have required a temperate pathway across the southern Indian Ocean, rather than the cold temperate one from South America via Antarctica (Raven and Axelrod, 1974).

Opportunities for biotic interchanges arose during the Pleistocene when Indonesia, like the rest of the world, experienced major falls in sea level with each major glacial period in the north. A fall in sea level of about 130 m was fairly general during the maximum glaciations. (There is, however, a record of a terrace at −200 m put at an age equal to or greater than 170000 yBP, from the Arafura Sea (Jongsma, 1970).) There was also regional variation within the Indonesian region, associated with continuing tectonic events. The extent of dry land during these periods of lowered sea-level is indicated in Fig. 1, showing the 2000 m contour. At this time Sumatra, Java, Bali and Borneo were continuous with Asia; land extended northwards towards the Philippines whilst the Makassar Strait was narrowed to half its present width. There was a junction between Sumbawa and Flores; extension of land around Sulawesi in addition to a considerable extension of the Sahul Shelf northwestwards from Australia towards Timor. Colonization routes, such as those suggested by Birdsell (1977) for aboriginal man, would have been available (see inset in Fig. 1). Climatic, and hence vegetational, changes have also been a major factor in the complex distribution patterns of plants and vertebrates in modern Indonesia (see also, Whitmore, 1981; Granbrook, 1981).

CONTEMPORARY BIOTAS WITHIN THE ASIAN–AUSTRALIAN INTERCHANGE ZONE

The Indo–Australian Region can best be considered in terms of four components (Fig. 1): (a) Asian mainland; (b) Sundaland (Malaysia and the Greater Sunda Ialands), the area of land lying on the Sunda Shelf; (c)

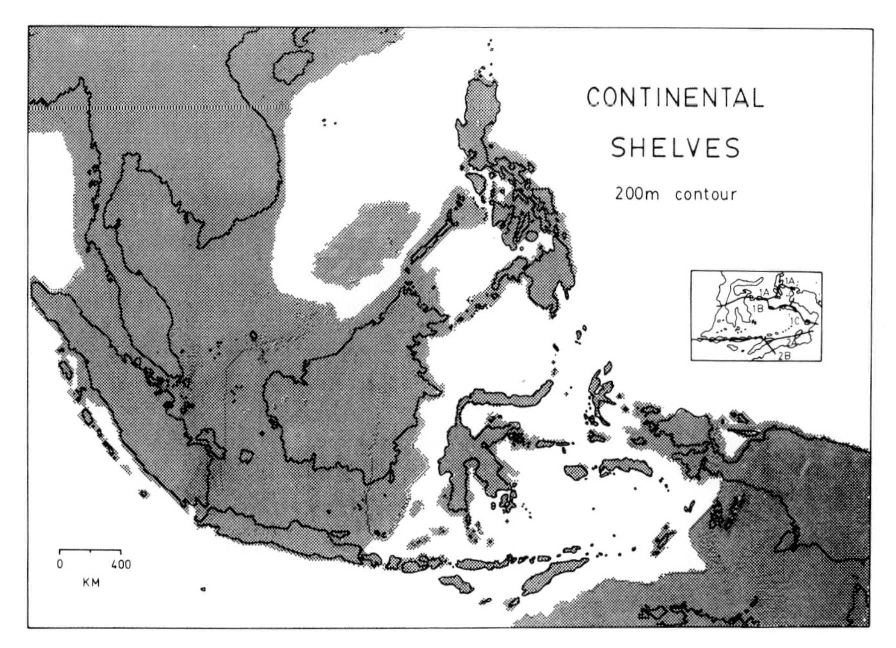

Fig. 1. Continental shelves, southeast Asia and Australia, 200 m contour (data taken from Times Atlas and other sources). Inset: possible colonization routes at times of lower sea-level, taken from Birdsell (1977).

Australo-Papua; (d) the intermediate zone between (b) and (c). This latter, embracing Sulawesi, the Lesser Sunda Islands, and the island zone between Sulawesi and New Guinea, is insular and biotically impoverished. The name Wallacea has been given to it (Dickerson *et al.*, 1928; Rensch, 1936); this name will be used here as a general term, but not, as extended by some authers, to include the Philippines.

Early recognition that (i) the main Oriental fauna extended as far east as Borneo and Java–Bali and (ii) the Australo–Papua fauna does not extend far beyond the limits of the Australian continental shelf, led to an early focus on faunal barriers. Pre-eminent amongst these latter was Wallace's Line and no account of Indo–Australian biogeography can be complete without reference to its more intriguing aspects.

Wallace's Line, that in the eyes of many came to assume almost mythical importance, separates off the rich Oriental freshwater fish, amphibian, bird and mammal faunas from the impoverished ones of Sulawesi and the Lesser Sunda Islands (with their partial Australo–

Papuan affinities) to the east. The transition is sharp; the water barrier is only 15 miles wide (between Bali and Lombok) to 28 miles wide (between Borneo and Sulawesi). Wallace (1860), in his classic essay *On the zoological geography of the Malay Archipelago*, after reviewing existing data highlighted the barrier with these remarks:

> The western and eastern islands of the archipelago belong to regions more distinct and contrasted than any other of the great zoological divisions of the globe. South America and Africa, separated by the Atlantic, do not differ so widely as Asia and Australia.

further:

> We may consider it established that the Strait of Lombok (separating Bali and Lombok) marks the limit and abruptly separates two of the great zoological regions of the globe.

Wallace did not name the line; Huxley (1868) did that.

Wallace's Line found wide acceptance amongst 19th and 20th century biologists (Mayr, 1944a), particularly those who had worked on both sides of it. Enthusiastic advocates included Haeckel (1893), who wrote: "Crossing the narrow but deep Lombok Strait we go with a single step from the Present Era to the Mesozoicum." Years later, Wallace found a further defender in Rensch (1930) who, after field studies in Java–Bali and Lombok, wrote concerning the avifaunas on either side of Lombok Strait, provided his now much quoted impressions:

> And the difference is indeed quite extraordinary! Much more conspicuous than I would have ever imagined. As soon as I entered the woods on a small native trail a whole chorus of strange bird songs greets me—in fact, among the real songsters there is not a single one with which I was familiar. One surprise follows the other. The very species that are most common on Bali, are absent on the islands to the east. The most characteristic bird of these woods is a green barbet . . . it belongs to the family Capitonidae which is entirely absent on Lombok! The woodpeckers also, which are represented on the islands farther east by a single species only are found on Bali in five different species. On the other hand I missed a whole number of species of birds which are characteristic for the islands visited previously.
>
> (Rensch, 1930, quoted by Mayr, 1944a).

The euphoria of such advocates has not been uniformly shared. As a biogeographic feature Wallace's Line is only of modest significance to plants and insects. Even some early workers, including ornithologists, suggested it to be entirely imaginary: see the summary in Meyer and

Wiglesworth (1898). After noting that numbers of protagonists and antagonists of Wallace's Line were about equal, these authors suggested that many writers had not independently assessed the evidence but had simply followed the acknowledged authorities:

> There can be no doubt that in our present state of knowledge it is premature to define the problem for solution, however interesting and suggestive it may be, and that it is, therefore, a waste of time to speculate on it with the help of an up-and-down system for the islands and continents, just as required. It is characteristic of an inadequate hypothesis that it is always in need of a new one which should sustain it, and as geology and palaeontology are as yet powerless to guide us, we must restrict ourselves to zoology, though we know that here also our knowledge is defective in a high degree.

For earlier lists of the Indonesian biogeographical literature relative to Wallace's Line, see Sarasin (1901), Pelseneer (1904), and Meyer and Wiglesworth (1898); later relevant authors included Mayr (1944a), Simpson (1977), Whitmore (1981), and many others.

Much of this early discussion and argument on the biogeography of Indonesia stemmed from a desire to find the "ultimate line" separating the Asian and Australian biotas. Conflicting evidence was provided by different groups. Simpson (1977) provided a recent assessment of the various "biogeographic lines" advanced, and the bases on which they were described. Amongst them were: (1) a modification to the original Wallace's Line by Huxley; (2) that of the ornithologist Sclater whose line of separation lay east of Sulawesi; (3) Lydekker's Line (representing a boundary between the Australian Region and an Austro–Malayan Region between the former and the Oriental Region); (4) that of Weber (Fig. 1). Weber's Line, like that of Wallace, has been subject to modification by later workers (see Simpson, 1977) and actually represents the line of faunal balance between Oriental and Asian elements (Mayr, 1944a). Mayr (*op. cit.*) regards Lydekker's Line as equivalent to the limits of the Australo–Papuan mainland fauna (Fig. 1).

Wallace's Line, the real significance of which is to demarcate a rich and diversified mainland Asian biota from a impoverished insular one, is of great significance in Amphibia, primary division freshwater fish (i.e. those intolerant of salty or brackish water—see Myers, 1951; Darlington, 1980), mammals (Raven, 1935), and birds. It is, however, only of modest or limited importance as a division line in plants (Balgooy, 1971; van Steenis, 1979) and insects (Gressitt, 1961). In these groups, the really

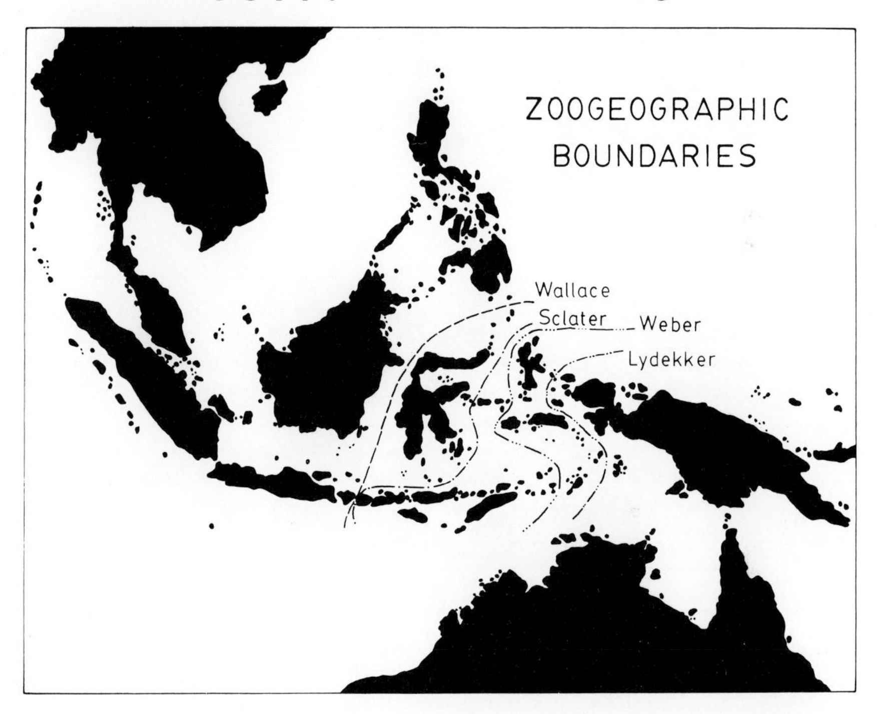

Fig. 2. Zoogeographic boundaries that have been postulated for the Indonesian region. Redrawn from Mayr (1944a), and Simpson (1977). (See also recent discussions in Whitmore, 1981.)

profound separation is of Australian elements from those of New Guinea and Malesia. Makassar Strait, whilst significant, is a lesser barrier (see Figs 2–3).

REPRESENTATION OF AVIAN TAXONOMIC GROUPS

Major studies of Asian–Australian biogeography include Mayr (1944a), Scrivenor *et al.* (1943) and Holloway and Jardine (1968). Various other workers have commented on wider aspects in the course of analysing specific groups, e.g. Medway (1972), Groves (1975), Goodwin (1979) and Auffenberg (1980). Other recent writers have suggested interpretations relative to the newer concept of plate tectonics (Raven and Axelrod, 1974; Raven, 1979; Audley-Charles, 1981; Whitmore, 1981).

The major work on Indonesian bird geography remains that of Mayr (1944a), and it is thanks to his astute analysis that we enjoy a reasonably

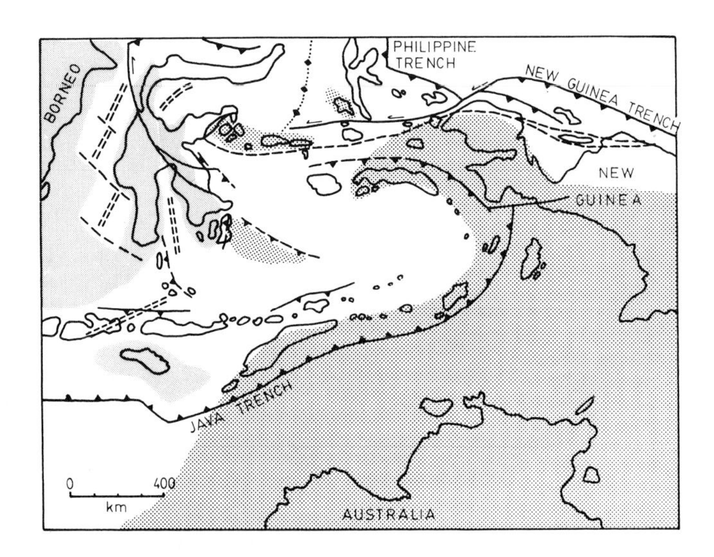

Fig. 3. Recent tectonics map for eastern Indonesia (adapted from Hamilton (1979:Fig. 53). A major subduction trench extends from west to east through the region immediately to the south of Sumatra, Java and Timor, then loops north and west around the Banda Sea, representing the dipping of the edge of the Australian–New Guinea section of the Indo–Australian plate following its movement up from the south. The Pacific plate is encroaching on the region from the east. Lesser trenches and subjuction zones occur within the region, especially relative to Sulawesi. Darkened areas represent continental crust, that of Australia–New Guinea being of Precambrian and Paleozoic age, that of Borneo and Sulawesi Cretaceous and Paleogene. Sections to the east of Sulawesi are of Precambrian or Paleozoic age and represent detached portions of New Guinea rafted westwards along the Sorong Fault. Sumba is exceptional amongst the Lesser Sunda Islands in being on continental strata; since this is of Cretaceous–Paleogene age an Asian, rather than Australian, origin is suggested (vide by Hamilton, 1979). Timor, Ceram, Tenimbar, and associated islands are composed of melange, scraped up from the sea floor. The inner arc of the Lesser Sundas (Flores, etc), are volcanic. Sea floor spreading is occurring between Borneo and Sulawesi, and between the arms of Sulawesi. Makassar Strait is variously regarded as having originated in the Miocene (Hamilton, 1979), or Pleistocene (Katili, 1978). Audley-Charles (1981), however, suggests that western Sulawezi may have been intermittently linked with Borneo by land during the Cenozoic and Quaternary.

balanced perspective. Mayr's conclusions were derived from a quantitative analysis of the avifaunas of the various islands. Tables I–III show the avian families present and numbers of genera and species for the major islands and segments between Malaya and New Guinea–Australia. The islands are separated into a northern tier (Table II) and a southern tier (Table III). Figs 4–7 show in diagrammatic form the changeover in 11 different avian families, including some equally represented in both the Oriental and Australian regions and others that are either entirely or predominantly Asian or Australian.

The data have been built up from various sources as follows: Malaya, Sumatra, Java, and Bali (Chasen, 1936, with some modifications from King *et al.*, 1975, and other sources); Borneo (Smythies, 1960); Sulawesi (Streseman 1939, 1940, 1941); the Moluccas (Bemmel, 1948); Lombok, Sumbawa, and Flores (Rensch, 1931); Timor and Sumba (Mayr, 1944b); New Guinea (Mayr, 1941, Rand and Gilliard, 1968); Australia (R.A.O.U. Checklists; Pizzey, 1980; and other sources). At the eastern end of the island chain, divisions of "Northwest Australia", "Eastern Australia", Vogelkop, and "Eastern New Guinea" are introduced to bring out the richness of the Australian Region in its areas of greatest diversity. Thus, whilst "Northwestern Austrialia" is dry and the Vogelkop small and semi-isolated, "Eastern New Guinea" and "Eastern Australia" are highly diversified environments.

The analysis is restricted to breeding land and freshwater birds. There is a minor degree of arbitrariness to the counts since it is uncertain how many of the species of water-birds, and others that disperse northwards from Australia, breed in the islands. The Palearctic Region contributes a great many species of wintering ducks, hawks, flycatchers, warblers and others, to Southeast Asia, Sundaland, and Wallacea. A few have established breeding races. The validity of excluding migrants from such faunal analyses as the present has recently been questioned in the case of central America (Keast, 1980), many of the migrants spending seven months in their wintering grounds and maintaining niches there that remain vacant after they leave.

Again, because the basic regional lists from the different segments and islands were compiled at different times (Flores and Lombok in 1931; Timor in 1944; Borneo in 1961) there are necessarily differences in the state of the taxonomies. To bring all the lists completely into line would entail a complete taxonomic revision of the avifauna of the whole area.

Table I. Avifauna composition, islands between Malay Peninsular and New Guinea

	Malay Peninsula		*Borneo*		*Sulawesi*		*Halmahera*		*Buru*		*Ceram*		*Vogelkop*		*Eastern New Guinea*	
Dromaiidae			—	—	—	—	—	—	—	—	—	—	—	—	—	—
Casuariidae	—	—	—	—	—	—	—	—	—	—	—	—	1	3	1	3
Podicipedidae	1	1	1	1	1	1	1	1	1	1	1	1	1	1	1	1
Pelecanidae	1	1	—	—	—	—	—	—	—	—	—	—	—	—	—	—
Phalacrocoracidae	1	1	1	3	1	1	1	2	1	2	1	2	1	1	1	3
Anhingidae	1	1	1	1	1	1	—	—	1	1	—	—	—	—	1	1
Ardeidae	8	13	6	8	10	11	7	7	7	7	7	8	6	6	9	9
Ciconiidae	1	1	2	2	1	1	—	—	—	—	—	—	—	—	1	1
Threskiornithidae	—	—			—	—	—	—	—	—	—	—	1	1	1	1
Anatidae	4	4	1	1	3	3	2	2	4	4	3	3	3	4	6	8
Pandionidae	1	1	1	1	1	1	1	1	1	1	1	1	1	1	1	1
Accipitridae			11	15	14	17	7	11	6	8	5	7	8	12	12	17
Falconidae			2	3	1	3	1	2	1	2	1	2	1	1	2	5
Megapodiidae	—	—	1	1	1	1	2	2	2	2	2	2	3	4	3	4
Phasianidae	13	16	13	14	2	2	1	1	1	1	1	1	—	—	2	2
Turnicidae	1	1	—	—	1	2	—	—	—	—	—	—	—	—	1	1
Pedionomidae	—	—	—	—	—	—	—	—	—	—	—	—	—	—	—	—
Rallidae	6	12	7	9	8	11	5	5	5	5	5	5	10	11	11	12
Cruidae	1	1	—	—	—	—	—	—	—	—	—	—	—	—	1	1
Otididae					—	—	—	—	—	—	—	—	—	—	1	1
Jacanidae	2	2	2	2	1	1	1	1	1	1	—	—	—	—	1	1
Burhinidae													1	1	1	1
Rostratulidae	1	1	1	1	—	—	—	—	—	—	—	—	—	—	—	—
Charadriidae	2	2	1	1	—	—	—	—	—	—	—	—	—	—	2	2

continued

Table I: continued (2)

	Malay Peninsula		*Borneo*		*Sulawesi*		*Halmahera*		*Buru*		*Ceram*		*Vogelkop*		*Eastern New Guinea*	
Recurvirostridae					—	—	—	—	—	—	—	—	—	—	1	1
Scolopacidae	2	2	1	1	—	—	—	—	—	—	—	—	1	1	1	1
Glareolidae	1	1	1	1	—	—	—	—	1	1	—	—	—	—	—	—
Columbidae	7	17	9	19	8	16	9	15	9	13	7	12	13	32	14	39
Psittacidae	3	4	4	5	4	6	8	8	11	11	9	11	20	29	19	34
Cuculidae	14	24	11	20	8	10	5	7	5	6	5	6	8	12	8	14
Strigidae	7	13	5	9	2	3	2	3	2	3	2	2	2	3	2	6
Tytonidae	2	2	1	1	1	2	—	—	—	—	1	1	1	1	1	4
Podargidae	1	3	1	6	—	—	—	—	—	—	—	—	1	2	1	2
Caprimulgidae	2	3	2	4	2	4	1	1	1	1	1	1	2	2	2	2
Aegothelidae	—	—	—	—	—	—	1	1	—	—	—	—	1	3	1	4
Apodidae	3	9	5	11	2	4	1	2	1	2	1	2	1	3	1	3
Hemiprocnidae					1	1	1	1	1	1	1	1	1	1	1	1
Trogonidae	1	4	1	6	—	—	—	—	—	—	—	—	—	—	—	—
Alcedinidae	5	10	5	9	5	9	4	10	4	5	4	6	8	13	10	17
Meropidae	2	4	2	2	2	2	—	—	—	—	—	—	—	—	1	1
Coraciidae	2	2	1	1	2	2	1	1	—	—	—	—	1	1	1	1
Bucerotidae	6	9	6	8	2	2	1	1	—	—	1	1	1	1	1	1
Capitopnidae	6	10	7	9	—	—	—	—	—	—	—	—	—	—	—	—
Picidae	14	24	13	18	21	21	—	—	—	—	—	—	—	—	—	—
Eurylaimidae	6	7	5	8	—	—	—	—	—	—	—	—	—	—	—	—
Pittidae	1	6	1	6	1	2	1	2	1	2	1	2	1	2	1	2
Menuridae	—	—	—	—	—	—	—	—	—	—	—	—	—	—	—	—
Atrichornithidae	—	—	—	—	—	—	—	—	—	—	—	—	—	—	—	—

continued

Table I: continued (3)

	Malay Peninsula		Borneo		Sulawesi		Halmahera		Buru		Ceram		Vogelkop		Eastern New Guinea	
Alaudidae	—	—	1	1	1	1	—	—	—	—	—	—	—	—	1	1
Hirundinidae	2	3	1	1	1	2	1	1	1	1	1	1	1	1	1	1
Motacillidae	3	3	1	1	1	1	—	—	—	—	—	—	—	—	1	2
Campephagidae	3	8	6	10	3	8	3	5	2	2	2	4	4	11	4	13
Chloropseidae	2	5	2	3	—	—	—	—	—	—	—	—	—	—	—	—
Pycnonotidae	14	31	4	23	—	—	1	1	1	1	1	1	—	—	—	—
Dicruridae	4	6	1	4	2	2	1	1	1	1	1	1	2	2	2	2
Oriolidae	1	4	1	4	1	1	1	1	1	1	1	1	1	1	2	4
Corvidae	3	5	5	7	2	2	1	2	1	1	1	1	2	2	2	2
Paridae	2	2	1	1	—	—	—	—	—	—	—	—	—	—	—	—
Sittidae	1	2	1	1	—	—	—	—	—	—	—	—	—	—	—	—
Timaliinae	22	46	16	34	4	4	—	—	—	—	—	—	1	2	3	5
Orthonychinae	—	—	—	—	—	—	—	—	—	—	—	—	3	5	3	5
Muscicapinae	18	28	8	27	9	14	3	8	5	10	5	9	12	34	12	43
Pachycephalinae	1	1	1	2	3	3	1	1	1	2	1	2	6	13	6	16
Falcunculinae	—	—	—	—	—	—	—	—	—	—	—	—	1	1	1	1
Turdinae	12	19	7	12	4	4	—	—	1	1	2	2	1	1	4	5
Sylviinae	10	17	6	11	5	6	2	2	5	5	3	3	2	2	4	5
Malurinae	—	—	—	—	—	—	—	—	—	—	—	—	4	5	4	5
Acanthizinae	—	—	—	—	—	—	—	—	—	—	—	—	3	12	4	18
Neosittidae	—	—	—	—	—	—	—	—	—	—	—	—	1	1	1	2
Climacteridae	—	—	—	—	—	—	—	—	—	—	—	—	1	1	1	1
Meliphagidae	—	—	—	—	—	—	—	—	—	—	—	—	16	32	19	48
Ephthianuridae	—	—	—	—	—	—	—	—	—	—	—	—	—	—	—	—

continued

Table 1: continued (4)

	Malay Peninsula		*Borneo*		*Sulawesi*		*Halmahera*		*Buru*		*Ceram*		*Vogelkop*		*Eastern New Guinea*	
Artamidae	—	—	1	1	1	2	1	1	1	1	1	1	1	2	1	2
Laniidae	4	7	1	1	2	2	1	1	—	—	1	1	—	—	1	1
Sturnidae	2	3	2	2	6	8	1	2	1	2	2	3	3	5	3	3
Nectariniidae	6	17	4	16	3	4	1	2	1	2	1	2	1	2	1	2
Dicaeidae	4	11	2	11	1	4	1	1	1	1	1	1	7	8	8	9
Pardalotidae	—	—	—	—	—	—	—	—	—	—	—	—	—	—	—	—
Zosteropidae	1	1	3	3	2	6	1	1	2	3	2	5	1	3	1	3
Ploceidae	4	8	2	4	2	6	2	3	1	1	2	2	2	3	3	11
Grallinidae	—	—	—	—	—	—	—	—	—	—	—	—	1	1	1	1
Corcoracidae	—	—	—	—	—	—	—	—	—	—	—	—	—	—	—	—
Cracticidae	—	—	—	—	—	—	—	—	—	—	—	—	1	2	2	4

Table II. Avifaunal Composition, genera and species, breeding land and freshwater birds Islands between Malay Peninsula and Australia

	Malay Peninsula		Sumatra		Java		Bali		Lombok		Sumbawa		Flores		Sumba		Timor		Northwest Australia		Eastern Australia	
Dromaiidae	—	—	—	—	—	—	—	—	—	—	—	—	—	—	—	—	—	—	1	1	1	1
Casuariidae	—	—	—	—	—	—	—	—	—	—	—	—	—	—	—	—	—	—	—	—	1	1
Podicipedidae	1	1	1	1	1	1	1	1	1	1	—	—	1	1	1	1	1	1	1	2	1	3
Pelecanidae	1	1	1	1	1	1	—	—	—	—	—	—	—	—	—	—	—	—	1	1	1	1
Phalacrocoracidae	1	2	1	2	1	1	—	—	1	2	—	—	1	1	1	1	1	1	1	3	1	4
Anhingidae	1	1	1	1	1	1	—	—	—	—	1	1	—	—	—	—	—	—	1	1	1	1
Ardeidae	8	13	8	13	8	13	6	8	7	7	6	7	6	8	7	8	8	9	7	11	9	14
Ciconiidae	1	1	1	1	1	1	—	—	1	1	1	1	—	—	—	—	1	1	1	1	1	1
Threskiornithidae	—	—	—	—	—	—	—	—	—	—	—	—	—	—	—	—	—	—	2	4	2	5
Anatidae	4	4	4	4	4	4	2	2	1	2	2	2	1	1	1	1	1	1	8	10	13	17
Pandionidae	1	1	1	1	1	1	1	1	1	1	—	—	—	—	1	1	—	—	1	1	1	1
Accipitridae	11	15	10	15	10	13	5	5	5	5	4	5	5	7	6	6	8	8	12	15	12	16
Falconidae	2	2	2	2	2	2	1	2	1	3	1	2	1	3	1	1	1	2	1	4	1	4
Megapodiidae	—	—	—	—	1	1	—	—	1	1	1	1	1	1	1	1	—	—	1	1	3	3
Phasianidae	13	15	9	11	4	5	1	2	1	1	2	2	3	3	2	2	2	2	1	3	1	3
Turnicidae	1	2	1	1	1	2	1	1	1	1	1	2	1	2	1	1	1	1	1	4	1	6
Pedionomidae	—	—	—	—	—	—	—	—	—	—	—	—	—	—	—	—	—	—	—	—	1	1
Rallidae	6	9	6	9	6	8	3	3	4	4	4	4	5	6	4	4	5	5	6	7	9	13
Heliornithidae	1	1	—	—	—	—	—	—	—	—	—	—	—	—	—	—	—	—	—	—	—	—
Cruidae	—	—	—	—	—	—	—	—	—	—	—	—	—	—	—	—	—	—	1	1	1	2
Burhinidae	—	—	—	—	—	—	—	—	—	—	—	—	—	—	—	—	—	—	1	1	1	1
Otididae	—	—	—	—	—	—	—	—	—	—	—	—	—	—	—	—	—	—	1	1	1	1
Jacanidae	1	1	1	1	1	1	—	—	—	—	1	1	1	1	1	1	1	1	1	1	1	1

continued

Table II: continued (2)

	Malay Peninsula	*Sumatra*	*Java*	*Bali*	*Lombok*	*Sumbawa*	*Flores*	*Sumba*	*Timor*	*Northwest Australia*	*Eastern Australia*
Rostratulidae	1 1	1 1	1 1	— —	— —	1 1	— —	— —	— —	1 1	1 1
Charadriidae	1 3	1 3	1 2	1 1	— —	1 1	1 1	— —	1 1	2 3	2 6
Recurvirostridae	— —	— —	— —	— —	— —	— —	— —	— —	— —	2 2	2 2
Scolopacidae	— —	— —	— —	— —	— —	— —	— —	— —	— —	— —	— —
Glareolidae	1 1	1 1	1 1	— —	— —	— —	— —	— —	— —	1 1	— —
Columbidae	7 18	7 14	7 13	6 10	9 14	9 14	7 13	8 10	10 14	7 13	10 16
Psittacidae	3 4	3 4	2 2	3 3	2 2	2 2	6 6	4 4	4 6	8 14	20 37
Cuculidae	7 20	7 18	7 16	6 7	3 3	4 4	4 4	3 3	5 6	5 5	5 11
Strigidae	5 10	4 8	4 8	4 4	1 1	1 2	2 3	1 1	1 1	1 3	1 4
Tytonidae	1 1	1 1	1 1	1 1	1 1	— —	1 1	1 1	1 1	1 3	1 4
Podargidae	1 4	1 4	1 1	— —	— —	— —	— —	— —	— —	1 1	1 3
Aegothelidae	— —	— —	— —	— —	— —	— —	— —	— —	— —	1 1	1 1
Caprimulgidae	2 5	2 4	1 4	1 2	1 2	1 2	1 1	1 2	1 2	1 2	1 2
Hemiprocnidae											
Apodidae	5 9	5 8	4 7	3 3	1 2	1 1	1 1	1 2	1 2	— —	1 2
Trogonidae	1 6	1 6	1 2	— —	— —	— —	— —	— —	— —	— —	— —
Alcedinidae	5 12	5 11	5 11	4 7	5 6	5 6	4 5	3 5	2 3	3 7	5 9
Meropidae	2 4	2 4	1 3	1 2	1 1	1 1	1 1	1 1	1 1	1 1	1 1
Coraciidae	2 2	2 2	1 1	1 1	1 1	1 1	1 1	— —	1 1	1 1	1 1
Bucerotidae	6 10	6 10	3 3	2 2	— —	— —	— —	1 1	— —	— —	— —
Capitonidae	2 9	3 11	2 6	3 3	— —	— —	— —	— —	— —	— —	— —
Indicatoridae	1 1	1 1								— —	— —
Picidae	13 25	11 23	11 15	4 5	1 1	1 1	1 1	— —	— —	— —	— —
Eurylaimidae	4 4	4 4	1 1	— —	— —	— —	— —	— —	— —	— —	— —

continued

Table II: continued (3)

	Malay Peninsula	*Sumatra*	*Java*	*Bali*	*Lombok*	*Sumbawa*	*Flores*	*Sumba*	*Timor*	*Northwest Australia*	*Eastern Australia*
Pittidae	1 5	1 5	1 4	— —	1 1	1 1	1 1	1 1	1 1	— —	— —
Menuridae	— —	— —	— —	— —	— —	— —	— —	— —	— —	— —	1 2
Atrichornithidae	— —	— —	— —	— —	— —	— —	— —	— —	— —	— —	1 1
Alaudidae	— —	— —	1 1	1 1	1 1	1 1	1 1	1 1	1 1	1 1	1 1
Hirundinidae	3 3	3 3	2 3		1 2	1 1	1 2	1 1	1 1	1 1	3 4
Motaciliidae	1 2	1 2	1 2	— —	1 1	1 1	1 1	1 1	1 1	1 1	1 1
Campephagidae	4 11	4 10	3 7	3 5	2 2	3 3	3 3	3 3	3 3	2 4	2 7
Chloropseidae	2 6	2 5	2 2	— —	— —	— —	— —	— —	— —	— —	— —
Pycnonotidae	3 14	3 14	3 15	2 3	1 1	— —	— —	— —	— —	— —	— —
Dicruridae	2 6	1 5	1 4	1 3	1 2	1 1	1 1	1 1	1 1	1 1	1 1
Oriolidae	1 4	1 5	1 3	1 1	1 1	1 1	1 1	1 1	2 2	2 3	2 3
Corvidae	6 7	4 5	3 4	2 3	1 1	1 1	1 2	1 1	1 1	1 2	1 4
Paridae	1 1	1 1	1 1	1 1	1 1	1 1	1 1	1 1	— —	— —	— —
Sittidae	1 2	1 2	1 2	— —	— —	1 1	— —	— —	— —	— —	— —
Timaliinae	16 45	14 35	10 16	2 2	1 1	1 1	1 1	— —	— —	1 2	1 2
Orthonychinae	— —	— —	— —	— —	— —	— —	— —	— —	— —	1 1	3 6
Muscicapinae	11 22	10 18	8 13	7 7	4 4	7 7	10 11	9 10	5 5	9 14	13 30
Pachycephalinae	1 1	1 1	1 1	1 2	1 1	1 2	1 2	1 1	1 2	3 7	3 8
Turdinae	7 14	6 11	5 10	3 3	4 5	2 3	3 4	1 1	3 3	— —	2 3
Sylviinae	6 10	6 9	6 10	5 6	5 6	3 4	3 3	3 3	13 14	4 6	4 7
Malurinae	— —	— —	— —	— —	— —	— —	— —	— —	— —	2 5	4 7
Acanthizinae	1 1	1 1	— —	— —	— —	— —	— —	— —	1 1	3 8	12 27
Neosittidae	— —	— —	— —	— —	— —	— —	— —	— —	— —	1 1	1 1
Climacteridae	— —	— —	— —	— —	— —	— —	— —	— —	— —	1 2	1 3

continued

Table II: continued (4)

	Malay Peninsula	*Sumatra*	*Java*	*Bali*	*Lombok*	*Sumbawa*	*Flores*	*Sumba*	*Timor*	*Northwest Australia*	*Eastern Australia*
Meliphagidae	— —	— —	— —	— —	3 3	3 3	3 3	3 3	4 6	13 23	21 60
Ephthianuridae	— —	— —	— —	— —	— —	— —	— —	— —	— —	1 22	1 3
Artamidae	— —	1 1	1 1	1 1	1 1	1 1	1 1	1 1	1 1	1 3	1 5
Laniidae	1 2	1 1	1 1	2 2	1 1	1 1				—	—
Sturnidae	5 8	4 6	4 4	2 2	2 3	2 2	2 2	1 1	1 1	— —	1 1
Nectariniidae	4 16	4 13	4 10	3 3	1 1	2 3	2 3	2 2	1 1	— —	1 1
Dicaeidae	2 8	2 7	2 7	1 4	1 2	3 3	3 4	2 2	2 3	1 1	1 1
Pardalotidae	— —	— —	— —	— —	— —	— —	— —	— —	— —	1 1	1 3
Zosteropidae	1 2	1 1	1 1	1 2	1 2	4 7	3 7	1 2	2 3	1 1	1 3
Ploceidae	4 7	2 5	4 9	4 7	5 8	4 8	4 7	3 5	5 7	4 10	5 10
Grallinidae	— —	— —	— —	— —	— —	— —	— —	— —	— —	1 1	1 1
Corcoracidae	— —	— —	— —	— —	— —	— —	— —	— —	— —	1 1	1 2
Cracticidae	— —	— —	— —	— —	— —	— —	— —	— —	— —	2 9	3 7
Ptiloworhynchidae	— —	— —	— —	— —	— —	— —	— —	— —	— —	1 1	6 9
Paradiseidae	— —	— —	— —	— —	— —	— —	— —	— —	— —	— —	2 3

Table III. Surface area, maximum altitude and avifauna composition of the major Islands, Indo–Australian region

Island	*Area* (km^2)	*Maximum Height* (m)	*Families*	*Genera*	*Species*	*Endemic Genera*	*Endemic Species*	*% Species of Western Origin*
Malay Peninsula	142 439		58	216	429	1		98
Sumatra	473 570	3522	58	189	359	1	5	97·8
Java	125 615	3407	58	165	287	1	16	97·5
Borneo	293 000	3804	57	209	386	5	29	96
Sulawesi	181 286	2825	47	160	220	16	84	72
Halmahera	17 779	1412	39	88	120	4 } 1	15	70
Buru	9064	2251	38	94	115	2 }	8	70
Ceram	18 623	2798	38	88	117	1	8	70
Vogelkop			52	175	323			22
East New Guinea		4662	65	221	422			16
Bali	5560	3105	43	99	129	1		87
Lombok	4729	3489	46	83	101	—		72·5
Sumbawa	15 448	2850	46	90	111	—	14	68·0
Flores	17 150	2203	44	96	119	—		63·0
Sumba	11 153	2100	42	78	89	—	8	56·5
Timor	32 000	1977	42	95	113	(1)	22 (11 on island)	50·4
N.W. Australia			63	152	239	—		15
E. Australia		2067	71	221	433			12

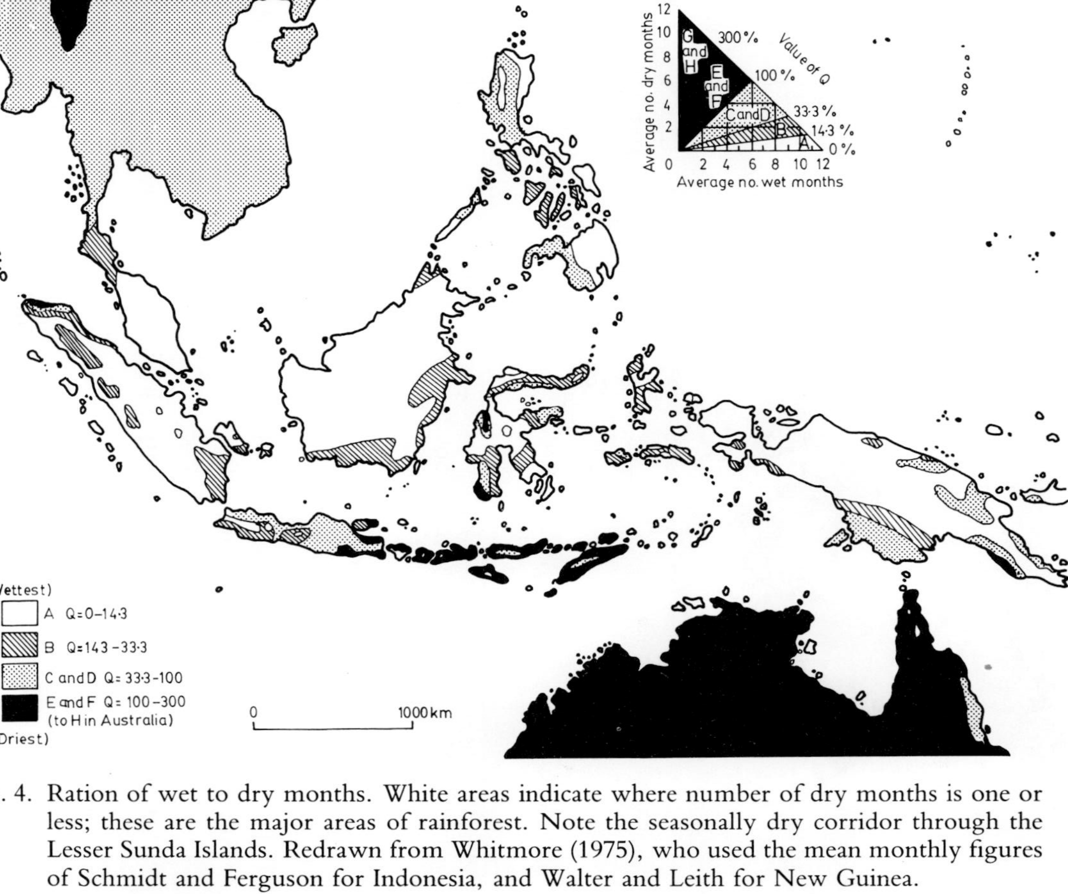

Fig. 4. Ration of wet to dry months. White areas indicate where number of dry months is one or less; these are the major areas of rainforest. Note the seasonally dry corridor through the Lesser Sunda Islands. Redrawn from Whitmore (1975), who used the mean monthly figures of Schmidt and Ferguson for Indonesia, and Walter and Leith for New Guinea.

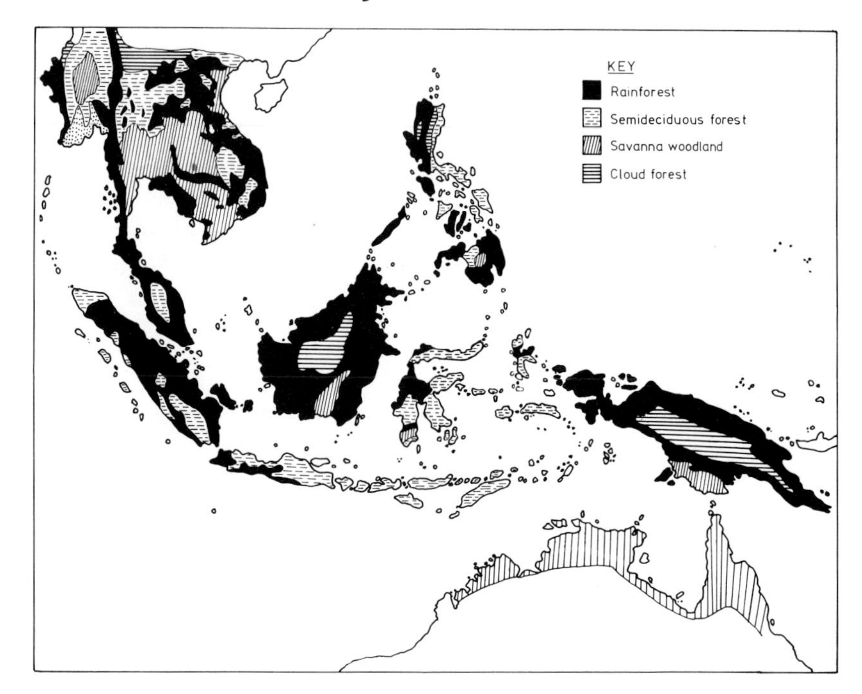

Fig. 5. Major vegetation zones of Southeast Asia and Australia–New Guinea. Based on the map of Steenis (1938), and Nix and Kalma (1972). Note the eastern and western rainforest blocks and intermediate sections of semi-deciduous forest.

This has not been possible here. Adding or eliminating a few genera and species would not materially affect the overall conclusions.

1. Families that are Entirely, or Mostly, Oriental

These are the Phasianidae, Heliornithidae, Trogonidae, Bucerotidae, Capitonidae, Picidae, Eurylaimidae, Chloropseidae, Pycnonotidae, Paridae, Timaliinae, Sylviinae. Diversity is high at both the generic and species levels in Malaya, Sumatra, and Borneo, with a precipitous drop off in diversity thereafter; see Fig. 6 (Phasianidae, Picidae) and Fig. 7 (Timaliinae, Sylviidae). Sometimes such families (Trogonidae, Eurylaimidae, Capitonidae) do not extend beyond Wallace's Line. In other cases, a few species may extend through the Lesser Sundas (Timaliinae), reach New Guinea (Bucerotidae), or reach Australia (Dicaeidae). Yet elsewhere, there has been minor secondary radiation in New Guinea and/or Australia (Phasianidae, Sylviinae).

2. Families that are Entirely or Predominantly Australian★

These are the Dromaiidae, Casuariidae, Megapodiidae, Pedionomidae, Podargidae, Aegothelidae, Menuridae, Atrichonrithidae, with the "subfamilies" Ornithonychinae, Pachycephalinae, Malurinae, Acanthizinae, Neosittidae, Climacteridae, Meliphagidae, Ephthianuridae, Artamidae, Pardalotidae, Grallinidae, Corcoracidae, Cracticidae, Ptilonorhynchidae, Paradiseidae. Some of these are restricted to the southern limits of Australia proper (Menuridae, Atrichornithidae), the continental mainland of Australia (Ephthianuridae), or the limits of the Australo–Papuan continental shelf (Casuaridae; Neosittidae; Climacteridae). Others (Meliphagidae) penetrate Wallacea and the Lesser Sundas in modest numbers. A few reach Bali and the Celebes but not beyond. There has been modest penetration of the Philippines by Australian parrots (Cacatuinae, within the Psittacidae) and New Guinea pigeons (*Ptilinopus*). Finally, single species of some four families penetrate through to the Asian mainland (*Gerygone* in the Acanthizinae; *Pachycephala* in the Pachycephalinae). A megapode reaches the Andamans.

Examples of Australian distribution patterns are the Meliphagidae, Malurinae and Acanthizinae (Fig. 8).

3. Families Well-Developed in Both Regions

Various families have radiated both in the Oriental and Australian Regions, different groups being commonly involved. Examples are the Columbidae (Fig. 6) and Alcedinidae (Fig. 9). The Australian Region is phenomenally rich in pigeons, kingfishers and parrots, with many striking genera; in each case the greatest diversity in the world occurs there. The endemic genera can rightly be regarded as an element quite distinct from their Oriental counterparts.

A slightly different situation is represented by the turnicid quails, emerald cuckoos (*Chrysococcyx* and other genera) and estrildine finches (Ploceidae). These groups extend from Africa to Australia and have undergone a modest but significant radiation at each end of their range. Interchange here must have been intermediate to late, but must have

★For Australian avian fossil record see Rich and van Tets (1982).

preceded such later colonizations as that (west–east) of the hornbills, *Dicaeum*, Nectariniids, and others.

4. Cosmopolitan Bird Groups, Comparative Representation in the Oriental and Australian Regions

Large water birds, such as cormorants, herons, rails, stilts and ducks, large birds of prey (hawks, owls) and other cosmopolitan groups such as nightjars, swifts, cuckoos and swallows, are richly developed both in the Oriental and Australian regions (see Ardeidae, Fig. 9). There is in between a modest drop in diversity (numbers of breeding species are necessarily reduced in the dry islands of Wallacea). In the case of the large water-birds (e.g. herons), whose survival depends on mobility, various species are shared between the Oriental Region and Australia. Both regions, however, have their endemic anatids (5 genera in the case of Australia) and, as noted, the Oriental duck fauna is annually supplemented by migrants from the Palearctic. (Rich, 1975, has suggested that some endemic genera in the non-passerines may belong to the "old southern" element). Australia and New Guinea are likewise rich in endemic genera of hawks.

5. The Papuan Element

The Papuan avifauna, as noted, represents a semi-independent radiation from that on the Australian mainland, and New Guinea has a very large number of endemic genera. The tropical, and some cases insular, adaptations of some Papuan groups have enabled them to make major penetrations into the island groups to the north, east, and west. Interesting examples are to be found in the pigeons (*Ptilinopus* and *Ducula*). The former, small-bodied, arboreal fruit-pigeons, commonly green in colour, are represented by eight endemic New Guinea species, which obviously makes this island the centre of radiation for the group (distribution maps in Goodwin, 1970). Extending outwards is a rich fauna of island endemics (Fig. 10). The Philippines have four species, two restricted to the more northern islands. Three are restricted to the Moluccas, and two to Sulawesi. Four (including *Megaloprepia* in *Ptilinopus*) penetrate Australia to the south. There are three Fijian endemics, one extending to the Marianas in the northwest, and a series of insular endemic species in the eastern Pacific. One wider ranging

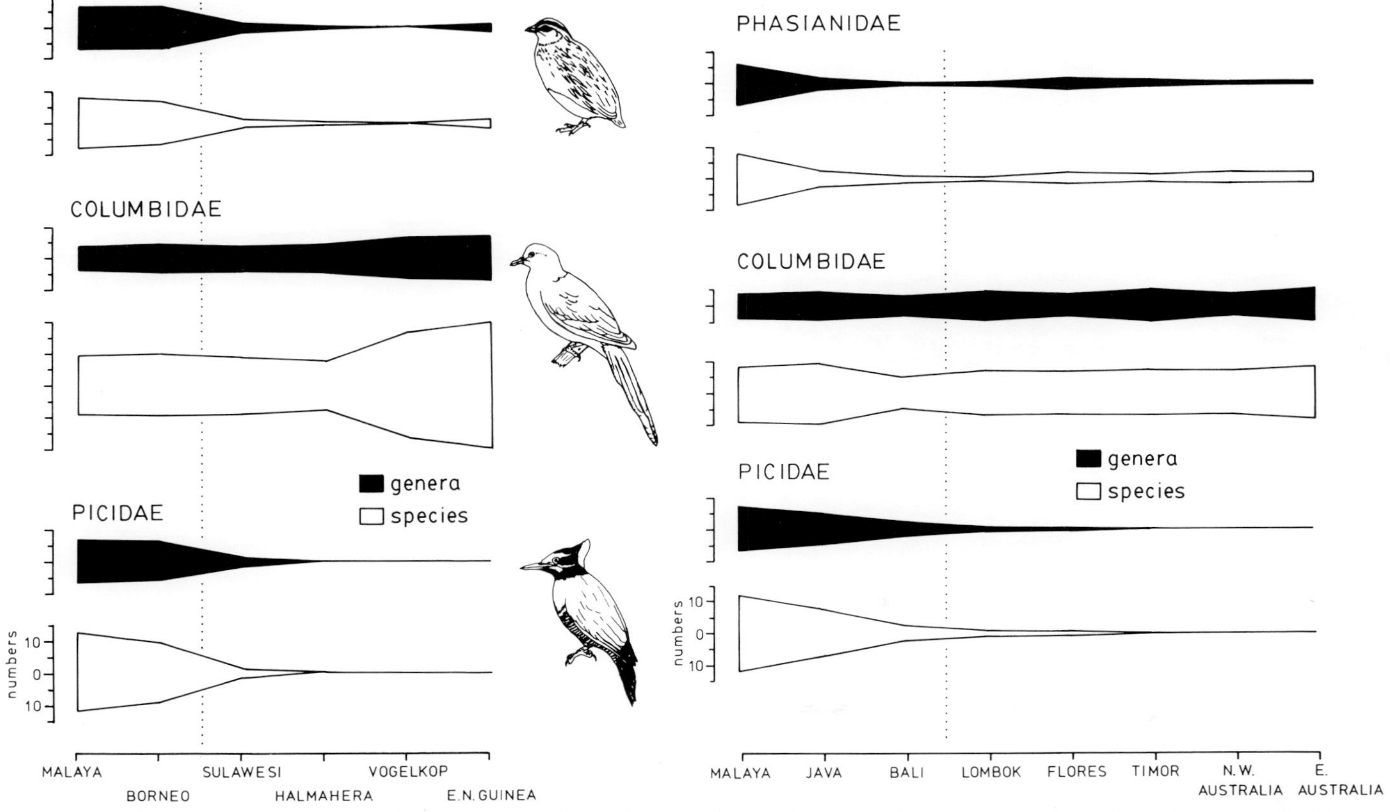

Fig. 6. Numbers of genera and species of Phasianidae, Columbidae, and Picidae; Malaya to eastern New Guinea (north), and Malaya to Australia (south). Dotted line represents Wallace's Line.

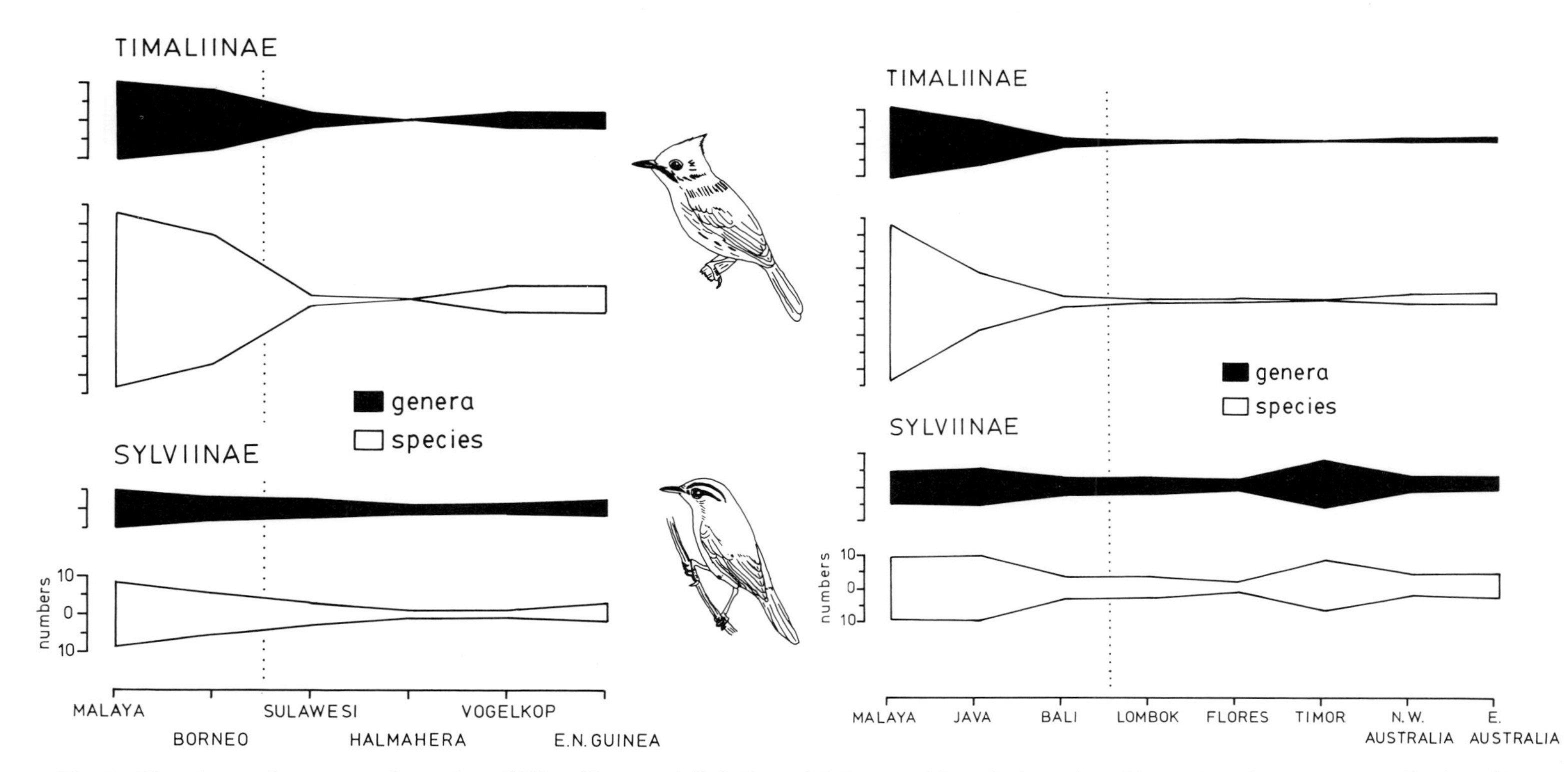

Fig. 7. Numbers of genera and species of Timaliinae and Sylviinae; Malaya to New Guinea (north), and Malaya to Australia (south).

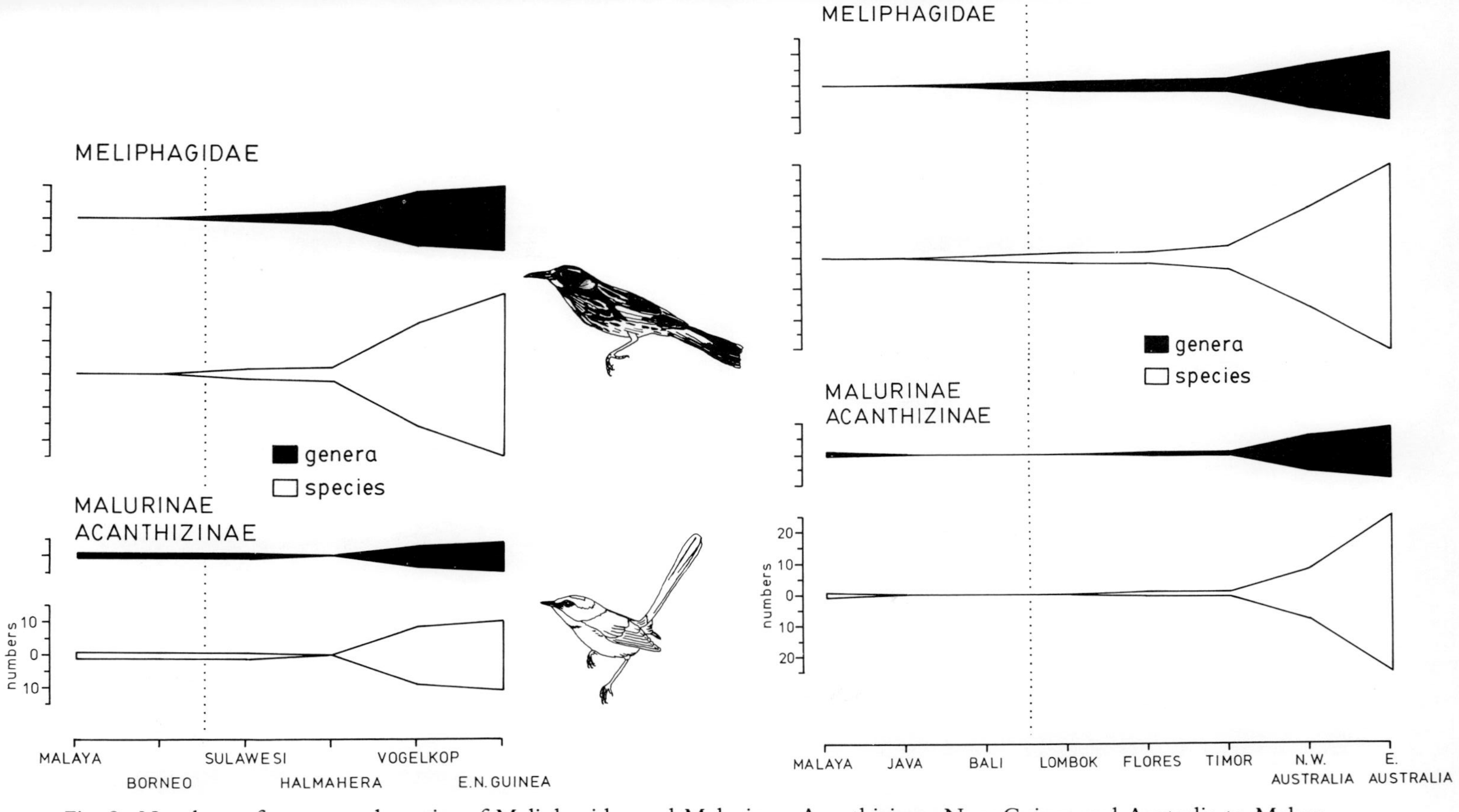

Fig. 8. Numbers of genera and species of Meliphagidae and Malurinae–Acanthizinae; New Guinea and Australia to Malaya.

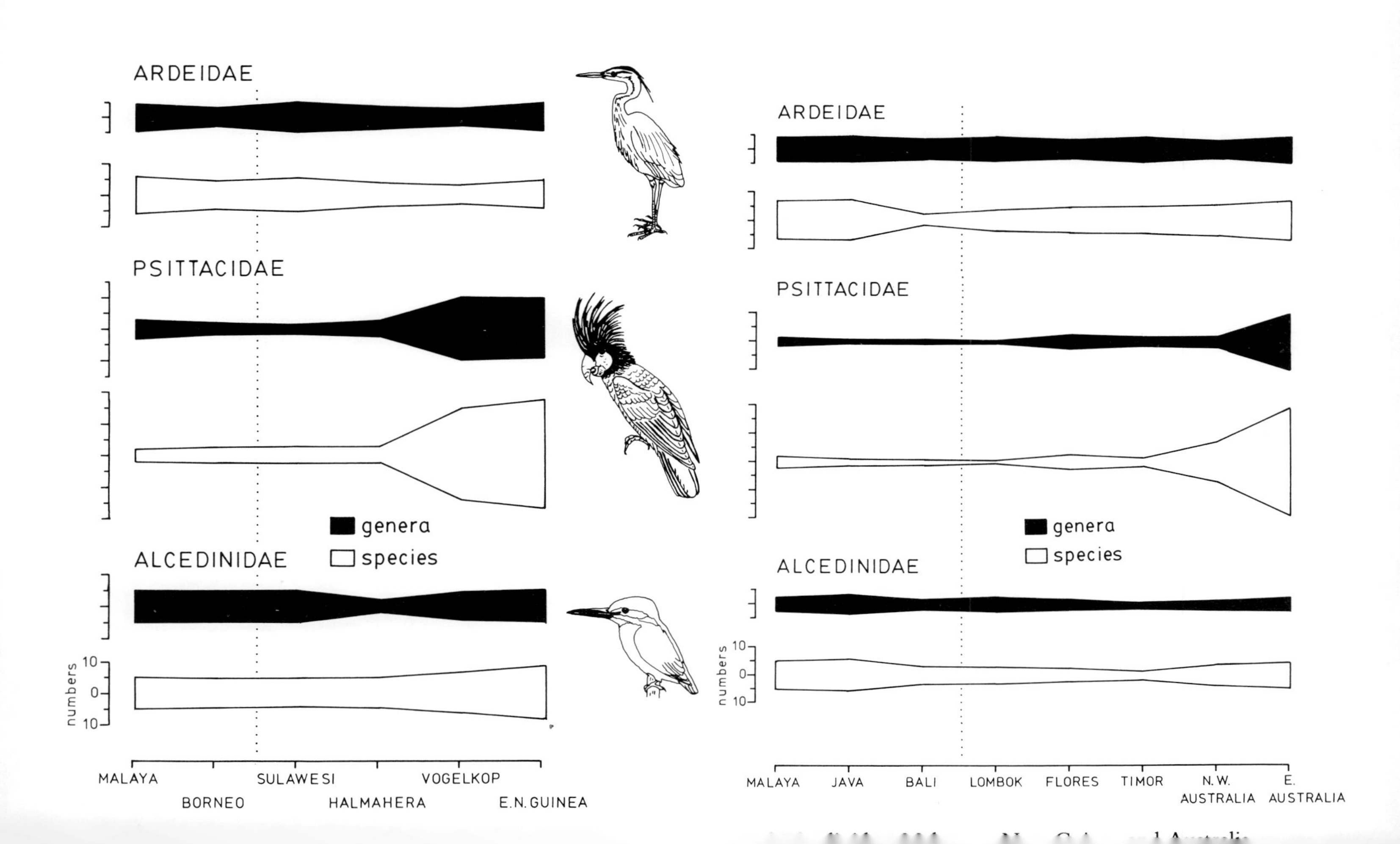
ARDEIDAE
PSITTACIDAE
ALCEDINIDAE
genera
species
numbers
10
0
10
MALAYA
BORNEO
SULAWESI
HALMAHERA
VOGELKOP
E.N.GUINEA
ARDEIDAE
PSITTACIDAE
ALCEDINIDAE
genera
species
numbers
10
0
10
MALAYA
JAVA
BALI
LOMBOK
FLORES
TIMOR
N.W. AUSTRALIA
E. AUSTRALIA

species occurs in Malaya, Sumatra and Borneo, and another on Sumatra and Java. Various evolutionary stages are to be seen in *Ptilinopus*, including endemics that are approaching the stage of differentiation characteristic of genera. Others represent strongly differentiated subspecies, e.g. the Wallacean races of *P. regina* (Fig. 11). In Australia, isolated populations of *Ptilinopus* (*Megaloprepia*) *magnifica* in isolated rainforest blocks from north to south show striking size variation. Finally, the genus contains species that extend over considerable distances with little or no differentiation. *P. superbus*, extending from eastern Australia to the Moluccas (not shown on Fig. 10), is an example. These are obviously cases of recent range spread.

6. *General*

Despite the opportunities for interchange provided by the approach of Australia to Asia in the Miocene, the Oriental and Australian regions retain avifaunas that are significantly different. Each retains a series of distinctive families. Some Oriental ones reach Wallace's Line from the west, but a few stop short of it. Many of the endemic Australian families do not extend beyond the continental limits. A few reach Wallace's Line from the east and even penetrate it. Of overriding importance, in all this, are obvious ecological adaptations and dispersive capacities.

7. *Avian Speciation in the Indo–Australian Region*

The islands of Sundaland, Wallacea and the Pacific, are major centres of speciation, and in many groups there are here more differentiated populations per given area than anywhere else in the world. As noted, Java is rich in endemic races; Timor has 67 endemic races, Ceram 47, Buru 44, the Sula Islands 34. Speciation in the southwest Pacific, where such species as *Pachycephala pectoralis*, *Petroica multicolor*, *Rhipidura rufifrons*, *Halcyon sancta* and *H. chloris* have large numbers of insular races, has been the basis of many papers by Mayr (listed in Keast, 1981). Such insular forms, in that they represent intermediate stages in the speciation process, are incipient species. As has been previously pointed out (Keast, 1961), however, most of these have little evolutionary future and will be eliminated if there is subsequent junction with better-adapted mainland stocks.

There are many cases of speciation "between" Asia and Australia.

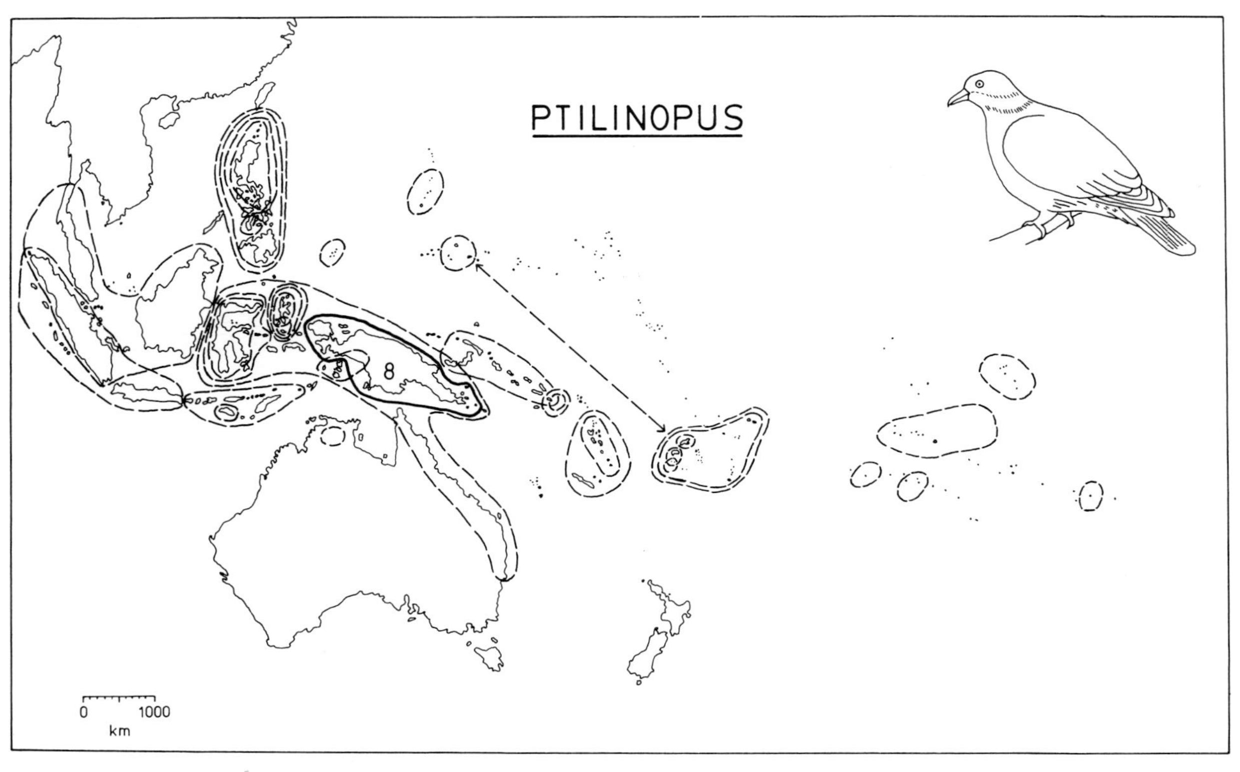

Fig. 10. Distribution of fruit-pigeons of the genus *Ptilinopus*. The genus is centred on New Guinea which has eight endemic species. Insular species occur throughout Indonesia and the Pacific.

Some examples are given in Figs 11–14. The examples chosen are interesting because they not only include simple situations of allopatric forms differentiating in isolation, but secondary extinction of intermediate populations. *Chalcophaps* is a genus of small-bodied, green, ground-feeding pigeons that inhabit shoreline thickets and the rain forest floor. There are two species of which one, *C. indica*, extends from India to northern and eastern Australia, with an isolated population in

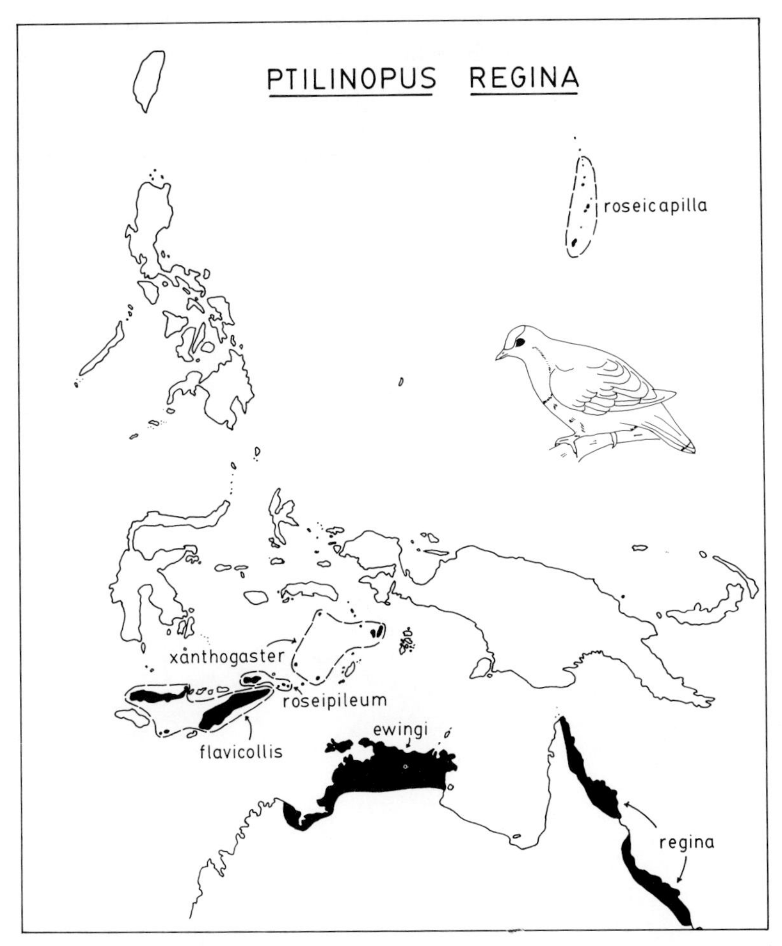

Fig. 11. Range of the fruit-pigeon *Ptilinopus regina*. Raciation has occurred in the northeastern Australia relative to the northwest, and islands to the northwest of the continent. An isolated form occurs many hundreds of miles to the north in the Marianas.

eastern New Guinea (range map adapted from Goodwin, 1970). A second species, *C. stephani*, occurs throughout New Guinea, the Bismarcks, Solomons, and islands to the east, with an isolated population on Sulawesi. This pattern would be consistent with any one of several explanations, such as: (a) original differentiation in the west (*indica*) and east (*stephani*), respectively; (b) an original species (*indica*) and a later New Guinea derivative (*stephani*) that secondarily spread out to Sulawesi and subsequently became replaced by *indica*—in the intermediate islands, in this case, the Australian and eastern New Guinea populations of *indica* would be relict ones; (c) a western origin for *indica*, invasion of Australia from the northwest and then of eastern New Guinea from Australia. Diamond (1975:418) has demonstrated that where these two species co-occur in New Guinea they replace each other in different habitats.

An equally complex distributional history is shown by the small brown cuckoo-doves of the genus *Macropygia* (*M. amboinensis* group).

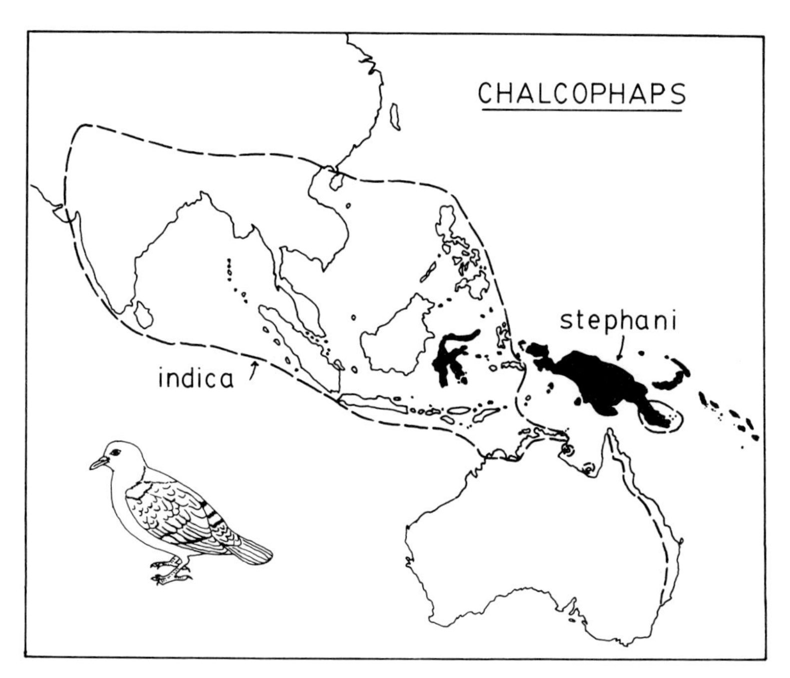

Fig. 12. Distribution of pigeons of the genus *Chalcophaps*. The two species have partially overlapping ranges.

The genus is most diversified in New Guinea, with three species. The larger-bodied *M. ambionensis* superspecies is represented by two distinctively-barred species in New Guinea–Sulawesi (*amboinensis*) and Timor (*magna*), and a peripheral form in the Andamans (*rufipennis*). A plain uniform dark brown form (in which some juveniles show faint and partial barring) occurs from the Philippines through northern Borneo to Sumatra and Java, and has an isolated population in eastern Australia (range map redrawn from Mayr, 1944b). Whilst these insular populations have slight differences, the whole of *phasianella* is remarkably uniform and strongly suggestive of a late-dispersing group. Again no ready explanation for this distribution pattern suggests itself. Is the

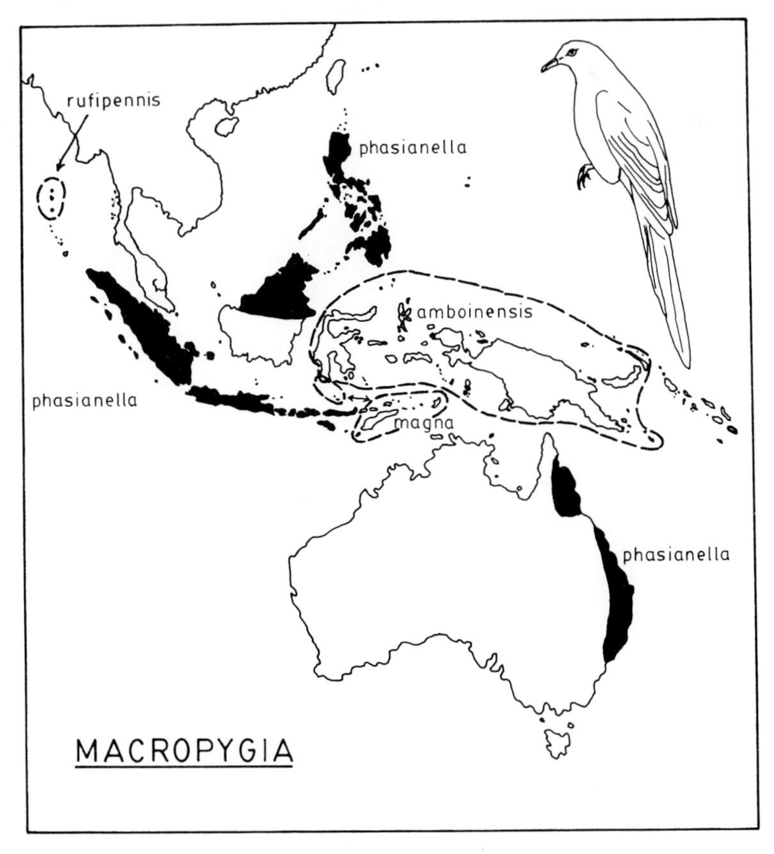

Fig. 13. Distribution of cuckoo-doves of the *Macropygia amboinensis* superspecies. *M. phasianella* has a curiously disjunct range.

Australian *phasianella* a late colonist from the northwest, that subsequently became extinct in northwestern Australia and then was secondarily displaced in Timor by colonizers from New Guinea? Or, is it relictual in Australia from an originally continuously-ranging ancestor? For the latter to be the case, a remarkable degree of evolutionary conservatism would be demanded.

A further problematical distribution pattern is shown by the remote Mariannas race of *Ptilinopus regina* (Fig. 11), the species otherwise being confined to Australia, the Lesser Sundas and the adjacent islands.

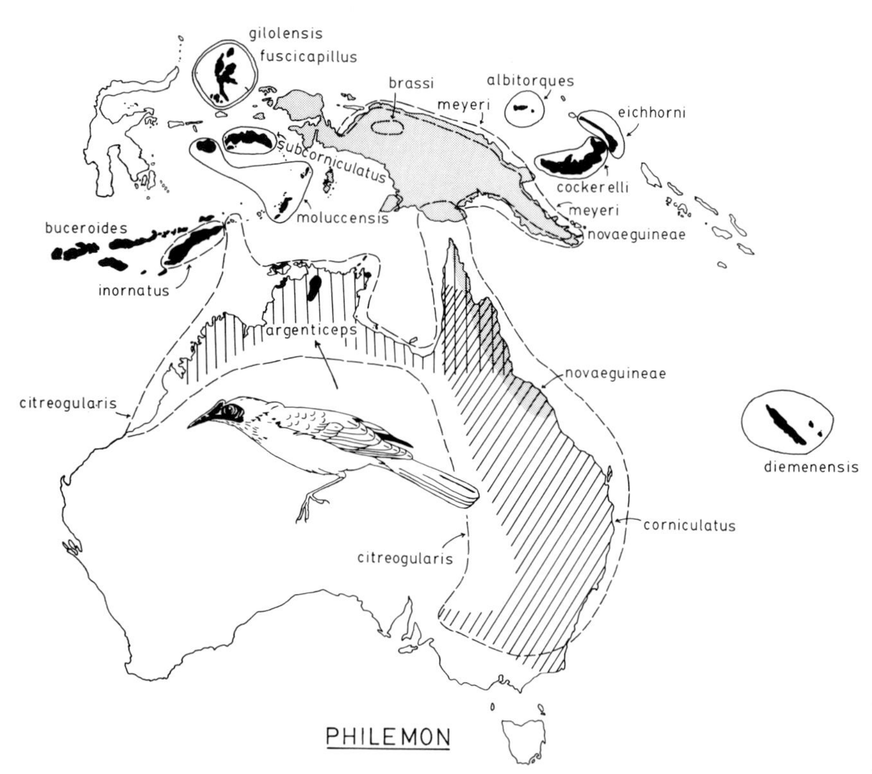

Fig. 14. Distribution of large honeyeaters (Meliphagidae) of the genus *Philemon*. The genus is characterized by a series of very distinct insular isolates that must have differentiated following colonization. The most primitive (generalized) species is *P. meyeri* of New Guinea. Australia has four species. Speciation by double invasion has occurred in the Moluccas and northern Australia. *Philemon* is one of the genera of Meliphagidae that have mounted successful penetrations of Wallacea.

Obviously, the stock has secondarily been eliminated from the intermediate islands.

Thus, patterns of distribution, isolation and differentiation may be complicated. Natural and evolutionary processes are dynamic, and climatic and competitive background patterns change.

Honeyeaters (Meliphagidae) of the genus *Philemon* provide an interesting example of insular speciation in the islands of Wallacea. This large-bodied Australian genus extends to the Lesser Sundas and Moluccas, the islands northeast of New Guinea, and New Caledonia to the east. There is a large number of distinctive insular species. The most primitive species is regarded by Mayr (1944b) as *P. meyeri* of New Guinea; this has a relative in Australia (*P. citreogularis*). There are two species in the Moluccas. Australia has five species, the distinctive bare-headed *corniculatus* in the east, with *buceroides* and *novaeguinea* (*yorki*) being outlying populations of northern forms. *P. argenticeps* of northern Australia is derived from an earlier colonizing wave of one of these stocks (Keast, 1961).

Philemon fills the niche of a large-bodied insectivore—nectarivore in the northern islands, an adaptive zone substantially similar to that of the Old World Oriolidae that are obviously later invaders of Wallacea, New Guinea and Australia from the west. Where orioles are absent *Philemon* may be convergent with them in colour pattern (Diamond, 1982).

ECOLOGICAL COUNTERPARTS AND THE SUB-DIVISION OF ECOLOGICAL OPPORTUNITIES BETWEEN ORIENTAL AND AUSTRALIAN SPECIES

One of the most promising areas for future research is examination of the regional and insular avifaunas of the Malayan–Australian Region to determine how the major adaptive zones (ecological opportunities) have been divided up in areas of mixed Oriental and Australian avifaunas. Clearly it can be assumed that ecological interactions are one major factor in limiting range spreads but how important it has been in establishing contemporary distribution patterns, cannot be ascertained without studies of the avian communities concerned *in situ*. Lincoln's (1975) analysis of bird communities in Bali and Lombok noted broad role differences between Asian and Australian forms, plus habitat differences. He was not, however, able to undertake an ecological niche analysis. Some examples of competitive displacement of Asian and

Australian ecological counterparts in the overlap zone are on record, e.g. between sunbirds (*Nectariniidae*) and the small, morphologically similar honeyeaters of the genus *Myzomela* (Ripley, 1959). The distribution patterns he noted suggested that any given island was capable of supporting only one or the other. This subject has since been taken up by Diamond (1975), who has demonstrated that replacement patterns of this type are characteristic of an extensive insular area morph of New Guinea.

Broadly speaking, it would appear that in the overlap zone Australian forms either occupy distinct niches or occupy marginal habitats. Note, for example, that the two most successful colonizers to reach the Asian mainland, *Gerygone* and *Pachycephala*, are inhabitants of either coastal mangroves or secondary growth (King *et al.*, 1975). (These authors described the habitat of *Pachycephala* as "mangroves, casuarinas, adjacent scrub, sometimes garden".) Megapodes are largely insular and beachedge dwellers in the western part of their range. Honeyeaters, cockatoos and wood-swallows, that evolved in the open terrain of Australia, are highly mobile and adapted for travelling long distances. Mobility is obviously advantageous to granivorous and nectarivorous species, since these resources are likely to be patchily and seasonally distributed. *Artamus* obviously evolved in Australia in the absence of endemic swifts and against a poor competitor background of swallows (Australia has only four species). Many *Artamus* spp. are wide-ranging and adapted to open terrain and shoreline, in insular colonization. So their far penetration of Indonesia where there are endemic swifts, is surprising, the more so since the few swallows present are migrants and occur only seasonally.

An analysis of how the major ecological opportunities are divided between Asian and Australian elements, from a series of islands where the former is dominant, through islands where proportions are equal, to yet others where the latter predominate, is a logical extension to this study. Are the Asian elements "stronger" in some adaptive zones and Australian ones in others? What attributes give superiority in each case? What form is competition taking?

It would be appropriate here to test out the "assembly rules" ideas of Diamond (1975) against this richer background. Again, the area lends itself to the testing of the "taxon cycles" concept of Rickleffs and Cox (1972)—i.e. that island colonists start off as generalists progressively becoming more specialized and narrowly adapted.

It is, of course, obvious that there is the need for a massive new series of avifaunal surveys throughout Indonesia. Most of our basic distributional and biogeographic data are well out of date. It must be reassembled into a modern framework before processes can be better understood.

ASIAN–AUSTRALIAN BIOGEOGRAPHY, VICARIANCE OR DISPERSAL

Vicariance and dispersal have both figured prominently in the origin and development of the Indonesian biota. Obviously, the former has played a major role in the faunas of the continental islands but the faunas of the Lesser fundas and Moluccas have largely arisen through over-water dispersal.

The Australo–Papuan and Asian biotas are basically very different, having come from different world stocks. Yet the degree to which this difference has been maintained has varied with the groups and their dispersal capacity. Australia maintains unique marsupial—monotreme mammal and amphibian faunas; its turtles are likewise distinct from that of Asia, and so on. By contrast, the Australian avifauna, whilst still characterized by high endemism in families and genera, clearly shows major infusions of new elements by long-distance dispersal. A considerable proportion of Australian genera and species entered from the north. Proof of the continued entry by dispersion comes from the number of undifferentiated species (or races of Asian species) that represent recent colonists. The Australian bird and bat-rodent faunas have also been built up by progressive colonization from the northwest (Simpson, 1961).

As further proof of the efficacy of over-water colonization, the close similarities of the avifaunas of the Lesser Sundas Islands and of the Moluccas should be noted. The same set of basic types is present on each. By contrast, rainforest dwelling timaliids, and others whose wings indicate poor flying capacities, are largely restricted to the continental islands; their penetration of even Sulawesi is non-existent or minimal. In terms of over-water dispersal, note also the readiness with which the inhabitants of these islands can be allocated to western or eastern groups. They have arisen, in other words, from colonization from both directions.

Underlying today's distribution patterns in the Wallacean region, hence, are historic and contemporary ecological factors. Dispersal has been a major biogeographic process.

ACKNOWLEDGEMENT

The review was written whilst the writer was the holder of a Canadian National Scientific and Engineering Research Council Grant and the writer would like to thank this body for its support. The diagrams were the work of Karen Brown.

REFERENCES

Archer, M. (1981). A review of the origins and radiations of Australian mammals. *In* "Ecological Biogeography of Australia" (A. Keast, ed.), pp. 1435–1742. W. Junk, The Hague.

Audley-Charles, M.G. (1975). The Sumba Fracture: a major discontinuity between eastern and western Indonesia. *Tectonophysics* **26**, 213–328.

Audley-Charles, J.G. and D.A. Hooijer (1973). Relations of Pleistocene migrations of pygmy Steegodonts to island are tectonics in Eastern Indonesia. *Nature, Lond.* **241**, 197–198.

Audley-Charles, M.G. (1981). Geological history of the region of Wallace's Line. *In* "Wallace's Line and Plate Tectonics" (T.C. Whitmore, ed.), pp. 24–35, Clarendon Press, Oxford.

Auffenberg, W. (1980). The herpetofauna of Komodo, with notes on adjacent areas. *Bull. Fla. St. Mus. biol. Sci.* **25**, 39–156.

Axelrod, D.I. and P.H. Raven (1981). Paleobiogeography and origin of the New Guinea flora. *In* "Ecology and Biogeography in New Guinea" (L. Gressit, ed.), W. Junk, The Hague (in press).

Balgooy, M.M.J. (1971). Plant-Geography of the Pacific. *Blumea, Suppl.* **6**, 22.

Balgooy, M.M.J. (1976). Phytogeography. *In* "New Guinea Vegetation" (K. Paijmans, ed.), pp. 1–22. Australian National University Press, Canberra.

Batchelor, B.C. (1979). Discontinuously rising Lake Cainozoic eustatic sea-levels with special reference to Sundaland, South-west Asia. *Geologie Mijnb.* **58**, 1–20.

Beadle, N.C.W. (1981). Origins of the Australian angiosperm flora. *In* "Ecological Biogeography of Australia" (A. Keast, ed.), pp. 391–406, W. Junk, The Hague.

Beaufort, L.F. de (1913). Fishes of the eastern part of the Indo–Australian Archipelago, with remarks on its zoogeography. *Bijkdr. Dierk.* **19**, 95–164.

Beaufort, L.F. de (1951). *Zoogeography of the land and inland waters.* Sidgwick and Jackson, London.

Bemmel, A.C.G. van (1948). A faunal List of the birds of the Moluccan Islands *Treubia* **19**, 323–402.

Birdsell, J.B. (1977). The calibration of a paradigm for the first peopling of Greater Australia. *In* "Sunda and Sahul: prehistoric studies in southeast Asia, Melanesia and Australia" (J. Allen, J. Golson, and R. Jones, eds), pp. 113–167. Academic Press, London and New York.

Chasen, F.N. (1936). A handlist of Malaysian Birds. *Bull. Raffles Mus.* **11**, 1–389.

Cogger, H.G. and Heatwole, H. (1981). The Australian reptiles: origin biogeography, distribution patterns and island evolution. *In* "Ecological Biogeography of Australia" (A. Keast, ed.), pp. 731–1376. W. Junk, The Hague.

Cranbrook, the Earl (1981). The vertebrate faunas. *In* "Wallace's Line and Plate Tectonics" (T.C. Whitmore, ed.), pp. 57–69, Clarendon Press, Oxford.

Darlington, P.J. jr. (1971). The carabid beetles of New Guinea. Part IV. General considerations; analysis and history of fauna; taxonomic supplement. *Bull. Mus. comp. Zool.* **142**, 129–337.

Darlington, P.J. jr. (1980). "Zoogeography: the Geographical Distribution of Animals". Robert E. Krieger: Huntington, New York (Reprint Edition).

Delacour, J. (1947). *Birds of Malaysia.* Macmillan, New York. 1–382.

Diamond, J.M. (1975). Assembly of species communities. *In* "Ecology and Evolution of Communities" (M.L. Cody and J.M. Diamond, eds), pp. 342–444. Harvard University Press, Cambridge, Mass.

Diamond, J.M. (1982). Mimicry of friarbirds by orioles. *Auk.* **99**, 187–196.

Dickerson, R.E. *et al.* (1928). Distribution of life in the Philippines. *Philip. J. Sci.* **21**, 1–322.

Dow, D.B. (1977). A geological synthesis of Papua New Guinea. *Bull. Bur. Miner. Resour. Geol. Geophys. Aust.* **201**, 1–41.

Good, R. (1964). "The Geography of Flowering Plants" (3rd edn). Longmans Green, London.

Goodwin, D. (1970). "Pigeons and Doves of the World". British Museum, (Natural History), London.

Goodwin, R.E. (1979). The bats of Timor: Systematics and ecology. *Bull. Am. Mus. nat. Hist.* **163**, 73–122.

Gressit, J.L. (1961). Problems in the zoogeography of Pacific and Antarctic insects. *Pacif. Insects Monogr.* **2**, 1–127.

Groves, C.P. (1975). Mammalian distribution and the palaeogeography of the Sunda shelf region in the late Quaternary. *Regional Conf. Geol. Min. Resour. S.E. Asia (Djakarta, 1971).*

Groves, C.P. (1976). The origin of the mammalian fauna of Sulawesi (Celebes). *Z. Saugetierk.* **41**, 201–216.

Groves, C.P. (1980). Speciation in *Macaca.* The View from Sulawesi. *In* "The Macaques: Studies in Ecology, Behaviour, and Evolution" (D.G. Lindburg, ed.), pp. 84–124. Van Nostrand, New York.

Haechek, E. (1893). Zur phylogenie der australischen fauna. *Denkschr. med.-naturw. Ges. Jena* **4**, 5.

Hallam, A. (1981). The relative importance of plate movements, eustasy and climate in controlling major biogeographic changes since the early Mesozoic.

In "Vicariance Biogeography" (G. Nelson and D.E. Nelson, eds), pp. 303–330. Columbia University Press, New York.

Hamilton, W.B. (1973). Tectonics of the Indonesian Region. *Bull. geol. Soc. Malaysia* **6**, 3–10.

Hamilton, W.B. (1979). Tectonics of the Indonesian region. *Prof. Pap. U.S. geol. Surv.* **1087**, 338 pp.

Holloway, J.D. and N. Jardine (1968). Two approaches to zoogeography: a study based on the distribution of butterflies, birds and bats in the Indo–Australia area. *Proc. Linn. Soc. Lond.* **179**, 153–188.

Hooijer, D.A. (1975). Quaternary mammals, west and east of Wallace's Line. *Neth. J. Zool.* **25**, 46–58.

Hooker, J.D. 1860 "The Botany of the Antarctic Voyage" III. "Flora Tasmaniae" I. "Introductory Essay." Lords Commission of the Admiralty, London.

Inger, R.F. (1966). The systematics of zoogeography of the Amphibia of Borneo. *Fieldiana, Zool.* **52**, 1–402.

Inger, R.F. and C.P. Kong (1962). The freshwater fishes of North Borneo. *Fieldiana, Zool.* **45**, 1–268.

Jongsma, D. (1970). Eustatic sea level changes in the Arafura Sea. *Nature, Lond.* **228**, 150–151.

Jongsma, W.J. and W. Gotham (1935). Die palaobotanischen Ergebnisse der Djambi-Expedition 1925. *Jaarb. Mijn. Ned. Oost-Indië 1930* **59** (2), 71–201.

Kalkman, C. 1955. A plant-geographic analysis of the Lesser Sunda Islands. *Acta bot. neerl.* **4**, 200–225.

Katili, J.A. (1975). Volcanism and plate tectonics in the Indonesian island area. *Tectonophysics* **26**, 165–188.

Katili, J.A. (1978). Past and present geotectonic position of Sulawesi. *Tectonophysics* **26**, 165–188.

Keast, A. (1961). Bird speciation on the Australian continent. *Bull. Mus. comp. Zool. Harv.* **123**, 303–495.

Keast, A. (1972). Australian mammals: zoogeography and evolution. *In* "Evolution, Mammals and Southern Continents" (A. Keast, F.C. Erk and B. Glass, eds), pp. 195–46. State University of New York, Albany.

Keast, A. (1973). Contemporary biotas and the separation sequence of the southern continents. *In* "Implications of Continental Drift to the Earth Sciences" (D.H. Tarling and S.K. Runcorn, eds), pp. 309–343. Academic Press, London and New York.

Keast, A. (1977). Historical biogeography of the marsupials. *In* "The Biology of Marsupials" (B. Stonehouse and G. Gilmore, eds), pp. 69–95. Macmillan, London.

Keast, A. (1980). The ecological basis and evolution of the nearctic–neotropical bird migration system. *In* "Migrant Birds in the Neotropics: ecology, behavior, distribution, and conservation" (A. Keast and E.S. Morton, eds), pp. 561–576. Smithsonian Insitution Press, Washington.

Keast, A. (1981). Origins and relationships of the Australian biota. *In*

"Ecological Biogeography of Australia" (A. Keast, ed.), pp. 1999–2050. W. Junk, The Hague.

Kemp, E.M. and W.K. Harris (1975). The vegetation of Tertiary islands on the Ninetyeast Ridge. *Nature, Lond.* **258**, 303–307.

Khan, A.M. (1976a). Palynology of Tertiary sediments from Papua New Guinea. I. New form genera and species from upper Tertiary sediments. *Aust. J. Bot.* **24**, 753–781.

Khan, A.M. (1976b). Palynology of Tertiary sediments from Papue New Guinea, II. Gymnosperm pollen from upper Tertiary sediments. *Aust. J. Bot.* **24**, 783–791.

King, B., M. Woodcock, and E.C. Dickinson (1975). "A Field Guide to the Birds of South-east Asia". Collins, London.

Koopmans, B.N. and P.H. Stauffer (1968). Glacial phenomena on Mount Kinabalu, Sabah. *Bull. geol. Surv. Malaysia Borneo Reg.* **8**, 25–35.

Laurie, E.M.O. and J.E. Hill (1954). "List of Land Mammals of New Guinea, Celebes and adjacent Islands, 1758–1952". British Museum (Natural History), London.

Lincoln, G.A. (1975). Bird counts either side of Wallace's Line. *J. Zool., Lond.* **177**, 349–361.

Lydekker, R. (1896). "A Geographical History of Mammals". Cambridge, University Press, Cambridge.

McDowall, R.W. (1981). The relationships of Australian freshwater fishes. *In* "Ecological Biogeography of Australia" (A. Keast, ed.), pp. 1251–1274. W. Junk, The Hague.

Mayr, E. (1941). "List of New Guinea Birds". American Museum of Natural History, New York.

Mayr, E. (1944*a*). Wallace's Line in the light of recent zoogeographic studies. *Q. Rev. Biol.* **19**, 1–14.

Mayr, E. (1944*b*). The birds of Timor and Sumba. *Bull. am. Mus. nat. Hist.* **83**, 127–194.

Mayr, E. (1944*c*). Timor and the colonization of Australia by birds. *Emu* **44**, 113–130.

Mayr, E. (1976). Discussion, symposium on the origins of Australasian avifauna, *Proc. Intern. Ornith. Congr., Canberra* **16**, 98.

Medway, Lord (1972). The Quarternary mammals of Malesia; a review. *In* "The Quarternary Era in Malesia" [Transactions of 2nd Aberdeen–Hull Symposium] (P. Ashton and M. Ashton, eds), pp. 63–98. Department of Geography, University of Hull, Hull.

Mertens, R. (1930). Die amphibien und reptilien der inseln Bali, Lombok, Sumbawa und Flores. *Abh. Senckenb. Naturforsch. Ges.*, **42**, 115–344.

Meyer, A.B. and Wigglesworth, L.W. (1898). "Birds of the Celebes".

Miller, J. (1972). Palynological evidence for change in geomorphology, climate, and vegetation in the Mio-Pliocene of Malesia. *In* "The Quarternary Era in Malesia" (P. Ashton and M. Ashton, eds.), pp. 6–34. Department of Geography, University of Hull, Hull.

Musser, G.G. and Boeadi (1980). A new genus of murid rodent from the Komodo Islands in Nusatenggara, Indonesia. *J. Mammal.* **61**, 395–413.

Myers, G.S. (1951). Fresh-water fishes and East Indian zoogeography. *Stanford ichthyol. Bull.* **4**, 11–21.

Nix, H.A. and Kalma, J.D. (1972). Climate as a dominant control in the biogeography of northern Australia and New Guinea. *In* "Bridge and Barrier: The Natural and Cultural History of Torres Strait" (D. Walker, ed.), pp. 61–92. Australian National University Press, Canberra.

Pilseneer, P. (1904). "La Ligne de Weber", limite zoologique de l'Asie et de L'Australia". *Bull. Acad. r. Belg. Cl. Sci.* **1904**, 1001–1022.

Pizzey, G. (1980). "A Field Guide to the Birds of Australia". Princeton University Press, Princeton, N.J.

Rand, A.L. and E.T. Gilliard (1968). "Handbook of New Guinea Birds". Natural History Press, Garden City, New York.

Raven, H.C. (1935). Wallace's Line and the distribution of Indo–Australian mammals. *Bull. Am. Mus. nat. Hist.* **68**, 179–293.

Raven, P.H. (1979). Plate tectonics and Southern Hemisphere biogeography. *In* "Tropical Botany" (K Larsen and L.B. Holm-Nielsen, eds), pp. 3–24. Academic Press, London and New York.

Raven, P.H. and D.I. Axelrod (1974). Angiosperm biogeography and past continental movements. *Ann. Mo. Bot. Gdn.* **61**, 539–673.

Rensch, B. (1930). "Eine biologische reise nach den Kleinen Sunda Inseln". Borntraeger, Berlin.

Rensch, B. (1931). Die vogelwelt von Lombok, Sumbawa, und Flores. *Mitt. Zool. Mus. Berl.* **17** (4), 451–537.

Rich, P.V. (1975). Antarctic dispersal routes, wandering continents, and the origin of Australia's non-passeriform avifauna. *Mem. natn. Mus. Vict.* **36**, 63–126.

Rich, P.V. and van Tets, G.F. (1982). Fossil birds of Australia and New Guinea: phylogenetic and biostratigraphic input. *In* "The Fossil Vertebrate Record of Australia" (P.V. Rich and E.M. Thompson, eds.), pp. 235–384. Publ. Monash Univ., Melbourne.

Richards, P.W. (1976). "The Tropical Rain Forest" (3rd edn). Cambridge University Press, Cambridge.

Rickleffs, R.E. and G.W. Cox (1972). Taxon cycles in the West Indian avifauna. *Am. Nat.* **106**, 195–219.

Ripley, S.D. (1959). Competition between sunbird and honeyeater species in the Moluccan Islands. *Am. Nat.* **93**, 127–132.

Rumney, G.R. (1968). "Climatology and the World's Climates". Macmillan, London.

Sarasin, P., and Sarasin, F. (1901). "Uber die geologische Geschichte der Insel Celebes suf Grund der Tierverbreitung". Wiesbaden.

Schuster, R.M. (1972). Continental movements. "Wallace's Line", and Indomalayan–Australasian dispersal of land plants: some eclectic concepts. *Bot. Rev.* **38**, 38–86.

Schuster, R.M. (1976). Plate tectonics and its bearing on the geographical origin and dispersal of Angiosperms. *In* "Origin and early evolution of Angiosperms" (C.B. Beck, ed.), pp. 48–138. Columbia University Press, New York.

Sclater, P.L. (1958). On the general geographical distribution of the Class Aves. *J. Linn Soc.* (*Zool.*) **37**, 130–145.

Scrivenor, J.B., Burkill, I.H., Smith, M.A., Corbet, A.S., Airey Shaw, H.K., Richards, P.W. and Zeuner, F.E. (1943). A discussion of the biogeographic division of the Indo–Australian Archipelago with criticism of the Wallace and Weber Lines and of any other dividing lines and with an attempt to obtain uniformity in the names used for the divisions. *Proc. Linn. Soc. Lond.* **154**, 120–165.

Simpson, G.G. (1961). Historical zoogeography of Australian mammals. *Evolution, Lancaster, Pa* **15**, 431–446.

Simpson, G.G. (1977). Too many lines: the limits of the Oriental and Australian zoogeographic regions. *Proc. Am. phil. Soc.* **121**, 107–120.

Smith, M. (1927). Contributions to the herpetology of the Indo–Australian Region, *Proc. zool. Soc. Lond.* **1927**, 199–225.

Smythies, B.E. (1960). "The Birds of Borneo". Oliver and Boyd, Edinburgh.

Speecht, R.L. (1981). Major vegetation formations of Australia. *In* "Ecological Biogeography of Australia" (A. Keast, ed.), pp. 163–298. W. Junk, The Hague.

Steenis, C.G.G.J. van (1979). Plant-geography of east Malesia. *Bot. J. Linn. Soc.* **79**, 97–178.

Stresemann, E. (1939–41). Die Vogel von Celebes. *J. Ornith.* **87**, 299–425; **88**, 1–135, 389–487; **89**, 1–102.

Tate, G.H.H. (1946). Geographic distribution of bats in the Australian Archipelago. *Am. Mus. Novit.* **1323**, 1–21.

Tyler, M.J. (1979). Herpetofaunal relationships of Australia and South America. *In* "The South American Herpetofauna: its origin, evolution and dispersal" (W.E. Duellman, ed.), *Monogr. Mus. nat. Hist. Univ. Kansas* **7**, 73–106.

Wallace, A.R. (1860). On the zoological geography of the Malay Archipelago. *J. Linn Soc.* (*Zool.*) **2**, 1104–1108.

Webb, L.J. and J.G. Tracey (1981). Australian rainforests: patterns and change. *In* "Ecological Biogeography of Australia" (A. Keast, ed.), pp. 605–694. W. Junk, The Hague.

Whitmore, T.C. (1975). "The Rain Forests of the Far East". Clarendon Press, Oxford.

Zaklinskaya, E.D. (1978). Palynological information for Late Pliocene–Pleistocene deposits recovered by deep-sea drilling in the region of the island of Timor. *Rev. Palaeobot. Palynol.* **26**, 227–241.

14 | Planarians, Plurality and Biogeographical Explanations

IAN R. BALL

Institute for Taxonomic Zoology, University of Amsterdam, Postbus 20125, 1000 HC Amsterdam, The Netherlands

Abstract: The development of comparative biology in recent years has taken an unusual turn. Phylogenetic systematics has become "transformed cladistics" with the result, so it seems, that evolution has become irrelevant to biological classification. At the same time some biogeographers have taken on the role of the alchemists of old and have sought infallible methods whereby base data may be transformed into golden theories of biotic relationships. But biogeography is not yet a formalized science. The biogeographer must concern himself with a full range of phenomena concerning chorology (area analysis), commonality (both continuous and disjunct), uniqueness (endemism), species mobility (active and passive) and speciation (both course and cause), and these facets are as important in themselves as any rigid rules of procedure for the erection of biogeographical theories. These topics are discussed with special reference to the study of Tricladida, or planarians. The terrestrial planarians are of relatively little interest to the historical biogeographer in that their distribution has been heavily influenced by man. Moreover, their taxonomic relationships are poorly known. The marine planarians, littoral organisms without larvae, show a high degree of endemicity, but almost certainly this is an artifact due to the small number of specialists that have worked upon them and thus it is of little theoretical interest at present. Some marine forms show amphi-atlantic and panaustral disjunctions at the species level. In contrast the freshwater planarians are well known and in general each of the major land masses has it own indigenous species; exceptions usually may be explained reasonably in terms of man's influence. Recent hypotheses concerning the origin and history of the freshwater planarians are critically reviewed.

Systematics Association Special Volume No. 23, "Evolution, Time and Space: The Emergence of the Biosphere", edited by R.W. Sims, J.H. Price and P.E.S. Whalley, 1983, pp. 409–430. Academic Press, London and New York.

> In biogeography perhaps more than in any other subject, a holistic approach is necessary. The solution of any biogeographic problem, no matter how small, requires the following of a great many varied lines of enquiry.
>
> E.C. Pielou (1979:v)

INTRODUCTION

Of the four fundamental principles of logic, the syllogism, the law of identity, the law of non-contradiction, and the law of the excluded middle, it is the latter alone that seems to have captured the imagination of biogeographers. The conflict between the so-called dispersalist and vicariance schools of biogeography (see Craw, 1978) has led to a hardening of positions, an avoidance of middle ground, which at times must seem unproductive to those who practise biogeography by studying real organisms rather than Euclidean diagrams. Croizat (1978, 1981), a vigorous advocate of the vicariance school, clearly believes that the competition between the two dogmas of dispersal and vicariance is such that one model will replace the other, the dispersalist model being the one marked for extinction. But while it is easy to agree with him that biogeography is on the threshold of change I suggest that the change is to be one of emphasis rather than one of content; the choice between the two viewpoints is not as rigid and clear-cut as Croizat and some of his followers seem to imply. It is my view that the conflict between the rival schools has become counter-productive; it is bogged down in what are essentially formalistic arguments often with little biological relevance.

A complete study of the biogeography of a group of organisms involves the consideration of five aspects, *viz.* chorology, commonality, uniqueness, species spread, and speciation.

By chorology I mean the general assessment of distribution patterns so as to recognize faunistic and floristic regions of the earth. The regions so recognized are generally agreed upon and provide a useful linguistic and descriptive aid to discussion. Chorology as such will not further be discussed here.

Commonality and uniqueness are aspects of that part of biogeography which may be considered simply as pattern analysis involving characters, populations, species, or higher taxa. Commonality includes continuous distributions, presumably a product of species spread, and disjuncts across barriers, variously argued as being products of dispersal or of vicariance of an ancestral biota. Uniqueness refers to the phenomenon of endemism, a phenomenon that has achieved some high importance in recent writings on biogeographical theory (Platnick and Nelson, 1978; Nelson and Platnick, 1978).

Species spread and speciation are aspects of process analysis whereby the science of biogeography becomes inextricably bound with the sciences of ecology on the one hand, and the phylogenetic aspects of evolution on the other. These processes have achieved a special importance in the writings of Croizat (e.g. 1964) who equates species movement with "translation in space" (during periods of "mobilism") and allopatric speciation with "differential form-making" (during periods of "immobilism"), the two together comprising for him the entire science of evolutionary biogeography. It is not clear whether "mobilism" and "immobilism" are to be construed as general statements covering particular periods of earth history but I would think that at any one time both "mobilism" and "immobilism" could be operational depending on the nature of the biota involved. The reality of "immobilism" is demonstrated by the fact that it is possible to write viable regional floras and faunas; and the concept of species spread, or "mobilism", is almost axiomatic for historical biogeography (Ball, 1979).

No biogeographer, whatever the group in which he is interested, can afford to ignore these topics. Yet while studies of all of them have played a role in the development of biogeography as a science, it cannot be denied that they have been variously emphasized in the course of time. Thirty years ago, for example, there was a general belief in the stability of oceans and continents with a consequent biogeographical emphasis on commonality and species mobility. More recently there has been a tendency to regard biology as relatively unimportant in biogeography and to consider the latter as an adjunct to geology, or as an enterprise principally concerned with the formal analysis of endemic distributions (Rosen, 1978; Platnick and Nelson, 1978).

In this paper, and since I was given the task of talking on planarians, I wish to consider these problems, and others, using a review of planarian biogeography to provide the examples. My intention is to illustrate the diversity of phenomena with which any biogeographer has to deal and the difficulties of collating individual data into valid universal statements, whilst at the same time hoping to stimulate work on a group of animals which I believe to be so interesting and valuable to the systematist and biogeographer. To assist the non-specialist with the names used, Table I is provided, based on current views on the classification of planarians.

Table I. Classification of the Tricladida.

1. DIPLONEURA
   - 1.1. Terricola (the land planarians)
2. HAPLONEURA
   - 2.1. Maricola (the marine planarians) e.g. *Procerodes*
   - 2.2. Paludicola (the freshwater planarians)
     - 2.2.1 Dugesioidea
       - 2.2.1.1. Dugesiidae e.g. *Dugesia, Girardia, Schmidtea, Neppia, Spathula, Cura, Eviella*
     - 2.2.2. Planarioidea
       - 2.2.2.1. Planariidae e.g. *Planaria*
       - 2.2.2.2. Dendrocoelidae

THE BIOGEOGRAPHY OF PLANARIANS

It is not the intention here to provide an exhaustive review of the geographical distribution of planarians, but rather to give an overview of the range of phenomena that are exhibited by the group so as to facilitate discussion of some broader issues. Included here are pattern analysis as exemplified by continuous, disjunct, and endemic distributions, and process analysis including species spread and speciation.

1. Pattern Analysis

In what follows under this heading it must be borne in mind that so far as aquatic planarians are concerned we are at a relatively primitive stage in their phylogenetic analysis (Ball, 1977a). Indeed, there have been attempts at strict phylogenetic analysis only for the Dugesiidae (Ball, 1974, 1975a), and there can be little doubt that substantial changes are to be expected in so far as phylogenetic principles come further to be applied (see Ball, 1981).

(a) Commonality—continuous distributions. There are no cosmopolitan species of marine or freshwater planarians. Most, indeed, are confined to particular continents where they may achieve a wide, more or less

continuous, distribution. Nor are there any cosmopolitan genera. Among the Dugesiidae the genus *Dugesia* s.s. has the widest continuous distribution, its species being found throughout the Palaearctic, and Oriental regions, and in Africa (Ball, 1975a: Fig. 6). The marine genus *Procerodes*, under current taxonomy, seems almost cosmopolitan in that species are found on the shores of western Europe, the Mediterranean, Pacific and Atlantic North America, South America and many southern hemisphere islands, but it is doubtful that the genus forms a monophyletic group (Ball, 1977a, 1977b). Within the freshwater triclads the most primitive family, the Dugesiidae, are cosmopolitan as a whole, the other families being confined to areas north of the Tethys geosyncline (Ball, 1974: Fig. 12).

(b) Commonality—disjunct distributions. As has been mentioned most species of aquatic triclads are confined to, or are endemic to, particular land masses and therefore species-level disjuncts are rare. Within the Paludicola both *Dugesia polychroa* and *D. tigrina* show amphi-Atlantic disjunctions but these almost certainly are due to anthropochore quantum dispersal (see below). In contrast the Maricola show a few species-level disjunctions, both amphi-Atlantic and pan-austral (Ball, 1975b, 1977b).

At the generic level the principal disjunctions are, for the Paludicola, in the southern hemisphere and involving South America, South Africa, and Australia/New Zealand as shown, for example by the dugesiid genera *Neppia* (Ball, 1975a: Fig. 5) and *Dugesia* (*Girardia*) (Ball, 1975a: Fig. 4). The only claimed strictly amphi-atlantic distribution, that of the genus *Planaria* (Ball, 1975a: Fig. 11) has been shown on anatomical and karyological grounds to be a false disjunct in that the genus is not monophyletic (Ball and Gourbault, 1978). An unusual disjunction shown by the genus *Cura* (eastern North America, and Notogaea) is also rendered suspect by recent anatomical, karyological, and ultrastructural studies which suggest a Pangaean vicariant origin of the Dugesiidae (Ball, 1977a, 1981; Gremigni, 1979) in contrast to the dispersal elements of an earlier hypothesis (Ball, 1974, 1975a); elements of species spread remain (see below). In this sense, then, *Cura foremanii* (eastern North America) and *C. pinguis* (Notogaea) may be considered relictual forms which do not themselves form a monophyletic group; an example of "wing dispersal" in the terminology of Croizat (1964).

(c) Uniqueness—endemic distributions. The study of endemics and their areas has played an important role in biogeography for a century and a half (see Nelson, 1978). The biogeographical and evolutionary theories of Willis (1922, 1940), and the analytical methods proposed by Platnick and Nelson (1978) rely heavily on a correct assessment of endemic forms. Notwithstanding their crucial importance, it must be acknowledged that the recognition of endemism depends upon negative evidence, that is, the failure to find the taxon in question outside the area or locality under consideration.

Endemics are of two types: true endemics, ones which have evolved *in situ* and not spread from the one area; and relictual endemics, or epibiotics, which are species that once had a range much wider than now known. Coope (this volume) has given many examples of epibiotics. Whilst it may be important to distinguish between endemics and epibiotics, it must be admitted that, in the absence of a fossil record, there are no sure ways to do this. Occasionally biological and anatomical studies, particularly of primitive features, coupled with karyological data can provide clues, as for some aquatic triclads (Ball, 1977a: 26–30).

Two of the most interesting areas of endemism for the aquatic triclads are the ancient Lakes Ochrid (Macedonia) and Baikal (Siberia). Both these lakes contain species flocks, the individuals often being of bizarre appearance (Stankovic, 1960; Kenk, 1978; Kozhov, 1963; Porfierieva, 1977). Unfortunately, most of the work on these endemic planarians has been of a basic taxonomic nature; there have been no ecological and evolutionary studies such as have been performed on the flocks of cichlid fishes in the African great lakes. Whatever the mechanisms of diversity in these two lakes, and however the species are partitioned, it seems more than likely that most, if not all, are true endemics.

True endemics may arise by autochthonous adaptive radiation or vicariance of an existing fauna. With respect to planarians interesting areas for studying this topic would be the Caribbean (Ball, 1971) and the Australian Alps (Hay and Ball, 1979). We have only begun to analyse the Caribbean freshwater triclads but all are *Dugesia* (*Girardia*) species and each island seems almost solely to harbour its own endemic form; thus the endemics are mostly monocentric. The exception is *Dugesia* (*Girardia*) *arimana* which is known from Trinidad, St Vincent, and Venezuela (Ball, 1980: Table 1). The tectonic history of this area is extremely complex and may well provide the explanation of current difficulties in analysing phylogenetically the Caribbean freshwater planarians.

In the Victorian Alps of south eastern Australia, endemic species of the genera *Spathula* and *Reynoldsonia* (Dugesiidae) have been described. For the most part, each mountain top has its own species, but one is bicentric and occurs on two mountains. Nevertheless, when sympatry occurs there is ecological separation, for each has its own habitat (Hay and Ball, 1979). It is noteworthy that the bicentric species, *Spathula agelaea*, occurs in intermittent streams and thus must have a drought resistant (dispersive?) stage in its life-cycle.

True endemics may also arise as a result of species spread into a new niche. It seems, for example, that freshwater planarians may be historically absent from many oceanic islands in the southern hemisphere and marine (Maricola) forms have been able to invade the vacant niche, whence speciation may occur. Thus, two marine planarians are endemic to the freshwaters of St Helena (Ball, 1977b) and it seems probable that an unusual triclad from phreatic waters in Tahiti has originated from marine ancestors which invaded this restricted habitat (Ball, 1977a).

Relictual endemics, epibiotics, are as indicated, hard to recognize; but there is evidence to suggest that one primitive Australian species, *Eviella hynesae* Ball (1977c), known only from a single locality, is a relictual form (Ball, 1977a). It has also been considered that certain unusual triclads from the waters of Mexican caves are marine relictual species (Mitchell and Kawakatsu, 1972). This possibility has similarly been suggested for species of Lake Baikal all of which have hitherto been assigned to the Dendrocoelidae (Ball, 1975b).

The high degree of endemism in aquatic triclads, especially freshwater forms, implying their great value to the biogeographer, makes it all the more regrettable that there has been so little phylogenetic and karyological work on these organisms over a broad geographical range.

2. Process Analysis

(a) Species spread. In the present review I shall consider species spread as comprising quantum dispersal, diffusion, and secular migration (see Pielou, 1979).

Quantum dispersal, known also as jump- or long-distance dispersal, is the movement of individual organisms over long distances, and across barriers, in a single lifetime, followed by successful colonization. Such dispersal is usually passive and it includes all the classic examples of anemochore, hydrochore, biochore, and anthropochore dispersal (see Udvardy, 1969). On arrival there must be suitable conditions, and a large enough population, for viability.

Diffusion is the movement of individuals across hospitable terrain over a period of many generations.

Secular migration is diffusion taking place so slowly that the diffusing species undergo appreciable evolutionary change in the process.

The role of species movement in biogeography is a contentious issue (Croizat *et al.*, 1974; McDowall, 1978; Craw, 1979). Vicariance biogeographers, including myself, have repeatedly decried the overemphasis on quantum dispersal in biogeographical studies, that is chance dispersal over pre-existing barriers, while allowing for dispersal, through diffusion or secular migration, as an explanation of the attainment of cosmopolitanism prior to the erection of barriers and vicariant disruption. One reason for the rejection of quantum dispersal explanations by vicariance biogeographers is that they believe such elements to be untestable within the Popperian framework of science. This may be so, but deductive explanations, although perhaps to be preferred, are not the only modes of explanation used in the sciences, and we must be cautious about confining ourselves to one prevailing philosophical approach. Nagel (1961), for example, has recognized four classes of scientific explanations, the deductive, probabilistic, teleological, and genetic. Most data on quantum dispersal would lie in the probabilistic category where usually the evidence is but circumstantial, even though occasionally overwhelming.

Quantum dispersed species may be expected to show immature ranges (that is, ranges not bounded by factors known to be limiting in the parental area), evidence of rapid expansion, and often r-strategy. Other characteristics may be important in relation to the presumed mode of transport.

As an example, we may consider *Dugesia polychroa*. This species is widespread in Europe and in 1969 was found in North America in Lake Champlain (New York), Bay of Quinte (Lake Ontario), and the St Lawrence River (Ontario) (Ball, 1969). These are interconnected shipping routes. Now the species extends from Quebec City to Toronto Harbour, where it is the dominant freshwater triclad in certain localities (Boddington and Mettrick, 1974); karyological evidence has shown it to be the vigorous triploid biotype-B (Benazzi and Benazzi-Lentati, 1976). The hypothesis that the species has recently been introduced into North America does not appear to be falsifiable in the Popperian sense, yet nonetheless the hypothesis is surely viable because of strong circumstantial evidence. It is also likely that the Canadian populations are undergoing an ecological shift (Boddington and Mettrick, 1974) and we may predict, therefore, that speciation will occur. If it does then we will have a taxonomic disjunction which, although resulting from the allopatric speciation of a founder population, has nothing to do with the Croizatian vision of vicariance.

Similar reasoning may also be used to explain the occurrence of *Dugesia* aff. *gonocephala* on the island of Madeira. Apart from the fact that we know historically that much of the flora and fauna of that island has been introduced it is noteworthy that this species has been found as an asexual triploid race, presumably an r-strategist, in but two rivers, both of which are known to have been artificially stocked with fish from the mainland of Portugal.

Other examples could be given (e.g. Ball and Fernando, 1970) but the ones above suffice to indicate that good circumstantial evidence for quantum dispersal in particular instances can be adduced, although it is noteworthy that in these particular cases only anthropochore dispersal is invoked. I know of no evidence, circumstantial or otherwise, for the quantum dispersal by hydrochore, anemochore or biochore means for aquatic triclads, although such evidence is known for other organisms (Udvardy, 1969; Gerlach, 1977).

The distributions of some marine planarians pose interesting problems. These benthic animals are unusual among marine littoral organisms in that they have no larval, dispersive, stage in their life-cycle, they cannot swim free nor have they ever been recorded in the plankton. Nevertheless, there are some species common to the Atlantic shores of North America and western Europe (Ball, 1975), and others are distributed on the shores of islands across the southern hemisphere (Ball, 1977b). Clearly these distributions could have come about as a result of continental drift, as has been suggested for amphi-Atlantic meiofaunal elements (Sterrer, 1973); since most planarians can go for long periods of time without food, passive dispersal in currents, perhaps attached to floating objects, cannot be ruled out, especially as the surrounding medium, seawater, is not inhospitable. No corroboration of this is yet available.

As stated above, vicariance biogeographers have tended to de-emphasize the role of quantum dispersal in biogeography, most explicitly in terms of problems dealing with disjunct, endemic, allopatric taxa (Nelson and Platnick, 1978). It does seem reasonable to caution that the automatic invocation of quantum dispersal events to explain observed disjunctions is often no real explanation; it obviates the need, and desire, to search for alternative, perhaps more meaningful, explanations (Ball, 1975a), but this is not to deny the reality of quantum dispersal in certain cases. Only its efficacy as the first chosen causal explanation is denied, as are also denied inductive leaps from one situation to another. Thus, because there is convincing circumstantial evidence that *Dugesia polychroa* has crossed the Atlantic this is no reason to assume or infer that, for example, the endemic *Dugesia seclusa* of the Crozet Islands has arrived there by trans-oceanic dispersal (cf. McDowall, 1973). Circumstantial evidence of the type discussed applies only to the particular case and a universal generalization cannot be obtained from it. The rejection of quantum dispersal explanations as initial explanations for observed disjunctions is only a methodological principle resulting from their lack of testability and an application of the heuristic principle of Popper that we select the most testable theories first. This is a point thoroughly misunderstood in the critique by McDowall (1978, 1980).

The continuing spread of *Dugesia polychroa* in North America, and of the introduced *D. tigrina* in Europe (Gourbault, 1969; van der Velde, 1975) provide examples of diffusion and there is also circumstantial evidence that some freshwater triclads in Britain are still extending their northwards ranges following the last glaciation (Reynoldson, 1966).

Secular migration is difficult to identify but perhaps the broad distribution of the polytypic *Dugesia tigrina* in North America and of the *D. gonocephala* species group in Palaearctic are in part the result of this process, although there appears to be no way of testing such a hypothesis at present. Such dispersal was implicitly assumed in my own earlier biogeographical writings in planarians (Ball, 1974; 1975a), although for some reason McDowall (1978) was unable to see this.

Much has been made by some workers of the criterion of dispersal provided by sympatric distributions (Croizat *et al.*, 1974; Rosen, 1978) but clearly this criterion fails in cases of non-allopatric speciation, which seem to be not uncommon in planarians, and probably many other organisms (White, 1978).

(b) Speciation.

> ... the mere fact that vicariant form-making is the rule of life tells us that immobilism is after all by far more important as a factor of evolution over space and time than mobilism.
>
> (Croizat, 1964:212).
>
> ... Caribbean biogeography is reviewed to illustrate within a particular geographic framework how evidence of plant and animal distribution may be interpreted without commitment to special assumptions other than allopatric speciation.
>
> (Rosen, 1975:43)

These two quotations demonstrate clearly the importance of allopatric speciation models to vicariance biogeography, and the point is made with equal emphasis in the papers of Nelson and of Platnick previously cited. Yet there is an increasing body of opinion arguing for alternative modes of speciation (White, 1973, 1978) which may well occur more commonly than is realized.

It has long been known that chromosomal changes, including translocations, aneuploidy, polyploidy with or without subsequent diploidization, and possibly hybridization, have played an important role in the evolution of freshwater triclads (Benazzi and Benazzi-Lentati, 1976). Many species and species groups are remarkably uniform in their karyotypes, being sexual diploids with or without the alternative of fission; for them it is probable that the allopatric model applies. Others, however, show divisions into races or biotypes characterized by various kinds of chromosomal differences, sexual and/or asexual reproduction, pseudogamy and so forth. In many of these cases, it seems unlikely that the allopatric model can apply. Some examples will illustrate the points.

Elizabeth de Vries (in preparation) has discovered some unusual polymorphisms in *Dugesia* "*gonocephala*" from the region of Montpellier in southwest France. A triploid asexual race, 3n = 24, occurs in most of the streams of the area. In one stream, however, there were also sexual diploids, 2n = 16, living sympatrically with asexual triploids showing a unique chromosomal polymorphism in one chromosome of one triplet; this triplet consisted of one acrocentric and two metacentric chromosomes. In the laboratory, some of these asexual individuals developed sexual organs, whereupon it was seen that their reproductive anatomy is quite distinct from that of the normal diploids. The differences are such as to convince us that the two sympatric forms are physiologically and cytologically isolated and it seems reasonable to assume that we have here a case of sympatric speciation.

Another interesting chromosomal polymorphism was noted independently by de Vries and by J. Baguñá in certain populations of *Dugesia* (*Schmidtea*) *mediterranea*, a species known from Catalonia, Sardinia, Corsica, and Sicily. On its original description, Benazzi *et al.* (1975) argued for a microplate fragmentation theory to explain its peculiar distribution and this has now received support from the discovery of the species in Mallorca (Ball, unpublished). Furthermore, certain populations, characterized by unusual reproductive behaviour and by a chromosomal polymorphism in the third pair of elements, are distributed in such a way as to conform with some geologists' views on the sequence of break-up of the relevant microplate (de Vries and Baguñá, in preparation). This study is continuing, but it does seem that here is a pattern of distribution agreeing well with classical vicariant concepts of allopatric speciation combined with earth history.

The role of hybridization in the evolution of planarians is less easy to elucidate. There is evidence that some biotypes within the *Dugesia gonocephala* group have arisen by crossing (Benazzi and Benazzi-Lentati, 1976) and the polyploid *Phagocata velata* of eastern North America, a species not known in the diploid condition, is probably an allopolyploid form (Gourbault and Ball, unpublished). Hybridization in nature is, of course, hard to detect but enzyme studies using iso-electric focussing revealed that, in some Dutch and French localities where *Polycelis nigra* and *P. tenuis* occur together, the two species are in genetic contact (Biersma and Wijsman, 1981); the two species interbreed under laboratory conditions (Benazzi, 1963).

It is sometimes claimed that the species flocks of ancient lakes must have arisen sympatrically although in general allopatric explanations have prevailed (White, 1973, 1978). Kozhov (1963), however, made a case for the sympatric model in Lake Baikal and this was accepted by White (1978). Although much work needs to be done, from the above and from the data summarized by Benazzi and Benazzi-Lentati (1976) and by White (1978), it seems clear that non-allopatric modes of speciation are of significance in the evolution of planarians. This has implications for the study of their biogeography.

GENERAL DISCUSSION—BIOGEOGRAPHICAL EXPLANATIONS

This review of the range of phenomena exhibited by aquatic planarians has been undertaken with the avowed intent of illuminating the quotation at the head of this paper. Within the fields outlined of both pattern analysis (commonality and uniqueness) and process analysis (species spread and speciation) the variety of attributes is such that it is surely not yet possible to apply a single, all-embracing explanatory principle of historical biogeography. There is no doubt that specialists in other groups of organisms are facing the same dilema, as has been well evidenced in another recent symposium volume (Nelson and Rosen, 1981), although the relative emphases of the various factors vary from group to group.

As a specialist in planarians, I am faced with problems of allopatric, sympatric, and hybrid speciation. I am sure that in some cases quantum dispersal has occurred. I suspect that some speciation has been multiple and not dichotomous. I know that diffusion occurs and I suspect that secular migration has occurred. I believe that, in some cases, I can recognize both true endemics and epibiotics. And perhaps there are 1001 other problems which I have not yet encountered. With this range before me what universal principles am I to apply, and who is to decide which of all these aspects is worthy of my immediate attention?

Croizat (1964) attempted to provide the universal principle with what has now become his slogan, *viz*. earth and life evolve together. This is undoubtedly true but unfortunately it is not the panacea some think it to be. The naïve belief that so-called "generalized tracks" as statistical measures of biotic relationships were more than preliminary rough estimates of the latter (Croizat *et al*., 1974; Rosen, 1975) and can provide the universal principles that we seek is now surely discredited. Severely criticised by both Ball (1975a) and McDowall (1978) this viewpoint now seems to have been abandoned by even its most ardent supporters (cf. Rosen, 1978; Platnick and Nelson, 1978) and, moreover, it can no longer be accepted that sympatry is always evidence of dispersal. But with the abandonment of this position, historical biogeography has developed in rather different ways as regards both method and purpose.

In a recent review, Patterson (1981a) discussed modern trends in historical biogeography making critical assessments of what he termed Ball's method, Rosen's 1976 method, Keast's method and the method of Platnick and Nelson (1978) and Rosen (1978). His inference that I proposed a distinct methodology for solving problems of historical biogeography I find somewhat embarrassing, in that this is something that I have never myself claimed. Patterson's criticisms, therefore, are for the most part irrelevant to what I was trying to say.

The stress on "method" that so many biogeographers now apply is apparently based on the belief that there is one biogeography, with a single unifying principle, that can be discovered by the correct logical procedures. I believe this to be entirely mistaken; there are no rules for erecting hypotheses that can be said, *a priori*, to be infallible in their results[1]. In this sense, the sense that testability is what counts, method and invention are equivalent in the erection of hypotheses, and I see nothing in Patterson's critique to counter this claim. He stated (1981a:452) that a sound biogeographical hypothesis comes from a deductive choice among known alternatives, but does not say from whence cometh these alternatives, especially the first! Now, I concur with Colles (1969) that the process of arriving at an hypothesis is of interest in that it relates to the productivity of science, but it has nothing to do with its validity. If we like we may seek hypotheses in a crystal ball. What really matters is that they be well corroborated. That is why I have laid great stress on the testability of biogeographical theories rather than on preconceptions of how we should arrive at them (Ball, 1975a, 1979)[2].

What I tried to do in my earlier paper was to stimulate critical thinking so as to avoid what may be termed category errors. In this I was influenced, in part, by Cassirer's theory of scientific knowledge (see Kaufman, 1949). Within the physical sciences, Cassirer recognized a hierarchy of types of statements which was intended to avoid the problems associated with the "vicious circle principle" in the analysis of empirical science, and for which Russell's famous "theory of types" was the intended solution in logic (see Chihara, 1973). It cannot be claimed that Cassirer's categories fit precisely the current situation in historical biogeography, but in distinguishing between descriptive, narrative, and analytical biogeography (Ball, 1975a), I was attempting to make the same sort of hierarchical distinctions for biogeographical statements with the intention of neutralizing pointless arguments that crossed, illegitimately, the boundaries within the hierarchy. Thus:

> I wish it to be understood that I am not suggesting that the different viewpoints of biogeography outlined throughout this paper are necessarily on trial. What I am saying is that in advocating a hypothesis to account for observed distributional data we must be clear into what category our explanation falls and proceed accordingly.
>
> (Ball, 1975a:422).

This is the limit of my so-called "method". Attempts to provide biogeographical explanations that are testable in the Popperian sense would be analytical by my criterion. Most dispersal hypotheses still fall under the narrative phase. There is still much work to be done at the descriptive level. Whereas all attempts to formulate broad principles must be based on the analytical level, this in no way can mean that all endeavours in descriptive and narrative biogeography are a waste of time or misguided. There may well be, indeed are, problems, methods, approaches appropriate to each level in the hierarchy. One studies dispersal, and expects different generalizations, in a way quite different to that appropriate for the analysis of endemic distributions.

Mayr's (1965, 1976) claim that inferences in biogeography are based upon the proper evaluation of three sets of information, the relative ages of various taxa, the determination of the dispersal capacities of taxa, and the distribution of related taxa, is one with which it is hard to disagree. But in his analysis of faunal components, Mayr stressed the various colonizer (dispersal) elements to the detriment of what he called autochthonous adaptive radiation, which latter seems to me to be very similar to, if not identical with, Croizat's vicariance[3].

It seems clear to me that the composition of a modern biota is likely to be the product of several speciation (vicariant and sympatric) and several dispersal events. The diversity of phenomena to be faced is such that it cannot be a simple question of dispersal or vicariance. Instead, it is more a question of refinement of the data so as to distinguish between the various dispersal and vicariant elements just as one has to distinguish between symplesiomorphy, synapomorphy and convergence, in assessing taxonomic similarity. Mayr (1965), with his bias towards stable continents and almost complete disregard of the role of vicariance, concentrated on the problem of delimiting the dispersal components, whilst Hennig (1966) tackled the vicariant components with his concept of multiple sister-group relationships. In the light of current stress on vicariance biogeography (Rosen, 1978; Platnick and Nelson, 1978), I emphasize my agreement that it is the vicariant components alone that, within limits, will enable us to make predictions concerning earth history, but such predictions are not the only, or even the principal, objectives of a science of biogeography. Again this is a point misunderstood by Patterson (1981a:453) when criticizing my hypothesis of planarian distributions.

The concept of multiple sister group relationships has undergone considerable revision in the hands of Platnick and Nelson (1978). They adopted the principle that, when faced with particular patterns of allopatric distributions, our first questions should be directed not to their cause but to whether or not there is conformity to a general pattern of relationships shown by the taxa endemic to the areas occupied. Apart from the rejection of initial investigations of causality, which would dispense with half the efforts of present day biogeographers, and the belief that correlation itself replaces or means causality, the problem of determining endemicity remains. Platnick (1977), for example, regarded cladograms as testable whereas phylogenetic trees are not, because the choice between trees rests on negative evidence, i.e. failure to find certain character distributions. Similarly, Nelson (1971) preferred cladograms of dichotomy rather than trichotomy because the latter rely on negative evidence[4]. Yet these positions seem curiously inconsistent with the stress now placed on endemic distributions, epibiotics notwithstanding, which distributions are indeed determined by negative evidence, i.e. failure to find the species in question in other geographical areas. Moreover, the question of the definition of an "area of endemism" arises for no clear definition has been given.

If, for example, I study the biotic relationships between western Europe, the Mediterranean, and North Africa, and remembering my primary interest in planarians, am I allowed to use the "endemism" of *Dugesia polychroa* in western Europe? The species does, after all, occur in North America, even though the evidence for its recent introduction there is overwhelming, as discussed above. Biogeographers are quite as justified in directing their first questions to cause as to possible correlations of endemicity with other groups, depending upon their interest. The question actually becomes one of deciding the exact area of endemism, i.e. the deleting of dispersal components, before a base for such correlations can be formed. I am aware, of course, that this brings us dangerously close once more to the "centre-of-origin" concept, albeit at the species level only, and involving "centre" in the broadest possible sense of the word.

Those who would make a revolution in biogeography, changing the content rather than the emphasis[5], would do well to remember the history of the "revolution" in physical science that took place at the beginning of this century. Einstein himself always stressed the continuity of the process of inquiry leading to the Special and General theories of Relativity and considered his theories as the perfection of classical physics rather than its destruction (Kaufman, 1949). Heisenberg

(1975) pointed out that at no time were the "inner-circle" attempting to make a revolution; indeed, on the contrary, they wished to preserve as much of the old physics as possible because, within the boundaries of its applicability, it had proved so universally useful.

This latter point, stemming from Heisenberg's ideas on closed and open theories in science, may be of value to biogeographers in our times of upheaval and debate, for often we pay too little attention to the boundary, limiting, conditions of our general theories. Thus, I would in no way question the logical validity and heuristic value of the proposals of Platnick and Nelson (1978) and Nelson and Platnick (1978) concerning the analysis of endemic distributions *whenever the conditions of their applicability exist*, i.e. when there are no complications raised by epibiotics, extinction, non-allopatric speciation and quantum dispersal. Yet all these phenomena clearly are a part of the history of life, they are therefore worthy of our attention. Indeed they must be studied if the principles of Platnick and Nelson (1978) are to be properly applied.

Kneale (1974) discussed the interesting point that science is about the frame of nature whereas history is about the content. Clearly the two biogeographers just discussed have done much to improve the frame of biogeography. But while vicariance biogeography may well be concerned with areas of endemism, their inter-relationships, and the geological factors that might have operated as causal agents (Nelson and Platnick, 1978:476) that does not exhaust the content of historical biogeography. Since biogeographers are perhaps as much historians as scientists, as also are geologists and cosmologists, they are as much concerned with individual cases, the content, as with the frame of nature. Or, to paraphrase Popper (1961:143), whereas theoretical sciences are mainly interested in finding and testing universal laws, the historical sciences take all kinds of universal laws for granted and are mainly interested in finding and testing singular statements. Paradoxical though it may sound, and the new rigour introduced by cladistics notwithstanding, the practising biogeographer must retain a pluralistic, although still rigorous, approach to his subject matter, if only because of the diversity of phenomena causing the distribution patterns with which he deals. The idea that such a pluralistic approach represents some weak compromise (cf. Patterson, 1981b) must be rejected; it is a reflection of the way things are!

ACKNOWLEDGEMENTS

Preparation of this lecture for the press was carried out under rather difficult circumstances and thus I beg the indulgence of both editors and readers. It would never have been done at all except for the privileges and facilities made available to me by the staff of Ziekenhuis Amsterdam Noord, to whom I am most grateful.

NOTES

1. My own earlier paper on planarian biogeography (Ball, 1975) contains a fundamental flaw (Ball, 1981) which no critic had observed simply, I believe, because they were more interested in my procedures than in my results.
2. The question of testability in biogeographical hypotheses is problematical. Many recent workers (e.g. Rosen, 1975, 1978; Patterson, 1981a) seem to believe that parsimony is a test in itself, rather than a methodological rule for selecting hypotheses for testing. This all too frequent misunderstanding has led to some rather curious positions. Thus, Rosen (1978) believed that distributions of sedentary organisms have the potential to falsify dispersal theories as applied to vagile organisms, a proposition which appears to me to be nonsense. Patterson's (1981a:449–453) views on the subject (see also notes 1, 4), in a critique of my work on planarians, are also incomprehensible to me. This is not the place for a detailed rejoinder but I state (1) Falsification of a cladogram may make a given biogeographical hypothesis irrelevant and therefore "falsifies" the latter. (2) That apparent falsifiers, perhaps stemming from the minimum acceptable number of internal inconsistencies of the hypothesis, may themselves be refuted is no criticism of the hypothesis; it is a normal method of procedure. All theories have some inconsistencies, especially cladistic ones, and these must be explained and tested; reliance on the principle of parsimony as a criterion of final acceptability is invalid. (3) New taxa can falsify both cladograms, by casting doubt on character-clines monophyly, etc., and biogeographical hypotheses. In the latter case, I would abandon my hypothesis if species of *Polycelis*, or *Planaria*, for example, were found in South America or Notogaea, and I was unable satisfactorily to explain these occurrences by quantum dispersal from the North. This is not simply a case of increasing a three-taxon statement to a four-taxon statement. (4) Both Patterson and Rosen have not avoided the circularity inherent in using geological data as a part of a biogeographical hypothesis and then using the hypothesis to make predictions concerning earth history. My hypothesis accepted certain geological data as given. It cannot therefore be used to make predictions concerning earth history. And if the geological data are wrong (e.g. concerning the Tethys sea) then the hypothesis is irrelevant and must be abandoned, wholly or in part.
3. Mayr (1976:556) lists six categories of faunal origins, *viz.* (1) authochthonous adaptive radiation, (2) continued single and (3) continued multiple origin

colonization, (4) fusion of two faunas, (5) successive adaptation, and (6) composite origin of all types. He also wrote (p. 563) that a fauna consists of faunal elements which differ in origin, age, and adaptation, and also (p. 553) that a fauna consists of unequal elements and no fauna can be fully understood until it is segregated into its elements and until one has succeeded in explaining the separate history of these elements. In this light it does not seem to be correct to say that all that he means by "unequal elements" are species with unique histories of dispersal as Nelson (1978:288) has charged.

4. In both phylogenetic systematics and biogeography the choice for dichotomies is no more than a methodological principle. Patterson (1981a), following Platnick (1977), believed that a trichotomy represents an unresolved problem; neither seems to consider that it may also represent a genuine trichotomous relationship.
5. Thus, Croizat (1981:517) "In summary, the line is by now sharply and finally drawn: either the Darwinians bury me, or I them".

REFERENCES

Ball, I.R. (1969). *Dugesia lugubris* (Tricladida, Paludicola), a european immigrant into North American freshwaters. *J. Fish. Res. Bd. Canada* **30**, 389–394.

Ball, I.R. (1971). Systematic and biogeographical relationships of some *Dugesia* species (Turbellaria, Tricladida) from Central and South America. *Am. Mus. Novit.* **2472**, 1–25.

Ball, I.R. (1974). A Contribution to the phylogeny and biogeography of the freshwater triclads (Platyhelminthes, Turbellaria). *In* "Biology of the Turbellaria" (N.W. Riser and M.P. Morse, eds), pp. 339–401. McGraw-Hill, New York.

Ball, I.R. (1975a). Nature and formulation of biogeographical hypotheses. *Syst. Zool.* **24**, 407–430.

Ball, I.R. (1975b). Contributions to a revision of the marine triclads of North America: the monotypic genera *Nexilis*, *Nesion* and *Foviella* (Turbellaria, Tricladida). *Can. J. Zool.* **53**, 395–407.

Ball, I.R. (1977a). On the phylogenetic classification of aquatic planarians. *Acta zool. fenn.* **154**, 21–35.

Ball, I.R. (1977b). La faune terrestre de l'Ile de Sainte-Hélène 2, Turbellaria. *Ann. Mus. Roy. Afr. Centr., ser. 8°, Zool.* **220**, 492–511.

Ball, I.R. (1977c). A new and primitive retrobursal planarian from Australian fresh-waters (Platyhelminthes, Turbellaria, Tricladida). *Bijdr. Dierk.* **47**, 149–155.

Ball, I.R. (1979). Problemen in de historische Biogeografie. *Vkbld. Biol.* **21**, 384–388.

Ball, I.R. (1980). Freshwater planarians from Colombia. A revision of Fuhrmann's types. *Bijdr. Dierk.* **50**, 235–242.

Ball, I.R. (1981). The phyletic status of the Paludicola (Tricladida). *Hydrobiologia*, **84**, 7–11.

Ball, I.R. and Fernando, C.H. (1970). Freshwater triclads (Platyhelminthes, Turbellaria) from Anticosti Island. *Nat. can.* **97**, 331–336.

Ball, I.R. and Gourbault, N. (1978). The phyletic status of the genus *Planaria* (Platyhelminthes, Turbellaria, Tricladida). *Bijdr. Dierk.* **48**(1), 29–34.

Benazzi, M. (1963). Il problema sistematico delle *Polycelis* del gruppo *nigratenuis* alla luce di richerche citologiche e genetiche. *Monit. Zool. Ital.* **70–1**, 288–300.

Benazzi, M., Baguñá, J., Ballester, R., Puccinelli, I. and del Papa, R. (1975). Further contributions to the taxonomy of the "*Dugesia lugubris-polychroa* group" with description of *Dugesia mediterranea* n. sp. (Tricladida Paludicola). *Boll. Zool.* **42**, 81–89.

Benazzi, M. and Benazzi-Lentati, G. (1976). "Animal Cytogenetics". vol. 1. "Platyhelminthes". Gebruder Borntraeger, Berlin and Stuttgart.

Biersma, R. and Wijsman, H. (1981). Studies on the speciation of the European freshwater planarians *Polycelis nigra* and *Polycelis tenuis* based on the analysis of enzyme variation by means of iso-electric focussing. *Hydrobiologia*, **84**, 79–85.

Boddington, M.J. and Mettrick, D.F. (1974). The distribution, abundance, feeding habits and population biology of the immigrant triclad *Dugesia polychroa* (Platyhelminthes, Turbellaria) in Toronto Harbour, Canada. *J. anim. Ecol.* **43**, 681–699.

Chihara, C.S. (1973). "Ontology and the Vicious Circle Principle". Cornell University Press, Ithaca.

Colles, D.H. (1969). The phylogenetic fallacy revisited. *Syst. Zool.* **18**, 115–126.

Craw, R.C. (1978). Two biogeographical frameworks: implications for the biogeography of New Zealand. A review. *Tuatara* **23**, 81–114.

Craw, R.C. (1979). Generalized tracks and dispersal in biogeography: a response to R.M. McDowall. *Syst. Zool.* **28**, 99–107.

Croizat, L. (1964). "Space, Time, Form: The Biological Synthesis". Published by the author, Caracas.

Croizat, L. (1978). Deduction, induction, and biogeography. *Syst. Zool.* **27**, 209–213.

Croizat, L. (1981). Biogeography: Past, Present, and Future. *In* "Vicariance Biogeography. A Critique" (G. Nelson and D.E. Rosen, eds), pp. 501–523. Columbia University Press, New York.

Croizat, L., Nelson, G. and Rosen, D.E. (1974). Centers of origin and related concepts. *Syst. Zool.* **23**, 265–287.

Gerlach, S.A. (1977). Means of Meiofauna Dispersal. *Mikr. Meersboden* **61**, 89–103.

Gourbault, N. (1969). Expansion de *Dugesia tigrina* (Girard), planaire américaine introduite en Europe. *Annls. Limnol.* **5**, 3–7.

Gremigni, V. (1979). An ultrastructural approach to planarian taxonomy. *Syst. Zool.* **28**, 345–355.

Hay, D.A. and Ball, I.R. (1979). Contributions to the biology of freshwater planarians (Turbellaria) from the Victorian Alps, Australia. *Hydrobiologia* **62**, 137–164.

Heisenberg, W. (1975). "Across the Frontiers". Harper Torchbrooks, Harper and Row, New York.

Hennig, W. (1966). "Phylogenetic Systematics". University of Illinois Press, Urbana.

Kaufmann, F. (1949). Cassirer's Theory of Scientific Knowledge. *In* "The Philosophy of Ernst Cassirer" (P.A. Schilpp, ed.), pp. 183–213. Open Court Publishing Company, Illinois.

Kenk, R. (1978). The Planarians (Turbellaria: Tricladida Paludicola) of Lake Ohrid in Macedonia. *Smithson. Contr. Zool.* **280**, i–iv, 1–56.

Kneale, W.C. (1974). The demarcation of science. *In* "The Philosophy of Karl Popper" (P.A. Schilpp, ed.), pp. 205–217. Open Court Publishing Company, Illinois.

Kozhov, M.M. (1963). "Lake Baikal and its Life". Junk, The Hague.

Mayr, E. (1965). What is a fauna? *Zool. Jb.* (Syst.) **92**, 473–486.

Mayr, E. (1976). "Evolution and the Diversity of Life. Selected Essays". The Belknap Press of Harvard University Press, Mass.

McDowall, R.M. (1973). Zoogeography and taxonomy. *Tuatara* **20**, 88–96.

McDowall, R.M. (1978). Generalized tracks and dispersal in biogeography. *Syst. Zool.* **27**, 88–104.

McDowall, R.M. (1980). Freshwater fishes and plate tectonics in the southwest Pacific. *Palaeogeogr. Palaeoclimatol. Palaeoecol.* **31**, 337–351.

Mitchell, R.W. and Kawakatsu, M. (1972). A new family, genus, and species of cave-adapted planarian from Mexico (Turbellaria, Tricladida, Maricola). *Occ. Pap. Mus. Texas Tech. Univ.* **8**, 1–16.

Nagel, E. (1961). "The Structure of Science". Routledge and Kegan Paul, London.

Nelson, G. (1971). "Cladism" as a philosophy of classification. *Syst. Zool.* **20**, 373–376.

Nelson, G. (1978). From Candolle to Croizat: comments on the history of biogeography. *J. Hist. Biol.* **11**, 269–305.

Nelson, G. and Platnick, N.I. (1978). The perils of plesiomorphy: Widespread taxa, dispersal and phenetic biogeography. *Syst. Zool.* **27**, 474–477.

Nelson, G. and Rosen, D.E. (1981). "Vicariance Biogeography. A Critique". Columbia University Press, New York.

Patterson, C. (1981a). Methods of paleobiogeography. *In* "Vicariance Biogeography. A Critique" (G. Nelson and D.E. Rosen, eds), pp. 446–489. Columbia University Press, New York.

Patterson, C. (1981b). Biogeography: in search of principles. (Review). *Nature, Lond.* **291**, 612–613.

Pielou, E.C. (1979). "Biogeography". J. Wiley, New York.

Platnick, N.I. (1977). Cladograms, phylogenetic trees, and hypothesis testing. *Syst. Zool.* **26**, 438–442.

Platnick, N.I. and Nelson, G. (1978). A method of analysis for historical biogeography. *Syst. Zool.* **27**, 1–16.

Popper, K.R. (1961). "The Poverty of Historicism" (2nd edn, corrected). Routledge and Kegan Paul, London.

Porfijeva, N.A. (1977). ["Planarians of Lake Baikal".] Novosibirsk: Akademia Nauk SSSR [In Russian].

Reynoldson, T.B. (1966). The distribution and abundance of Lake-dwelling triclads—towards a hypothesis. *Adv. Ecol. Res.* **3**, 1–71.

Rosen, D.E. (1975). A vicariance model of Caribbean biogeography. *Syst. Zool.* **24**, 431–464.

Rosen, D.E. (1978). Vicariant patterns and historical explanation in biogeography. *Syst. Zool.* **27**, 159–188.

Stankovic, S. (1960). "The Balkan Lake Ohrid and its Living World". Junk, The Hague.

Stankovic, S. (1969). Turbellariés Triclades endémiques nouveaux du lac Ohrid. *Arch. Hydrobiol.* **65**, 413–435.

Sterrer, W. (1973). Plate tectonics as a mechanism for dispersal and speciation in interstitial sand fauna. *Neth. J. Sea Res.* **7**, 200–222.

Udvardy, M.D.F. (1969). "Dynamic Zoogeography with Special Reference to Land Animals". Van Nostrand Reinhold, New York.

van der Velde, G. (1975). The immigrant triclad flatworm *Dugesia tigrina* (Girard) (Platyhelminthes, Turbellaria). Range extension and ecological position in the Netherlands. *Hydrobiol. Bull.* **9**, 123–130.

White, M.J.D. (1973). "Animal Cytology and Evolution" (3rd edn). University Press, Cambridge.

White, M.J.D. (1978). "Modes of Speciation". W.H. Freeman, San Francisco.

Willis, J.C. (1922). "Age and Area". Cambridge University Press, Cambridge.

Willis, J.C. (1940). "The Course of Evolution". Cambridge University Press, Cambridge.

15 The Biogeography of Salamanders in the Mesozoic and Early Caenozoic: a Cladistic-Vicariance Model

A.R. MILNER

Department of Zoology, Birkbeck College, University of London, London WC1E 7HX, UK

Abstract: A cladistic-vicariance model of the phylogeny and biogeography of the amphibian order Urodela (the salamanders and newts) is presented. It is argued that the patterns of relationship and distribution of living salamander families are consistent with a single major dispersal in the early Mesozoic resulting in a cosmopolitan Laurasian fauna, followed by a sequence of vicariance events identifiable with palaeogeographical isolation events in the northern continents in the Mesozoic and early Caenozoic. The urodeles probably originated in the late Permian in the north temperate humid zone in northern Laurasia, spreading southwards across Laurasia by the early Jurassic when humid conditions prevailed more widely. Successive subdivision of Laurasia by the Turgai Sea, the Mid-Continental Sea and the Atlantic Ocean isolated non-paedomorphic populations which evolved endemically to give the hynobiids in east Asia, the dicamptodontids in western North America, the ambystomatids and plethodontids in eastern North America and the salamandrids in Europe. Later withdrawal of the Mid-Continental and Turgai Seas, together with the intermittently present Bering land bridge, permitted secondary dispersal of representatives of several families, resulting in sympatric overlap of family ranges. The families of paedomorphic urodeles probably evolved from non-paedomorphic forms by niche differentiation and are not the direct result of vicariance on a continental scale. The diversity of paedomorphic urodeles may, however, be an indirect consequence of vicariance amongst their non-paedomorphic ancestors.

Systematics Association Special Volume No. 23, "Evolution, Time and Space: The Emergence of the Biosphere", edited by R.W. Sims, J.H. Price and P.E.S. Whalley, 1983, pp. 431–468. Academic Press, London and New York.

INTRODUCTION

The living members of the amphibian order Urodela comprise about 320 species of salamanders, newts and their neotenous relatives, currently grouped into nine families (Table I). They are unique among the orders of living lower tetrapods in being largely restricted to the northern continents (Fig. 1). Furthermore, most families are either entirely or predominantly restricted to a single continent, the Sirenidae, Dicamptodontidae, Ambystomatidae, Amphiumidae and most Plethodontidae being restricted to North and Central America, while the Hynobiidae are restricted to Asia and most Salamandridae to Eurasia. However, either intercontinental dispersal or vicariance within cosmopolitan families has occurred, the Proteidae and Cryptobranchidae occurring in North America and Eurasia, while a few bolitoglossine plethodontids occur in South America and Europe and a few salamandrids occur in North Africa and North America.

Explanations for the primarily northern distribution of salamanders are usually based on an origin as north temperate/montane stream-dwellers (Schmalhausen, 1968) in Laurasia during the Mesozoic (Savage, 1973), although a tropical origin, followed by dispersal from and extinction in the tropics, has been suggested (Darlington, 1957). The absence of salamanders in most tropical regions has been ascribed to inability to compete with tropical frogs (Darlington, 1957) or lungfish larvae (Schmalhausen, 1968). The distribution of some families on more than one continent were originally explained in terms of dispersal over fixed continents and land bridges (Dunn, 1926; Noble, 1931). Comprehensive dispersalist models have been proposed since, suggesting centres of origin, evolution and dispersal of salamanders either in the palaeotropical region (Darlington, 1957) or North America (Naylor, 1980, 1981). Neither model is readily amenable to analysis and both rely heavily on negative evidence from the fossil record. Recent studies of individual families of living and fossil urodeles suggest rather, that the constituent families of the order evolved in different regions of the northern hemisphere (Tihen, 1958; Wake, 1966; Wake and Özeti, 1969; Estes, 1981). The colonization of South America and North Africa is agreed to be the result of Caenozoic events occurring later than the establishment of the modern families which must have been complete by the Palaeocene at the very latest. The fossil record is of value in delimiting the latest time of appearance of the modern families,

Table I. The families (and one higher grouping) of living urodeles, their phylogenetic status, their inferred region of origin and latest time of differentiation as a family (based on earliest appearance in the fossil record).

Hynobiidae	(paraphyletic?)	East Asia	No fossils (Estes, 1981)
Dicamptodontidae	(paraphyletic?)	Western North America (Tihen, 1958)	Lower Eocene (Estes, 1981)*
Ambystomatidae	(paraphyletic?)	Eastern North America (Tihen, 1958)	Lower Oligocene (Holman, 1968)
Salamandridae	(monophyletic)	Europe	Upper Palaeocene (Estes *et al.*, 1967)
Plethodontidae	(monophyletic)	Eastern North America (Wake, 1966)	Lower Miocene (Tihen and Wake, 1981)
Cryptobranchidae	(monophyletic)	North America (Naylor, 1981), East Asia	Upper Palaeocene (Naylor, 1981)
Sirenidae	(monophyletic)	North America	Upper Cretaceous (Estes, 1965)
Amphiumidae	(monophyletic)	North America	Upper Cretaceous (Estes, 1969b)
Proteida (both families)	(monophyletic)	Euramerican Laurasia	Upper Cretaceous (Estes, 1981)

* Earliest record is taken as *Chrysotriton* (Estes 1981); following Estes, the scapherpetontids are not considered to be dicamptodonts, but unlike Estes, I do not consider the *Bargmannia/Geyeriella* group to belong here either.

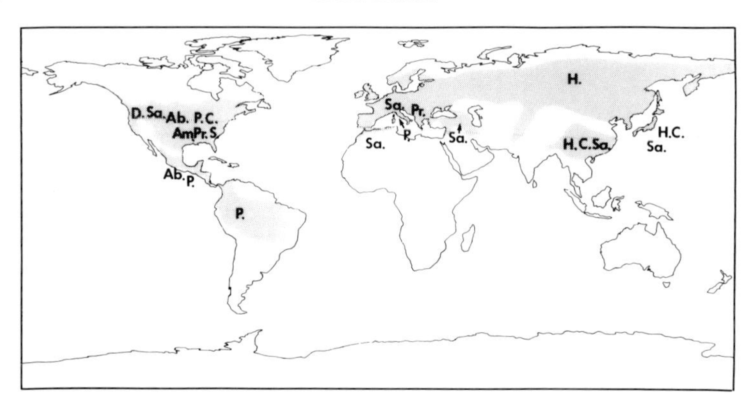

Fig. 1. Present day world map (cylindrical equidistant projection) showing general distribution of urodeles (stipple). The families present in each major continental area are indicated by abbreviations as follows: Ab = Ambystomatidae, Am = Amphiumidae, C = Cryptobranchidae, D = Dicamptodontidae, H = Hynobiidae, P = Plethodontidae, Pr = Proteidae, S = Sirenidae, Sa = Salamandridae. (Map copyright Cambridge University Press, 1980, taken from Smith, Hurley and Briden "Phanerozoic Paleocontinental World Maps".)

amphiumids and sirenids being known from the Upper Cretaceous (Estes, 1970) and several other families appear in the Palaeocene. However, in general the fossil record of urodeles is of limited use in elucidating the precise phylogeny and early biogeography of the group because: (1) the record is extremely poor prior to the Upper Cretaceous so that the period when family-level differentiation was occurring is hardly directly represented; (2) many fossil salamander taxa are known only from isolated bones of limited information content and uncertain relationship. However, the fossil record does suggest that the family-level differentiation of the group took place prior to the Upper Cretaceous, and for the modern distributions to be attained, much of the dispersal or vicariance must have occurred then.

Savage (1973) appears to have been one of the first authors to relate modern urodele distribution to a Mesozoic "Laurasian source-unit", this being the north temperate component of a Laurasian endemic amphibian fauna. He argued, with particular reference to the anuran component of the fauna, that it was subsequently modified by climatic changes rather than geographical events. However, several geographical isolation events (mostly marine transgressions) are known to have occurred within Laurasia during the Mesozoic and must have fragmented the Laurasian urodele fauna to give vicariant distribution patterns which may still persist. The morphological characteristics and geographical distributions of the families of salamanders may be subjected to cladistic and vicariance analysis based on the premise that at least some families or suprafamilial clades may have arisen by allopatric evolution following the fragmentation of an initially cosmopolitan Laurasian urodele fauna by geographical barriers. This study is an attempt to reconstruct the phylogeny and palaeogeography of salamanders by:

(i) briefly characterizing the nine families of living urodeles, their status as monophyletic or paraphyletic groups *sensu* Hennig, their earliest appearances as fossils, and their inferred regions of origin;

(ii) phylogeny reconstruction using the distribution of derived characters;

(iii) identifying the most probable time and, more speculatively, region of origin of the urodeles together with the time by which they were cosmopolitan across Laurasia;

(iv) identifying dichotomies in the phylogeny which are congruent in sequence and effect with palaeogeographical isolating events within Laurasia;

(v) identifying secondary dispersals, permitted by the subsequent breakdown of specific barriers, by the ensuing geographical sympatry of families.

ANALYSIS OF THE RELATIONSHIPS OF LIVING SALAMANDER FAMILIES

1. Units of Study—the Nine Families of Living Salamanders

The living urodeles are usually divided into nine families although there is little agreement about the higher relationships of these families. Prior to discussing their relationships, it may be useful to characterize each family in terms of its constituent forms, distribution, first appearance in the fossil record and its status as a monophyletic group *sensu* Hennig. In five families, the primitive condition of metamorphosis into a terrestrial adult is the characteristic life history. These "terrestrial" salamanders are discussed first.

(a) HYNOBIIDAE (*Hynobius*, *Salamandrella*, *Ranodon*, *Onychodactylus*, *Batrachuperus*, *Pachypalaminus*). Most are restricted to East Asia although

three species extend westwards, one as far as the Urals. No certain fossil hynobiids have been described (Estes, 1981). Diagnoses of this family are largely based on the absence of specializations of other families but the living Hynobiidae do share derived characters with the Cryptobranchidae (see node 2 of analysis and Dunn, 1922) and hence may be of monophyletic origin but paraphyletic with respect to the Cryptobranchidae.

(b) DICAMPTODONTIDAE (*Dicamptodon, Rhyacotriton*). Restricted to the coastal ranges of western North America. These have always been recognized as primitive ambystomatids or ambystomatid relatives, and were given subfamilial status by Tihen (1958), later elevated to family status by Edwards (1976). Edwards' family diagnosis combines primitive urodele characters with some derived ambystomatid characters but does not appear to include any unique characters, hence these two genera may represent a natural group or a grade. Tihen (1958) argued that they were endemic to western North America. The various fossils assigned to this family by Edwards (1976) and Estes (1981) mostly bear gradal rather than derived similarities to *Dicamptodon. Chrysotriton* (Estes, 1981) from the Eocene of North Dakota appears to be a true dicamptodontid but the relationships of the scapherpetontids and the *Bargmannia/Geyeriella* group are less certain.

(c) AMBYSTOMATIDAE (*Ambystoma, Rhyacosiredon*). Found in North America from Canada to Mexico. Tihen (1958) reviewed the relationships and distributions of the ambystomatids and concluded that the family originated in the Appalachians in the north-east USA. There are few fossil ambystomatids *sensu stricto*, the earliest being of Lower Oligocene age (Holman, 1968) and none from outside North America. The diagnosis of this family given by Tihen appears to be another combination of primitive urodele characters and a few derived characters shared with the Salamandridae. Neither Edwards (1976) nor Hecht and Edwards (1977) identified a unique ambystomatid character, so the family may represent a grade.

(d) SALAMANDRIDAE. This family comprises 14 living genera, the characters and relationships of which were reviewed by Wake and Özeti (1969). It is considered to be monophyletic. Seven genera occur in Europe, North Africa and Asia Minor, whilst fossils certainly attribut-

able to the Salamandridae occur in Europe in deposits from the Upper Palaeocene onwards (Hecht and Hoffstetter, 1962; Estes *et al.*, 1967). *Tylototriton*, known only as a fossil in Europe, survives in East Asia along with a probable clade of salamandrids made up of *Cynops*, *Pachytriton*, *Hypselotriton* and the less certainly related *Paramesotriton* (Wake and Özeti, 1969). A possible fossil salamandrid, *Procynops*, is known from the Miocene of China (Young, 1965). *Taricha* and *Notophthalmus* are restricted to North America where they are also known as fossils from the Oligocene onwards (Tihen, 1974; Naylor, 1979b). The salamandrids are widely agreed to have originated in Eurasia and the most parsimonious interpretation of Wake and Özeti's (*op cit.*: Fig. 7) suggested relationships is that they originated in Europe. The North American genera are certainly a derived subgroup of salamandrids which must have dispersed to North America in the early Caenozoic (Estes, 1970:146).

(e) PLETHODONTIDAE. This family contains many genera occurring in North America (Wake, 1966) and Central and South America (Wake and Lynch, 1976) while one North American genus *Hydromantes* also has species in Europe (Wake *et al.*, 1978). The only pre-Pleistocene plethodontid fossils are from the Lower Miocene of Montana (Tihen and Wake, 1981). The Plethodontidae is considered to be a monophyletic group. Those found outside North America belong to the highly specialized plethodontid tribe Bolitoglossini. These unusually terrestrial salamanders appear to have colonized Central America in the early Caenozoic and South America in the Pliocene (Wake and Lynch, 1976), while the genus *Hydromantes* appears to have dispersed to Eurasia from western North America and now survives in California and southern Europe (Wake, 1966; Estes, 1970; Wake *et al.*, 1978). Thus the plethodontids appear to be of North American origin and Wake (1966) argued that they originated in the north-east of North America in the Appalachians, basing this conclusion on the most parsimonious interpretation of their distribution in relation to their phylogeny, a small number of lineages having apparently dispersed to western North America.

The remaining four families comprise the specialized paedomorphic forms:

(f) CRYPTOBRANCHIDAE (*Cryptobranchus* and *Andrias*). Living crypto-

branchids occur in east Asia and in the Appalachians of North America. Fossil cryptobranchids are known from the Middle Oligocene to the Pliocene of Europe (Westphal, 1958) and from the Upper Palaeocene onwards in North America (Naylor, 1981). The family is uncontroversially monophyletic. The general distribution of cryptobranchids is consistent with an origin almost anywhere in the northern continents but it will be argued later in this paper that the derived similarities of cryptobranchids and hynobiids strongly indicate an Asian origin for the family.

(g) SIRENIDAE (*Siren* and *Pseudobranchus*). Restricted to southern and eastern North America. Fossil sirenids of Miocene and later age occur in the same area (Goin and Auffenberg, 1965), in addition there is an Upper Cretaceous sirenid *Habrosaurus* from Wyoming (Estes, 1965). The family is uncontroversially monophyletic and known exclusively from North America from its first appearance.

(h) AMPHIUMIDAE (*Amphiuma*). The single living genus is restricted to the south-east of North America. Fossil amphiumids are known from the Palaeocene of Wyoming (*Amphiuma*) and the Upper Cretaceous of Montana (*Proamphiuma*) (Estes, 1969b). Unknown outside North America and monophyletic.

(i) PROTEIDAE. This comprises the living genera *Proteus* from south-east Europe and *Necturus* from eastern North America. Fossil proteids include *Necturus* from the Upper Palaeocene of North America (Naylor, 1978) and *Mioproteus* from the Miocene of south-west Russia and Germany (Estes and Darevsky, 1977). The living Proteidae are widely but controversially, held to be monophyletic. Kezer, Seto and Pomerat (1965), Larsen and Guthrie (1974), Estes and Darevsky (1977) and Naylor (1978) have all argued for monophyly, with Hecht and Edwards (1976, 1977) dissenting. Monophyly is assumed here. The extinct family Batrachosauroididae (*Prodesmodon*, *Opisthotriton*, *Batrachosauroides* and *Palaeoproteus* appears to be the sister-group of the Proteidae (Estes, 1975; Naylor, 1978, 1979a). This family is known from the Cretaceous to Miocene of North America and the Palaeocene and Eocene of Europe. I follow Estes (1975) in using the subordinal taxon Proteida to comprise the Proteidae and Batrachosauroididae. The distribution of the Proteida

is consistent with an origin in Euramerican Laurasia prior to the Upper Cretaceous.

2. Character Distribution

The only comprehensive cladistic analyses of urodele relationships are those of Edwards (1976) and Hecht and Edwards (1977), part of the latter having been critically commented on by Naylor (1978). Hecht and Edwards summarized several earlier concepts of urodele relationship and attempted to express them in cladogram form. Edwards' analysis used 13 characters while Hecht and Edwards used 18 characters which they analysed in weighted and unweighted form producing several alternative results. Due to Hecht and Edwards' inability to resolve some of the problems of relationship, a further analysis is attempted below using additional characters in order to produce a working hypothesis of salamander relationships. The characters used are, for reasons of space, summarized in Table II and the resultant cladogram depicted in Fig. 2. The characters used are those which are useful as synapomorphies between two or more families and not those which demonstrate monophyly of any one family. Monophyly is assumed for six families and an indeterminate monophyly/paraphyly is assumed for the Hynobiidae and the Ambystomatidae. The Dicamptodontidae is treated as paraphyletic because of the uncertainty of the relationships of *Rhyacotriton*.

The resulting cladogram in Fig. 2 corresponds to one of Hecht and Edwards' favoured cladograms (*op. cit.*: Fig. 6a) in much of its structure and implied relationships. It differs only in the transfer of the Plethodontidae from being the sister-family of the Ambystomatidae to being the sister-family of the Amphiumidae, and the less certain removal of *Rhyacotriton* from the Dicamptodontidae to a basal position in the amphiumid-plethodontid clade. The plethodontid-amphiumid relationship suggested here was proposed by both Reed (1920) and Dunn (1922) on the basis of the ontogeny of the middle ear, and it receives support from the immunological work of Salthe and Kaplan (1966) and the behavioural study of Salthe (1967). Salthe and Kaplan (1966: Fig. 2) also found that the closest relatives of *Amphiuma* after the plethodontids appeared to be *Rhyacotriton* and *Dicamptodon* based on serological evidence, and this also relates well to the cladogram proposed here. The association of *Rhyacotriton* with the plethodontids

Table II Derived characters used to define nodes 1–8 in the cladogram of urodele family relationships in Fig. 2. Where no source is cited, information is taken from Edwards (1976) or Hecht and Edwards (1977). Although the nodes are defined by the derived conditions, some of the sources describe the contrasting primitive conditions.

NODE 1.	A.	Interventricular septum absent (present in *Siren*) (Putnam, 1975).
	B.	Periotic canal joins periotic cistern close to or at fenestra ovalis (Lombard, 1977:120).
	C.	Periotic canal has a horizontal course and is straight or shows flexures (Lombard, 1977:121).
	D.	*Adductor mandibulae internus pseudotemporalis superficialis* extends to the top of the skull roof or back to the cervical vertebrae (muscle is small and inserts on sides of skull roof in sirenids) (Carroll and Holmes, 1980).
	E.	Anterior glomeruli of kidney reduced or absent (retained in sirenids) (Goin and Goin, 1971).
	F.	Scapulocoracoid is a single ossification (coracoid persists in sirenids) (Goin and Goin, 1971).
	G.	At least one pair of cloacal glands present (no cloacal glands in sirenids).
	H.	Haploid chromosome number reduced to 31 or less (32 in *Pseudobranchus*).
NODE 2.	A.	First ceratobranchial and first epibranchial fused into a single cartilage rod (Dunn, 1922).
	B.	*Pubotibialis* muscle fused to *puboischiotibialis* (Dunn, 1922).
NODE 3.	A.	Elaboration of cloacal glands into three pairs of glands.
	B.	Haploid chromosome number reduced to 19 or less (Morescalchi, 1975).
	C.	Microchromosomes absent (also in sirenids and one hynobiid) (Morescalchi, 1975).
NODE 4.	A.	Haploid chromosome number reduced to 14 or less to give a "symmetrical" karyotype with all chromosomes metacentric or subtelocentric (Morescalchi, 1975).
	B.	Postsacral spinal nerves pass through intravertebral foramina while post-atlanteal presacral nerves are intervertebral (Edwards 1976—condition 6III′) or more derived conditions (6III″ or 6III‴). Because of the node 5,6 characters, the more primitive amphiumid condition here assumed to be a derived reversal.
	C.	Second epibranchial absent in post-metamorphic individuals.
	D.	Nasal ossification develops from lateral anlagen only (in other forms it develops from 2 anlage-hynobiids, the mesial anlage-sirenids or is absent-proteids) (Jurgens, 1971).

Table II continued

NODE 5.	A.	Loss of true operculum (i.e. the structure formed by the "pinching off" of the region of the otic capsule, caudal to fenestra ovalis, which carries the insertion of the *levator scapulae*). The operculum is absent or replaced by a neomorph, see node 6.
	B.	Pelvis lacks the ypsiloid cartilage (also lost in sirenids and proteids).
	C.	Reduction of the pterygoids (*Rhyacotriton* and *Amphiuma*) or loss of pterygoids (plethodontids).
NODE 6.	A.	Neomorph "operculum" fused to columella of stapes and attached to otic capsule by the *isthmus fenestralis*—a column of tissue from the ventrocephalic margin of the fenestra ovalis (Reed, 1920; Monath, 1965). Larsen (1963, unpub. thesis) suggested that this is the footplate of the columella substituting for the lost operculum.
	B.	Exoccipital fused to prootic + opisthotic to give single otic-occipital ossification (also node 8)
	C.	Non-appearance or early fusion of the angular (retained in *Rhyacotriton*, Srinivasachar, 1962).
	D.	Lachrymal completely absent.
NODE 7.	A.	Enlargement of *levator scapulae* muscle so that its posterior insertion occupies the entire anterior edge of the suprascapula (Monath 1965). In most salamandrids a more derived condition occurs in which the muscle is divided into two posterior heads which insert widely over the suprascapula and scapulocoracoid.
	B.	Vomerine teeth replaced anterolaterally or laterally (Regal, 1966: Fig. 2, subfigs 1,2a,2b,5). (Hynobiids, plethodontids and *Rhyacotriton* show the presumably primitive condition of mesial or antero-mesial replacement).
	C.	Loss or fusion of the angular.
NODE 8.	A.	Post-atlanteal spinal nerves all exit through intravertebral foramina except for the first 3 (most Ambystomatidae) or 2 (Salamandridae and some Ambystomatidae) (Edwards, 1976). Occurs convergently in Sirenidae (2) and Plethodontidae (3).
	B.	Exoccipital fused to prootic + opisthotic to give a single otic-occiptial ossification (also node 6).
	C.	Lachrymal fuses with prefrontal to give a compound prefrontal bearing a nasolachrymal duct (Jurgens, 1971:110).

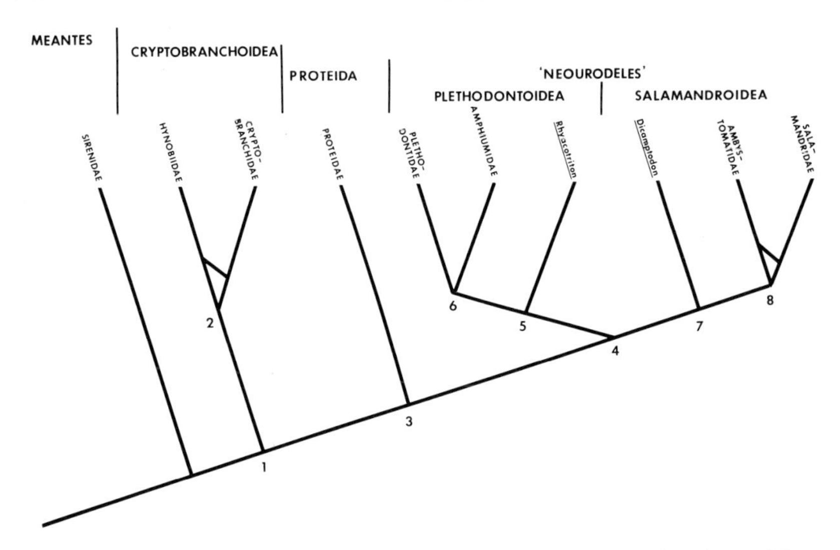

Fig. 2. Cladogram depicting the relationships of the nine families of living urodeles. Nodes 1–8 as diagnosed in Table II. Names for the sister-taxa of the Meantes and the Cryptobranchoidea are not proposed, no pre-existing names being appropriate. The term "neourodeles" is used for the sister-group of the Proteida and the superfamilial terms Plethodontoidea and Salamandroidea are used in slightly different contexts from previous works. The genera *Dicamptodon* and *Rhyacotriton* are depicted separately.

and amphiumids is the least satisfactory part of this hypothesis of relationships. Some of the shared characters are those of upland stream-dwellers occurring convergently in stream-dwelling hynobiids and salamandrids, and it is possible that *Rhyacotriton* is a specialized stream-dwelling relative of *Dicamptodon*. Regal (1966) considered the dicamptodonts to represent a basal grade of the "ambystomatoids" (including plethodontids but excluding salamandrids) and noted the similarities of *Rhyacotriton* to the plethodontids based on primitive and adaptive features, the latter being potential shared derived characters. The monophyly of the Dicamptodontidae has not been strongly argued and the only possible unique derived character for *Rhyacotriton* and *Dicamptodon* in Edwards' diagnosis (1976:327) is the phalangeal reduction in the fourth toe, but even this appears intermittently in the ambystomatids (Tihen, 1958). In conclusion, I do not consider the position of

Rhyacotriton in this scheme of relationships to be more than possible but I cannot perceive any more satisfactory relationship for it.

3. Chronology of the Dichotomies

The earliest appearances of the undoubted members of some families in the fossil record does permit latest dates of dichotomies to be established. Most informative is the presence of the amphiumid *Proamphiuma* in the Upper Cretaceous. By reference to Fig. 2, it can be seen that not only did the plethodontid-amphiumid (node 6) dichotomy occur prior to the Upper Cretaceous but also the preceding dichotomies at nodes 5,4,3,1 and the sirenid–non-sirenid dichotomy. The dichotomies at nodes 2 and 8 had occurred by the Upper Palaeocene when the earliest cryptobranchids and salamandrids appear in the fossil record. The latest date for the node 7 dichotomy is uncertain and depends on whether any Upper Cretaceous forms attributed to the Dicamptodontidae can be demonstrated to have a derived similarity to *Dicamptodon*. Thus, most of the differentiation of the urodeles into modern families had taken place by the Upper Cretaceous and some of it undoubtedly significantly earlier. However, prior to the Upper Cretaceous, no urodeles can be certainly associated with modern families, the principal earlier forms being the Prosirenidae from several localities, *Karaurus* from the Upper Jurassic of the USSR, *Comonecturoides* (a femur) from the Upper Jurassic of North America and *Hylaeobatrachus* from the Wealden of Belgium (Estes, 1981). None of these shows obvious relationship to the living families. Thus the fossil record provides little direct information about the differentiation of the urodele families in space and time, and it is necessary to turn to modern material in an attempt to construct a model of the early biogeography of urodeles.

TIMES OF ORIGIN AND COSMOPOLITANISM IN LAURASIA

The earliest certain urodeles are *Comonecturoides* from the Upper Jurassic of North America (Hecht and Estes, 1960) and *Karaurus* from the Upper Jurassic of Kazakhstan (Ivachnenko, 1978). These two specimens do not provide a significant basis for any inference about urodele origins or geography beyond the implication of widespread distribution in Laurasia by the Upper Jurassic. However, several forms of indirect evidence permit some controlled speculation on the time and region of origin of the urodeles together with the time at which they became widespread.

1. Time of Origin

Most hypotheses of amphibian relationships treat the Anura and the Urodela as sister-groups, in which circumstance the presence of the proanuran *Triadobatrachus* in the Lower Triassic of Madagascar implies the existence of urodeles or at least a urodele stem-group in the Lower Triassic. No obvious urodele relatives are known from the taxonomically rich small amphibian assemblages of the Permo-Carboniferous and it appears that few of the "lissamphibian" skeletal characters had arisen at that time. The most plausible estimate for the time of origin of the urodeles as a distinct group is therefore the Upper Permian. Salthe and Kaplan (1966) suggested a Permian date for the urodele-anuran dichotomy based on serological distance.

2. Region of Origin

The concept of centres of origin has recently been the subject of much legitimate criticism as being inherently speculative and not susceptible to analysis (Croizat *et al.*, 1974; Platnick and Nelson, 1978; Rosen, 1978). However, the postulated origin of urodeles in the Permo-Triassic during the existence of the supercontinent of Pangaea together with their apparent subsequent scarcity in Gondwanaland until the late Caenozoic, begs a fundamental question about their region of origin. As already noted, a Laurasian origin can be inferred from the generalized track and climatic limitations of living forms, the South American and African forms being attributable to Caenozoic dispersals. The Mesozoic and Caenozoic fossil records are almost entirely consistent with this with two exceptions noted below. But prior to being widespread in Laurasia in the Upper Jurassic, were the urodeles geographically restricted to a "region of origin" within northern Pangaea and hence unable to penetrate extensively into southern Pangaea prior to the separation of the northern and southern continents later in the Mesozoic? This problem can be speculatively answered by reference to the biology of primitive living urodeles and the distribution of climates in the Permo-Triassic world. Schmalhausen (1968) argued that the primitive urodele life-style was montane stream-dwelling in a cool temperate climate,

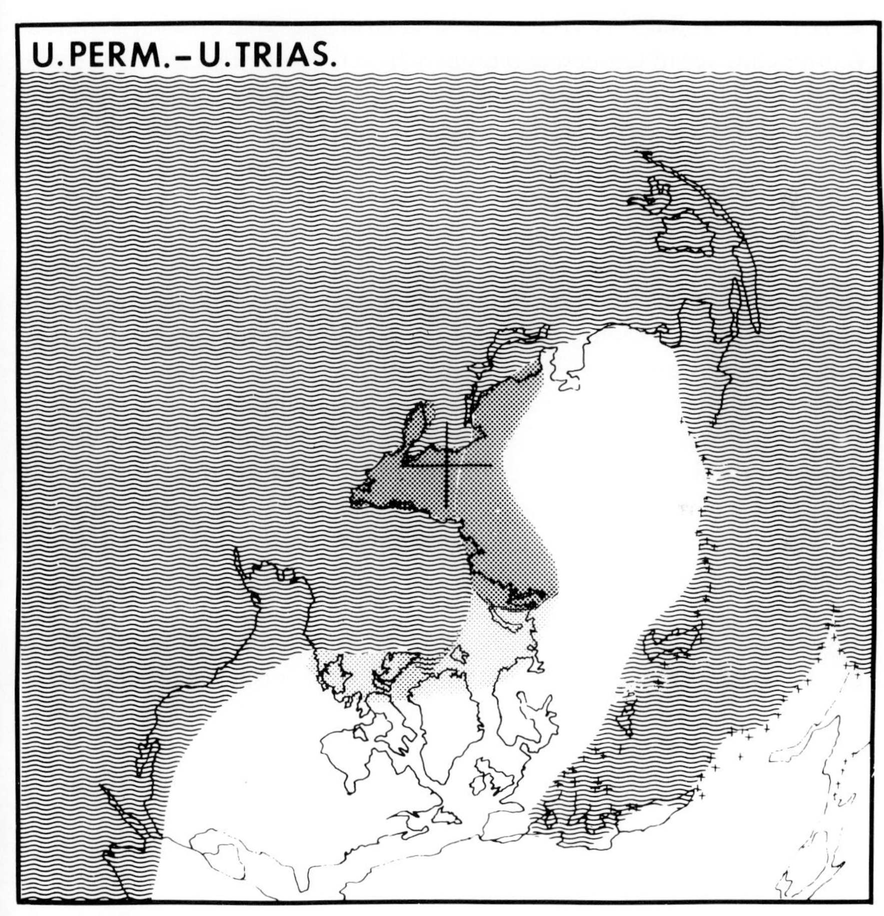

Fig. 3. Upper Triassic north polar projection (Lambert equal area) depicting north temperate humid belts hypothesized as the areas of endemic origin of the urodeles during the Permo-Triassic. Dark stipple = Upper Permian north temperate humid belt. Light stipple = extension of belt by Upper Triassic. Humid belts after Robinson (1973). (Map copyright Cambridge University Press, 1980, taken from Smith *et al.* "Phanerozoic Paleocontinental World Maps".

citing the life-styles of most hynobiids, the cryptobranchids, the dicamptodontids, the primitive plethodontids and the primitive salamandrids. The lowland neotenous forms and the terrestrial plethodontids are agreed to be secondarily derived forms with "secondary" lifestyles. Thus the most parsimonious hypothesis of early urodele ecology is that the Permo-Triassic stem-urodeles were restricted to cool humid climates in temperate regions where they were probably stream-dwellers. Their subsequent occurrence in Laurasia implies an origin in the north temperate zone of Pangaea. Robinson's (1973) analysis of the climatic zonation of the world in the Upper Triassic gives a valuable indication of the extent of the northern high latitude humid belt in which urodeles might be expected to be able to have existed. Urodeles, if recognizably present in the Upper Permian, would have been restricted to the north-east region of Asia as indicated by dark stipple in Fig. 3 based on Robinson (1973: Fig. 13). By the Upper Triassic, the high latitude humid belt extended further around the north coast of Laurasia (light stipple in Fig. 3 based on Robinson *op. cit.*: Fig. 10). Southern Euramerican Laurasia sustained a dry tropical climate at this time and urodeles may have remained restricted to the northern region of Laurasia until the early Jurassic when more humid climates prevailed in the northern continents and a cosmopolitan Laurasian urodele fauna could have become established.

I suggest therefore, that after the coalescence of Angara and Euramerica early in the Permian, an unspecified group of Euramerican amphibians (either temnospondyl or microsaur) dispersed over Angara and, at the northern end of their range, adopted to the cool humid conditions there as the urodele stem-group, initially isolated from tropical amphibians by the dry tropical belt across much of European Laurasia.

3. Cosmopolitanism

As noted above, certainly attributable urodele fossils are known from the Upper Jurassic of Kazakhstan and Wyoming, implying that urodeles were widespread across Laurasia by then. By the Upper Jurassic, the marine incursions collectively known as the Turgai Sea (Fig. 4) had already divided Laurasia north–south and as the two Upper Jurassic urodeles occur one on each side of the Turgai Sea, the group must have been widespread earlier in the Jurassic. Thus a cosmopolitan distribution of urodeles in Laurasia in the early Jurassic is not just a theoretical prerequisite for vicariance analysis of urodele biogeography but appears actually to have occurred. It must be noted that the maximum range during the Mesozoic embraced northern Gondwanaland as evidenced by *Ramonellus* from the Lower Cretaceous of Israel (Nevo and Estes, 1969) and an undescribed urodele from the Cretaceous of Niger (De Broin *et al.*, 1974). This demonstrates a Mesozoic penetration into Gondwanaland without living descendants which presumably occurred via the Jurassic European–African land bridge (Cox, 1975).

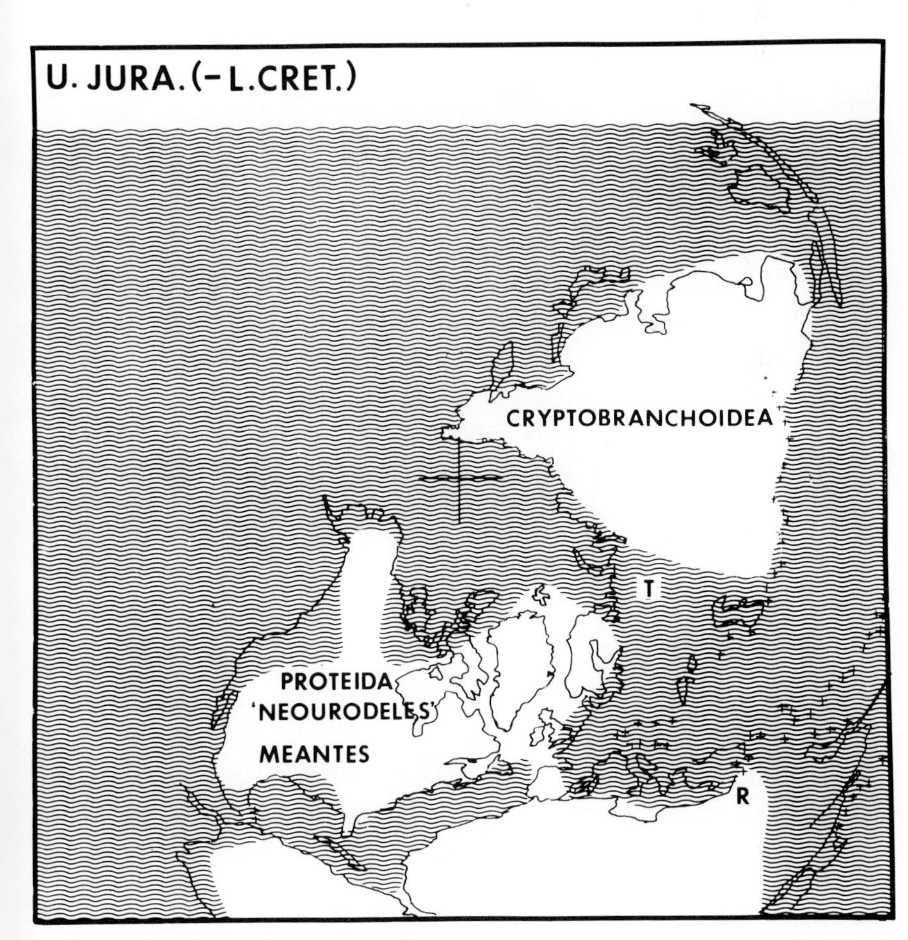

Fig. 4. Upper Jurassic north polar projection (Lambert equal area) depicting vicariance resulting from the isolation produced by the Turgai Sea. R = *Ramonellus* locality in Lower Cretaceous of Israel. T = Turgai Sea. The presence of the Meantes throughout Laurasia would be consistent with the vicariance model presented here but there is no evidence for their presence in east Asia. (Map copyright Cambridge University Press, 1980, taken from Smith *et al.* "Phanerozoic Paleocontinental World Maps".)

BIOGEOGRAPHICAL ANALYSIS AND HISTORICAL MODEL

A basis for reconstructing the biogeographical history of the urodeles is now established, with a base-point of cosmopolitanism across Laurasia by the Middle Jurassic, a reconstructed phylogeny made up of a sequence of dichotomous branches leading to ten living units (Fig. 2) and centres of endemism for each of these units (Fig. 5, Table 1). The palaeogeographical history of Laurasia from the Mesozoic onwards, as discussed here, is based on the relative continental positions as proposed by Smith *et al.* (1980) and the appearance and disappearance of marine barriers as reviewed by Hallam (1981).

Of the nine dichotomies forming the cladogram (Fig. 5), four (nodes 1,5,7,8) are congruent in both sequence and effect with the development of marine barriers across Laurasia from the mid-Jurassic onwards as summarized by Hallam (1981). Node 1 implies the separation of an

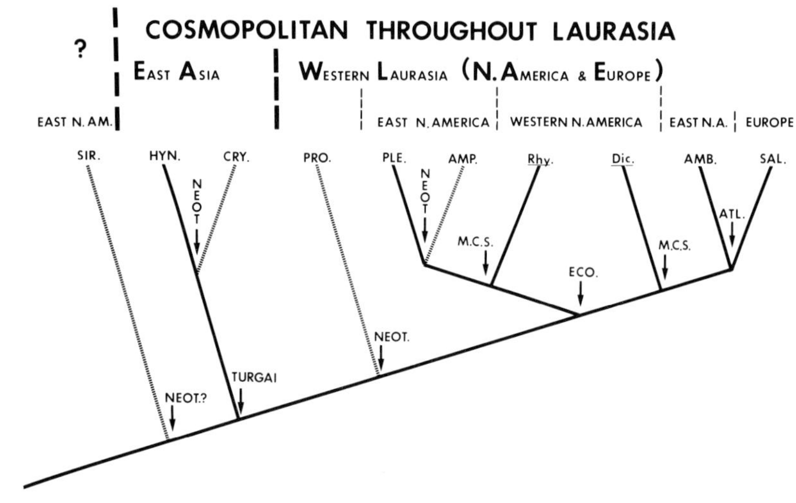

Fig. 5. Cladogram as in Fig. 2 but depicting the proposed geographical, ontogenetic or ecological bases for the dichotomies. The geographical isolation events are depicted by the abbreviations: TURGAI = Turgai Sea (from 150 Ma), MCS = Mid-Continental Sea (from c100 Ma), ATL = Atlantic Ocean (from 50–40 Ma). Dates from Hallam (1981). Other abbreviations are NEOT = dichotomy between terrestrial and paedomorphic groups, ECO = dichotomy based on ecological divergence. The nested sets of continental areas within Laurasia indicate the areas of endemism resulting from vicariance.

Asian area from a Euramerican area and corresponds to the Turgai marine transgression at about 150 Ma (Hallam, 1981). Nodes 5 and 7 imply separation of Western North America from Eastern North America and Europe and correspond to the Mid-Continental Sea which appeared at about 100 Ma. Node 8 corresponds to separation of Eastern North America from Europe and is more difficult to date. Although the North Atlantic was not completely open as an ocean until 40 Ma, it was preceded by shelf seas with the result that only intermittent terrestrial continuity appears to have occurred from 80 Ma onwards. In all cases, it is possible that isolation of urodele populations particularly upland forms, preceded the formation of a complete marine barrier which represents the latest date of isolation. This sequence of dichotomies is congruent with some of those determined for Laurasian freshwater bony fish by Patterson (1981). The area cladograms (Patterson *op. cit.*: Fig. 20.17B,D,H) for *Lepisosteus*, *Amia*, the hiodontids and some esocoids correspond to the salamander area cladogram derived from nodes 1,5,7 and 8.

Four of the remaining five dichotomies (sirenid–non-sirenid, and nodes 2,3,6) are between a clade/family of paedomorphic urodeles and their terrestrial sister-group. None of these are geographically consistent with vicariance events produced by isolation on a continental scale, in fact all are consistent with geographically sympatric origins above a local level. In view of the circumstances in which facultative and permanent paedomorphosis can occur within living ambystomatid, dicamptodontid and plethodontid genera, this is not surprising. In species of all these families, local populations may be permanently, facultatively or non-neotenous depending on the immediate environment, suggesting that the reproductive isolation of neotenic forms can develop locally.

The ninth dichotomy (node 4) cannot be attributed to paedomorphosis and has no obvious geographical basis. It may be speculatively attributed to ecological differentiation into two distinct habitats/lifestyles within one geographical area and is here regarded as ecological or local geographical allopatry.

To modify the resultant partly vicariance-based pattern in order to generate the present-day family distributions, it is necessary to postulate several secondary dispersals. These are invoked (i) where sympatry of two or more terrestrial families demands that is must have occurred, (ii) where geographical barriers are known to have broken down or, (iii) where a derived member of a clade which is largely restricted to one area, occurs over a wider area. The nine dichotomies are discussed below individually and sequentially with their deduced causes and geographical effects. This is followed by discussion of the subsequent dispersals, to give a narrative sequence of events which make up this model of the history of the group.

1. The Sirenid–Non-sirenid Dichotomy

The sirenids appear to be the neotenous sister-group of the remaining living urodeles, implying an origin prior to the subdivision of Laurasia by the Turgai Sea in the mid-Jurassic (see node 1 below). As noted previously, sirenids are only certainly known from North America from the Upper Cretaceous onwards. They may have arisen endemically in, and remained restricted to, western Laurasia in the Mesozoic but their relationship to other urodeles suggests that even when they first appear in the fossil record, they are already relicts of an early Mesozoic "sidebranch" of neotenous urodeles which may have been cosmopolitan across Laurasia. Because they are hypothesized as having differentiated prior to the Turgai transgression, it would in no way violate this model if sirenids were found in the Mesozoic of Eurasia. They may even have extended their range into Gondwanaland as the more primitive pro-sirenids (*Ramonellus* from the Lower Cretaceous of Israel) have done (Nevo and Estes, 1969).

2. The Node 1 Dichotomy

The first recognizable subdivision of Laurasia in the Mesozoic was by the Turgai Sea, a continental sea which functioned as a marine barrier from the mid-Jurassic (150 Ma) to the early Palaeocene and again for most of the Eocene, dividing Laurasia into two terrestrial regions, an east Asian region and a Euramerican region (Fig. 4). The node 1 dichotomy between the cryptobranchoids and the proteid + neourodele group is geographically consistent with being the vicariant result of this subdivision of Laurasia. The Cryptobranchoidea, a clade definable by derived characters (Dunn, 1922), is still almost entirely restricted to central and east Asia, only *Salamandrella keyserlingii* and some cryptobranchid species being known from outside this area. The Proteida and most neourodeles, both living and fossil, occur in Euramerican Laurasia or contacting southern continents, with only some subgroups of the Salamandridae occurring in east Asia. The initial dichotomy would have been between terrestrial urodeles with a gradal similarity to hynobiids and dicamptodontids. The Upper Jurassic *Karaurus* broadly represents this grade but is more primitive, Estes (1981) having identified primitive conditions in *Karaurus* which make it the sister-taxon of all other urodeles, living and fossil. In conclusion, the Turgai incursion appears to have been followed by vicariance, giving two urodele populations whose descendants have remained morphologically and substantially geographically distinct to the present day.

3. The Node 2 Dichotomy

As discussed earlier, the Hynobiidae may represent the grade from which *Cryptobranchus* evolved rather than being the sister-group of the Cryptobranchidae. As the two families share a complex of derived characters and as almost all hynobiids and some *Cryptobranchus* occur in Asia, the most economical hypothesis must be that the cryptobranchoid clade evolved within Asia. Hence *Cryptobranchus* arose within Asian Laurasia, effectively as a large paedomorphic hynobiid and subsequently dispersed across the northern hemisphere, first appearing in the fossil record in North America in the Upper Palaeocene (Naylor, 1981) and in Europe in the Middle Oligocene (Westphal, 1958). The possible dispersal routes and chronology are discussed later. The cryptobranchids must have differentiated from the hynobiids between the mid-Jurassic (150 Ma) when the cryptobranchoids were first isolated, and the mid-Palaeocene (60 Ma) when they first appear as fossils. Naylor (1981) tentatively argued a North American origin for the cryptobranchids based largely on the geography of the oldest fossil cryptobranchid in the Palaeocene of Saskatchewan. For this theory to be compatible with cryptobranchoid monophyly, it would be necessary for a hynobiid-like cryptobranchoid to disperse from Asia to North America, evolve to the cryptobranchid morphology and then disperse back again. It is more economical to assume that cryptobranchids evolved in Asia where all their nearest relatives are, their absence in pre-Pleistocene deposits in that continent merely reflecting a general absence of studied microvertebrate assemblages in that area. Naylor's literalistic interpretation of the fossil record could only conceivably be justified on the basis of comprehensive assemblages of associated salamander skeletons from ecologically diverse circumstances, from every substage of the late Mesozoic and early Caenozoic from every northern continent.

4. *The Node 3 Dichotomy*

No primitive relicts of the stem-terrestrial urodeles of Euramerican Laurasia survive, the neotenous Proteida and the largely terrestrial "neourodeles" being sister-groups. This dichotomy probably occurred in the late Jurassic after the Turgai incursion and was well established by the Upper Cretaceous as evidenced by the fossil record. The proteids and their stem/sister-group, the batrachosauroidids were widespread in Euramerican Laurasia in the late Cretaceous and early Caenozoic, and the living *Necturus* and *Proteus* appear to be disjunct relicts of this former distribution (Fig. 4). All neourodele families appear to have originated within Euramerican Laurasia with fossils such as *Scapherpeton* and *Lisserpeton* from the Upper Cretaceous being possible stem-neourodeles.

It appears that, considered on a continental scale, the neotenous proteids and cryptobranchids arose sympatrically with the neourodeles and hynobiids respectively. The node 2 and 3 dichotomies are suggested as having originated by relatively local isolation of populations in permanent water bodies, giving rise to paedomorphic species and ultimately clades of widespread urodeles. It is possible that the Cryptobranchidae and the Proteida were able to arise separately only as result of the isolation produced by the Turgai Sea and are hence indirect products of vicariance.

5. *The Node 4 Dichotomy*

This is the single dichotomy which cannot be correlated with either geographical barriers or paedomorphosis. One sister-group is the plethodontoid clade of primarily small specialized upland stream-dwellers, but including more derived subgroups of terrestrial (bolitoglossines) and aquatic (amphiumids) forms. This clade appears to have originated in North American Euramerica and not extended to European Euramerica, perhaps because of an absence of continuity of uplands across eastern Euramerica in the Mesozoic. The other sister-group is the salamandroid clade consisting primarily of larger less specialized stream, pond and leaf-litter dwellers extending more widely into lowland areas and being initially cosmopolitan across Euramerica (Fig. 6). This

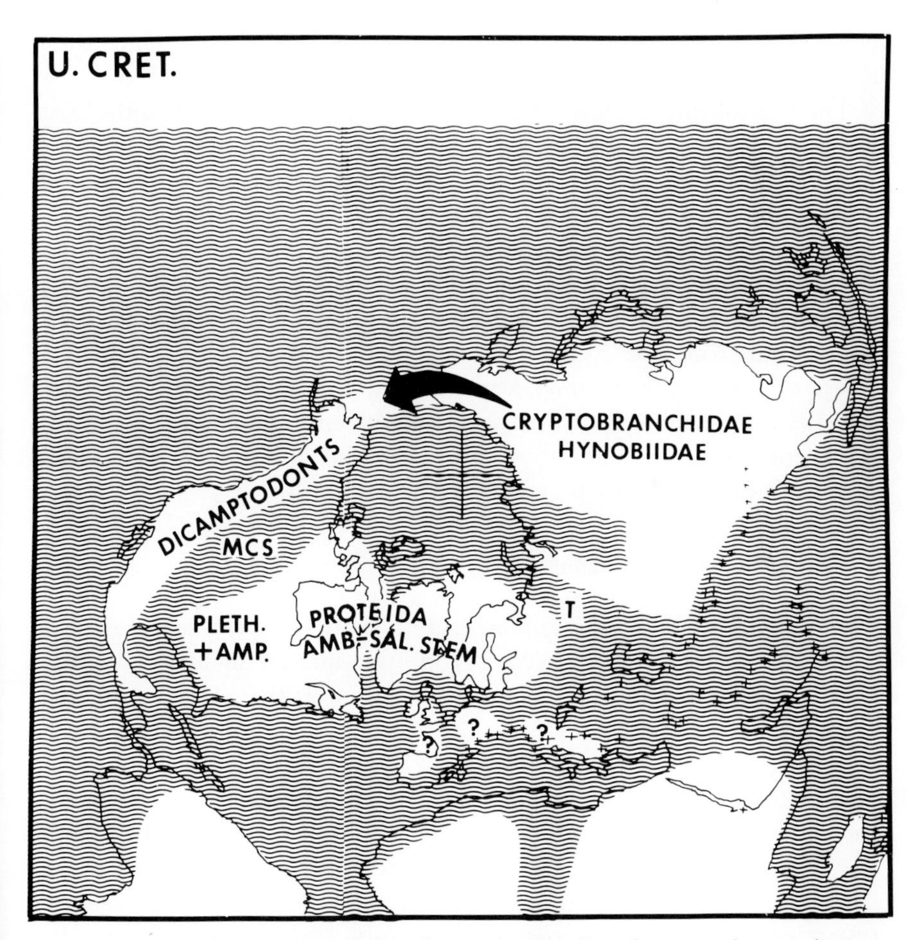

Fig. 6. Upper Cretaceous north polar projection (Lambert equal area) depicting vicariance resulting from the isolation produced by the Mid-Continental Sea. MCS = Mid-Continental Sea. T = Turgai Sea. AMB.-SAL. STEM = ambystomatid-salamandrid stem group, AMP. = Amphiumidae, PLETH. = Plethodontidae. Arrow depicts proposed dispersal route of *Cryptobranchus* in late Cretaceous or early Palaeocene. ? = European islands with unknown urodele fauna possibly including stem-salamandrids. (Map copyright Cambridge University Press, 1980, taken from Smith *et al.* "Phanerozoic Paleocontinental World Maps".)

dichotomy must also have occurred after the Turgai transgression (150 Ma) but well before the Upper Cretaceous (100 Ma) by which time the Mid-Continental Sea had further subdivided these clades.

The relationship of *Rhyacotriton* to the other plethodontoids is the least certain of those proposed here, being based on functional adaptations which are known to have evolved convergently within the Salamandridae and the Hynobiidae. It remains possible that *Rhyacotriton* is a dicamptodontid in the derived sense and that small stream-dwelling specialists have arisen independantly in montane "islands" in Asia (*Onychodactylus*, a hynobiid), Europe (*Chioglossa*, a salamandrid), eastern North America (plethodontines) and western North America (*Rhyacotriton*). If *Rhyacotriton* were made the sister-taxon of *Dicamptodon*, the congruence of vicariance-related dichotomies and the sequence of marine barriers would not be affected and hence its precise placement is not critical to this discussion.

6. The Node 5 and 7 Dichotomies

The second major geographical subdivision of terrestrial Laurasia occurred in the mid-Cretaceous with the marine incursions across North American Laurasia, collectively known as the Mid-Continental Sea, which completely divided the continent from north to south between 100 and 70 Ma (Hallam, 1981, Fig. 7). These shallow seas fluctuated but reached their maximum extent in the early Campanian, thereafter withdrawing southwards. The next dichotomies within both the Plethodontoidea and the Salamandroidea (as used here) can be correlated with this isolation. Since *Rhyacotriton* and *Dicamptodon* are restricted to the western Coastal Range in North America, they can be considered as the western relicts of the resultant vicariance, being the sister-groups of the plethodontid + amphiumid and ambystomatid + salamandrid clades respectively. The barrier formed by the Mid-Continental Sea may also have resulted in vicariance and divergent evolution within the sirenids, proteids and batrachosauroidids but too little is known about the fossil forms on either side of the Mid-Continental Sea to be able to assess whether this occurred.

7. The Node 6 Dichotomy

This appears to have been a paedomorphosis-based dichotomy occurring in geographical sympatry in eastern North America. The Caenozoic and living amphiumids, the desmognathine plethodontids and many of the plethodontines occur only in eastern North America implying that both families arose endemically in that area. The dichotomy had occurred by the late Cretaceous when the Mid-Continental Sea was retreating, as evidenced by the presence of an amphiumid in the Upper Cretaceous of Montana. The dichotomy presumably occurred earlier in the Upper Cretaceous, following as it does the node 5 dichotomy which is attributed to vicariance generated by the Mid-Continental Sea.

8. The Node 8 Dichotomy

The third major subdivision of terrestrial Laurasia was the progressive splitting of the continent into the North American and Eurasian continents by the sea-floor spreading producing the North Atlantic Ocean (Fig. 7). The completion of the oceanic barrier at 45 Ma is the latest possible geographical isolating event which could have formed the basis for this dichotomy, but effective isolation had already occurred at this time. The node 8 dichotomy is between the Ambystomatidae (uncertainly monophyletic but exclusively North American) and the Salamandridae (monophyletic and believed to be of European origin). The European and North American mammal faunas show permanent divergence from the Middle Eocene onwards (Szalay and McKenna, 1971) but the earliest European salamandrids are Upper Palaeocene. It appears that the passage of the relatively mobile mammals between North America and Europe was intermittent in the Palaeocene and early Eocene and it must be presumed that the less mobile salamanders were unable to disperse at the same times as the mammals. The ambystomatid-salamandrid dichotomy can have been no later than Middle Palaeocene and may have occurred in the late Cretaceous or early Palaeocene. However the possibility exists that the centre of endemism of salamandrids was not the principal north European landmass but one of the central or south European islands which were present in the late Cretaceous and early Caenozoic. The Pan-Euramerican proteids and batrachosauroidids were also isolated as disjunct populations by the same marine barriers and are represented by different genera in the two continents in the Caenozoic.

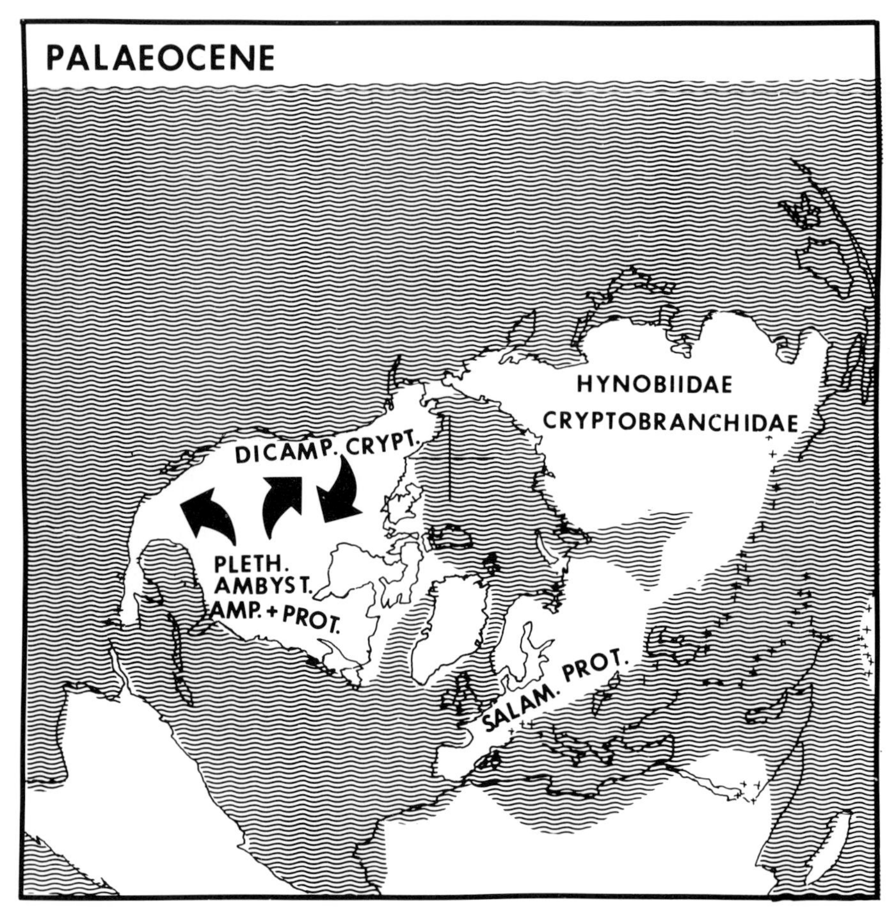

Fig. 7. Palaeocene north polar projection (Lambert equal area) depicting vicariance resulting from the developing Atlantic Ocean, together with post-vicariance dispersal in North America following withdrawal of the Mid-Continental Sea. The temporary isthmus of land across the south Turgai Sea appears to have had no biogeographical consequences for urodeles. AMBYST. = Ambystomatidae, AMP. = Amphiumidae, DICAMP. = *Dicamptodon* and *Rhyacotriton*, PLETH. = Plethodontidae, PROT. = Proteidae, SALAM. = Salamandridae. (Map copyright Cambridge University Press, 1980, taken from Smith *et al.* "Phanerozoic Paleocontinental World Maps".

9. Disappearance of Barriers and Subsequent Dispersals

By the late Cretaceous/early Caenozoic, it may be inferred that the fragmentation of the urodele fauna of Laurasia was at its maximum and that the modern family-level diversity had been generated largely by vicariance and paedomorphosis. In the model proposed here, the hynobiids and cryptobranchids were present in east Asia, the salamandrids and some proteids in Europe, the ambystomatids, plethodontids, amphiumids, sirenids and some proteids in eastern North America and *Dicamptodon* and *Rhyacotriton* relatives and further proteids in western North America. The possibility remains that sirenids were more widespread as relicts.

However, even before the Atlantic Ocean barrier was finally established, other barriers were disappearing and one new intercontinental connection was intermittently established. The sequence of disappearance of barriers together with the dispersals attributable to the opportunities presented by these events, is given below in chronological order. These secondary dispersals are only invoked where sympatry of terrestrial families demands such an explanation or where a single form within a clade occurs substantially outside the range of the rest of the clade. The time of disappearance of a barrier marks the earliest possible date of dispersal while the earliest fossil representative of a group in a colonized area provides the latest date for dispersal. Clearly then, the discovery of earlier fossil material of a given group in a given continent can necessitate significant revision of the chronology of a dispersal.

10. Initial Closure of the Bering Straits (Fig. 6).

As already discussed, the phylogenetic evidence strongly indicates an Asian origin for the cryptobranchids and the appearance of a cryptobranchid in the Upper Palaeocene of North America (Naylor, 1981) necessitates a pre-Upper Palaeocene dispersal, the most likely route being the Bering land connection. The earliest evidence for this area being a route for the dispersal of terrestrial vertebrates is the appearance of some eucosmodontid and taeniolabidid multituberculate mammals in North America in the late Cretaceous, the apparent centre of endemism and diversification of these families being Asia (Kielan-Jaworowska, 1974). Kielan-Jaworowska argued that the apparent dispersal of only a few mammal families from Asia to North America indicated a "sweepstakes" route, perhaps rafting. The appearance of cryptobranchids in the Palaeocene of North America is consistent with such a one-way dispersal of terrestrial vertebrates but for cryptobranchids to have dispersed, there must, of course have been a continuity of fresh water, i.e. a network of river-systems.

The temporary withdrawal of the Turgai Sea for part of the Palaeocene (Russell, 1975: Fig. 4) and the intermittent movement of mammals between North America and Europe during the Palaeocene allows the alternative possibility of a Palaeocene "window" when cryptobranchids might have reached North America from Asia via Europe. However, not only are no cryptobranchids recorded from Europe prior to the Oligocene but Russell (1975) noted that such mammalian faunal evidence as there is, does not demand interchange with Asia prior to the early Eocene. The Palaeocene connection between Asia and Europe appears to have been relatively tenuous and intermittent as does the Europe to North America connection, and the probability of cryptobranchids dispersing across both as well as the distances in between, seems less probable than a shorter dispersal via the Bering land route.

11. Withdrawal of the Mid-Continental Sea (Fig. 7)

Towards the end of the Cretaceous, between 65 and 70 Ma, the Mid-Continental Sea withdrew southwards (Hallam, 1981). Some subgroups of ambystomatids, plethodontids and amphiumids into western North America where the former two families are still sympatric with *Dicamptodon* and *Rhyacotriton*, relicts of the earlier isolation (Fig. 7). The westward dispersal of amphiumids as evidenced by *Proamphiuma* from the Upper Cretaceous of Montana (Estes, 1969b) appears to have been an ephemeral extension of their range. Having reached western North America, both ambystomatids and plethodontids were also able to disperse southwards into central America for the first time in the early Caenozoic (Savage, 1966; Wake and Lynch, 1976). The desmognathine plethodontids have remained restricted to eastern North America but others such as the genera *Plethodon* and *Aneides* and the ambystomatid genus *Ambystoma* still occur on both sides of North America. Subsequent climatic and vegetational changes during the later Caenozoic have resulted in an almost complete arid barrier across central North America isolating eastern and western salamander faunas, only *Ambystoma tigrinum* maintaining an unbroken distribution across North America. Two west-to-east dispersals appear to have followed the withdrawal of the Mid-Continental Sea, namely those of the cryptobranchids, now occurring in the Appalachians and the Ozarks, and the dicamptodontids which reached North Dakota in the Eocene (*Chrysotriton* Estes, 1981).

12. Withdrawal of the Turgai Sea (Fig. 8)

The Turgai Sea complex withdrew partially in the Palaeocene, only to reappear in the early Eocene before disappearing permanently at the end of the Eocene (Russell, 1975). There is no evidence for salamander dispersal between Europe and Asia prior to the Oligocene and hence during the earlier withdrawal of the Turgai, but the discovery of pre-Oligocene salamandrids in Asia or pre-Oligocene cryptobranchids or hynobiids in Europe would necessitate invoking the Palaeocene European–Asian corridor. The Salamandridae diversified considerably in Europe during the Eocene and reached Asia by the Miocene (Young, 1965), but their presence in the Upper Oligocene of North America (van Frank, 1955) implies their presence in Asia in the Oligocene as well. Wake and Özeti's (1969: Fig. 7) cladogram of salamandrid relationships requires three lineages to disperse into east Asia after the withdrawal of the Turgai Sea: (i) *Tylototriton*; (ii) a clade made up of the remaining Asian salamandrids; and (iii) the stem lineage which ultimately dispersed to North America to give rise to *Taricha* and *Notophthalmus*. The former two groups occur sympatrically with hynobiids in Asia, the latter with ambystomatids in North America. Western dispersal was also made possible by the withdrawal of the Turgai Sea as evidenced by the appearance of cryptobranchids in the fossil record of Europe in the middle Oligocene (Westphal, 1958). They became extinct in the Pliocene. *Salamandrella keyserlingii* is the only living hynobiid to extend westwards across the area once occupied by the Turgai Sea.

13. The Bering Land Connection in the Caenozoic (Fig. 8)

Apart from the appearance of cryptobranchids in North America, the Bering connection must be invoked to explain two other intercontinental dispersals, both of which appear to have taken place by the late

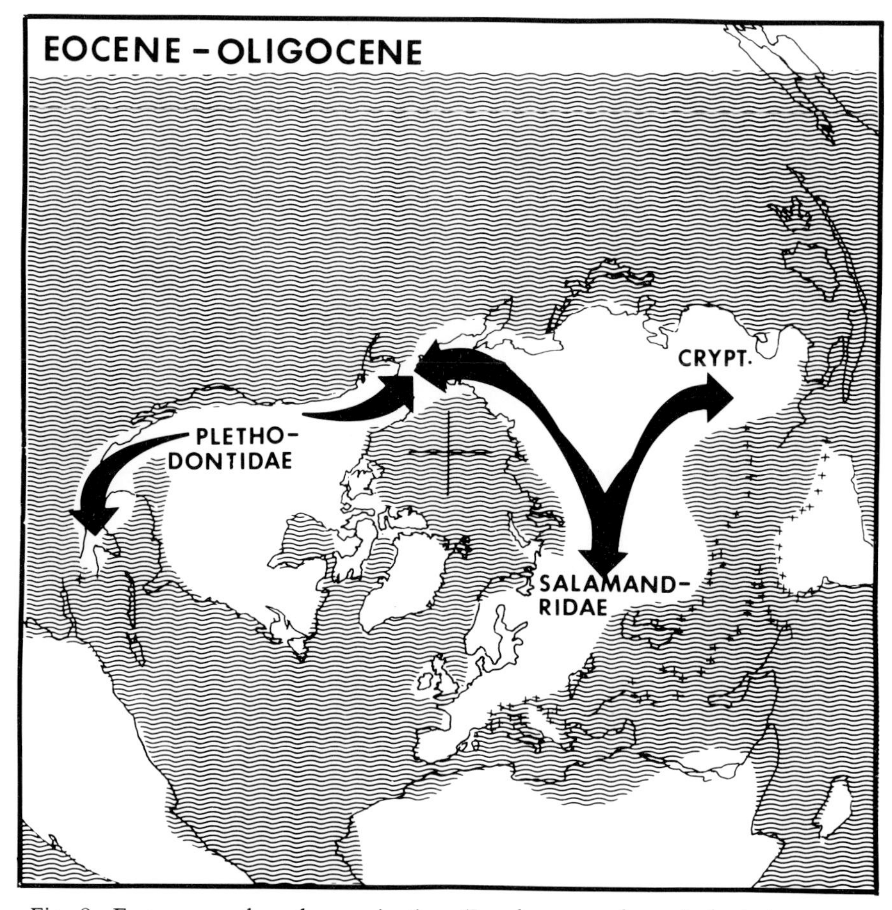

Fig. 8. Eocene north polar projection (Lambert equal area) depicting post-vicariance dispersals of the early Caenozoic involving the Cryptobranchidae, Plethodontidae and Salamandridae. The Turgai Sea finally withdrew at the end of the Eocene. The genera involved in the dispersals are discussed in the text. (Map copyright Cambridge University Press, 1980, taken from Smith *et al.* "Phanerozoic Paleocontinental World Maps".)

Oligocene (Fig. 8). One is the dispersal of salamandrids to North America, the living North American genera *Taricha* and *Notophthalmus* appearing in the Oligocene and Miocene respectively (van Frank, 1955; Tihen, 1974; Naylor, 1979b). The two genera form a monophyletic group (Wake and Özeti, 1969; Morescalchi, 1975) within the Salamandridae and therefore may be assumed to represent the descendants of a single population dispersing across the Bering route. The only North America to Eurasia dispersal to be invoked is that of the plethodontid *Hydromantes* which is represented by very similar species in California, south-east France, Italy and Sardinia. The species of *Hydromantes* form a natural group characterized by a unique tongue structure (Wake, 1966; Lombard and Wake, 1977), and Estes (1970) and Wake *et al.* (1978) have argued on palaeogeographical and immunological grounds for an Oligocene dispersal post-dating the formation of the Atlantic Ocean but pre-dating the isolation of the Sardinian uplands in the Miocene and hence necessarily via the Bering connection. The presence of *Hydromantes* and the other bolitoglossines in the west but not the east of North America is obviously helpful to this hypothesis. It is interesting to note that the only plethodontid successfully to colonize any part of Eurasia is a purely terrestrial form, montane rather than stream-dwelling and without a larval phase. The failure of plethodontids to reach European Laurasia in the late Cretaceous or early Palaeocene may be inferred from the presence of the plethodontid-convergent salamandrids such as *Chioglossa* filling the specialized stream-dwelling niche in Europe, certainly from the Miocene onwards (Estes and Hoffstetter, 1976) and probably earlier. In Asia, stream-dwelling specialists such as the hynobiid *Onychodactylus* and the salamandrid *Pachytriton* are present. The niche for a specialized stream-dweller was thus not available for plethodontids in Eurasia but at least one terrestrial niche apparently was.

14. The Iberia–Morocco Isthmus

The primarily European species *Salamandra salamandra* and *Pleurodeles waltl* also occur in the Atlas Mountains in north-west Africa where there is an indigenous species *Pleurodeles poireti*. The former two species occur in the Iberian peninsula and their distributions suggest a recent terrestrial connection between Iberia and north-west Africa. *Salamandra* and *Pleurodeles* presumably dispersed to North Africa in the late Miocene between 7 and 5·5 Ma when an isthmus between Iberia and Africa cut off the Mediterranean region from the Atlantic Ocean (Hsü, 1972). At about 5.5 Ma, the Straits of Gibraltar were formed, the Mediterranean region was inundated and the connection between Iberia and North Africa severed.

15. The Panama Isthmus

The final establishment of a permanent isthmus between Central and South America is believed to have taken place in the early Pliocene and the presence of a montane corridor along this isthmus permitted the bolitoglossine plethodontid genera *Oedipina* and *Bolitoglossa* to disperse into South America (Wake and Lynch, 1976).

THE DIVERSIFICATION OF THE URODELES

Three major conclusions concerning the family-level diversification of urodeles derive from this cladistic-vicariance model of urodele relationships and distributions, and each may be contrasted with the view of urodele history recently put forward by Naylor (1980, 1981).

(i) The dichotomies which gave rise to the five families of living "terrestrial" urodeles are mostly attributable to geographical isolation events in Laurasia during the Mesozoic, and some of the diversity of neotenous urodele families is indirectly attributable to these events. Naylor (1980:122) has recently suggested that the radiation of urodeles is partly attributable to selective pressures associated with the coevolution of insects and angiosperms during the Cretaceous. This speculation might be appropriate as an explanation for some of the convergence in feeding specializations within some terrestrial urodele families but it does not account for the diversity and distribution of terrestrial families nor does it explain the congruences between the dichotomies in urodele diversification and the barriers within Laurasia.

(ii) Nearly all families of living urodeles are hypothesized as having differentiated between the Middle Jurassic (150 Ma) when isolation events commenced, and the early Palaeocene (60 Ma) by which time all dichotomies had occurred. The Sirenidae is the only living family which may have differentiated prior to the Mid-Jurassic, and the ambystomatid-salamandrid dichotomy is the only one which might have been as late as the Palaeocene. Most isolation events and, by implication, most dichotomies, took place between 150 and 100 Ma. This is similar to the time-scale suggested by Naylor (1980:123) who proposed that much of the radiation of urodeles occurred in the pre-Campanian Cretaceous (130–80 Ma). However, Naylor's conclusion was based on the inferred correlation of salamander diversification with those of angiosperms and insects, together with a literal interpretation of the fossil record. The former observation is highly speculative and the latter is based on negative evidence. There are very few small vertebrate assemblages from the Jurassic or the pre-Campanian Cretaceous and hence few faunas from which the urodeles are seen to be absent. The absence of various urodele families from pre-Campanian rocks is seen here as a deficiency of the fossil record, not an argument for late Cretaceous diversification of salamanders.

(iii) There was no centre of evolution and dispersal of urodele families in any area less than Laurasia. The families differentiated in the areas where they still predominantly occur and only a few dispersals need be hypothesized. Naylor (1980) suggested that much of the evolution of the Urodela occurred in North America but the Cryptobranchoidea (Hynobiidae and Cryptobranchidae) appear to be Asian endemics, the Salamandridae evolved in Europe, the Proteidae originated as a Euramerican group and the Sirenidae may be a relictual Laurasian group. The remaining families, however, appear to be of North American origin.

CONCLUSIONS

A hypothesis of relationships for the nine families of living urodeles is presented in cladogram form and it is suggested that four of the nine dichotomies in the cladogram result from vicariance, being congruent in sequence and effect with geographical isolation events resulting from the development of marine barriers within Laurasia. The other dichotomies appear to have been ecological isolation events, four of which are characterized by paedomorphosis in one branch of the dichotomy. However, although the latter dichotomies are not attributable directly to geographical isolation, the evolution of four distinct groups of paedomorphic urodeles may, in part, be an indirect result of vicariant evolution of the terrestrial salamanders.

The geographical history of urodeles and their diversification at a family level can be modelled largely in terms of subdivision of a cosmopolitan early–mid-Jurassic fauna of terrestrial salamanders. The sequential appearance of the Turgai Sea, the Mid-Continental Sea and the north Atlantic Ocean between the mid-Jurassic and the early Caenozoic is consistent in sequence and effect with the dichotomies producing most of the primarily terrestrial salamander clades and explains the Mesozoic diversification of salamanders more precisely than general ecological/evolutionary explanations. Most of the dichotomies must have occurred prior to the late Cretaceous, over a period which is barely represented in the fossil record of urodeles. The later withdrawal of continental seas and the appearance of the Bering connection and the Iberian–Moroccan and Panamanian isthmuses has permitted several post-vicariance dispersals to occur.

This model of salamander historical geography is essentially an economical model invoking the minimal number of dispersal events to account for the distribution of living forms and their unequivocal fossil relatives. I have no doubt that the actual historical events were more complex than this model but there can be no justification for invoking further complexity unless one of the testable components of the model is falsified. The hypotheses of relationships and biogeography proposed here are testable: (a) by further examination of character distribution in living urodeles and hence reevaluation of their relationships; (b) by further analysis of palaeogeographical events and patterns; and (c) by discovery of further determinate fossil urodele material from new localities and horizons particularly prior to the Palaeocene, thus permitting more accurate assessment of the rôles of dispersal and extinction in the production of the modern distribution patterns. Possibly this model of urodele distribution in space and time may be of use in evaluating the states and relationships of the many fragmentary fossil urodeles, some of which are still effectively classified independantly of living forms.

ACKNOWLEDGEMENTS

I should like to thank Richard Estes for invaluable discussions and information on fossil urodeles and for commenting on an earlier draft of this paper. I should also like to thank J.J. Hooker for helpful discussions on Caenozoic geography, my wife Angela for assistance in the preparation of the figures of this paper, and Richard Packer for photographic work associated with the figures.

REFERENCES

Broin, F. de, Buffetaut, E., Koeniguer, J.C., Rage, J.C., Russell, D., Taquet, P., Vergnaud-Grazzini, C. and Wenz, S. (1974). La faune de Vertébrés continentaux du gisement d'In Beceten (Sénonien du Niger). *C. r. hebd. Séanc. Acad. Sci., Paris* D: **279**, 469–472.

Carroll, R.L. and Holmes, R. (1980). The skull and jaw musculature as guides to the ancestry of salamanders. *Zool. J. Linn. Soc.* **68**, 1–40.

Cox, C.B. (1975). Vertebrate palaeodistributional patterns and continental drift. *J. Biogeogr.* **1**, 75–94.

Croizat, L., Nelson, G. and Rosen, D.E. (1974). Centers of origin and related concepts. *Syst. Zool.* **23**, 265–287.

Darlington, P.J., jr. (1957). "Zoogeography: the geographical distribution of animals". J. Wiley, New York.

Dunn, E.R. (1922). The sound-transmitting apparatus of salamanders and the phylogeny of the Caudata. *Am. Nat.* **56**, 418–427.

Dunn, E.R. (1926). "The salamanders of the family Plethodontidae". Smith College, Northampton, Mass.

Edwards, J.L. (1976). Spinal nerves and their bearing on salamander phylogeny. *J. Morph.* **148**, 305–328.

Estes, R. (1965). Fossil salamanders and salamander origins. *Am. Zool.* **5**, 319–334.

Estes, R. (1969a). The Batrachosauroididae and Scapherpetontidae, Late Cretaceous and Early Cenozoic salamanders. *Copeia* **1969**, 225–234.

Estes, R. (1969b). The fossil record of amphiumid salamanders. *Breviora* **332**, 1–11.

Estes, R. (1970). Origin of the recent North American lower vertebrate fauna: an inquiry into the fossil record. *Forma Functio* **3**, 139–163.

Estes, R. (1975). Lower Vertebrates from the Fort Union Formation, late Paleocene. Big Horn Basin, Wyoming. *Herpetologica* **31**, 365–385.

Estes, R. (1981). "Handbuch der Paläoherpetologie". Teil 2. "Gymnophiona, Caudata". G. Fischer, Stuttgart.

Estes, R. and Darevsky, I. (1977). Fossil amphibians from the Miocene of the North Caucasus, U.S.S.R. *J. palaeont. Soc. India* **20**, 164–169.

Estes, R., Hecht, M. and Hoffstetter, R. (1967). Paleocene amphibians from Cernay, France. *Am. Mus. Novit.* **2295**, 1–25.

Estes, R. and Hoffstetter, R. (1976). Les Urodèles du Miocène de la Grive-Saint-Alban (Isère, France). *Bull. Mus. natn. Hist. nat., Paris*, sér. 3, **398**, 297–343.

Goin, C.J. and Auffenberg, W. (1955). The fossil salamanders of the family Sirenidae. *Bull. Mus. comp. Zool. Harv.* **113**, 497–514.

Goin, C.J. and Goin, O.B. (1971). "Introduction to Herpetology" (2nd edn) W.H. Freeman, San Francisco.

Hallam, A. (1981). Relative importance of plate movements, eustasy and climate in controlling major biogeographical changes since the early Mesozoic. *In* "Vicariance Biogeography" (G. Nelson and D.E. Rosen, eds), pp. 303–340. Columbia, New York.

Hecht, M.K. and Edwards, J.L. (1976). The determination of parallel or monophyletic relationships: the proteid salamanders—a test case. *Am. Nat.* **110**, 653–677.

Hecht, M.K. and Edwards, J.L. (1977). The methodology of phylogenetic inference above the species level. *In* "Major Patterns in Vertebrate Evolution" (M.K. Hecht, P.C. Goody and B.M. Hecht, eds), pp. 3–51. Plenum, New York and London.

Hecht, M.K. and Estes, R. (1960). Fossil amphibians from quarry Nine. *Postilla* **46**, 1–19.

Hecht, M.K. and Hoffstetter, R. (1962). Note préliminaire sur les amphibiens et les squamates du Landenien Supérieur et du Tongrien de Belgique. *Bull. Inst. r. Sci. nat. Belg.* **38**, 1–30.

Holman, J.A. (1968). Lower Oligocene amphibians from Saskatchewan. *Q. Jl. Fla. Acad. Sci.* **31**, 273–289.

Hsü, K.J. (1972). When the Mediterranean dried up. *Sci. American* **227**, 26–36.

Ivachnenko, M.F. (1978). Urodelans from the Triassic and Jurassic of Soviet central Asia. *Paleont.J.* **1978**, 362–368.

Jurgens, J.D. (1971). The morphology of the nasal region of Amphibia and its bearing on the phylogeny of the group. *Annale Univ. Stellenbosch* **46A** (2), 1–146.

Kezer, J., Seto, T. and Pomerat, C.M. (1965). Cytological evidence against parallel evolution of *Necturus* and *Proteus*. *Am. Nat.* **99**, 153–158.

Kielan-Jaworowska, Z. (1974). Migrations of the Multituberculata and the late Cretaceous connections between Asia and North America. *Ann. S. Afr. Mus.* **64**, 231–243.

Larsen, J., jr. (1963). "The cranial osteology of neotenic and transformed salamanders and its bearing on interfamilial relationships". PhD thesis, University of Washington. [Not seen, cited by Hecht and Edwards (1977).]

Larsen, J.H., jr. and Guthrie, D.J. (1974). Parallelism in the Proteidae reconsidered. *Copeia* **1974**, 635–643.

Lombard, R.E. (1977). Comparative morphology of the inner ear in salamanders (Caudata: Amphibia). *Contrib. Vert. Evoln* **2**, 1–140.

Lombard, R.E. and Wake, D.R. (1977). Tongue evolution in the lungless salamanders, family Plethodontidae. II. Function and evolutionary diversity. *J. Morph.* **153**, 39–80.

Monath, T. (1965). The opercular apparatus of salamanders. *J. Morph.* **116**, 149–170.

Morescalchi, A. (1975). Chromosome evolution in the caudate Amphibia. *Evol. Biol.* **8**, 339–387.

Naylor, B.G. (1978). The earliest known *Necturus* (Amphiba, Urodela), from the Paleocene Ravenscrag formation of Saskatchewan. *J. Herpetol.* **12**, 565–569.

Naylor, B.G. (1979a). The Cretaceous salamander *Prodesmodon* (Amphibia: Caudata). *Herpetologica* **35**, 11–20.

Naylor, B.G. (1979b). A new species of *Taricha* (Caudata: Salamandridae), from the Oligocene John Day Formation of Oregon. *Can. J. Earth Sci.* **16**, 970–973.

Naylor, B.G. (1980). Radiation of the Amphibia Caudata: are we looking too far in the past? *Evol. Theory* **5**, 119–126.

Naylor, B.G. (1981). Cryptobranchid salamanders from the Paleocene and Miocene of Saskatchewan. *Copeia* **1981**, 76–86.

Nevo, E. and Estes, R. (1969). *Ramonellus longispinus*, an early Cretaceous salamander from Israel. *Copeia* **1969**, 540–547.

Noble, G.K. (1931). "The Biology of the Amphibia". McGraw-Hill, New York.

Patterson, C. (1981). The development of the North American fish fauna—a problem of historical biogeography. *In* "The Evolving Biosphere—(Chance, change and challenge)" (P.L. Forey, ed.), pp. 265–282. Cambridge University Press, Cambridge.

Platnick, N.I. and Nelson, G. (1978). A method of analysis for historical biogeography. *Syst. Zool.* **27**, 1–16.

Putnam, J.L. (1975). Septation in the ventricle of the heart of *Siren intermedia*. *Copeia* **1975**, 773–774.

Reed, H.D. (1920). The morphology of the sound-transmitting apparatus in caudate Amphibia and its phylogenetic significance. *J. Morph.* **33**, 325–387.

Regal, P. (1966). Feeding specializations and the classifications of terrestrial salamanders. *Evolution, Lancaster, Pa* **20**, 392–407.

Robinson, P.L. (1973). Palaeoclimatology and continental drift. *In* "Implications of Continental Drift to the Earth Sciences" (D.H. Tarling and S.K. Runcorn, eds), pp. 451–476. Academic Press, London and New York.

Rosen, D.E. (1978). Vicariant patterns and historical explanation in biogeography. *Syst. Zool.* **27**, 159–188.

Russell, D.E. (1975). Paleoecology of the Paleocene–Eocene transition in Europe. *In* "Approaches to Primate Paleobiology" (F. Szalay, ed.), *Contr. Primatol.* **5**, 28–61.

Salthe, S.N. (1967). Courtship patterns and the phylogeny of the urodeles. *Copeia* **1967**, 100–117.

Salthe, S.N. and Kaplan, N.O. (1966). Immunology and rates of enzyme evolution in the Amphibia in relation to the origins of certain taxa. *Evolution, Lancaster, Pa.* **20**, 603–616.

Savage, J.M. (1966). The origins and history of the Central American herpetofauna. *Copeia* **1966**, 719–766.

Savage, J.M. (1973). The geographic distribution of frogs: patterns and predictions. *In* "Evolutionary Biology of the Anurans" (J.L. Vial, ed.), pp. 351–445. University of Missouri Press, Columbia, Missouri.

Schmalhausen, I.I. (1968). "The Origin of the Terrestrial Vertebrates". Academic Press, New York and London.

Smith, A.G., Hurley, A.M. and Briden, J.C. (1980). "Phanerozoic paleocontinental world maps". Cambridge University Press, Cambridge.

Srinivasachar, H.R. (1962). Development and morphology of the skull of *Rhyacotriton olympicus olympicus* Gaige (Amphibia, Urodela, Ambystomidae). *Morph. Jahrb.* **103**, 263–302.

Szalay, F.S. and McKenna, M.C. (1971). Beginning of the age of mammals in Asia: the late Paleocene Gashato fauna, Mongolia. *Bull. Am. Mus. nat. Hist.* **144**, 269–319.

Tihen, J.A. (1958). Comments on the osteology and phylogeny of ambystomatid salamanders. *Bull. Fla St. Mus.* **3**, 1–50.

Tihen, J.A. (1974). Two new North American salamandrids. *J. Herpetol.* **8**, 211–18.

Tihen, J.A. and Wake, D.B. (1981). Vertebrae of plethodontid salamanders from the Lower Miocene of Montana. *J. Herpetol.* **15**, 35–40.

van Frank, R. (1955). *Palaeotaricha oligocenica*, new genus and species, an Oligocene salamander from Oregon. *Breviora* **45**, 1–12.

Wake, D.B. (1966). Comparative osteology and evolution of the lungless salamanders, family Plethodontidae. *Mem. Sthn. Calif. Acad. Sci.* **4**, 1–111.

Wake, D.B. and Lynch, J.F. (1976). The distribution, ecology and evolutionary history of plethodontid salamanders in tropical America. *Nat. Hist. Mus. Los Angeles Co., Sci. Bull.* **25**, 1–65.

Wake, D.B., Maxson, L.R. and Wurst, G.Z. (1978). Genetic differentiation, albumin evolution and their biogeographic implications in plethodontid salamanders of California and southern Europe. *Evolution, Lancaster, Pa.* **32**, 529–539.

Wake, D.B. and Özeti, N. (1969). Evolutionary relationships in the family Salamandridae. *Copeia* **1969**, 124–137.

Westphal, F. (1958). Die Tertiaren und Rezenten eurasiatischen Riesensalamander (genus *Andrias*, Urodela, Amphibia). *Palaeontographica* **110**, 20–92.

Young, C.C. (1965). On the first occurrence of the fossil salamanders from the Upper Miocene of Shantung, China. *Acta paleont. sin.* **13**, 457–459.

16 Vicariance and Cladistics: Historical Perspectives with Implications for the Future

G. NELSON

Department of Ichthyology, American Museum of Natural History, New York, N.Y. 10024, USA

Abstract: The history of biogeography features a long-continuing conflict between creation myth and empirical science. From creation myth came the notions of centre of origin and dispersal, which were, and continue to be, employed in biogeography as "explanatory" devices. Empirical conflicts began at least as early as Candolle's (1820) determination of botanical regions, and continue with the modern concern for the interrelationships of areas of endemism. Biogeographical classification has been predominantly artificial, as exemplified by the schemes of Sclater (1858) and Wallace (1876) and the many subsequent commentaries upon them. Possibilities for natural classification have been often ignored, but they are currently emphasized by vicariance biogeography. Prospects for the future, already outlined by Croizat, depend upon the actual degree of geographical congruence of cladograms of different groups of organisms.

> Experience is . . . the name everyone gives their mistakes.
>
> Oscar Wilde.

> Our whole problem is to make the mistakes as fast as possible.
>
> John Archibald Wheeler.

If Wheeler is right and if his problem can be overcome, one popular notion (that mistakes are a waste of time) is probably wrong, for the

Systematics Association Special Volume No. 23, "Evolution, Time and Space: The Emergence of the Biosphere", edited by R.W. Sims, J.H. Price and P.E.S. Whalley, 1983, pp. 469–492. Academic Press, London and New York.

time involved (what is wasted) would be minimal. In any event, one might reckon human knowledge and its progressive development in terms of the mistakes known, through subsequent discovery, to be such. I am not sure that this trial-and-error measure of progress, "Wheeler's Law", would suffice for human knowledge in all its forms, but I am sure that it suffices, to some degree at least, for science.

A case in point is Darwin's experiences with the Galapagos birds. These he collected in a casual fashion, apparently in the beliefs that the differences in the bird fauna from island to island were trifling, and that the bird fauna as a whole represented, for the most part, mere varieties of mainland species. The story according to Grinnell (1974:260–261) is as follows:

> During their visit to the Galápagos Archipelago in the autumn of 1835, Charles Darwin and Robert Fitzroy, the young captain of the *Beagle*, had fallen into one of their numerous disputes, this time over the origin of a group of brownish-black birds resembling finches. Darwin surmised that their ancestors had been blown out from the mainland, and that successive generations had adapted themselves to their new environment by their own resources. Having done so, they should, he believed, be considered only varieties, rather than new species, and hence were not worth collecting carefully. Fitzroy, on the other hand, believed them to be new, hitherto undiscovered, species and had a special collection made for himself. Fitzroy had noted that the shape of the beak varied from group to group such that each was created for the type of island on which it had been placed. "This appears to be one of those admirable provisions of Infinite Wisdom by which each created thing is adapted to the place of which it was intended," he wrote in the account of the voyage, and concluded therefrom that these species were created by God.[1]

According to Grinnell it was only later, after their return to London, and after John Gould determined that the birds comprise several distinct species new to science that the birds (and the islands themselves) assumed a special significance. Fitzroy's collection, unlike Darwin's, had proper locality data, so using it as a reference Darwin was able later to sort some of his own specimens to locality.

[1] However interesting an interpretation, Grinnell's account is not as fully supported by published information as it might seem. The quote from Fitzroy is a case in point. The passage occurred between the following two sentences in Fitzroy's narrative (1839:503): "All the small birds that live on these lava-covered islands have short beaks, very thick at the base, like that of a bull-finch. . . . In picking up insects, or seeds which lie on hard iron-like lava, the superiority of such beaks over delicate ones, cannot I think, be doubted; but there is, perhaps, another object in their being so strong and wide." Fitzroy continued with an account of "an old bird in the act of supplying three young ones with drink, by squeezing the berry of a tree into their mouths."

Had Darwin intensely collected, and properly labelled, the birds, without preconceptions as to what among them might be a variety as opposed to a species, he might never have had cause to reflect further upon them. Thus, Gould's taxonomic treatment might never have disturbed Darwin at all, any more than it disturbed Fitzroy. The preconceptions, in short, eventually brought Darwin into conflict with the "facts" of Gould's taxonomy, which in Grinnell's words "threw Darwin into intellectual turmoil" (1974:261). The preconceptions, and the casual approach to collecting that followed from them, were thereby exposed as mistakes, which led Darwin into further investigation of the causal factors. Had the history been otherwise, we today probably would have the misfortune of knowing the birds as "Fitzroy's finches."

It has sometimes been remarked that Darwin's *Origin of Species* is mistitled, because the book does not treat that subject. But in a way the book does, if we understand the term "species" to relate, in part, to the Galapagos birds, to Darwin's discovery of his mistaken notions about their taxonomy, to the origin of those notions, and, in addition, to the origin of alternative notions. Darwin's view of the matter, characteristically, was briefly stated; his words, penned no one knows exactly when after his return to London, are as follows (1887, 1:276):

> In July opened first notebook on Transmutation of Species. Had been greatly struck from about the previous March on character of South American fossils, and species on Galapagos Archipelago. These facts (especially latter), origin of all my views.

Considering the field of biogeography from this point of view, I confess to a certain poverty of material. However much I read the literature of biogeography, I almost never encounter a first-hand admission of mistake. I do not mean that mistakes are not there to be found, but only that, when they are presented as such, they are always, or almost always, the other fellow's.

There is not much that can be called biogeography before the time of Linnaeus. Although he was never preoccupied with the subject, Linnaeus accepted the biblical version of creation but pondered the nature and location of the garden of Eden (1781:77):

> To enter into the remainder of this subject with as much brevity as possible, I think myself not greatly in danger of error in laying down the following proposition: *That the Continent in the first ages of the world lay*

> *immersed under the sea, except a single island in the midst of this immense ocean; where all animals lived commodiously, and all vegetables were produced in the greatest luxuriance.*

To satisfy the ecological requirements both of alpine and of tropical biota, Linnaeus considered how both kinds of organisms could survive on a single island (1781:90):

> First let us conceive Paradise situated under the Equator; and nothing further is requisite to demonstrate the possibility of these two indispensible conditions, than supposing a very lofty mountain to have adorned its beautiful plains.

He allowed that there was a problem (1781:94):

> Here we foresee that a difficulty will be started by some, how such a number of vegetables, such immense woods and thickets, those multitudes of flowers that cover every field and meadow, can have been disseminated over the world by single plants.

So he considered how (1781:95):

> from a single central spot, a plant of a given species may be so disseminated as to be found in all parts of the world.

His discussion concluded (1781:113–114):

> We have seen the Winds, the Rains, the Rivers, the Sea, Heat, Animals, Birds, the structure of Seeds, and Seed Vessels, the peculiar Natures of Plants, and even Ourselves, contribute to this great work.—I have shewn, that any one single plant alone would have been able to have covered the face of the globe: I have demonstrated that the dry land has always been increasing, and dilating itself; and therefore once was infinitely less than it is at this present:—I have traced back the orders of animals and vegetables, and found them to terminate in individuals created by the hand of God.

Naïve as these ideas may seem today, they were embodied in a diagram, admittedly hypothetical, published one hundred years later by Haeckel, which purported to show the "Monophyletic Origin . . . of the 12 Races of Man from Lemuria over the Earth" (Fig. 1).

Haeckel differed from Linnaeus with respect to notions of creation and evolution, and also with respect to the number of individuals involved in the origin of whatever species might have inhabited "paradise" (1876, 2:304):

> The whole of the celebrated dispute, as to whether the human race is descended from a single pair or not, rests upon a completely false way of putting the question. It is just as senseless as the dispute as to whether all

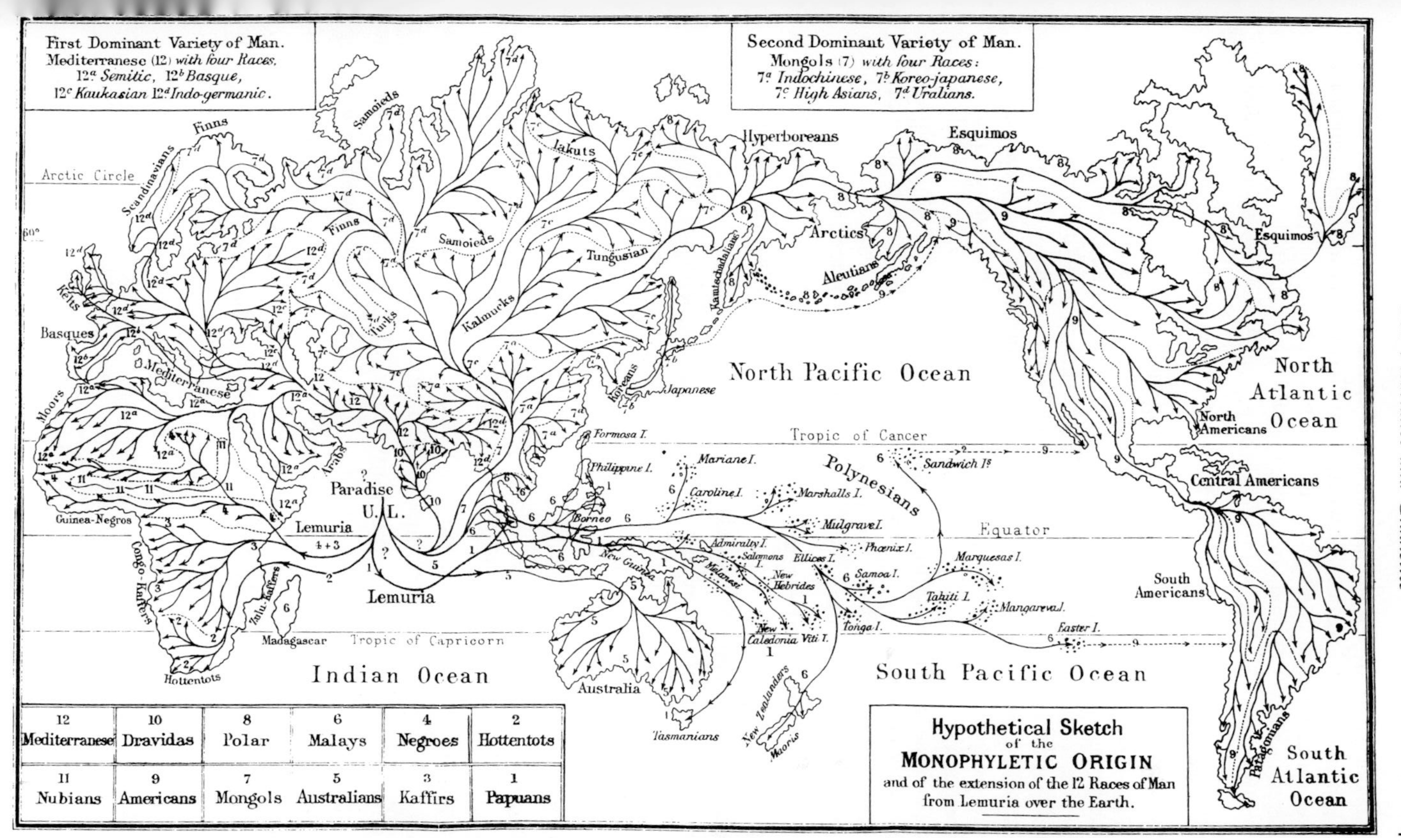

Fig. 1. Centre of origin/dispersal interpretation of humans. After Haeckel (1876: Pl. 20).

> sporting dogs or all race-horses are descended from a single pair. We might with equal justice ask whether all Germans or all Englishmen are "descended from a single pair", etc. A "first human pair," or "first man", has in fact never existed, any more than there ever existed a first pair or a first individual of Englishmen, Germans, race-horses, or sporting dogs.

With respect to the map itself, Haeckel asserted that (1876, 2:326):

> there are a number of circumstances (especially chorological facts) which suggest that the primaeval home of man was a continent now sunk below the surface of the Indian Ocean, which extended along the south of Asia, as it is as present (and probably in direct connection with it), towards the east, as far as further India and the Sunda Islands; towards the west, as far as Madagascar and the south-eastern shores of Africa. We have already mentioned that many facts in animal and vegetable geography render the former existence of such a south Indian continent very probable. . . . Sclater has given this continent the name of Lemuria, from the Semi-apes which were characteristic of it. By assuming this Lemuria to have been man's primaeval home, we greatly facilitate the explanation of the geographical distribution of the human species by migration.

One might emphasize the differences in the "explanations" of Linnaeus and Haeckel: that Linnaeus invoked miraculous creation; and Haeckel, evolution by means of natural selection. But what if we emphasize the similarity in conception of what is being "explained"? Consider what the map portrays, which fits Linnaeus' notions as well as Haeckel's. "Explain" the map in any way but it does not thereby change its character. With respect to the "explanations" themselves, I confess that I don't see them as all that different. "Miraculous creation" and "evolution by means of natural selection" I see for all practical purposes as merely different names for what was, and remains, unknown. Although evolution by means of natural selection is regarded in some quarters today as the more scientific "explanation," because it allows us to believe that we have some knowledge of the process, it has the additional defect of allowing us, if we so choose, to believe that we have knowledge as concrete and decisive as the pious faith of the creationist. The point was better made some years ago by Thompson (1963:566) whose words fit equally the viewpoints of the creationist and, as he intended, the selectionist:

> The reader may be completely ignorant of biological processes yet he feels that he really understands and in a sense dominates the machinery by which the marvellous variety of living forms has been produced.

Around the turn of the century, Haeckel revised this map because he had decided that Lemuria was an unlikely centre of origin for humans (Fig. 2). He wrote (1907, 2:437):

> this hypothesis—formerly advocated by me also—has of late years been opposed by such weighty considerations, especially from a geological point of view, that we must for the present give it up. In this case, of all the various parts of the earth, in which the "Paradise", or place of origin of the human species, might be looked for—by far the most likely which remains is Southern Asia, and, indeed, the western part, Further India.

With the trunk of Haeckel's tree transplanted from Lemuria to the *terra firma* a little northward of it, a trifling adjustment, I pass over the next half-century to Darlington's *Principal Existing Routes of Dispersal of Cold-Blooded Vertebrates* (Fig. 3). This figure was republished nine years later as "the main pattern of dispersal of vertebrates", not merely the cold-blooded ones but all of them. To clarify the geography of this figure, one need only consider Darlington's view of the "apparent history of dispersal of fresh-water fishes" (Fig. 4). Geographically it is quite clear and quite familiar, for, aside from its "explanation," it contains the same concepts that Linnaeus gave to us 200 years ago, and that Haeckel gave to us again 100 years later. This brief survey of zoological distribution closes with *Homo sapiens* and a figure entitled "topology of the minimum-evolution tree uniting fifteen human populations: constructed on the basis of blood group alleles" (Fig. 5). Again I stress the similarity in conception, rather than the details that differentiate between the "explanations" of one and another author. It is easy, after all, to find what seems to be mistakes of detail. It is not so easy to find the flaw, if indeed there is one, in the basic concept. Nor is it easy to explain why, through 200 years, there has been one basic concept entertained by authors so diverse, and calculated by them to portray correctly the distribution of organisms of even greater diversity. Only one conclusion is possible: Wheeler's Law has not been working very well in this area of science. In other words, progress has been minimal.

On the botanical side, matters are somewhat more complicated. By way of an introduction, it is useful to quote Takhtajan on the problem of flowering plants (1969:137):

> We must now consider the geographical location of that "isolated area" from which, in Darwin's opinion, the angiosperms eventually, as a result of geographical changes, managed to escape and then spread over the whole world.

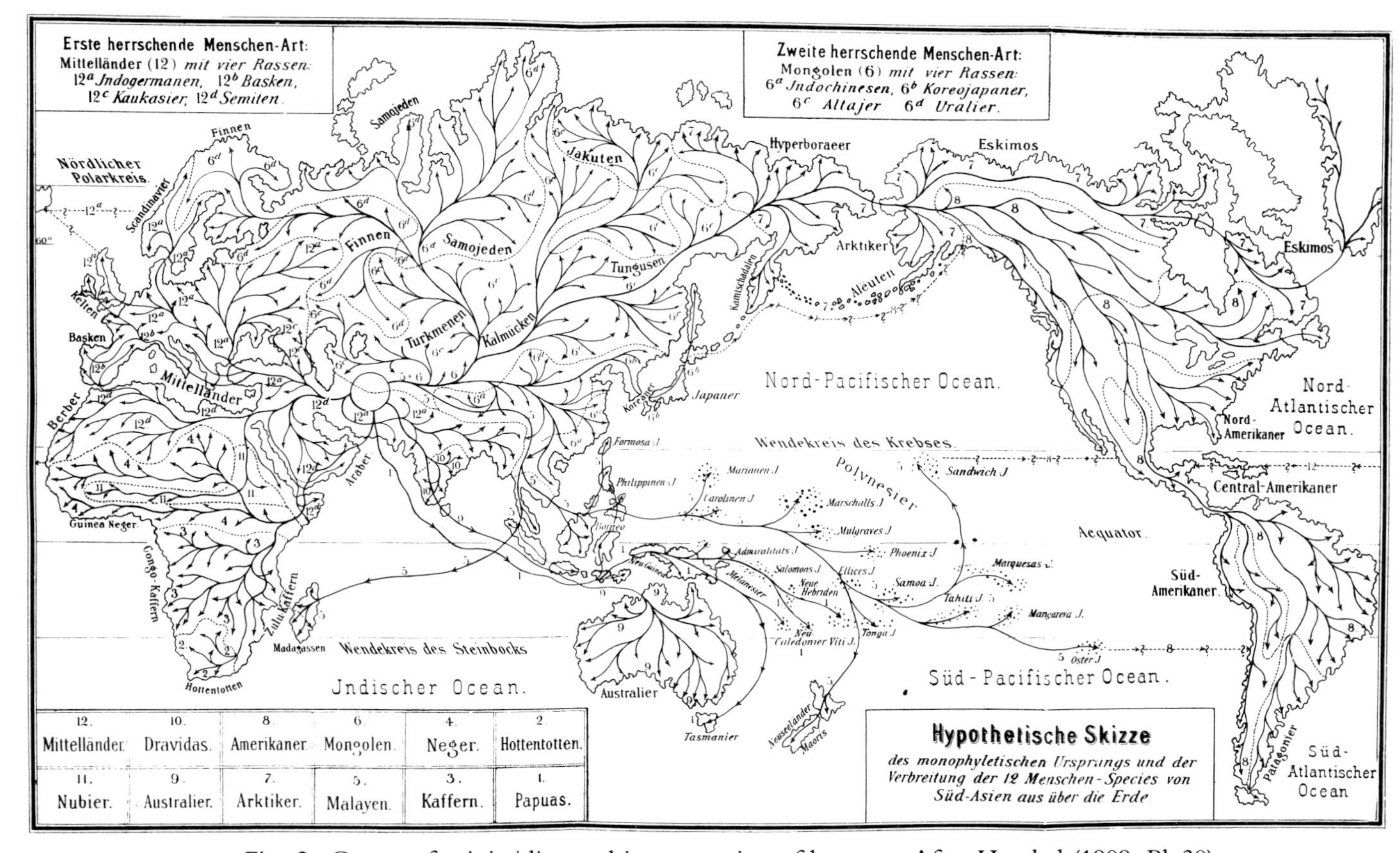

Fig. 2. Centre of origin/dispersal interpretation of humans. After Haeckel (1909: Pl. 30).

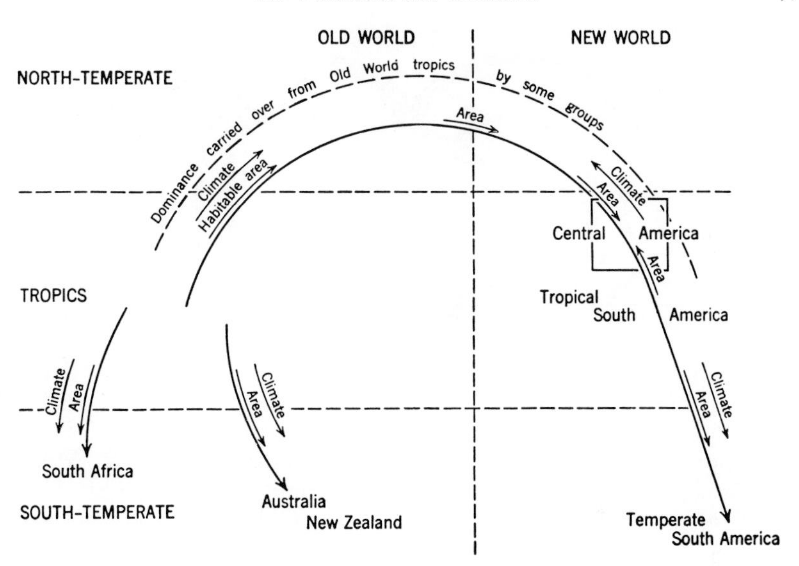

Fig. 3. Diagram of main dispersal pattern. [From *Q. Rev. Biol.* **23**, 112, Fig. 4 (1948).] Diagram is intended to suggest origin of dominant groups in main Old World tropics and dispersal along three main routes toward smaller areas and less favourable climates. Small areas show the expected effects of area and climate on direction of dispersal at critical points. After Darlington (1957: Fig. 72).

After reviewing the various possibilities he concluded (1969:142):

> It is here, in eastern and south-eastern Asia, Australasia and Melanesia, that the cradle of the angiosperms must be sought. As the whole pattern of the geographical distributions and connections of the primitive angiosperms (especially the families of the order *Magnoliales*) shows, it was this part of the world which had been, if not the birthplace, then at least the original centre of the wide-spread Cretaceous expansion of the angiosperms; and this could hardly have been very far from their birthplace.

Again we have the same basic concept.

Admittedly the sample is somewhat biased, in that authors have not been considered who believed that the "cradle" of angiosperms may have been, for example, in the far north (or the far south), or the centre of origin of humans may have been in Africa (or some other place). These different schemes have not been considered because they are but variations on a theme, and are played as easily as Haeckel's variations on Lemuria. The theme is that of the centre of origin of some taxon, wherever that might have been, and dispersal from there to other places.

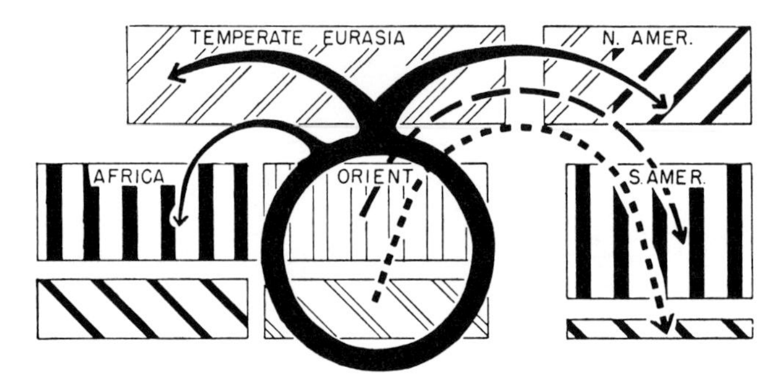

Fig. 4. Apparent history of dispersal of fresh-water fishes. Simplified diagram to show how existing distribution may have been produced by the following situation and events: zonal distribution of old non-ostariophysan fish faunas (diagonal bars) in the main Old World tropics and the north-temperate zone; rise of primitive Ostariophysi (characins and catfishes—vertical bars) in the main Old World tropics; movement of fractions of the old tropical fauna to South America, perhaps via the edge of the north-temperate zone (broken arrows), followed by multiplication of primitive Ostariophysi in South America; and radiation from the tropical Orient of dominant Cypriniformes (solid circle and arrows), causing extinction (white bars) of many older fishes in the Orient, temperate Eurasia, and western North America, and survival (solid bars) of the old tropical fish fauna in Africa and (modified) in South America and of the old north-temperate fauna in eastern North America. After Darlington (1957: Fig. 18).

Variations on the theme are played by grasping the trunk of the branching diagram displayed on the map, uprooting it, and transplanting it to another plot, while keeping the branch tips in place in the geographical areas actually occupied by the Recent members of the taxon.

The "minimum-evolution tree" is different from the other diagrams in lacking an identifiable trunk with roots so it cannot therefore be transplanted from one centre of origin to another. Its purpose is not to show a centre of origin, but rather some notion of interrelationship, primarily of the diverse groups of humans, and secondarily of the geographical areas occupied by the groups. Still the central axis of the diagram may be moved around on the map, in which case certain of the lines of the diagram would be stretched and others compressed. But elasticity as such is no problem. The point is illustrated, in your mind's eye, by lifting the main axis at some point so pulling the diagram off the

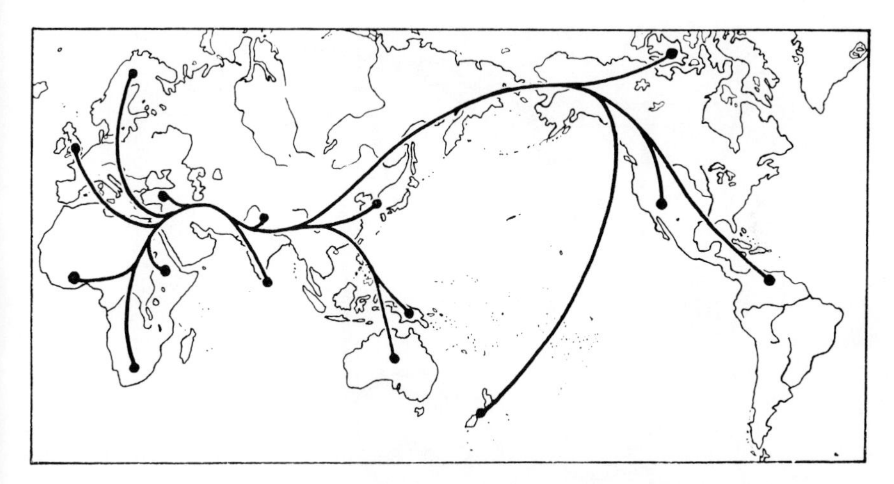

Fig. 5. Topology of the minimum-evolution tree uniting fifteen human populations; constructed on the basis of the frequency of blood-group alleles. After Edwards and Cavalli-Sforza (1964: Fig. 1).

surface of the map but letting the lines stretch as they will with the branch tips attached. From that lofty perspective, observe how the diagram relates the areas.

That we recognize some general notion of biogeography enables us to question its validity, and in particular to ask what sort of information might indicate it to be false, if it is. Here we confront various levels of generality. At one level we may consider the evidence that would require a trunk to be planted in one area rather than another. There is a large literature on this subject, and I have no doubt that problems of this sort can and do absorb whatever time and energy a person is willing to give them. But I do not see that the problems lead anywhere; they might just as well, all of them, be dancing on pins. Alternatively, we may deal with trunkless diagrams. But what do we do with them? What can they teach us?

One diagram of this sort purports to show the interrelationships of taxa that occupy the different parts of the globe (Croizat, 1958; Fig. 6). It is more complex than the others and, unlike most of them, is without trunk and roots. Its purpose is not to show centres of origin, and indeed none is shown. Croizat titles it "The main channels of dispersal (plants and animals alike) of 'modern' life, so called." Of interest are the "main nodes controlling the course of important 'tracks' on an intercontinental

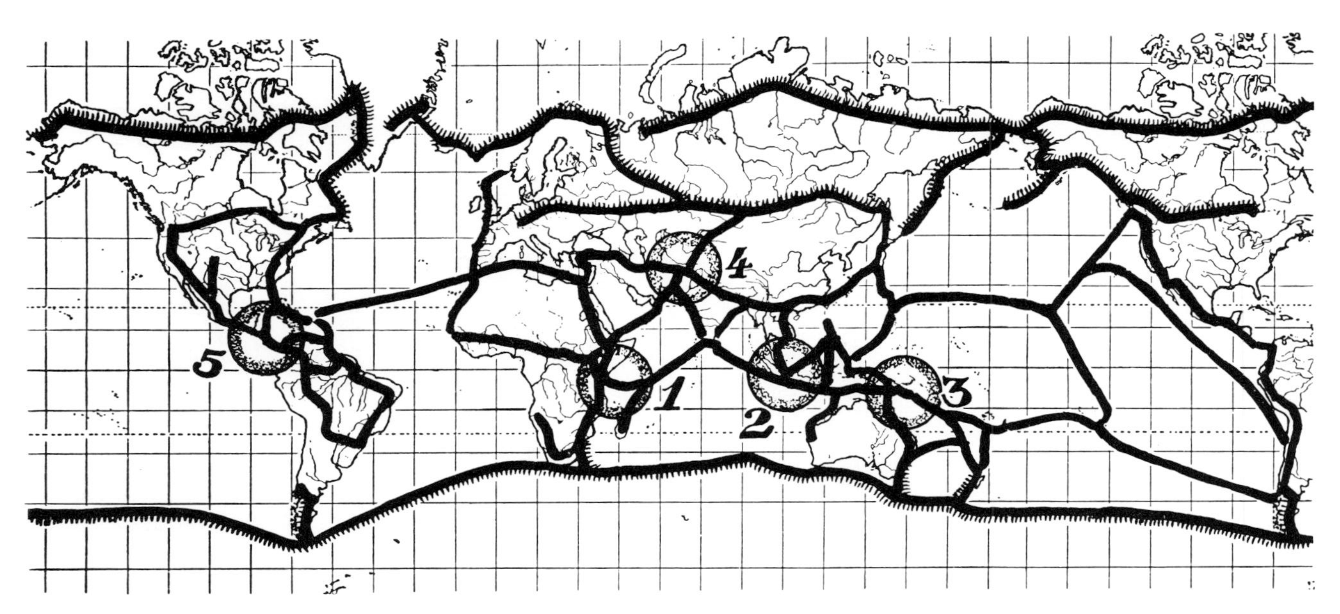

Fig. 6. The main channels of dispersal (plants and animals alike) of "modern" life, so called. Hatched in the high north and high south, respectively, are "tracks" that could with some property be identified as *boreal* ("Holarctic", "Palaearctic", "Nearctic") and as *austral* ("Antarctic", "Old Oceanic", etc.). Set out within stippled circles 1, 2, 3, 4, 5 are the very main nodes controlling the course of important "tracks" on an intercontinental scale. The map of the new World to the left stresses the "tracks" toward the Atlantic; that to the right those toward the Pacific. After Croizat (1958: Fig. 259).

scale." The nodes are numbered 1 through 5. I doubt not that Croizat considers this a reasonable summary, at a fairly general level, of organismic distribution that we can see today. For reasons of his own he uses the term "dispersal" to mean distribution, and not migration from some centre or origin.

The complexity of this pattern, if I may term it that, Croizat considers a decisive falsification of notions, like those I have already mentioned, that illustrate a centre of origin for some taxon or taxa. In other words, the centre of origin in Croizat's view of modern life is not less than the world at large. In his view the main outlines fit the tectonic structure of the earth as closely as, say, a glove fits the hand within it.

Areas of particular complexity, his "nodes", exist, in his view, because their tectonic history extends concurrently in many directions in space through time. They may be regarded as zones of tectonic convergence or compression, possibly of continental collision or rapprochement. The New World illustrates this aspect for like other major continental areas, it has a complex pattern of relationships. Node 5, for example, involves relationships with Africa, by way of the Atlantic, and with Australasia, by way of a more complex network spanning the Pacific. This is seen by repeating the "mind's eye" experiment on this diagram, here best performed by simultaneously grasping the diagram at three points, one each for the tropical Atlantic, Indian, and Pacific Oceans.

To some extent Croizat's notions at this global level are supported by the facts and theories pertaining to continental drift. Nevertheless, Croizat may be wrong as there are three sorts of information that might indicate his views to be false. The first is the age of taxa that exemplify this global pattern of interrelationships. Such taxa and their differentiation are predicted to be old, that is, mid-Mesozoic. This prediction conflicts with what is often asserted to be already known on the basis of the fossil record. In Croizat's view, some, and perhaps most if not all, of these assertions are false; that is, they are mistakes due to misinterpretation of what fossil record there is. The second is the compositions of the groups of taxa, plant and animal alike, that exemplify the trans-Pacific network of relationships. Such taxa are predicted to be differentiated in response to tectonic developments roughly comparable to those having occurred in the Indian and Atlantic Oceans. This interpretation which amounts to asserting that the Pacific in some sense was born in the mid-Mesozoic, conflicts with what is often asserted to be already known on the basis of geological and geophysical data. In Croizat's view, some of these assertions are mistakes and, like the others, are exposed as such by the biological evidence of Recent distribution. To sum up the prospects for these two sorts of information: Wheeler's Law has more numerous possibilities, that is wider application, in paleontology and geophysics than it has in biogeography.

The third sort of information falls within the domain of cladistic systematics, as considered in the geographical dimension. I begin with what Good (1974) called a "Map of the World Showing Floristic Regions" (Fig. 7). At a fairly coarse level the 37 regions delineated may be understood as areas of endemism of land plants. (The term "at a fairly coarse level" is employed because within each of most of these regions there are sub-areas of endemism more local in character.) The problem arises from the relationships of the areas, which may be considered as the sum of the interrelationships of the taxa endemic to them. In a simple way the relationship between two species, for example one endemic in Brazil with a close relative in tropical west Africa, may be portrayed by a line drawn between areas 27 and 12. Adding other lines for other relationships increases the complexity of the pattern. The basic question is the following: precisely what is the complex pattern? Good's figure specified the areas but not their interrelationships. Croizat's figure specified relationships, but the areas are not specified so precisely. Good listed 37 regions. By my count, all but about 7 have exact counterparts in the pioneering work of this sort by Candolle in papers published in 1820 and 1838: the exceptions are regions 10, 18, 22, 32, 33, 34, 37. Candolle treated regions 10 and 12 together as tropical west Africa, and regions 17 and 18 together as continental India with its archipelago; region 22 was unmentioned; Australia was not subdivided; while among the south oceanic islands Candolle mentioned only Tristan da Cunha, which he considered a region by itself.

One finds Candolle's regions in other places, for example in Wallace's (1876) *Geographical Distribution of Animals* (Table I). Again the correspondence is not exact, but of the 24 "subregions" listed by Wallace, only three by my count have no good counterparts in Candolle: Ceylon, Austromalaya, and Rocky Mountains. Wallace considered the 24 as subregions of more inclusive regions that date from Sclater (1858). In the arrangement there is a sort of classification of lower taxa (subregions) into higher taxa (regions), in accordance with Wallace's notions of their relationships. Characteristic of his treatment are the items in the

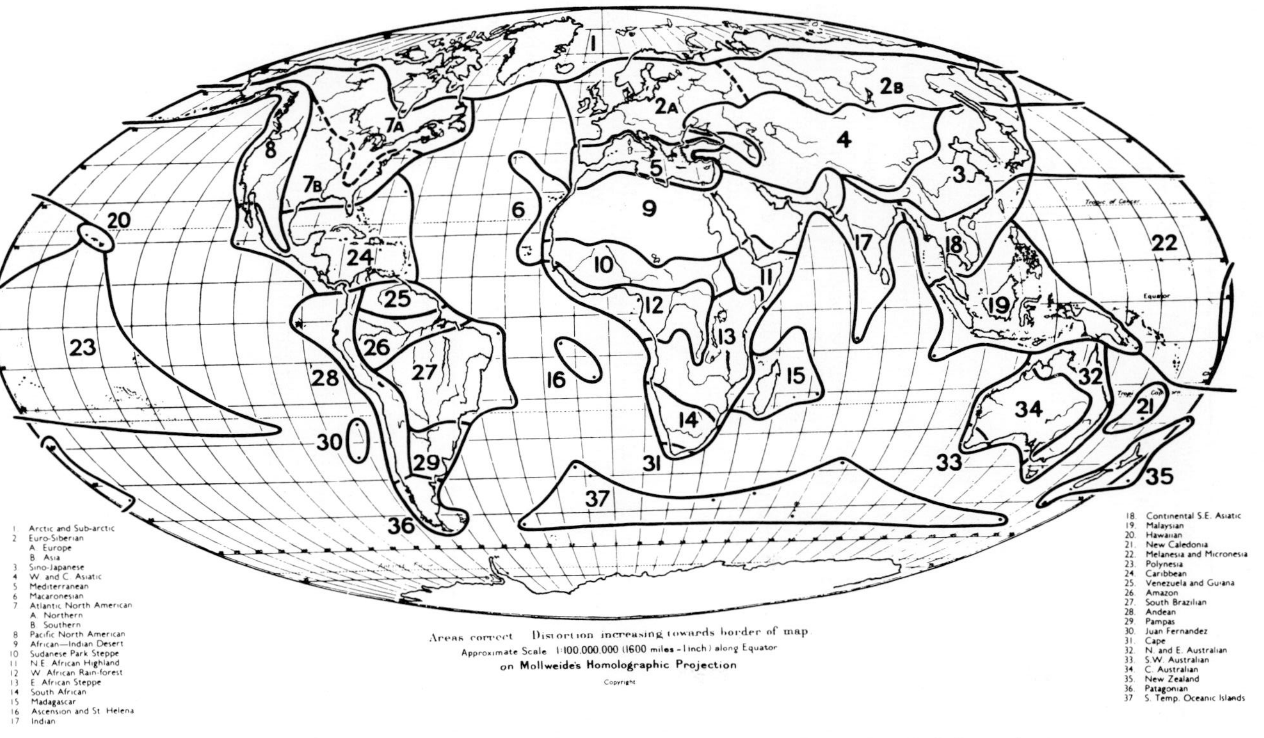

Fig. 7. Map of the World showing Floristic Regions. After Good (1974: Pl. 4).

Table I. Regions and sub-regions. After Wallace (1876:81–82).

Regions	*Subregions*	*Remarks*
I. Palaearctic.....	1. North Europe	
	2. Mediterranean (or S. Eu.)	Transition to Ethiopian.
	3. Siberia.	Transition to Nearctic.
	4. Manchuria (or Japan)	Transition to Oriental.
II. Ethiopian......	1. East Africa.	Transition to Palaearctic.
	2. West Africa.	
	3. South Africa.	
	4. Madagascar.	
III. Oriental	1. Hindostan (or Central Ind.)	Transition to Ethiopian.
	2. Ceylon.	
	3. Indo-China (or Himalayas)	Transition to Palaearctic.
	4. Indo-Malaya.	Transition to Australian.
IV. Australian.....	1. Austro-Malaya.	Transition to Oriental.
	2. Australia.	
	3. Polynesia.	
	4. New Zealand.	Transition to Neotropical.
V. Neotropical...	1. Chili (or S. Temp. Am.)	Transition to Australian.
	2. Brazil.	
	3. Mexico (or Trop. N. Am.)	Transition to Nearctic.
	4. Antilles.	
VI. Nearctic	1. California.	
	2. Rocky Mountains.	Transition to Neotropical.
	3. Alleghanies (or East U.S.)	
	4. Canada.	Transition to Palaearctic.

"Remarks" section, where certain sub-regions are said to be transitions to other regions.

Classification as an expression of relationship has seldom, if ever, been an exact procedure, but in the cladistic sense classification can be exact, conditions permitting. One may ask, therefore, if Wallace's six regions are natural ones in a cladistic sense? Apparently not, to judge from Wallace's account (1876, 1:80):

> The twenty-four sub-regions here adopted were arrived at by a careful consideration of the distribution of the more important genera, and of the materials, both zoological and geographical, available for their determination; and it was not till they were almost finally decided on, that they were found to be equal in number throughout all the regions—four in each. As this uniformity is of great advantage in tabular and diagrammatic

> presentations of the distribution of the several families, I decided not to disturb it unless very strong reasons should appear for adopting a greater or less number in any particular case. Such, however have not arisen. . . .

Wallace's account expressed a concern with uniformity of division, a concern also of the quinarian approach to classification, which was of some interest in Britain during the middle years of the 19th century. Wallace was aware of this approach, and in an earlier paper argued against it within the context of zoological classification (1855:187): "We shall thus find ourselves obliged to reject all those systems of classification . . . which fix a definite number for the divisions of each group."

Wallace's classification of regions lends itself to a quinarian portrayal, wherein the "transitions" from one region to another are the lines between circles representing regions (Fig. 8). If we disregard the circles

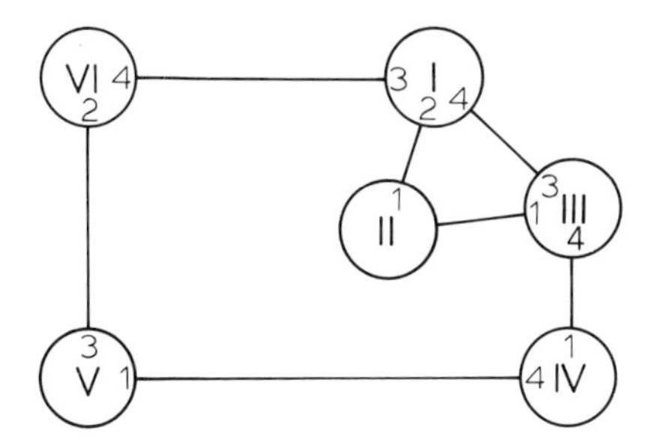

Fig. 8. A quinarian arrangement of Wallace's table (cf. Table 1).

entirely, and pay attention only to the "transitions," we have a system of relationships that is surprisingly modern with respect to some of its elements (Fig. 9): in particular the Chile–New Zealand and India–Africa groupings. Wallace did not pursue this possibility of natural classification. The reasons, so far as I can determine, are two. First, he had no theoretical notion of natural classification as it might have been applied to geography, and he therefore adopted what, for all intents and purposes, is a quinarian theory. Second, in the development of his approach he looked for physical rather than biological clues.

In Wallace's view he did not assume geographical stability but proved that stability was true (1905, 2:403):

> The general permanence of oceanic and continental areas was first taught by Professor J.D. Dana, the eminent American geologist, and again by Darwin in his "Origins of Species"; but I am, I believe, the only writer

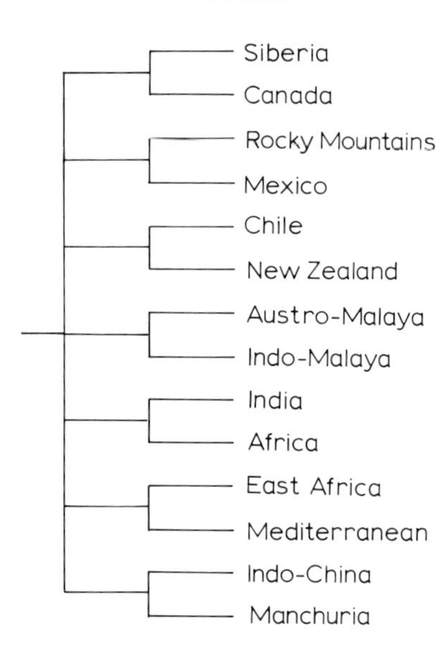

Fig. 9. An area-cladogram based on the "transitions" of Wallace's table (cf. Table 1).

> who has brought forward a number of other considerations, geographical and physical, which, with those of previous writers, establish the proposition on almost incontrovertible grounds. . . . The doctrine may be considered as the only solid basis for any general study of the geographical distribution of animals, and it is for this reason that I have made it the subject of my careful consideration.

Wallace's overall result by today's standards is largely, if not purely, artificial. He, of course, thought it natural, but he had a very broad notion of naturalness (1894:613):

> There is thus, in my opinion, no question of who is *right* and who is *wrong* in the . . . grouping of these regions, or of determining what are the *true* primary regions. All proposed regions are, from some points of view, natural, but the whole question of their grouping . . . is one of convenience and of utility . . .

Such notions have always been the hallmark of artificiality in classification. Their effect is to retard progress through their immunity to falsification. In this way they render inoperative Wheeler's Law. Such is the price of incontrovertibility.

Some of these points are further illustrated in Fig. 10. In A are shown the six regions and the "transitions" listed in Wallace's table (Table I).

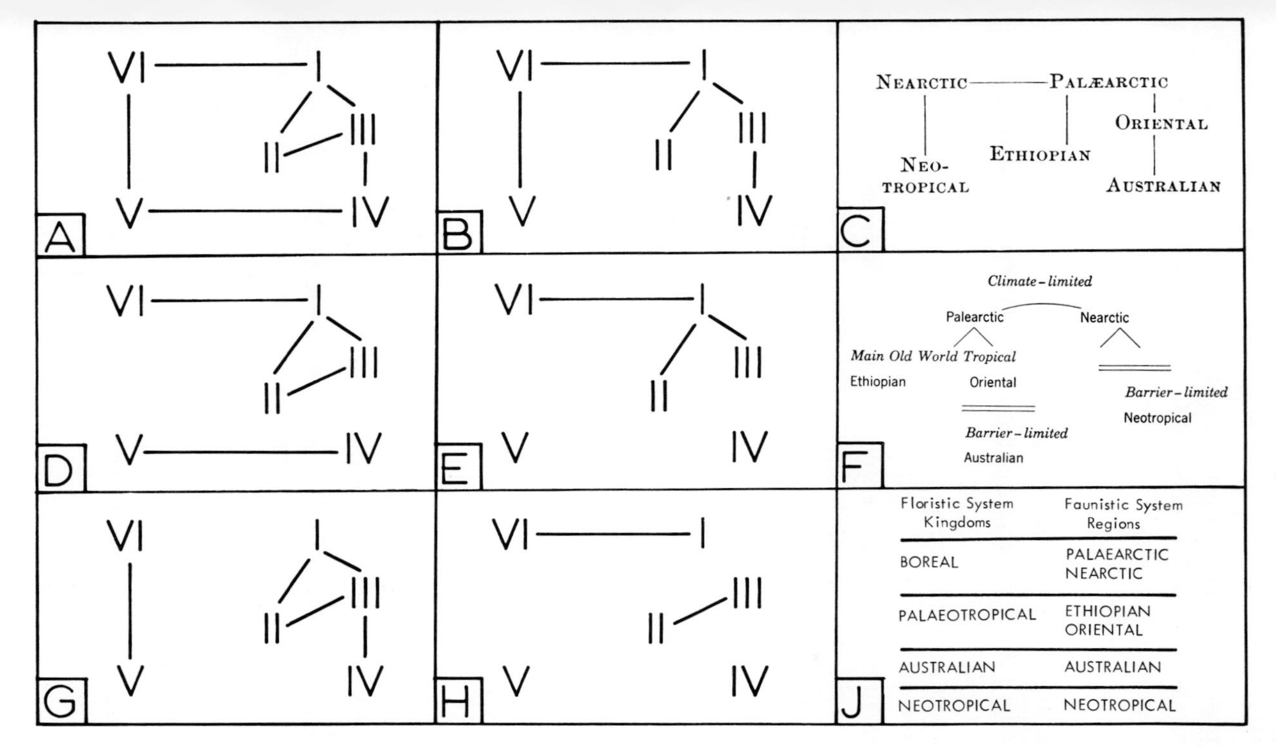

Fig. 10. A, Simplified representation of Wallace's table (cf. Table 1, Fig. 8). B, Derivation of Wallace's diagram of area-relationships (cf. C). C. Wallace's "arrangement of the regions", after Wallace (1881:52). D, Simplified representation of Huxley's (1868) classification of regions. E. Derivation of Darlington's diagram of area-relationship (cf. F). F, Darlington's "Diagrammatic arrangement of the faunal regions", after Darlington (1957: Fig. 45). G, Simplified representation of Sclater's (1858) classification of regions. H, Derivation of Udvardy's "Floristic System" (cf. J). J, Comparison of floristic and faunistic systems, redrawn from Udvardy (1975:9).

In C is a diagram later published by Wallace, about which he stated (1881:52): "The following arrangement of the regions will indicate their geographical position, and to a considerable extent, their relation to each other." In B is shown the derivation of C from A, which is achieved simply by omission of two lines. If the lines mean anything at all, surely their omission is something of a loss. Seemingly, Wallace omitted them because they cross areas of deep ocean between Africa and India, and between Australia and South America; in other words, the lines pass over what Wallace thought were significant and permanent barriers to dispersal.

Darlington applied the same sort of reason in his account of 1957, shown in F, when he decided that the next most impressive barriers to dispersal were likewise worth emphasizing. The derivation of his scheme, shown in E, has two lines fewer than Wallace's. Thus, emphasis is achieved through omission. The derivation specifies Darlington's classification of regions (the main part of the world, Megagea, I + II + III + VI; Notogea, IV; Neogea, V).

That I focus on Wallace and Darlington reflects my bias as a zoologist. Their notions are commonly encountered in the zoological literature. I don't know the botanical literature so well, but Udvardy's recent review, shown in J, purports to be the current situation in botany, which is poorer yet by one line, as indicated in the derivation H. We see there the "kingdoms" of the "Floristic System": Boreal, I + VI; Palaeotropical, II + III; Australian, IV; and Neotropical, V.

There are other variations on this theme, which are played by omitting one or another line of Wallace's original notion (A). Two others, both published prior to Wallace, are shown in D and G. In D is Huxley's (1868) division into the North World (Arctogaea, I + II + III + VI) and the South World (Notogaea, IV + V). In G is Sclater's (1858) division into the New World (Neogaea, V + VI) and Old World (Palaeogaea, I + II + III + IV). Interestingly, adding D and G together gives A as a result, and Wallace may have arrived at his notion through some such process of addition of what had come before him.

That there is a logical relation between these variations does not mean that they have any real substance. Artificial classifications always have some logical thread that binds them together. Yet the thread is invariably undone by empirical considerations, and these schemes are no exception. A minimum of distributional information is enough to dissolve them all into areas of local endemism, which at this level are approximated well enough by Candolle's regions. A consideration of the interrelationships of the endemic areas, based on the information available to Wallace (Table 1) gives a picture similar to Croizat's (Fig. 6). The endemic areas group together by ocean basins not by continental geography.

Wallace had some appreciation of this empirical connection. He wrote (1894:612):

> The South European, the Malayan, the Brazilian, or the South African faunas and floras, are constantly referred to, because those districts are really characterised by distinct assemblages of animals and plants. . . .

The "South European", the "Malayan", the "Brazilian", and the "South African" are for the most part Candollean regions. These regions and their interrelationships Wallace ignored as a matter of principle. The principle is convenience. He wrote further (1894:612): "But such areas as these are too small and too numerous to enable us to express the broader features of the distribution of animals. . . ." After arguing that Sclater's six regions suffice for his purposes, Wallace then suggested that the Candollean regions (Wallace's subregions) be effectively wiped from the map (1894:613):

> I would suggest, therefore, that for the present, at all events, no definite named subdivisions should be attempted, but that the continental portion of each region be subdivided by the use of the terms north, south, east, west, and central, with their combinations where required.

With a uniform division into north, south, east, west, and central, Wallace achieved a numerically quinarian arrangement. The problem, however, is not with number *per se*, but the empirical connection of the classification—the areas of local endemism—"those districts [that] are really characterised by distinct assemblages of animals and plants." A biological analog is the argument that, because there are so many species, some lesser number of families is convenient, which, when established, would allow the numerous species to be dispensed with altogether. If the families prove to be artificial assemblages, based only on consensus, we are left with nothing.

I have elsewhere dealt with the history of biogeography (Nelson, 1978; Nelson and Platnick, 1980, 1981) and it is clear that the problem of biogeographical classification is due to the failure to recognize the fundamentals of biological classification. In reality, the controversy is very simple at root, and stems from a confusion of the principles governing artificial and natural classification, which were distinguished long ago by Linnaeus and others. The distinction itself is simple. A natural, in contrast to an artificial, classification has an empirical connection that allows for falsification. The obscurity that attends the failure to make this distinction clouds the Anglo-American literature back to the time of Ray, and even then the situation was not altogether clear. It is no coincidence that the phenetic movement in systematics, which I believe has only deepened the gloom, was at heart an Anglo-American enterprise, or that quinarianism is seriously advocated today from that quarter without being recognized as such (McNeill, 1979).

Future prospects, as I see them, stem from two possibilities, one methodological and one empirical. The methodological approach will employ cladistics, which promises a simplification of theory that may help systematics to become intelligible to, and appreciated by, biologists generally. The impact may cause significant change in some areas, and I have in mind evolutionary theory in particular, which in its traditional form seems doomed to extinction. The empirical approach concerns the use of biogeographical data as viewed in the cladistic aspect, which promises to be more orderly than would be expected. Croizat has tirelessly informed us that "distribution forever repeats". If so, the repetition must appear as geographic congruence of cladograms for different groups of organisms. Enough congruence has so far been found to suggest that in the coming years this general problem will attract more, rather than less, interest, and to some extent will revive the study of systematics and give it greater coherence.

One area of general interest is the Pacific Basin and its history. With its many now-isolated islands, the Pacific has been the stronghold of the dispersalist tradition in biogeography, and remains so today, seemingly strengthened by the revival of continental-drift theory. Croizat, however, is not alone in affirming that the coherence of biological distribution in and around the Pacific belies the usual story of this ocean as an old and stable barrier to dispersal (Nelson and Rosen, 1981).

The Pacific holds the answer to a fundamental question concerning biogeographical classification, namely whether the continental areas—the nearctic, the neotropics, etc.—are real units, or whether they are geological and biological conglomerates of smaller and more local units with diverse histories (e.g. Ben-Avraham *et al.*, 1981). Biogeographically, the question has never really been much explored except by Croizat but means are now available to do so. The means are cladistics and the interrelationships of areas of endemism, i.e. "vicariance biogeography".

ACKNOWLEDGEMENTS

I am grateful to the English Speaking Union (New York Branch), The Systematics Association, and the American Museum of Natural History for the financial support that enabled me to attend this symposium.

REFERENCES

Ben-Avraham, Z., A. Nur, D. Jones and A. Cox (1981). Continental accretion: from oceanic plateaus to allochthonous terranes. *Science, N.Y.* **213**, 47–54.

Candolle, A.-P. de (1820). Géographie botanique. *Dict. Sci. Nat.* **18**, 359–422. F.G. Levrault, Strasbourg and Paris.

Candolle, A.-P. de (1838). "Statistique de la Famille des Composées". Treuttel and Würtz, Paris and Strasbourg.

Croizat, L. (1958). "Panbiogeography", published by the author. Caracas.

Darlington, P.J., jr. (1957). "Zoogeography: the geographical distribution of animals". J. Wiley and Sons, New York.

Darwin, F. (ed.) (1887). "The life and letters of Charles Darwin, including an autobiographical chapter". J. Murray, London.

Edwards, A.W.F. and Cavalli-Sforza, L.L. (1964). Reconstruction of evolutionary trees. *In* "Phenetic and phylogenetic classification" (V.H. Heywood and J. McNeill, eds), pp. 67–76. Systematics Association, London.

Fitzroy, R. (1839). "Narrative of the surveying voyages of His Majesty's Ships Adventure and Beagle, between the years 1826 and 1836, describing their examinations of the southern shores of South America, and the Beagle's circumnavigation of the globe", vol. 2. Colburn, London.

Good, R. (1974). "The Geography of the Flowering Plants" (4th edn). Longman, London.

Grinnell, G. (1974). The rise and fall of Darwin's first theory of transmutation. *J. Hist. Biol.* **7**, 259–273.

Haeckel, E. (1876). "The History of Creation, or the development of the earth and its inhabitants by the action of natural causes". Appleton, New York.

Haeckel, E. (1907). "The History of Creation, or the development of the earth and its inhabitants by the action of natural causes" (4th edn; from the eighth German edition). Appleton, New York.

Haeckel, E. (1909). "Natürliche Schöpfungs-Geschichte. Gemeinverständliche wissenschaftliche Vorträge über die Entwickelungslehre" (11th edn). G. Reimer, Berlin.

Huxley, T.H. (1868). On the classification and distribution of the *Alectoromorphae* and *Heteromorphae*. *Proc. Zool. Soc. Lond.* **1868**, 294–319.

Linnaeus, C. (1781). "Selected dissertations from the Amoenitates Academicae". London.

McNeill, J. (1979). Structural value: a concept used in the construction of taxonomic classifications. *Taxon* **28**, 481–504.

Nelšon, G. (1978). From Candolle to Croizat: comments on the history of biogeography. *J. Hist. Biol.* **11**, 269–305.

Nelson, G., and N. Platnick (1980). A vicariance approach to historical biogeography. *BioScience* **30**, 339–343.

Nelson, G., and N. Platnick (1981). "Systematics and Biogeography: cladistics and vicariance". Columbia Univeristy Press, New York.

Nelson, G., and D.E. Rosen (eds) (1981). "Vicariance Biogeography: a critique". Columbia University Press, New York.

Sclater, P.L. (1858). On the general geographical distribution of the members of the class Aves. *J. Proc. Linn. Soc. Zool.* **2**, 130–145.

Takhtajan, A. (1969). "Flowering Plants: origin and dispersal". Smithsonian Institution, Washington D.C.

Thompson, W.R. (1963). A critique of the Darwinian theory of evolution. *Studia Entomologica*, N.S. **6**, 563–574. [Reprinted from the "Introduction" to the "Origin of Species", Everyman's Library no. 811. 1958. Dent and Sons, London.]

Udvardy, M.D.F. (1975). A classification of the biogeographical provinces of the world. *Internat. Un. Conserv. Nat. Nat. Res., Occ. Pap.* **18**, 1–48.

Wallace, A.R. (1855). On the law which has regulated the introduction of new species. *Ann. Mag. nat. Hist.*, ser. 2, **16**, 184–196.

Wallace, A.R. (1876). "The Geographical Distribution of Animals". Macmillan, London.

Wallace, A.R. (1881). "Island Life or, the phenomena and causes of insular faunas and floras including a revision and attempted solution of the problem of geological climates". Harper, New York.

Wallace, A.R. (1894). What are zoological regions? *Nature, Lond.* **49**, 610–613.

Wallace, A.R. (1905). "My Life: a record of events and opinions". Dodd, Mead, New York.

Editor-in-Chief, Special Volume Series

D. L. HAWKSWORTH PhD DSc FLS FIBiol

Commonwealth Agricultural Bureaux

Systematics Association Publications

1. BIBLIOGRAPHY OF KEY WORKS FOR THE IDENTICATION OF THE BRITISH FAUNA AND FLORA *3rd edition* (1967)
   Edited by G. J. Kerrich, R. D. Meikle and N. Tebble
   Out of print
2. FUNCTION AND TAXONOMIC IMPORTANCE (1959)
   Edited by A. J. Cain
3. THE SPECIES CONCEPT IN PALAEONTOLOGY (1956)
   Edited by P. C. Sylvester-Bradley
4. TAXONOMY AND GEOGRAPHY (1962)
   Edited by D. Nichols
5. SPECIATION IN THE SEA (1963)
   Edited by J. P. Harding and N. Tebble
6. PHENETIC AND PHYLOGENETIC CLASSIFICATION (1964)
   Edited by V. H. Heywood and J. McNeil
   Out of print
7. ASPECTS OF TETHYAN BIOGEOGRAPHY (1967)
   Edited by C. G. Adams and D. V. Ager
8. THE SOIL ECOSYSTEM (1969)
   Edited by H. Sheals
9. ORGANISMS AND CONTINENTS THROUGH TIME (1973)†
   Edited by N. F. Hughes

Published by the Association

Systematics Association Special Volumes

1. THE NEW SYSTEMATICS (1940)
   Edited by Julian Huxley (Reprinted 1971)
2. CHEMOTAXONOMY AND SEROTAXONOMY (1968)*
   Edited by J. G. Hawkes
3. DATA PROCESSING IN BIOLOGY AND GEOLOGY (1971)*
   Edited by J. L. Cutbill
4. SCANNING ELECTRON MICROSCOPY (1971)*
   Edited by V. H. Heywood
5. TAXONOMY AND ECOLOGY (1973)*
   Edited by V. H. Heywood

* Published by Academic Press for the Systematics Association

† Published by the Palaeontological Association in conjunction with the Systematics Association

6. THE CHANGING FLORA AND FAUNA OF BRITAIN (1974)*
   Edited by D. L. Hawksworth
7. BIOLOGICAL IDENTIFICATION WITH COMPUTERS (1975)*
   Edited by R. J. Pankhurst
8. LICHENOLOGY: PROGRESS AND PROBLEMS (1976)*
   Edited by D. H. Brown, D. L. Hawksworth and R. H. Bailey
9. KEY WORKS TO THE FAUNA AND FLORA OF THE BRITISH ISLES AND NORTHWESTERN EUROPE (1978)*
   Edited by G. J. Kerrich, D. L. Hawksworth and R. W. Sims
10. MODERN APPROACHES TO THE TAXONOMY OF RED AND BROWN ALGAE (1978)*
    Edited by D. E. G. Irvine and J. H. Price
11. BIOLOGY AND SYSTEMATICS OF COLONIAL ORGANISMS (1979)*
    Edited by G. Larwood and B. R. Rosen
12. THE ORIGIN OF MAJOR INVERTEBRATE GROUPS (1979)*
    Edited by M. R. House
13. ADVANCES IN BRYOZOOLOGY (1979)*
    Edited by G. P. Larwood and M. B. Abbot
14. BRYOPHYTE SYSTEMATICS (1979)*
    Edited by G. C. S. Clarke and J. G. Duckett
15. THE TERRESTRIAL ENVIRONMENT AND THE ORIGIN OF LAND VERTEBRATES (1980)*
    Edited by A. L. Panchen
16. CHEMOSYSTEMATICS: PRINCIPLES AND PRACTICE (1980)*
    Edited by F. A. Bisby, J. G. Vaughan and C. A. Wright
17. THE SHORE ENVIRONMENT: METHODS AND ECOSYSTEMS (2 Volumes) (1980)*
    Edited by J. H. Price, D. E. G. Irvine and W. F. Farnham
18. THE AMMONOIDEA (1981)*
    Edited by M. R. House and J. R. Senior
19. BIOSYSTEMATICS OF SOCIAL INSECTS (1981)*
    Edited by P. E. Howse and J.-L. Clément
20. GENOME EVOLUTION (1982)*
    Edited by G. A. Dover and R. B. Flavell
21. PROBLEMS OF PHYLOGENETIC RECONSTRUCTION (1982)*
    Edited by K. A. Joysey and A. E. Friday
22. CONCEPTS IN NEMATODE SYSTEMATICS (1983)*
    Edited by A. R. Stone, H. M. Platt and L. F. Khalil
23. EVOLUTION, TIME AND SPACE: THE EMERGENCE OF THE BIOSPHERE (1983)*
    Edited by R. W. Sims, J. H. Price and P. E. S. Whalley
24. PROTEIN POLYMORPHISM: ADAPTIVE AND TAXONOMIC SIGNIFICANCE (1983)*
    Edited by G. S. Oxford and D. Rollinson

* Published by Academic Press for the Systematics Association